Advances in Experimental Medicine and Biology

Volume 894

Advances in Experimental Medicine and Biology presents multidisciplinary and dynamic findings in the broad fields of experimental medicine and biology. The wide variety in topics it presents offers readers multiple perspectives on a variety of disciplines including neuroscience, microbiology, immunology, biochemistry, biomedical engineering and cancer research.

Advances in Experimental Medicine and Biology has been publishing exceptional works in the field for over 30 years and is indexed in Medline, Scopus, EMBASE, BIOSIS, Biological Abstracts, CSA, Biological Sciences and Living Resources (ASFA-1), and Biological Sciences. The series also provides scientists with up to date information on emerging topics and techniques.

2014 Impact Factor: 1.958

More information about this series at http://www.springer.com/series/5584

Pim van Dijk • Deniz Başkent • Etienne Gaudrain
Emile de Kleine • Anita Wagner • Cris Lanting
Editors

Physiology, Psychoacoustics and Cognition in Normal and Impaired Hearing

Editors
Pim van Dijk
Deniz Başkent
Etienne Gaudrain
Emile de Kleine
Anita Wagner
Cris Lanting

Department of Otorhinolaryngology / Head and Neck Surgery
University of Groningen
University Medical Center Groningen
Groningen, The Netherlands

ISSN 0065-2598 ISSN 2214-8019 (electronic)
Advances in Experimental Medicine and Biology
ISBN 978-3-319-25472-2 ISBN 978-3-319-25474-6 (eBook)
DOI 10.1007/978-3-319-25474-6

Library of Congress Control Number: 2016930406

Springer Cham Heidelberg New York Dordrecht London

Printed on acid-free paper

This Springer imprint is published by the registered company Springer International Publishing AG part of Springer Nature.
The registered company address is: Gewerbestrasse 11, 6330 Cham, Switzerland

Preface

This volume constitutes the Proceedings of the 17th International Symposium on Hearing (ISH), held from 15 to 19 June 2015 in Groningen, The Netherlands. This meeting continued a great tradition of conferences that started in 1969 and has been held every 3 years. It was the second ISH meeting that took place in Groningen.

The emphasis for the ISH series has traditionally been on bringing together researchers from basic research on physiological and perceptual processes of the auditory system, including modeling. However, this tradition has also meant usually excluding great research on other topics related to hearing sciences, such as hearing impairment, hearing devices, as well as cognitive auditory processes. During the ISH 2015, we have aimed to expand to include such relatively new research, while still continuing with the ISH tradition of basic science. As a result, we had a great program that covered many exciting, and some contemporary, topics from a wide range of disciplines related to auditory sciences.

Following the format of the previous ISH meetings, all chapters were edited by the organizing committee, and these were made available to participants prior to the meeting. The timetable allowed ample time for discussions. The chapters are organized according to broad themes, and their order reflects the order of presentations at the meeting.

We would also like to acknowledge everyone who has contributed to ISH 2015. We would like to thank the Koninklijke Nederlandse Akademie van Wetenschappen (KNAW), Heinsius Houbolt Foundation, The Oticon Foundation, Cochlear Inc., Med-El, Phonak, the Gemeente Groningen, the Provincie Groningen and the Research Institute for Behavioral and Cognitive Neurosciences of the University of Groningen for the financial support. Also thanks to Jellemieke Ekens and Els Jans of the Groningen Congres Bureau for their logistic support. Thanks to Jeanne Clarke, Nawal El Boghdady, Enja Jung, Elouise Koops, Annika Luckmann, Jefta

Fig.1 Conference participants in front of the Familiehotel in Paterswolde

Saija and Mirjan van Dijk for their help during the meeting. Finally we would like to thank all authors and participants for their scientific contributions, and for the lively discussions.

Organising Committee
Pim van Dijk
Deniz Başkent
Emile de Kleine
Etienne Gaudrain
Anita Wagner
Cris Lanting
Ria Woldhuis

This edition of the ISH was supported by

- Royal Netherlands Academy of Arts and Sciences (KNAW)
- Oticon Fonden
- Heinsius-Houbolt Fonds
- Phonak
- Med-El
- Cochlear
- City of Groningen
- Province of Groningen
- Research School of Behavioral and Cognitive Neurosciences
- University of Groningen
- University Medical Center Groningen

Previous ISH Meetings and Books

ISH 2012
Cambridge, United-Kingdom: *Basic Aspects of Hearing, Physiology and Perception*. Edited by B.C.J. Moore, R.D. Patterson, I.M. Winter, R.P. Carlyon, H.E. Gockel. Springer: New York.

ISH 2009
Salamanca, Spain: *The Neurophysiological Bases of Auditory Perception*. Edited by E.A. Lopez-Poveda, A.R. Palmer, R. Meddis. Springer: New York.

ISH 2006
Cloppenburg, Germany: *Hearing-From Sensory Processing to Perception*. Edited by B. Kollmeier, G. Klump, V. Hohmann, U. Langemann, M Mauermann, S. Uppenkamp, J. Verhey. Springer: New York.

ISH 2003
Dourdan, France: *Auditory signal processing: physiology, psychoacoustics, and models*. Edited by D. Pressnitzer, A. de Cheveigné, S. McAdams, L. Collet. Springer: New York.

ISH 2000
Mierlo, Netherlands: *Physiological and psychophysical bases of auditory function*. Edited by D.J. Breebaart, A.J.M. Houtsma, A. Kohlrausch, V.F. Prijs, R. Schoonhoven. Shaker Publishing: Maastricht.

ISH 1997
Grantham, United-Kingdom: *Psychophysical and physiological advances in hearing*. Edited by A.R. Palmer, A. Rees, A.Q. Summerfield, R. Meddis. Whurr Publisher: London.

ISH 1994
Irsee, Germany: *Advances in Hearing Research*. Edited by G. A. Manley, G. M. Klump, C. Koppl, H. Fastl, H. Oeckinghaus. World Scientific: Singapore.

ISH 1991
Carcans, France: *Auditory Physiology and Perception*. Edited by Y. Cazals, L. Demany and K. Horner. Pergamon: Oxford.

ISH 1988
Groningen, Netherlands: *Basic issues in hearing*. Edited by H. Duifhuis, J.W. Horst, H.P. Wit. Academic Press: London.

ISH 1986
Cambridge, United-Kingdom: *Auditory frequency selectivity*. Edited by B.C.J. Moore and R.D. Patterson. Plenum Press: New York.

ISH 1983
Bad Nauheim, F.R. Germany: *Hearing – physiological bases and psychophysics*. Edited by R. Klinke, R. Hartmann. Springer Verlag: Berlin.

ISH 1980
Noordwijkerhout, Netherlands: *Psychophysical, physiological and behavioural studies in hearing*. Edited by G. van der Brink, F.A. Bilsen. Delft University Press: Delft.

ISH 1977
Keele, United-Kingdom: *Psychophysics and physiology of hearing*. Edited by E.F. Evans, J.P. Wilson. Academic Press: London.

ISH 1974
Tutzing, F.R. Germany: *Facts and models in hearing*. Edited by E. Zwicker, E. Terhardt. Springer Verlag: Berlin.

ISH 1972
Eindhoven, Netherlands: *Hearing theory*. Edited by B.L. Cardozo. IPO: Eindhoven.

ISH 1969
Driebergen, Netherlands: *Frequency analysis and periodicity detection in hearing*. Edited by R. Plomp, G.F. Smoorenburg. Sijthoff: Leiden.

Contents

Contributors

Michael A. Akeroyd MRC Institute of Hearing Research, Nottingham, UK

Martin Andermann Section of Biomagnetism, Department of Neurology, University of Heidelberg, Heidelberg, Germany

Go Ashida Cluster of Excellence "Hearing4all", University of Oldenburg, Oldenburg, Germany

Andreas Büchner Dept. of Otolaryngology and Hearing4all, Medical University Hannover, Hannover, Germany

Deniz Başkent Department of Otorhinolaryngology / Head and Neck Surgery, University of Groningen, University Medical Center Groningen, Groningen, The Netherlands

Graduate School of Medical Sciences (Research School of Behavioural and Cognitive Neurosciences), University of Groningen, Groningen, The Netherlands

Oliver Behler Medizinische Physik, Carl von Ossietzky Universität Oldenburg, Oldenburg, Germany

Sharath Bennur Department of Otorhinolaryngology, University of Pennsylvania, Philadelphia, USA

Frédéric Berthommier Grenoble Images Parole Signal Automatique-Lab, Speech and Cognition Department, CNRS, Grenoble University, Grenoble, France

Virginia Best Department of Speech, Language and Hearing Sciences, Boston University, Boston, MA, USA

Rainer Beutelmann Cluster of Excellence "Hearing4all", Animal Physiology and Behaviour Group, Department for Neuroscience, University of Oldenburg, Oldenburg, Germany

Charlotte M. de Blecourt Graduate School of Medical Sciences (Research School of Behavioural and Cognitive Neurosciences), University of Groningen, Groningen, The Netherlands

Kristen Bond Auditory Behavioral Research Laboratory, Communication Sciences and Disorders, University of Wisconsin-Madison, Madison, WI, USA

Yves Boubenec Laboratoire des Systèmes Perceptifs, CNRS UMR 8248, Paris, France

Département d'études cognitives, Ecole normale supérieure PSL Research University, Paris, France

Jörg M. Buchholz National Acoustic Laboratories, Macquarie University, Sydney, NSW, Australia

Angela Canas UCL Ear Institute, London, UK

Robert P. Carlyon MRC Cognition and Brain Sciences Unit, Cambridge, UK

Laurel H. Carney Departments of Biomedical Engineering, Neurobiology & Anatomy, Electrical & Computer Engineering, University of Rochester, Rochester, NY, USA

Catherine Carr Department of Biology, University of Maryland, College Park, MD, USA

Yves Cazals LNIA-CNRS, UMR 7260, Aix-Marseille Univ., Marseille, France

An-Chieh Chang Auditory Behavioral Research Laboratory, Communication Sciences and Disorders, University of Wisconsin-Madison, Madison, WI, USA

Kate Christison-Lagay Neuroscience Graduate Group, University of Pennsylvania, Philadelphia, USA

Martin Coath Cognition Institute and School of Psychology, Plymouth University, Plymouth, UK

Yale E. Cohen Department of Otorhinolaryngology, University of Pennsylvania, Philadelphia, USA

Department of Neuroscience, University of Pennsylvania, Philadelphia, USA

Department of Bioengineering, University of Pennsylvania, Philadelphia, USA

Nigel P. Cooper Erasmus MC, Rotterdam, Nederland

William Cross The Bionics Institute of Australia, Melbourne, Australia

Department of Medicine, The University of Melbourne, Melbourne, Australia

John F. Culling School of Psychology, Cardiff University, Cardiff, UK

Helen Dare School of Psychology, Cardiff University, Cardiff, UK

Torsten Dau Hearing Systems Group, Department of Electrical Engineering, Technical University of Denmark, Kongens Lyngby, Denmark

Steven van de Par Acoustics Group, Cluster of Excellence "Hearing4All", Carl von Ossietzky University, Oldenburg, Germany

Aneeka Degun UCL Ear Institute, London, UK

Susan L. Denham Cognition Institute and School of Psychology, Plymouth University, Plymouth, UK

Giovanni M. Di Liberto Trinity College Institute of Neuroscience, Trinity College Dublin, Dublin, Ireland

School of Engineering, Trinity College Dublin, Dublin, Ireland

Trinity Centre for Bioengineering, Trinity College Dublin, Dublin, Ireland

Mathias Dietz Medizinische Physik, Universität Oldenburg, Oldenburg, Germany

Pim van Dijk Department of Otorhinolaryngology / Head and Neck Surgery, University of Groningen, University Medical Center Groningen, Groningen, The Netherlands

Graduate School of Medical Sciences (Research School of Behavioural and Cognitive Neurosciences), University of Groningen, Groningen, The Netherlands

Ruijuan Dong Otolaryngology—Head & Neck Surgery, Beijing Tongren Hospital, Beijing Institute of Otolaryngology, Capital Medical University, Beijing, China

Markus Drexl German Center for Vertigo and Balance Disorders (IFB), Department of Otorhinolaryngology, Head and Neck Surgery, Grosshadern Medical Centre, University of Munich, Munich, Germany

Vit Drga Ear Institute, University College London (UCL), London, UK

Bernhard Englitz Laboratoire des Systèmes Perceptifs, CNRS UMR 8248, Paris, France

Département d'études cognitives, Ecole normale supérieure PSL Research University, Paris, France

Department of Neurophysiology, Donders Centre for Neuroscience, Radboud Universiteit Nijmegen, Nijmegen, The Netherlands

Almudena Eustaquio-Martín INCYL, IBSAL, Universidad de Salamanca, Salamanca, Spain

Christian Füllgrabe MRC Institute of Hearing Research, Nottingham, UK

Shigeto Furukawa Human Information Science Laboratory, NTT Communication Science Laboratories, NTT Corporation, Atsugi, Japan

Robert Gürkov German Center for Vertigo and Balance Disorders (IFB), Department of Otorhinolaryngology, Head and Neck Surgery, Grosshadern Medical Centre, University of Munich, Munich, Germany

Frederick J. Gallun National Center for Rehabilitative Auditory Research, VA Portland Health Care System, Portland, OR, USA

Attigodu Chandrashekara Ganesh Grenoble Images Parole Signal Automatique-Lab, Speech and Cognition Department, CNRS, Grenoble University, Grenoble, France

Adam M. Gifford Neuroscience Graduate Group, University of Pennsylvania, Philadelphia, USA

Helen Glyde National Acoustic Laboratories, Macquarie University, Sydney, NSW, Australia

Hedwig E. Gockel MRC Cognition and Brain Sciences Unit, Cambridge, UK

Nicolas Grimault Centre de Recherche en Neurosciences de Lyon, CNRS UMR 5292, Université Lyon 1, Lyon, France

Wilko Grolman Department of Otorhinolaryngology and Head & Neck Surgery, University Medical Center Utrecht, Utrecht, The Netherlands

Gábor P. Háden Institute of Cognitive Neuroscience and Psychology, Research Centre for Natural Sciences, Hungarian Academy of Sciences, Budapest, Hungary

Nicholas Haywood UCL Ear Institute, London, UK

Erica L. Hegland Purdue University, West Lafayette, IN, USA

Marcel van der Heijden Erasmus MC, Rotterdam, Nederland

Antje Heinrich MRC Institute of Hearing Research, Nottingham, UK

Michael G. Heinz Department of Speech, Language, & Hearing Sciences, Purdue University, West Lafayette, IN, USA

Weldon School of Biomedical Engineering, Purdue University, West Lafayette, IN, USA

Inseok Heo Auditory Behavioral Research Laboratory, Communication Sciences and Disorders, University of Wisconsin-Madison, Madison, WI, USA

Diana Herzog Phonak AG, Staefa, Switzerland

Hongmei Hu Medizinische Physik, Universität Oldenburg, Oldenburg, Germany

Hamish Innes-Brown Bionics Institute, Melbourne, Australia

Research Group Experimental Oto-laryngology, KU Leuven, Leuven, Belgium

Toshio Irino Faculty of Systems Engineering, Wakayama University, Wakayama-city, Japan

Anoop Jagadeesh Medizinische Physik and Cluster of Excellence Hearing4all, Department of Medical Physics and Acoustics, Oldenburg University, Oldenburg, Germany

Kasey M. Jakien National Center for Rehabilitative Auditory Research, VA Portland Health Care System, Portland, OR, USA

Esther Janse Centre for Language Studies, Radboud University, Nijmegen, HD, The Netherlands

Donders Institute for Brain, Cognition and Behavior, Nijmegen, GL, The Netherlands

Max Planck Institute for Psycholinguistics, Nijmegen, AH, The Netherlands

Niels S. Jensen Eriksholm Research Centre, Snekkersten, Denmark

Walt Jesteadt Boys Town National Research Hospital, Omaha, USA

Philip X Joris Laboratory of Auditory Neurophysiology, University of Leuven, Leuven, Belgium

Sean D. Kampel National Center for Rehabilitative Auditory Research, VA Portland Health Care System, Portland, OR, USA

Hideki Kawahara Faculty of Systems Engineering, Wakayama University, Wakayama-city, Japan

Richard Kempter Institute for Theoretical Biology, Department of Biology, Humboldt-Universität zu Berlin, Berlin, Germany

Gerald Kidd Department of Speech, Language and Hearing Sciences, Boston University, Boston, MA, USA

Duck O. Kim Department f Neuroscience, University of Connecticut Health Center, Farmington, CT, USA

Sjaak F.L. Klis Department of Otorhinolaryngology and Head & Neck Surgery, University Medical Center Utrecht, Utrecht, The Netherlands

Georg M. Klump Cluster of Excellence "Hearing4all", Animal Physiology and Behaviour Group, Department for Neuroscience, University of Oldenburg, Oldenburg, Germany

Sarah Knight MRC Institute of Hearing Research, Nottingham, UK

Birger Kollmeier Medizinische Physik, Universität Oldenburg, Oldenburg, Germany

Eike Krause German Center for Vertigo and Balance Disorders (IFB), Department of Otorhinolaryngology, Head and Neck Surgery, Grosshadern Medical Centre, University of Munich, Munich, Germany

Heather A. Kreft Department of Otolaryngology, University of Minnesota—Twin Cities, Minneapolis, MN, USA

Bianca Krumm Cluster of Excellence "Hearing4all", Animal Physiology and Behaviour Group, Department for Neuroscience, University of Oldenburg, Oldenburg, Germany

Shigeyuki Kuwada Department f Neuroscience, University of Connecticut Health Center, Farmington, CT, USA

Björn Lübken Department of Experimental Audiology, Otto von Guericke University Magdeburg, Magdeburg, Germany

Edmund C. Lalor Trinity College Institute of Neuroscience, Trinity College Dublin, Dublin, Ireland

School of Engineering, Trinity College Dublin, Dublin, Ireland

Trinity Centre for Bioengineering, Trinity College Dublin, Dublin, Ireland

Ulrike Langemann Cluster of Excellence "Hearing4all", Animal Physiology and Behaviour Group, Department for Neuroscience, University of Oldenburg, Oldenburg, Germany

Dave R. M. Langers Department of Otorhinolaryngology / Head and Neck Surgery, University of Groningen, University Medical Center Groningen, Groningen, The Netherlands

Graduate School of Medical Sciences (Research School of Behavioural and Cognitive Neurosciences), University of Groningen, Groningen, The Netherlands

Cris Lanting Department of Otorhinolaryngology / Head and Neck Surgery, University of Groningen, University Medical Center Groningen, Groningen, The Netherlands

Graduate School of Medical Sciences (Research School of Behavioural and Cognitive Neurosciences), University of Groningen, Groningen, The Netherlands

Søren Laugesen Eriksholm Research Centre, Snekkersten, Denmark

Mathieu Lavandier Laboratoire Génie Civil et Bâtiment, ENTPE, Université de Lyon, Vaulx-en-Velin, France

Jennifer Lawlor Laboratoire des Systèmes Perceptifs, CNRS UMR 8248, Paris, France

Département d'études cognitives, Ecole normale supérieure PSL Research University, Paris, France

Thibaud Leclère Laboratoire Génie Civil et Bâtiment, ENTPE, Université de Lyon, Vaulx-en-Velin, France

Jungmee Lee Auditory Behavioral Research Laboratory, Communication Sciences and Disorders, University of Wisconsin-Madison, Madison, WI, USA

Katharina Liebner Cluster of Excellence "Hearing4all", Animal Physiology and Behaviour Group, Department for Neuroscience, University of Oldenburg, Oldenburg, Germany

Ruth Litovsky Waisman Center, The University of Wisconsin-Madison, Madison, USA

Dongxin Liu Otolaryngology—Head & Neck Surgery, Beijing Tongren Hospital, Beijing Institute of Otolaryngology, Capital Medical University, Beijing, China

Glenis Long Speech-Language-Hearing Sciences, Graduate Center of City University of New York, New York, NY, USA

Enrique A. Lopez-Poveda INCYL, IBSAL, Dpto. Cirugía, Facultad de Medicina, Universidad de Salamanca, Salamanca, Spain

Robert Lutfi Auditory Behavioral Research Laboratory, Communication Sciences and Disorders, University of Wisconsin-Madison, Madison, WI, USA

Olivier Macherey LMA-CNRS, UPR 7051, Aix-Marseille Univ., Centrale Marseille, Marseille, France

Jeremy Marozeau Bionics Institute, Melbourne, Australia

Department of Electrical Engineering, Technical University of Denmark, Kongens Lyngby, Denmark

Torsten Marquardt UCL Ear Institute, London, UK

Christine R. Mason Department of Speech, Language and Hearing Sciences, Boston University, Boston, MA, USA

Toshie Matsui Faculty of Systems Engineering, Wakayama University, Wakayama-city, Japan

Manfred Mauermann Medizinische Physik and Cluster of Excellence Hearing4all, Department of Medical Physics and Acoustics, Oldenburg University, Oldenburg, Germany

David McAlpine UCL Ear Institute, London, UK

Thomas McColgan Institute for Theoretical Biology, Department of Biology, Humboldt-Universität zu Berlin, Berlin, Germany

Colette McKay The Bionics Institute of Australia, Melbourne, Australia

Department of Medical Bionics, The University of Melbourne, Melbourne, Australia

David McShefferty MRC/CSO Institute of Hearing Research—Scottish Section, Glasgow, UK

Brian C.J. Moore Department of Experimental Psychology, University of Cambridge, Cambridge, UK

Fiona Murray School of Psychology, Plymouth University, Plymouth, UK

Misaki Nagae Faculty of Systems Engineering, Wakayama University, Wakayama-city, Japan

Marjolijn M.B. Nagelkerke Department of Otorhinolaryngology and Head & Neck Surgery, University Medical Center Utrecht, Utrecht, The Netherlands

Stephen T. Neely Boys Town National Research Hospital, Omaha, USA

Waldo Nogueira Dept. of Otolaryngology and Hearing4all, Medical University Hannover, Hannover, Germany

Atsushi Ochi Department of Otolaryngology, Faculty of Medicine, University of Tokyo, Tokyo, Japan

Andrew J. Oxenham Departments of Psychology and Otolaryngology, University of Minnesota—Twin Cities, Minneapolis, MN, USA

Carina Pals Department of Otorhinolaryngology / Head and Neck Surgery, University of Groningen, University Medical Center Groningen, Groningen, The Netherlands

Roy D. Patterson Department of Physiology, Development and Neuroscience, Centre for the Neural Basis of Hearing, University of Cambridge, Cambridge, UK

Department of Physiology, Development and Neuroscience, University of Cambridge, Cambridge, UK

Julie H. Pedersen Eriksholm Research Centre, Snekkersten, Denmark

Patrycja Piktel Medizinische Physik and Cluster of Excellence Hearing4all, Department of Medical Physics and Acoustics, Oldenburg University, Oldenburg, Germany

Christopher J. Plack School of Psychological Sciences, The University of Manchester, Manchester, UK

Filip M. Rønne Eriksholm Research Centre, Snekkersten, Denmark

Dyan Ramekers Department of Otorhinolaryngology and Head & Neck Surgery, University Medical Center Utrecht, Utrecht, The Netherlands

Lars Riecke Department of Cognitive Neuroscience, Maastricht University, Maastricht, The Netherlands

Thilo Rode HörSys GmbH, Hannover, Germany

Stuart Rosen UCL Speech, Hearing & Phonetic Sciences, London, UK

André Rupp Section of Biomagnetism, Department of Neurology, University of Heidelberg, Heidelberg, Germany

Anastasios Sarampalis Department of Psychology, University of Groningen, Groningen, The Netherlands

Graduate School of Medical Sciences (Research School of Behavioural and Cognitive Neurosciences), University of Groningen, Groningen, The Netherlands

Mark Sayles Department of Speech, Language, & Hearing Sciences, Purdue University, West Lafayette, IN, USA

Laboratory of Auditory Neurophysiology, Katholieke Universiteit Leuven, Leuven, Belgium

Mark Sayles Centre for the Neural Basis of Hearing, The Physiological Laboratory, Department of Physiology, Development and Neuroscience, Cambridge, UK

Laboratory of Auditory Neurophysiology, Campus Gasthuisberg, Leuven, Belgium

Odette Scharenborg Centre for Language Studies, Radboud University, Nijmegen, HD, The Netherlands

Donders Institute for Brain, Cognition and Behavior, Nijmegen, GL, The Netherlands

Reinhold Schatzer Institute of Mechatronics, University of Innsbruck, Innsbruck, Austria

Juliane Schmidt Centre for Language Studies, Radboud University, Nijmegen, HD, The Netherlands

Esther Schoenmaker Acoustics Group, Cluster of Excellence "Hearing4All", Carl von Ossietzky University, Oldenburg, Germany

Jean-Luc Schwartz Grenoble Images Parole Signal Automatique-Lab, Speech and Cognition Department, CNRS, Grenoble University, Grenoble, France

Konrad E. Schwarz R&D MED-EL GmbH, Innsbruck, Austria

Abd-Krim Seghouane Department of Electrical and Electronic Engineering, The University of Melbourne, Melbourne, Australia

Adnan Shah The Bionics Institute of Australia, Melbourne, Australia

Department of Electrical and Electronic Engineering, The University of Melbourne, Melbourne, Australia

Shihab Shamma Laboratoire des Systèmes Perceptifs, CNRS UMR 8248, Paris, France

Département d'études cognitives, Ecole normale supérieure PSL Research University, Paris, France

Neural Systems Laboratory, University of Maryland, College Park, MD, USA

Philipp Spitzer R&D MED-EL GmbH, Innsbruck, Austria

Thomas Stainsby Cochlear Technology Centre, Mechelen, Belgium

Arkadiusz Stasiak Centre for the Neural Basis of Hearing, The Physiological Laboratory, Department of Physiology, Development and Neuroscience, Cambridge, UK

Christophe Stoelinga Auditory Behavioral Research Laboratory, Communication Sciences and Disorders, University of Wisconsin-Madison, Madison, WI, USA

Joshua S. Stohl MED-EL Corporation, Durham, NC, USA

Stefan B. Strahl R&D MED-EL GmbH, Innsbruck, Austria

Elizabeth A. Strickland Purdue University, West Lafayette, IN, USA

Jayaganesh Swaminathan Department of Speech, Language and Hearing Sciences, Boston University, Boston, MA, USA

David Théry Laboratoire Génie Civil et Bâtiment, ENTPE, Université de Lyon, Vaulx-en-Velin, France

Andrea C. Trevino Boys Town National Research Hospital, Omaha, USA

Renee Tsongas Bionics Institute, Melbourne, Australia

Joji Tsunada Department of Otorhinolaryngology, University of Pennsylvania, Philadelphia, USA

Jaime Undurraga UCL Ear Institute, London, UK

Stefan Uppenkamp Medizinische Physik and Cluster of Excellence Hearing4All, Carl von Ossietzky Universität Oldenburg, Oldenburg, Germany

Steven van de Par Acoustics group, Cluster of Excellence "Hearing4All", Carl von Ossietzky Universität Oldenburg, Oldenburg, Germany

Filiep Vanpoucke Cochlear Technology Centre, Mechelen, Belgium

Jesko L. Verhey Department of Experimental Audiology, Otto von Guericke University Magdeburg, Magdeburg, Germany

Sarah Verhulst Medizinische Physik and Cluster of Excellence Hearing4all, Department of Medical Physics and Acoustics, Oldenburg University, Oldenburg, Germany

Huib Versnel Department of Otorhinolaryngology and Head & Neck Surgery, University Medical Center Utrecht, Utrecht, The Netherlands

Deborah Vickers UCL Ear Institute, London, UK

Anita Wagner Department of Otorhinolaryngology / Head and Neck Surgery, University of Groningen, University Medical Center Groningen, Groningen, The Netherlands

Graduate School of Medical Sciences (Research School of Behavioural and Cognitive Neurosciences), University of Groningen, Groningen, The Netherlands

Hermann Wagner Institute for Biology II, RWTH Aachen, Aachen, Germany

Michael K. Walls Department of Speech, Language, & Hearing Sciences, Purdue University, West Lafayette, IN, USA

Shuo Wang Otolaryngology—Head & Neck Surgery, Beijing Tongren Hospital, Beijing Institute of Otolaryngology, Capital Medical University, Beijing, China

William M. Whitmer MRC/CSO Institute of Hearing Research—Scottish Section, Glasgow, UK

Lutz Wiegrebe Division of Neurobiology, Dept. Biology II, University of Munich, Martinsried, Germany

Blake S. Wilson Duke University, Durham, NC, USA

István Winkler Institute of Cognitive Neuroscience and Psychology, Research Centre for Natural Sciences, Hungarian Academy of Sciences, Budapest, Hungary

Institute of Psychology, University of Szeged, Szeged, Hungary

Ian M. Winter Centre for the Neural Basis of Hearing, The Physiological Laboratory, Department of Physiology, Development and Neuroscience, Cambridge, UK

Aron Woźniak Department of Otorhinolaryngology / Head and Neck Surgery, University of Groningen, University Medical Center Groningen, Groningen, The Netherlands

Graduate School of Medical Sciences (Research School of Behavioural and Cognitive Neurosciences), University of Groningen, Groningen, The Netherlands

Robert D. Wolford MED-EL Corporation, Durham, NC, USA

Li Xu Communication Sciences and Disorders, Ohio University, Athens, OH, USA

Tatsuya Yamasoba Department of Otolaryngology, Faculty of Medicine, University of Tokyo, Tokyo, Japan

Ifat Yasin Ear Institute, University College London (UCL), London, UK

Johannes Zaar Hearing Systems Group, Department of Electrical Engineering, Technical University of Denmark, Kongens Lyngby, Denmark

Luo Zhang Otolaryngology—Head & Neck Surgery, Beijing Tongren Hospital, Beijing Institute of Otolaryngology, Capital Medical University, Beijing, China

Xin Zhou The Bionics Institute of Australia, Melbourne, Australia

Department of Medical Bionics, The University of Melbourne, Melbourne, Australia

Effects of Age and Hearing Loss on the Processing of Auditory Temporal Fine Structure

Brian C. J. Moore

Abstract Within the cochlea, broadband sounds like speech and music are filtered into a series of narrowband signals, each of which can be considered as a relatively slowly varying envelope (ENV) imposed on a rapidly oscillating carrier (the temporal fine structure, TFS). Information about ENV and TFS is conveyed in the timing and short-term rate of nerve spikes in the auditory nerve. There is evidence that both hearing loss and increasing age adversely affect the ability to use TFS information, but in many studies the effects of hearing loss and age have been confounded. This paper summarises evidence from studies that allow some separation of the effects of hearing loss and age. The results suggest that the monaural processing of TFS information, which is important for the perception of pitch and for segregating speech from background sounds, is adversely affected by both hearing loss and increasing age, the former being more important. The monaural processing of ENV information is hardly affected by hearing loss or by increasing age. The binaural processing of TFS information, which is important for sound localisation and the binaural masking level difference, is also adversely affected by both hearing loss and increasing age, but here the latter seems more important. The deterioration of binaural TFS processing with increasing age appears to start relatively early in life. The binaural processing of ENV information also deteriorates somewhat with increasing age. The reduced binaural processing abilities found for older/hearing-impaired listeners may partially account for the difficulties that such listeners experience in situations where the target speech and interfering sounds come from different directions in space, as is common in everyday life.

Keywords Frequency discrimination · Envelope · Binaural processing · Interaural phase discrimination · Envelope processing · Intelligibility · Processing efficiency · Sound localization · Pitch

B. C. J. Moore (✉)
Department of Experimental Psychology, University of Cambridge, Downing Street, CB2 3EB Cambridge, UK
e-mail: bcjm@cam.ac.uk

P. van Dijk et al. (eds.), *Physiology, Psychoacoustics and Cognition in Normal and Impaired Hearing,* Advances in Experimental Medicine and Biology 894,
DOI 10.1007/978-3-319-25474-6_1

1 Introduction

Within the healthy cochlea, broadband sounds are decomposed into narrowband signals, each of which can be considered as a relatively slowly varying envelope (ENV) imposed on a rapidly oscillating carrier (the temporal fine structure, TFS). Information about ENV and TFS is conveyed in the timing and short-term rate of nerve spikes in the auditory nerve (Joris and Yin 1992). Following Moore (2014), a distinction is made between the physical ENV and TFS of the input signal (ENV_p and TFS_p), the ENV and TFS at a given place on the basilar membrane (ENV_{BM} and TFS_{BM}), and the neural representation of ENV_{BM} and TFS_{BM} (ENV_n and TFS_n). This paper reviews studies that separate the effects of hearing loss and age on the auditory processing of ENV and TFS.

2 Effects of Age

2.1 Monaural Processing of TFS

It is widely believed that the difference limen for frequency (DLF) of pure tones depends on the use of TFS information for frequencies up to 4–5 kHz (Moore 1973), and perhaps even up to 8 kHz (Moore and Ernst 2012). If so, the DLF at low and medium frequencies provides a measure of TFS processing. Abel et al. (1990) and He et al. (1998) compared DLFs for young and older subjects with normal hearing (audiometric thresholds ≤20 dB HL from 0.25 to 4 kHz). In both studies, DLFs were larger by a factor of 2–4 for the older than for the younger subjects, suggesting that there is an effect of age. However, performance on other tasks (duration discrimination for Abel et al. and intensity discrimination for He et al.) was also poorer for the older group, suggesting that there may be a general effect of age that leads to reduced "processing efficiency". The effect of age cannot be attributed to reduced frequency selectivity, since auditory filters do not broaden with increasing age when absolute thresholds remain normal (Lutman et al. 1991; Peters and Moore 1992).

Monaural TFS processing has also been assessed using the TFS1 test (Hopkins and Moore 2007; Moore and Sek 2009). This requires discrimination of a harmonic complex tone (H) from an inharmonic tone (I), created by shifting all components in the H tone up by a fixed amount in hertz, Δf. The value of Δf is adaptively varied to determine the threshold. Both tones are passed through a fixed bandpass filter centred on the higher (unresolved) components, and a background noise is used to mask combination tones and components falling on the skirts of the filter. Evidence suggesting that performance on this test reflects sensitivity to TFS rather than the use of excitation-pattern cues is: (1) Performance on the TFS test does not worsen with increasing level, except when the level is very high (Moore and Sek 2009, 2011; Marmel et al. 2015); (2) Randomizing the level of the individual components would be expected to impair the ability to use excitation-pattern cues, but this has very little effect on performance of the TFS1 test (Jackson and Moore 2014); (3) Differences in excitation level between the H and I tones at the threshold value of Δf, estimated using the H and I tones as forward maskers, are too small to be usable (Marmel et al. 2015).

Moore et al. (2012) used the TFS1 test with centre frequencies of 850 and 2000 Hz to test 35 subjects with audiometric thresholds ≤20 dB HL from 0.25 to 6 kHz and ages from 22 to 61 years. There were significant correlations between age and the thresholds in the TFS1 test ($r=0.69$ and 0.57 for the centre frequencies of 850 and 2000 Hz, respectively). However, TFS1 scores at 850 Hz were correlated with absolute thresholds at 850 Hz ($r=0.67$), even though audiometric thresholds were within the normal range, making it hard to rule out an effect of hearing loss.

Füllgrabe et al. (2015) eliminated confounding effects of hearing loss by testing young (18–27 years) and older (60–79 years) subjects with *matched* audiograms. Both groups had audiometric thresholds ≤20 dB HL from 0.125 to 6 kHz. The TFS1 test was conducted using centre frequencies of 1 and 2 kHz. The older subjects performed significantly more poorly than the young subjects for both centre frequencies. Performance on the TFS1 task was not correlated with audiometric thresholds at the test frequencies. These results confirm that increased age is associated with a poorer ability to use monaural TFS cues, in the absence of any abnormality in the audiogram. It is hard to know whether this reflects a specific deficit in TFS processing or a more general reduction in processing efficiency.

2.2 Monaural Processing of ENV

Füllgrabe et al. (2015) also assessed sensitivity to ENV cues by measuring thresholds for detecting sinusoidal amplitude modulation imposed on a 4-kHz sinusoidal carrier. Modulation rates of 5, 30, 90, and 180 Hz were used to characterize the temporal-modulation-transfer function (TMTF). On average, thresholds (expressed as $20\log_{10}m$, where m is the modulation index) were 2–2.5 dB higher (worse) for the older than for the younger subjects. However, the shapes of the TMTFs were similar for the two groups. This suggests that increasing age is associated with reduced efficiency in processing ENV information, but not with reduced temporal resolution for ENV cues. Schoof and Rosen (2014) found no significant difference in either processing efficiency or the shape of TMTFs measured with noise-band carriers between young (19–29 years) and older (60–72 years) subjects with near-normal audiograms. It is possible that, with noiseband carriers, amplitude-modulation detection is limited by the inherent fluctuations in the carrier (Dau et al. 1997), making it hard to measure the effects of age.

2.3 Binaural Processing of TFS

The binaural processing of TFS can be assessed by measuring the smallest detectable interaural phase difference (IPD) of a sinusoidal carrier relative to an IPD of 0°, keeping the envelope synchronous at the two ears. A task of this type is the TFS-LF test (where LF stands for low-frequency) (Hopkins and Moore 2010b; Sek and Moore 2012). Moore et al. (2012) tested 35 subjects with audiometric thresholds ≤20 dB HL from 0.25 to 6 kHz and ages from 22 to 61 years. The TFS-LF test was used at center frequencies of 500 and 850 Hz. There were significant correlations between age and IPD thresholds at both 500 Hz ($r=0.37$) and 850 Hz ($r=0.65$). Scores on the TFS-LF task were not significantly correlated with absolute thresholds at the test frequency. These results confirm those of Ross et al. (2007) and Grose and Mamo (2010), and indicate that the decline in sensitivity to binaural TFS with increasing age is already apparent by middle age.

Füllgrabe et al. (2015) used the TFS-LF task with their young and older groups with matched (normal) audiograms, as described above. The older group performed significantly more poorly than the young group at both centre frequencies used (500 and 750 Hz). The scores on the TFS-LF test were not significantly correlated with audiometric thresholds at the test frequencies. Overall, the results indicate that the ability to detect changes in IPD worsens with increasing age even when audiometric thresholds remain within the normal range.

The binaural masking level difference (BMLD) provides another measure of binaural sensitivity to TFS. The BMLD is the difference in the detection threshold between a condition where the masker and signal have the same interaural phase and level relationship, and a condition where the relationship is different. Although BMLDs can occur as a result of differences in ENV_n for the two ears, it is generally believed that the largest BMLDs result from the use of TFS_n. Because the BMLD represents a *difference* between two thresholds, it has the advantage that differences in processing efficiency across age groups at least partly cancel. Several researchers have compared BMLDs for young and older subjects with (near-) normal hearing (Pichora-Fuller and Schneider 1991, 1992; Grose et al. 1994; Strouse et al. 1998). All showed that BMLDs (usually for a 500-Hz signal frequency) were smaller for older than for young subjects, typically by 2–4 dB.

In summary, binaural TFS processing deteriorates with increasing age, even when audiometric thresholds are within the normal range.

2.4 Binaural Processing of ENV

King et al. (2014) measured the ability to discriminate IPD using amplitude-modulated sinusoids. The IPDs were imposed either on the carrier, TFS_p, or the modulator, ENV_p. The carrier frequency, f_c, was 250 or 500 Hz and the modulation frequency was 20 Hz. They tested 46 subjects with a wide range of ages (18–83 years) and degrees of hearing loss. The absolute thresholds at the carrier frequencies of 250 and 500 Hz were not significantly correlated with age. Thresholds for detecting IPDs in ENV_p were positively correlated with age for both carrier frequencies (r=0.62 for f_c=250 Hz; r=0.58, for f_c=500 Hz). The correlations remained positive and significant when the effect of absolute threshold was partialled out. These results suggest that increasing age adversely affects the ability to discriminate IPDs in ENV_p, independently of the effects of hearing loss.

3 Effects of Cochlear Hearing Loss

3.1 Monaural Processing of TFS

Many studies have shown that cochlear hearing loss is associated with larger than normal DLFs. For a review, see Moore (2014). However, in most studies the hearing-impaired subjects were older than the normal-hearing subjects, so some (but probably not all) of the effects of hearing loss may have been due to age. DLFs for hearing-impaired subjects are not correlated with measures of frequency selectivity (Tyler et al. 1983; Moore and Peters 1992), confirming that DLFs depend on TFS information and not excitation-pattern information.

Several studies have shown that cochlear hearing loss is associated with a greatly reduced ability to perform the TFS1 test or similar tests (Hopkins and Moore 2007, 2010a, 2011). Many hearing-impaired subjects cannot perform the task at all. Although the effects of hearing loss and age were confounded in some of these studies, the performance of older hearing-impaired subjects seems to be much worse than that of older normal-hearing subjects, suggesting that hearing loss *per se* adversely affects monaural TFS processing.

3.2 Monaural Processing of ENV

Provided that the carrier is fully audible, the ability to detect amplitude modulation is generally not adversely affected by cochlear hearing loss and may sometimes be better than normal, depending on whether the comparison is made at equal sound pressure level or equal sensation level (Bacon and Viemeister 1985; Bacon and Gleitman 1992; Moore et al. 1992; Moore and Glasberg 2001). When the modulation is clearly audible, a sound with a fixed modulation depth appears to fluctuate more for an impaired ear than for a normal ear, probably because of the effects of loudness recruitment (Moore et al. 1996).

3.3 Binaural Processing of TFS

Many studies of the effects of hearing loss on discrimination of interaural time differences (ITDs) are confounded by differences in age between the normal-hearing and hearing-impaired subjects. One exception is the study of Hawkins and Wightman (1980). They measured just noticeable differences in ITD using low-frequency (450–550 Hz) and high-frequency (3750–4250 Hz) narrow-band noise stimuli. The ITD was present in both TFS_p and ENV_p, except that the onsets and offsets of the stimuli were synchronous across the two ears. Performance was probably dominated by the use of TFS cues for the low frequency and ENV cues for the high frequency. They tested three normal-hearing subjects (mean age 25 years) and eight subjects with hearing loss (mean age 27 years). The two groups were tested both at the same sound pressure level (85 dB SPL) and at the same sensation level (30 dB SL). ITD thresholds were significantly higher for the hearing-impaired than for the normal-hearing subjects for both signals at both levels. These results suggest that cochlear hearing loss can adversely affect the discrimination of ITD carried both in TFS_p and in ENV_p.

The study of King et al. (2014), described above, is relevant here. Recall that they tested 46 subjects with a wide range of ages (18–83 years) and degrees of hearing loss. Thresholds for detecting IPDs in TFS_p were positively correlated with absolute thresholds for both carrier frequencies (r=0.45 for f_c=250 Hz; r=0.40, for f_c=500 Hz). The correlations remained positive and significant when the effect of age was partialled out. Thus, hearing loss adversely affects the binaural processing of TFS.

Many studies have shown that BMLDs are reduced for people with cochlear hearing loss, but again the results are generally confounded with the effects of age. More research is needed to establish the extent to which BMLDs are affected by hearing loss *per se*.

3.4 *Binaural Processing of ENV*

Although the results of Hawkins and Wightman (1980), described in Sect. 3.3, suggested that hearing loss adversely affected the discrimination of ITD carried in ENV_p, King et al. (2014) found that thresholds for detecting IPDs in ENV_p were not significantly correlated with absolute thresholds. Also, Léger et al. (2015) reported that thresholds for detecting IPDs in ENV_p were similar for subjects with normal and impaired hearing (both groups had a wide range of ages). Overall, the results suggest that hearing loss does not markedly affect the ability to process binaural ENV cues.

4 Summary and Implications

The monaural and binaural processing of TFS is adversely affected by both increasing age and cochlear hearing loss. The efficiency of processing monaural ENV information may be adversely affected by increasing age, but the temporal resolution of ENV cues appears to be unaffected. The processing of binaural ENV information is adversely affected by increasing age. Cochlear hearing loss does not markedly affect the processing of monaural or binaural ENV information.

The effects of age and hearing loss on TFS processing may partially explain the difficulties experienced by older hearing-impaired people in understanding speech in background sounds (Hopkins and Moore 2010a, 2011; Neher et al. 2012; Füllgrabe et al. 2015).

Acknowledgments This work was supported by grant G0701870 from the MRC (UK). I thank Tom Baer, Hedwig Gockel, and Pim van Dijk for helpful comments.

References

Abel SM, Krever EM, Alberti PW (1990) Auditory detection, discrimination and speech processing in ageing, noise sensitive and hearing-impaired listeners. Scand Audiol 19:43–54

Bacon SP, Gleitman RM (1992) Modulation detection in subjects with relatively flat hearing losses. J Speech Hear Res 35:642–653

Bacon SP, Viemeister NF (1985) Temporal modulation transfer functions in normal-hearing and hearing-impaired subjects. Audiology 24:117–134

Dau T, Kollmeier B, Kohlrausch A (1997) Modeling auditory processing of amplitude modulation. I. Detection and masking with narrowband carriers. J Acoust Soc Am 102:2892–2905

Füllgrabe C, Moore BCJ, Stone MA (2015) Age-group differences in speech identification despite matched audiometrically normal hearing: contributions from auditory temporal processing and cognition. Front Aging Neurosci 6(347):1–25

Grose JH, Mamo SK (2010) Processing of temporal fine structure as a function of age. Ear Hear 31:755–760

Grose JH, Poth EA, Peters RW (1994) Masking level differences for tones and speech in elderly listeners with relatively normal audiograms. J Speech Hear Res 37:422–428

Hawkins DB, Wightman FL (1980) Interaural time discrimination ability of listeners with sensori-neural hearing loss. Audiology 19:495–507

He N, Dubno JR, Mills JH (1998) Frequency and intensity discrimination measured in a maximum-likelihood procedure from young and aged normal-hearing subjects. J Acoust Soc Am 103:553–565

Hopkins K, Moore BCJ (2007) Moderate cochlear hearing loss leads to a reduced ability to use temporal fine structure information. J Acoust Soc Am 122:1055–1068

Hopkins K, Moore BCJ (2010a) The importance of temporal fine structure information in speech at different spectral regions for normal-hearing and hearing-impaired subjects. J Acoust Soc Am 127:1595–1608

Hopkins K, Moore BCJ (2010b) Development of a fast method for measuring sensitivity to temporal fine structure information at low frequencies. Int J Audiol 49:940–946

Hopkins K, Moore BCJ (2011) The effects of age and cochlear hearing loss on temporal fine structure sensitivity, frequency selectivity, and speech reception in noise. J Acoust Soc Am 130:334–349

Jackson HM, Moore BCJ (2014) The role of excitation-pattern and temporal-fine-structure cues in the discrimination of harmonic and frequency-shifted complex tones. J Acoust Soc Am 135:1356–1570

Joris PX, Yin TC (1992) Responses to amplitude-modulated tones in the auditory nerve of the cat. J Acoust Soc Am 91:215–232

King A, Hopkins K, Plack CJ (2014) The effects of age and hearing loss on interaural phase discrimination. J Acoust Soc Am 135:342–351

Léger A, Heinz MG, Braida LD, Moore BCJ (2015) Sensitivity to interaural time differences in envelope and fine structure, individually and in combination. In: ARO 38th Annual Midwinter Meeting (Abstract PS-703). ARO, Baltimore, MD

Lutman ME, Gatehouse S, Worthington AG (1991) Frequency resolution as a function of hearing threshold level and age. J Acoust Soc Am 89:320–328

Marmel F, Plack CJ, Hopkins K, Carlyon RP, Gockel HE, Moore BCJ (2015) The role of excitation-pattern cues in the detection of frequency shifts in bandpass-filtered complex tones. J Acoust Soc Am 137:2687–2697

Moore BCJ (1973) Frequency difference limens for short-duration tones. J Acoust Soc Am 54:610–619

Moore BCJ (2014) Auditory processing of temporal fine structure: effects of age and hearing loss. World Scientific, Singapore

Moore BCJ, Ernst SM (2012) Frequency difference limens at high frequencies: evidence for a transition from a temporal to a place code. J Acoust Soc Am 132:1542–1547

Moore BCJ, Glasberg BR (2001) Temporal modulation transfer functions obtained using sinusoidal carriers with normally hearing and hearing-impaired listeners. J Acoust Soc Am 110:1067–1073

Moore BCJ, Peters RW (1992) Pitch discrimination and phase sensitivity in young and elderly subjects and its relationship to frequency selectivity. J Acoust Soc Am 91:2881–2893

Moore BCJ, Sek A (2009) Development of a fast method for determining sensitivity to temporal fine structure. Int J Audiol 48:161–171

Moore BCJ, Sek A (2011) Effect of level on the discrimination of harmonic and frequency-shifted complex tones at high frequencies. J Acoust Soc Am 129:3206–3212

Moore BCJ, Shailer MJ, Schooneveldt GP (1992) Temporal modulation transfer functions for band-limited noise in subjects with cochlear hearing loss. Br J Audiol 26:229–237

Moore BCJ, Wojtczak M, Vickers DA (1996) Effect of loudness recruitment on the perception of amplitude modulation. J Acoust Soc Am 100:481–489

Moore BCJ, Vickers DA, Mehta A (2012) The effects of age on temporal fine structure sensitivity in monaural and binaural conditions. Int J Audiol 51:715–721

Neher T, Lunner T, Hopkins K, Moore BCJ (2012) Binaural temporal fine structure sensitivity, cognitive function, and spatial speech recognition of hearing-impaired listeners. J Acoust Soc Am 131:2561–2564

Peters RW, Moore BCJ (1992) Auditory filters and aging: filters when auditory thresholds are normal. In: Cazals Y, Demany L, Horner K (eds) Auditory physiology and perception. Pergamon, Oxford, pp 179–185

Pichora-Fuller MK, Schneider BA (1991) Masking-level differences in the elderly: a comparison of antiphasic and time-delay dichotic conditions. J Speech Hear Res 34:1410–1422

Pichora-Fuller MK, Schneider BA (1992) The effect of interaural delay of the masker on masking-level differences in young and old subjects. J Acoust Soc Am 91:2129–2135

Ross B, Fujioka T, Tremblay KL, Picton TW (2007) Aging in binaural hearing begins in mid-life: evidence from cortical auditory-evoked responses to changes in interaural phase. J Neurosci 27:11172–11178

Schoof T, Rosen S (2014) The role of auditory and cognitive factors in understanding speech in noise by normal-hearing older listeners. Front Aging Neurosci 6(307):1–14

Sek A, Moore BCJ (2012) Implementation of two tests for measuring sensitivity to temporal fine structure. Int J Audiol 51:58–63

Strouse A, Ashmead DH, Ohde RN, Grantham DW (1998) Temporal processing in the aging auditory system. J Acoust Soc Am 104:2385–2399

Tyler RS, Wood EJ, Fernandes MA (1983) Frequency resolution and discrimination of constant and dynamic tones in normal and hearing-impaired listeners. J Acoust Soc Am 74:1190–1199

Aging Effects on Behavioural Estimates of Suppression with Short Suppressors

Erica L. Hegland and Elizabeth A. Strickland

Abstract Auditory two-tone suppression is a nearly instantaneous reduction in the response of the basilar membrane to a tone or noise when a second tone or noise is presented simultaneously. Previous behavioural studies provide conflicting evidence on whether suppression changes with increasing age, and aging effects may depend on whether a suppressor above (high-side) or below (low-side) the signal frequency is used. Most previous studies have measured suppression using stimuli long enough to elicit the medial olivocochlear reflex (MOCR), a sound-elicited reflex that reduces cochlear amplification or gain. It has a "sluggish" onset of approximately 25 ms. There is physiological evidence that suppression may be reduced or altered by elicitation of the MOCR. In the present study, suppression was measured behaviourally in younger adults and older adults using a forward-masking paradigm with 20-ms and 70-ms maskers and suppressors. In experiment 1, gain was estimated by comparing on-frequency (2 kHz) and off-frequency (1.2 kHz) masker thresholds for a short, fixed-level 2-kHz signal. In experiment 2, the fixed-level signal was preceded by an off-frequency suppressor (1.2 or 2.4 kHz) presented simultaneously with the on-frequency masker. A suppressor level was chosen that did not produce any forward masking of the signal. Suppression was measured as the difference in on-frequency masker threshold with and without the suppressor present. The effects of age on gain and suppression estimates will be discussed.

Keywords Two-tone suppression · Medial olivocochlear reflex · MOCR · Endocochlear potential · Gain estimates · Duration

1 Introduction

We live in a noisy world, and understanding speech in noise is a common challenge. As people age, this task becomes increasingly difficult, and elevated thresholds alone cannot account for this increased difficulty (e.g., Pichora-Fuller et al. 1995). One explanation for this discrepancy may be attributed to the decrease in the en-

E. L. Hegland (✉) · E. A. Strickland
Purdue University, 715 Clinic Drive, West Lafayette, IN 47907, USA
e-mail: ehegland@purdue.edu

P. van Dijk et al. (eds.), *Physiology, Psychoacoustics and Cognition in Normal and Impaired Hearing,* Advances in Experimental Medicine and Biology 894,
DOI 10.1007/978-3-319-25474-6_2

docochlear potential that tends to occur with increasing age (e.g., Schuknecht et al. 1974). A decreased endocochlear potential negatively affects the function of inner and outer hair cells (OHCs) (e.g., Gates et al. 2002), and disruptions in OHC function could decrease cochlear nonlinearity (e.g., Schmiedt et al. 1980).

A sensitive measure of cochlear nonlinearity is two-tone suppression, e.g., a nearly instantaneous reduction in the basilar membrane (BM) response to one tone in the presence of a second tone. There is psychoacoustic evidence that suppression measured with suppressors above the signal frequency (high-side suppression) and below the signal frequency (low-side suppression) may show different aging effects. Several of these psychoacoustic studies have used a band-widening technique in a forward-masking paradigm to measure suppression (Dubno and Ahlstrom 2001a, b; Gifford and Bacon 2005). Narrow bandpass noise centred at the signal frequency was widened by adding noise bands above, below, or above and below the signal frequency. Suppression was measured as the difference in signal threshold between the narrowest and widest noisebands. Dubno and Ahlstrom (2001a, b) found less suppression with more noise above rather than below the signal frequency, and the older adults with normal hearing had less suppression than the younger adults. Older adults had minimal low-side suppression and absent high-side suppression (Dubno and Ahlstrom 2001a). In contrast, using the band-widening technique with noise added below the signal, Gifford and Bacon (2005) found no effect of age on suppression. Using tonal maskers and suppressors in a forward-masking paradigm, Sommers and Gehr (2010) found reduced or absent high-side suppression with increased age.

It is possible that previous psychoacoustic studies of suppression underestimated the amount of suppression by using maskers that were long enough to elicit the medial olivocochlear reflex (MOCR). The MOCR is a sound-elicited efferent feedback loop that reduces OHC amplification or gain (Cooper and Guinan 2006). Gain starts to be affected approximately 25 ms after elicitor onset. Evidence from physiological studies in animals suggests that elicitation of the MOCR may decrease or modify suppression (Winslow and Sachs 1987; Kawase et al. 1993). Previous studies of aging effects on suppression used masker and suppressor durations of 200–500 ms (Dubno and Ahlstrom 2001a, b; Gifford and Bacon 2005; Sommers and Gehr 1998, 2010). These durations were long enough to elicit MOCR-induced gain reductions and possibly reduce suppression estimates.

The goal of this study was to investigate the effect of age on estimates of two-tone suppression using shorter stimuli. Several prior studies of aging effects on suppression have used masker levels that had been previously shown to produce large amounts of suppression in young adults (Dubno and Ahlstrom 2001a, 2001b; Gifford and Bacon 2005; Sommers and Gehr 2010). In the present study, suppressor levels were chosen that did not produce forward masking of the signal, which can be presumed to produce minimal spread of excitation at the signal frequency place (Moore and Vickers 1997). If simultaneous masking is produced by spread of excitation and suppression, using a suppressor that does not produce forward masking should result in a measure of suppression rather than a mixture of suppression and excitation.

2 Methods

2.1 Participants

Participants included five younger adults between 18 and 30 years old (mean = 21.4; 3 female) and five older adults between 62 and 70 years old (mean = 65.4; 4 female). The average threshold for younger adults at 2 kHz was 5 dB HL (SD 3.54), and the average threshold for older adults at 2 kHz was 15 dB HL (SD 9.35). The Montreal Cognitive Assessment (MoCA; used with permission, copyright Z. Nasreddine) was administered, and all results were within the normal range (≥26/30).

2.2 Stimuli

The signal was a 10-ms, 2-kHz sinusoid with 5-ms onset and offset ramps and no steady state portion. This signal frequency has been used in previous studies on aging effects on suppression (Sommers and Gehr 1998, 2010; Dubno and Ahlstrom 2001a, 2001b; Gifford and Bacon 2005). The signal was presented at 10 dB SL and/or 50 dB SPL for each participant, and the masker level was adaptively varied to find threshold. Masker durations were 20 (short) and 70 ms (long). The 20-ms masker should be too short to elicit the MOCR, while the 70-ms masker should be long enough to elicit full gain reduction from MOCR elicitation (Roverud and Strickland 2010). For experiment 1, masker thresholds for the short and long on-frequency (2 kHz) and off-frequency (1.2 and 2.4 kHz) maskers were measured (Fig. 1). At the signal frequency place, the BM response to the 1.2-kHz masker should be approximately linear, even with MOCR stimulation (Cooper and Guinan 2006). The 2.4-kHz masker frequency was chosen because it has produced substantial suppression of a 2-kHz masker in previous studies with young adults (Shannon 1976; Sommers and Gehr 2010). Gain was estimated as the difference in threshold between the on-frequency masker and the 1.2-kHz masker (Yasin et al. 2014). For experiment 2, suppression was measured by presenting a short or long, 1.2- or 2.4-kHz suppressor simultaneously with the on-frequency masker and just

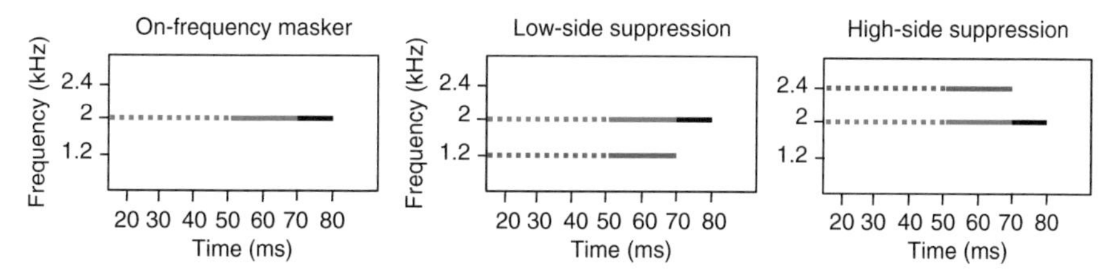

Fig. 1 Schematic of stimuli frequencies (kHz) and durations (ms). Masker durations were 20 ms (*solid lines*) and 70 ms (*solid + dashed lines*). Maskers were on-frequency (*blue*, 2 kHz) and off-frequency (*red*, 1.2 and 2.4 kHz)

prior to presentation of the 10-ms, 2-kHz signal. This will be referred to as the "combined masker." The masker and suppressor durations were always equal. The on-frequency masker level was adaptively varied to find threshold.

All stimuli were digitally produced with MATLAB™ software (2012a, The Math Works, Natick, MA). They were output via a Lynx II XLR sound card to a headphone buffer (TDT HB6) and presented to the right ear or the better ear of each participant through an ER-2 insert earphone (Etymotic Research, Inc., Elk Grove Village, IL).

2.3 Procedure

Participants sat in a sound-attenuating booth and listed to stimuli presented to their right ear or better ear. Masker threshold for a fixed signal level was measured using a three-interval forced-choice adaptive procedure. Participants were instructed to use a computer mouse or keyboard to choose the interval that sounded different from the others. The masker level was increased following one incorrect response and decreased following two correct responses. Following this procedure, an estimate of 70 % correct on the psychometric function is obtained (Levitt 1971). Masker threshold for a given trial was calculated as an average of the last eight of the 12 total reversals. Thresholds with a standard deviation of 5 dB or greater were not included in the final average. The order of the conditions was counterbalanced across the participants.

3 Results

3.1 Experiment 1

Thresholds for the short and long on-frequency (2 kHz) and off-frequency maskers (1.2 and 2.4 kHz) were obtained for signal levels fixed at 10 dB above each participant's threshold for the 2-kHz, 10-ms tone. Because the absolute signal level differed between younger and older participants, masker thresholds were also obtained for a 50-dB SPL signal, which was 10-dB SL for some older participants.

Estimates of cochlear gain were calculated as the difference in masker threshold between the 2-kHz and 1.2-kHz short maskers (Yasin et al 2014). Gain was determined separately for each signal level. A one-way ANOVA was conducted to compare the effect of age group on gain estimates at each signal level. There was a significant effect of age group on gain estimates at the 10-dB SL signal level [$F(1,8)=7.03$, $p=0.029$] and the 50-dB SPL signal level [$F(1,7)=6.86$, $p=0.034$]. Average amounts of gain are shown in Fig. 2. These results suggest that gain estimates were significantly decreased in older adults compared to younger adults.

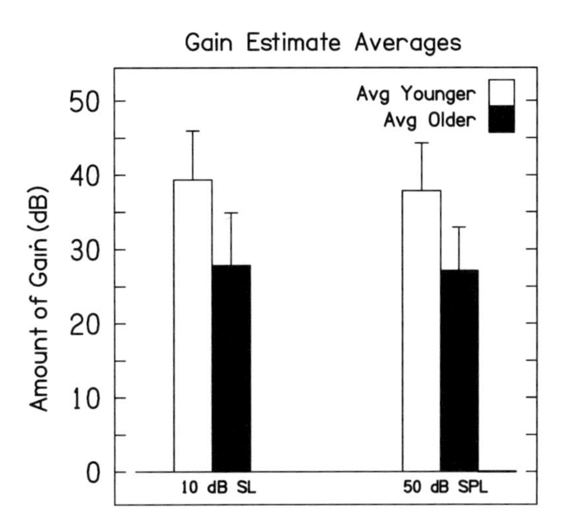

Fig. 2 Average amounts of gain for older and younger listeners at the 10-dB SL and 50-dB SPL signal levels

Correlation coefficients were computed between auditory threshold at 2 kHz and gain estimates at both signal levels, and neither of the correlations reached statistical significance.

3.2 Experiment 2

Suppression was measured in a forward-masking paradigm (see Fig. 1). Suppressor levels were chosen that were below the thresholds obtained in Experiment 1 to ensure that the suppressor did not produce any forward masking of the signal. For each participant, several suppressor levels were presented, and the level producing the greatest amount of suppression was used for analysis. The suppressor was an average of 4 dB below the off-frequency masker threshold obtained in Experiment 1 [*SD*=1.60]. For younger adults, suppressors were presented at an average of 74.50 dB SPL [*SD*=5.22] for low-side suppression and 74.23 dB SPL [*SD*=4.00] for high-side suppression. For older adults, suppressors were presented at an average of 67.60 dB SPL [*SD*=2.12] for low-side suppression and 75.20 dB SPL [*SD*=3.68] for high-side suppression. Compared to the 10 dB SL signal level, suppressors were on average 6.8 dB higher at the 50 dB SPL signal level for the younger adults and 2.7 dB higher for the older adults.

Average amounts of low-side and high-side suppression for younger adults (open bars) and older adults (filled bars) at the 10-dB SL and 50-dB SPL signal levels are shown in Fig. 3. With the short combined masker, all participants showed some low-side suppression, and all participants except for one older adult showed high-side suppression. With the long combined masker, 2/5 older adults had no measurable low-side suppression, and 3/5 older adults had no measurable high-side suppression. A one-way ANOVA was conducted to compare the effect of age group

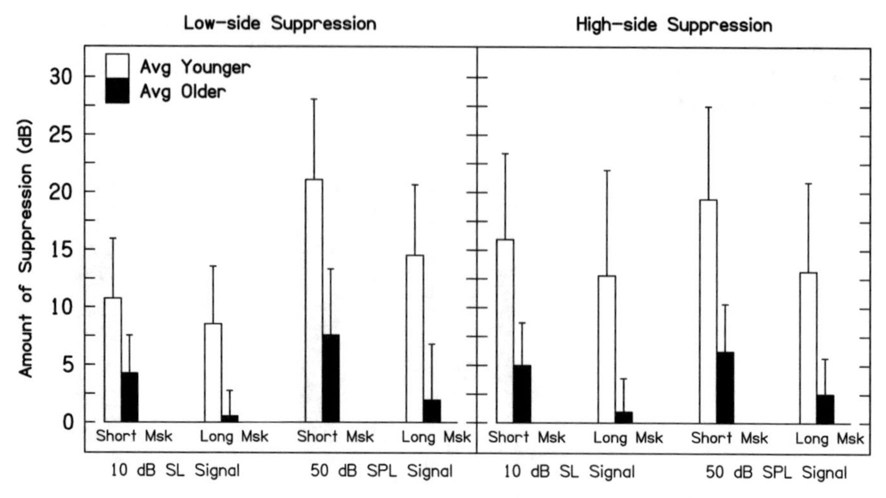

Fig. 3 Average suppression estimates for younger adults (*open bars*) and older adults (*filled bars*). Results are shown for the 1.2 kHz suppressor (*low-side suppression*), 2.4 kHz suppressor (*high-side suppression*), 20-ms combined masker (*short msk*), 70-ms combined masker (*long msk*), 10-dB SL signal level, and 50-dB SPL signal level

on suppression estimates. There was a significant effect of age group on suppression estimates at the $p<0.05$ level for all conditions tested. There was a significant effect of age group for low-side suppression with a short combined masker [10-dB SL signal: $F(1,8)=5.53, p=0.047$; 50-dB SPL signal: $F(1,8)=11.09, p=0.010$] and with a long combined masker [10-dB SL signal: $F(1,8)=10.57, p=0.012$; 50-dB SPL signal: $F(1,8)=12.94, p=0.007$]. There was a significant effect of age group for high-side suppression with a short combined masker [10-dB SL signal: $F(1,8)=8.66$, $p=0.019$; 50-dB SPL signal: $F(1,7)=10.37$, $p=0.015$] and with a long combined masker [10-dB SL signal: $F(1,8)=7.55$, $p=0.025$; 50-dB SL signal: $F(1,7)=7.99$, $p=0.026$]. This suggests that both high-side and low-side suppression estimates decreased with age.

A paired-samples t-test was conducted to compare suppression estimates obtained with short and long combined maskers. In all suppression conditions tested, there were significant differences at the $p<0.05$ level between suppression estimates obtained with short and long combined maskers. There was a significant difference between low-side suppression estimates obtained with short [$M=7.49$, $SD=5.34$] and long combined maskers [$M=4.54$, $SD=5.57$] at the 10-dB SL signal level [$t(9)=2.94, p=0.017$] and between short [$M=14.33$, $SD=9.33$] and long combined maskers [$M=8.25$, $SD=8.42$] at the 50-dB SPL signal level [$t(9)=3.41, p=0.008$]. There was a significant difference between high-side suppression estimates obtained with short [$M=10.46$, $SD=7.99$] and long combined maskers [$M=6.86$, $SD=8.95$] at the 10-dB SL signal level [$t(9)=3.83$, $p=0.004$] and between short [$M=12.08$, $SD=9.03$] and long combined maskers [$M=7.24$, $SD=7.66$] at the 50-dB SPL signal level [$t(9)=4.56, p=0.002$]. This suggests that suppression estimates were significantly greater when measured with a short combined masker.

It is possible that the decreased suppression in the older adults was due to their increased auditory thresholds. To determine the effect of auditory threshold on suppression estimates, correlation coefficients were computed between threshold in dB HL and suppression estimates from each of the eight conditions tested. Out of the eight suppression conditions, threshold at 2 kHz was negatively correlated with low-side suppression at the 50-dB SPL signal level measured with the short combined masker [$r(10)=-0.68$, $p=0.031$] and with the long combined masker [$r(10)=-0.67$, $p=0.33$]. This suggests that those with higher auditory thresholds had smaller low-side suppression estimates at the 50-dB SPL signal level. Otherwise, auditory threshold was not significantly correlated with either suppression or gain estimates.

4 Discussion

In the present experiment, several important differences were seen between younger adults and older adults. In Experiment 1, older adults were found to have less gain than younger adults, measured as the difference in on-frequency and off-frequency masker threshold for a fixed signal level (Yasin et al. 2014). The off-frequency, 1.2-kHz masker is understood to have a linear BM response at the signal frequency place, and the on-frequency, 2-kHz masker and signal have a compressive BM response at the signal frequency place (Oxenham and Plack 2000). Comparing these two conditions gives an estimate of cochlear gain around the signal frequency place. It is possible that the decrease in gain for the older adults was due to their slightly poorer auditory thresholds. However, correlation coefficients between auditory threshold and gain estimates did not reach significance. This measure of gain may reflect decreases in cochlear function that occur prior to elevations in auditory thresholds. Decreased endocochlear potential may have less of an effect on thresholds closer to the apex of the cochlea and a larger effect in the base (Mills et al. 2006) while still impacting cochlear nonlinearity.

Estimates of high-side and low-side suppression were significantly smaller for older adults than for younger adults. Most previous studies of aging effects on suppression have found decreased or absent suppression for older adults (Sommers and Gehr 1998, 2010; Dubno and Ahlstrom 2001a, b). In the present study, suppressor levels were chosen to not produce forward masking and to result in the greatest measure of suppression. Also, with the long combined masker, some older adults did not show high-side suppression, and some did not show low-side suppression. By using short combined maskers with individually-selected suppressor levels, accurate measurements of suppression can be achieved.

Both younger and older adults in the present study showed significantly more suppression measured with short combined maskers than with longer combined maskers. Several of the older participants, who had no measurable suppression with the longer combined masker, had a small amount of suppression measured with the shorter combined masker. When suppression is measured with combined maskers long enough to elicit MOCR-induced gain reduction, suppression may be underestimated. However, even when measured with the shorter combined maskers, older adults had less suppression than younger adults. Evidence from animal studies suggests that gain reduction from the MOCR and suppression interact, resulting in decreased suppression (Winslow and Sachs 1987; Kawase et al. 1993). The results of the present study support this idea.

Dubno and Ahlstrom (2001a) found a negative correlation between auditory thresholds and suppression estimates. In the present study, there was a negative correlation between auditory thresholds and suppression estimates for low-side suppression at the 50-dB SPL signal level. The correlation was not significant for low-side suppression at the 10-dB SL signal level or for high-side suppression. The suppression estimates in Dubno and Ahlstrom (2001a) were measured between the narrowest and widest bandwidth maskers. This measure would include elements of both high-side and low-side suppression. It is possible that auditory threshold is more strongly correlated with low-side suppression.

The present study found strong evidence for age-related decreases in both high-side and low-side two-tone suppression and in estimates of cochlear gain. These findings are consistent with a decreasing endocochlear potential that occurs with increasing age. The results also support the hypothesis that that suppression estimates may be underestimated when measured with long combined maskers, which would be consistent with a reduction in suppression due to the MOCR.

Acknowledgments This research was supported by a fellowship through NIH(NIDCD) T32 DC000030-21 and NIH(NIDCD) F31 DC014395 (first author), and NIH(NIDCD) R01 DC008327 (second author).

References

Cooper NP, Guinan JJ Jr (2006) Efferent-mediated control of basilar membrane motion. J Physiol 576(1):49–54

Dubno JR, Ahlstrom JB (2001a) Psychophysical suppression effects for tonal and speech signals. J Acoust Soc Am 110(4):2108–2119

Dubno JR, Ahlstrom JB (2001b) Psychophysical suppression measured with bandlimited noise extended below and/or above the signal: effects of age and hearing loss. J Acoust Soc Am 110(2):1058–1066

Gates GA, Mills D, Nam B, D'Agostino R, Rubel EW (2002) Effects of age on the distortion product otoacoustic emission growth functions. Hear Res 163(1–2):53–60

Gifford RH, Bacon SP (2005) Psychophysical estimates of nonlinear cochlear processing in younger and older listeners. J Acoust Soc Am 118(6):3823–3833
Kawase T, Delgutte B, Liberman MC (1993) Antimasking effects of the olivocochlear reflex. II. Enhancement of auditory-nerve response to masked tones. J Neurophysiol 70(6):2533–2549
Levitt H (1971) Transformed up-down methods in psychoacoustics. J Acoust Soc Am 49(2):467–477
Mills JH, Schmiedt RA, Schulte BA, Dubno JR (2006) Age-related hearing loss: a loss of voltage, not hair cells. Semin Hear 27(4):228–236
Moore BCJ, Vickers DA (1997) The role of spread of excitation and suppression in simultaneous masking. J Acoust Soc Am 102(4):2284–2290
Oxenham AJ, Plack CJ (2000) Effects of masker frequency and duration in forward masking: further evidence for the influence of peripheral nonlinearity. Hear Res 150:258–266
Pichora- Fuller MK, Schneider BA, Daneman M (1995) How young and old adults listen to and remember speech in noise. J Acoust Soc Am 97(1):593–608
Roverud E, Strickland EA (2010) The time course of cochlear gain reduction measured using a more efficient psychophysical technique. J Acoust Soc Am 128(3):1203–1214
Schmiedt RA, Zwislocki JJ, Hamernik RP (1980) Effects of hair cell lesions on responses of cochlear nerve fibers. I. Lesions, tuning curves, two-tone inhibition, and responses to trapezoidal-wave patterns. J Neurophysiol 43(5):1367–1389
Schuknecht HF, Watanuki K, Takahashi T, Belal AA Jr, Kimura RS, Jones DD, Ota CY (1974) Atrophy of the stria vascularis, a common cause for hearing loss. Laryngoscope 84(10):1777–1821
Shannon RV (1976) Two-tone unmasking and suppression in a forward-masking situation. J Acoust Soc Am 59(6):1460–1470
Sommers MS, Gehr SE (1998) Auditory suppression and frequency selectivity in older and younger adults. J Acoust Soc Am 103(2):1067–1074
Sommers MS, Gehr SE (2010) Two-tone auditory suppression in younger and older normal-hearing adults and its relationship to speech perception in noise. Hear Res 264:56–62
Winslow RL, Sachs MB (1987) Effect of electrical stimulation of the crossed olivocochlear bundle on auditory nerve response to tones in noise. J Neurophysiol 57(4):1002–1021
Yasin I, Drga V, Plack CJ (2014) Effect of human auditory efferent feedback on cochlear gain and compression. J Neurosci 34(46):15319–15326

Contributions of Coding Efficiency of Temporal-Structure and Level Information to Lateralization Performance in Young and Early-Elderly Listeners

Atsushi Ochi, Tatsuya Yamasoba and Shigeto Furukawa

Abstract The performance of a lateralization task based on interaural time or level differences (ITDs or ILDs) often varies among listeners. This study examined the extent to which this inter-listener variation could be accounted for by the coding efficiency of the temporal-structure or level information below the stage of interaural interaction. Young listeners (20s to 30s) and early-elderly (60s) listeners with or without mild hearing loss were tested. The *ITD, ILD, TIME*, and *LEVEL* tasks were intended to measure sensitivities to ITDs, ILDs, the temporal structure of the stimulus encoded by the neural phase locking, and the stimulus level, respectively. The performances of the *ITD* and *ILD* tasks were not significantly different between the age groups, while the elderly listeners exhibited significantly poorer performance in the *TIME* task (and in the *LEVEL* with a high-frequency stimulus only) than the young listeners. Significant correlations were found between thresholds for the *ILD* and *LEVEL* tasks with low- and high-frequency stimuli and for the *ITD* and *TIME* tasks for the high-frequency stimulus, implying peripheral coding efficiency as a major factor determining lateralization performance. However, we failed to find a correlation between the *ITD* and *TIME* tasks for the low-frequency stimulus, despite a large range of threshold values in the *TIME* task. This implies that in a low frequency region, the peripheral coding efficiency of the stimulus temporal structure is a relatively minor factor in the ITD-based lateralization performance.

Keywords Interaural time difference · Interaural level difference · Temporal structure · Intensity coding · Inter-individual variation · Aging

---

S. Furukawa (✉)
Human Information Science Laboratory, NTT Communication Science Laboratories, NTT Corporation, Atsugi, Japan
e-mail: furukawa.shigeto@lab.ntt.co.jp

A. Ochi · T. Yamasoba
Department of Otolaryngology, Faculty of Medicine, University of Tokyo, Tokyo, Japan
e-mail: ochia-tky@umin.ac.jp

T. Yamasoba
e-mail: tyamasoba-tky@umin.ac.jp

P. van Dijk et al. (eds.), *Physiology, Psychoacoustics and Cognition in Normal and Impaired Hearing,* Advances in Experimental Medicine and Biology 894,
DOI 10.1007/978-3-319-25474-6_3

1 Introduction

Interaural time and level differences (ITDs and ILDs) are the major cues for horizontal sound localization. Sensitivities to ITDs and ILDs, evaluated by lateralization tasks, often vary markedly among listeners. Lateralization performance based on ITDs and ILDs should reflect not only the listener's ability to compare time and level information, respectively, between ears but also the efficiency of encoding information about the temporal structure and intensity of stimuli at stages below binaural interactions in auditory processing. Our earlier study attempted to evaluate the relative contributions of these processing stages to the inter-listener variability in lateralization performance, by comparing individual listeners' monaural sensitivities to the temporal structure and intensity of a sound stimulus with their ITD and ILD sensitivities (Ochi et al. 2014). The results showed significant correlation of ILD discrimination thresholds with thresholds for monaural level-increment detection task. This could be interpreted as indicating that the inter-individual differences in ILD sensitivity could be (partially) accounted for by the level coding efficiency at stages before binaural interaction. Similarly, ITD discrimination thresholds were found to correlate with the listeners' sensitivities to the temporal structure of monaural stimuli, when the stimuli were in high frequency range (around 4000 Hz). However, we failed to find a positive correlation for stimuli in low-frequency range (around 1100 Hz).

The present study extends our earlier study (Ochi et al. 2014) by incorporating early-elderly listeners under essentially the same experimental settings. We adopted early-elderly listeners because generally they would exhibit deteriorated sensitivities to temporal structures of stimuli, while their audiometric thresholds remain within a normal to mildely-impaired range. We first examined the effects of age on the performance of individual tasks. We then analysed correlations, as in the earlier study, between task performances. It has been reported that sensitivities to the temporal structure and intensity of stimuli decline with age (e.g., Hopkins and Moore 2011). A population including young and elderly listeners would therefore exhibit a large variability of thresholds in the monaural tasks, which would lead to improved sensitivity of the correlation analyses and provide further insights as to the roles of monaural processing in ITD or ILD discrimination. Supplemental data were also obtained to evaluate underlying mechanisms for the monaural tasks.

As in the earlier study, we measured listeners' performances in four basic tasks, namely *ITD, ILD, TIME*, and *LEVEL* tasks, which would reflect sensitivities to ITDs, ILDs, the temporal structure, and the level change of stimuli, respectively. Low- and high-frequency stimuli were tested, which were centred at around 1100 and 4000 Hz, respectively. Supplementary experiments measured frequency resolution (*FRES* task), frequency discrimination limens (*FDISC* task), and (for low-frequency stimulus only) the discrimination threshold of Huggins pitch (*HUGGINS* task; Cramer and Huggins 1958).

2 Methods

2.1 Listeners

Forty-three adults participated in the experiment. All gave written informed consent, which was approved by the Ethics Committee of NTT Communication Science Laboratories. Those included 22 normal-hearing young listeners (referred to as the YNH group; 10 males and 12 females; 19–43 years old, mean 32.0) and 21 elderly listeners (11 males and 10 females; 60–70 years old, mean 63.0). The data from the YNH listeners have been represented in the earlier study (Ochi et al. 2014). In the analyses, the elderly listeners were further divided into normal-hearing (referred to as ENH) and hearing-impaired (EHI) groups. Listeners with audiometric thresholds of <30 dB HL at all the frequencies between 125 and 4000 Hz in both ears were classified as normal-hearing; otherwise, as (mildly) hearing-impaired. For the *FDISC* and *HIGGINS* tasks, a subset of YNH listeners (N=12) and all the elderly listeners participated. Table 1 summarizes the means and standard deviations of hearing levels obtained by pure-tone audiometry.

2.2 Stimuli

Stimuli were presented to the listener through headphones. Except for the binaural tasks (i.e., *ITD, ILD*, and *HUGGINS* tasks), the stimuli were presented to the right ear.

The main four tasks (namely, *ITD, ILD, TIME*, and *LEVEL* tasks) employed two types of stimuli, referred to as the low- and high-frequency stimuli, which were identical to those used in our earlier study (Ochi et al. 2014) and are thus only described briefly here. The stimuli were designed to assess the listener's ability to use information on the basis of neural phase-locking to the stimulus temporal structure, respectively, in the *ITD* and *TIME* tasks. Essentially the same stimuli were also used in the *ILD* and *LEVEL* tasks. The low-frequency stimulus was a spectrally shaped multicomponent complex (SSMC), which was a harmonic complex with a fundamental frequency (F_0) of 100 Hz. The spectral envelope had a flat passband and sloping edges ($5 \times F_0$ centered at 1100 Hz).The overall level of the complex was 54 dB SPL. Threshold equalizing noise, extending from 125 to 15,000 Hz,

Table 1 Mean and standard deviations of hearing levels for the three listener groups. Columns represent, from left to right, averages across all the frequencies, 1000-Hz tone, and 4000-Hz tone, respectively

	Hearing level (dB) mean ± standard deviation		
	125–4000 Hz	1000 Hz	4000 Hz
YNH (N=22)	8.1 ± 6.5	5.5 ± 5.6	3.1 ± 6.5
ENH (N=12)	13.0 ± 6.1	10.2 ± 7.0	11.7 ± 6.2
EHI (N=9)	21.1 ± 13.4	11.9 ± 7.1	35.3 ± 18.8

was added. The high-frequency stimulus was a "transposed stimulus," which was a 4-kHz tone carrier amplitude-modulated with a half-wave rectified 125-Hz sinusoid (Bernstein and Trahiotis 2002). It is considered that the auditory-nerve firing is phase locked to the modulator waveform, which provides the cue for judging the ITD and modulation rate of the stimulus. The overall level of the transposed stimulus was set to 65 dB SPL. A continuous, low-pass filtered Gaussian noise was added to prevent the listener from using any information at low spectral frequencies (e.g., combination tones).

Stimuli used for supplementary tasks (namely, *FRES* and *FDISC* tasks) involved tone-burst signals at frequencies of 1100 and 4000 Hz. Specifically to these two tasks, the low- and high-frequency stimuli refer to the 1100- and 4000-Hz tones, respectively. The frequency band of interest in the *HUGGINS* task (another supplementary task) was centred at 1100 Hz. Other details about the stimuli for the *FRES*, *FDISC*, and *HUGGINS* tasks are described in the next subsection.

2.3 Procedures

2.3.1 General Procedure

A two-interval two-alternative forced-choice (2I-2AFC) method was used to measure the listener's sensitivities to stimulus parameters. Feedback was given to indicate the correct answer after each response. The two-down/one-up adaptive tracking method was used to estimate discrimination thresholds.

2.3.2 Task Specific Procedures

2.3.2.1 *ITD* Task

In a 2I-2AFC trial, stimuli in the two intervals had ITDs of $+\Delta ITD/2$ and $-\Delta ITD/2$ μs. Each stimulus was 400-ms long, including 100-ms raised-cosine onset and offset ramps, which were synchronized between the two ears. The listeners were required to indicate the direction of the ITD change between the two intervals on the basis of the laterality of sound images.

2.3.2.2 *ILD* Task

Similarly to the *ITD* task, stimuli in the two intervals had ILDs of $+\Delta ILD/2$ and $-\Delta ILD/2$ dB. Each stimulus was 400-ms long, including 20-ms raised-cosine onset and offset ramps. The listeners were required to indicate the direction of the ILD change between the two intervals on the basis of the laterality of sound images.

2.3.2.3 *TIME* Task

For the low-frequency stimulus, the listeners were required to detect a common upward frequency shift (Δf Hz) imposed on the individual components of the SSMC with the spectral envelope remaining unchanged (Moore and Sek 2009). It was assumed that the listeners based their judgments on pitch changes, reflecting the temporal fine structure encoded as the pattern of neural phase locking. The "signal" and "non-signal" intervals in the 2I-2AFC method contained RSRS and RRRR sequences, respectively, where R indicates the original SSMC and S indicates a frequency-shifted SSMC. For the high-frequency stimulus, the listener's task was to discriminate the modulation frequencies of the transposed stimuli between f_m (=125 Hz) and $f_m+\Delta f$ Hz, referred to as R and S, respectively. Each R and S had a duration of 100 ms, including 20-ms raised-cosine ramps. The threshold was expressed as $\Delta f/f_0$ or $\Delta f/f_m$ for the low- or high-frequency stimuli, respectively. When adaptive tracking failed to converge within this limit, trials with a shift of $0.5F_0$ were repeated 30 times. In that case, the proportion of correct trials was converted to d', and then the "threshold" was derived on the assumption that d' is proportional to the frequency shift (Moore and Sek 2009).

2.3.2.4 *LEVEL* Task

In a 2I-2AFC trial, the listeners were required to indicate an interval containing a 400-ms-long SSMC or a transposed stimulus whose central 200-ms portion (including 20-ms raised-cosine ramps) was incremented in level by ΔL dB, while the other non-signal interval contained an original SSMC or a transposed stimulus.

2.3.2.5 *FRES* Task

The notched-noise masking method (Patterson et al. 1982) was employed to evaluate frequency selectivity. The signals were pure-tone busts centred at 1100 or 4000 Hz, and the maskers were notched noises with varying notch width (0, 0.05, 0.1, 0.2, 0.3, and 0.4 relative to the signal frequency). The spectrum level within the passband was 40 dB SPL. A rounded-exponential filter (Patterson et al. 1982) was fitted to the experimental data using a least-square fit. The equivalent rectangular bandwidth (*ERB*) was then derived from the parameters of the fitted filter.

2.3.2.6 *FDISC* Task

Frequency difference limens were measured with pure-tone bursts centred at 1100 and 4000 Hz. Similarly to the *TIME* task, the sequence of RRRR and RSRS was presented, R and S representing tone bursts with frequencies of f_c and $f_c+\Delta f$ Hz, respectively (f_c=1100 or 4000 Hz). Each tone burst had 200 ms of duration with 20 ms onset-offset ramps.

2.3.2.7 *HUGGINS* Task

Bandpass-filtered noises (passband: 250–4000 Hz) with a duration of 200 ms were used as stimuli. The noise was diotic except for a narrow frequency band centred around 1100 Hz (f_c) with an 18 % width around the centre frequency, on which an interaural phase transition was imposed. This stimulus elicits a sensation of pitch corresponding to f_c (Cramer and Huggins 1958). Similarly to the *TIME* task, the sequences of RRRR and RSRS were presented, with R and S representing tone bursts with frequencies of fc and $f_c + \Delta f$ Hz, respectively. The discrimination threshold was expressed as $\Delta f / f_c$.

2.4 Data Analyses

MATLAB with Statistical Toolbox was used for statistical analyses of the data. For the *ITD, TIME, LEVEL, FDISC*, and *HUGGINS* tasks, the analyses were performed with log-transformed threshold data.

3 Results

The left four columns of panels of Fig. 1 compare the performance of the four basic tasks (*ITD, ILD, TIME*, and *LEVEL*) between listener groups. For the low-frequency stimulus (the upper row of panels in Fig. 1), a one-way analysis of variance indicated a statistically significant effect of listener group in the *TIME* task only. Subsequent pair-wise comparisons indicated higher thresholds for the ENH and EHI listeners than for the YNH listeners. For the high-frequency stimulus (the lower rows of panels), a significant effect of listener group was found in the *ITD, TIME*, and *LEVEL* tasks. These results reflect higher thresholds for *EHI* than for *ENH* (*ITD* task); for ENH and EHI than for YNH (*TIME* task); and for ENH than for YNH (*LEVEL* task). The listener-group effect was not significant for the ILD task for either stimulus type. The figure also shows the data obtained in the supplementary experiments (*FRES, FDISC*, and *HUGGINS* tasks). Significant listener-group effects were found in the *HUGGINS* task and the *FDISC* task (for the high-frequency stimulus only).

The thresholds of individual listeners are compared between pairs of tasks in Fig. 2. The results show statistically significant positive correlations for the *ITD-ILD* and *ILD-LEVEL* pairs. The partial correlation coefficients by controlling the effect of age (YNH versus ENH/EHI) were also significant. A significant partial correlation coefficient was also found for the *ITD-TIME* pair with the high-frequency stimulus.

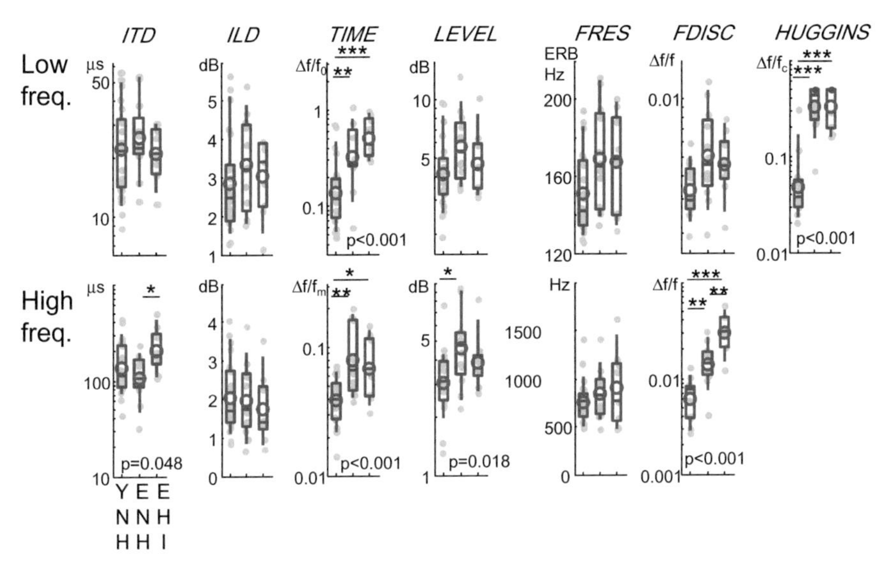

Fig. 1 Comparisons of thresholds among listener groups. Each column of panels represents one task, as labelled. The three columns of data within a panel correspond to, from *left* to *right*, YNH, ENH, and EHI, respectively. The 10th, 25th, 50th, 75th, and 90th percentiles are shown by the box and whisker plot. The *red* circle shows the mean, and the *grey* dots represent individual listeners' data. The task for which a significant effect of listener group was revealed by one-way analysis of variance is indicated by the *p*-values. *Asterisks* indicate group pairs for which a significant difference was indicated by a post hoc tast with Tukey's honestly significant difference criterion ($^*p<0.05$; $^{**}p<0.01$; $^{***}p<0.001$)

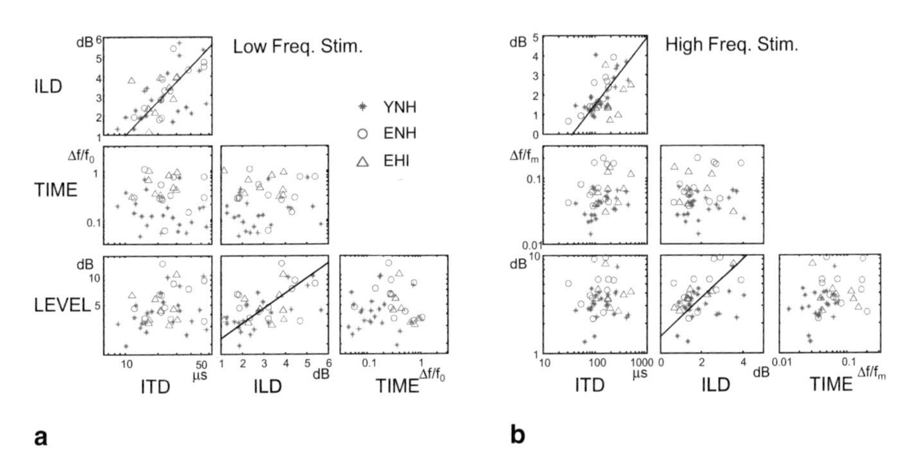

Fig. 2 Comparisons of individual listeners' thresholds between tasks for **a** low—and **b** high- frequency stimuli. Significant correlations were marked by *straight lines* obtained by orthogonal regressions. See also Table 2

Table 2 Pearson's correlation coefficients and p-values (in parentheses) for the data shown in Fig. 2. The second line of each entry indicates the values when the effect of age was partialled out. Significant correlations ($p<0.05$) were marked as italic characters

		ITD	ILD	TIME
Low Freq.	ILD	*0.59 (<0.001)* *0.59 (<0.001)*		
	TIME	−0.10 (0.520) −0.17 (0.283)	0.16 (0.318) 0.09 (0.572)	
	LEVEL	0.28 (0.069) 0.27 (0.079)	*0.53 (<0.001)* *0.52 (<0.001)*	0.03 (0.825) −0.16 (0.316)
High Freq.	ILD	*0.55 (<0.001)* *0.56 (<0.001)*		
	TIME	0.30 (0.053) *0.33 (0.031)*	0.08 (0.632) 0.16 (0.325)	
	LEVEL	0.12 (0.436) 0.12 (0.462)	*0.43 (0.004)* *0.52 (<0.001)*	0.28 (0.064) 0.08 (0.616)

4 Discussion

Elderly listeners (ENH and EHI) showed higher thresholds than young listeners (YNH) in the *TIME* task (with both low- and high-frequency stimuli) and the *HUGGINS* task. This confirms earlier findings that the sensitivity to temporal structure declines without accompanying elevated audiometric thresholds (e.g., Hopkins and Moore 2011). Thresholds for the *LEVEL* task also tended to be higher in the elderly than the young, consistently with previous finding (e.g., He et al. 1998). Despite these consistent declines in the performance of the monaural tasks with age, the present study failed to find significant age effect in the lateralization tasks (*ITD* and *ILD*).

The thresholds for the *ILD* and *LEVEL* tasks correlated for both the low- and high-frequency stimuli. This confirms the results of our earlier study (Ochi et al. 2014), suggesting that the efficiency of level coding in the auditory periphery is a major factor accounting for inter-individual variation of ILD sensitivity. The *ITD* task showed a correlation with the *TIME* task for the high-frequency stimulus; when the factors of age and hearing-impairments were controlled. This again is consistent with the finding of our earlier study (Ochi et al. 2014), suggesting that a listener's ITD sensitivity is well accounted for by the listener's sensitivity to temporal (envelope) structure.

Despite a relatively large number of participants and greater range of threshold values in the *TIME* task, however, we failed to find a correlation between the *ITD* and *TIME* tasks for the low-frequency stimulus. The thresholds for the *HUGGINS* task showed a significant positive correlation with those for the *TIME* task ($r=0.50$, $p=0.004$), but not with those for the *ITD* task ($r=0.11$, $p=0.553$). This suggests that the performances of the *TIME* and *HUGGINS* tasks capture inter-individual variation of the efficiency of temporal-structure processing, but that of the *ITD* task is determined primarily by other factors. The *HUGGINS* and *ITD* tasks are similar in that both require interaural comparison of temporal-structure information, but differ in the perceptual domain in which listeners are expected to respond (i.e., pitch versus laterality).

It should be noted, however, that the positive correlation found for the *TIME* and *HUGGINS* tasks was due predominantly to the consistent age effect in the both tasks (see Fig. 1). When the effects of age were partialled out, the correlation coefficient was not significantly different from zero ($r=-0.24$, $p=0.191$), implying that the positive correlation was due to underlying non-temporal factors that are sensitive to aging. A candidate for such a factor is frequency selectivity. It has been argued that (intermediately) resolved frequency components of the SSMC could contribute to the performance of the *TIME* task (Micheyl et al. 2010). Peripheral frequency resolution could influence the representation of interaural correlation across a frequency axis, which is the basis for the Huggins pitch. Indeed, *ERB* was significantly correlated with thresholds for the *TIME* and *HUGGINS* tasks (*ERB-TIME*: $r=0.30$, $p=0.047$; *ERB-HUGGINS*: $r=0.35$, $p=0.050$; the effects of age were not partialled out).

Acknowledgments This study was supported by an internal basic research fund of NTT Corporation.

References

Bernstein LR, Trahiotis C (2002) Enhancing sensitivity to interaural delays at high frequencies by using "transposed stimuli". J Acoust Soc Am 112(3):1026–1036

Cramer EM, Huggins WH (1958) Creation of pitch through binaural interaction. J Acoust Soc Am 30(5):413–417

He N, Dubno JR, Mills JH (1998) Frequency and intensity discrimination measured in a maximum-likelihood procedure from young and aged normal-hearing subjects. J Acoust Soc Am 103(1):553–565

Hopkins K, Moore BC (2011) The effects of age and cochlear hearing loss on temporal fine structure sensitivity, frequency selectivity, and speech reception in noise. J Acoust Soc Am 130(1):334–349

Micheyl C, Dai H, Oxenham AJ (2010) On the possible influence of spectral- and temporal-envelope cues in tests of sensitivity to temporal fine structure. J Acoust Soc Am 127(3):1809–1810

Moore BC, Sek A (2009) Development of a fast method for determining sensitivity to temporal fine structure. Int J Audiol 48(4):161–171

Ochi A, Yamasoba T, Furukawa S (2014) Factors that account for inter-individual variability of lateralization performance revealed by correlations of performance among multiple psychoacoustical tasks. Front Neurosci 8. doi:10.3389/fnins.2014.00027

Patterson RD, Nimmo-Smith I, Weber DL, Milroy R (1982) The deterioration of hearing with age: frequency selectivity, the critical ratio, the audiogram, and speech threshold. J Acoust Soc Am 72(6):1788–1803

Investigating the Role of Working Memory in Speech-in-noise Identification for Listeners with Normal Hearing

Christian Füllgrabe and Stuart Rosen

Abstract With the advent of cognitive hearing science, increased attention has been given to individual differences in cognitive functioning and their explanatory power in accounting for inter-listener variability in understanding speech in noise (SiN). The psychological construct that has received most interest is working memory (WM), representing the ability to simultaneously store and process information. Common lore and theoretical models assume that WM-based processes subtend speech processing in adverse perceptual conditions, such as those associated with hearing loss or background noise. Empirical evidence confirms the association between WM capacity (WMC) and SiN identification in older hearing-impaired listeners. To assess whether WMC also plays a role when listeners without hearing loss process speech in acoustically adverse conditions, we surveyed published and unpublished studies in which the Reading-Span test (a widely used measure of WMC) was administered in conjunction with a measure of SiN identification. The survey revealed little or no evidence for an association between WMC and SiN performance. We also analysed new data from 132 normal-hearing participants sampled from across the adult lifespan (18–91 years), for a relationship between Reading-Span scores and identification of matrix sentences in noise. Performance on both tasks declined with age, and correlated weakly even after controlling for the effects of age and audibility ($r=0.39$, $p\leq0.001$, one-tailed). However, separate analyses for different age groups revealed that the correlation was only significant for middle-aged and older groups but not for the young (<40 years) participants.

Keywords Aging · Audiometrically normal · Correlations · Cognition · Noise · Older listeners · Reading-span test · Speech intelligibility · Working-memory capacity · Young listeners · Matrix sentences

---

C. Füllgrabe (✉)
MRC Institute of Hearing Research, Science Road, Nottingham NG7 2RD, UK
e-mail: christian.fullgrabe@ihr.mrc.ac.uk

S. Rosen
UCL Speech, Hearing & Phonetic Sciences, 2 Wakefield Street, London WC1N 2PF, UK
e-mail: stuart@phon.ucl.ac.uk

P. van Dijk et al. (eds.), *Physiology, Psychoacoustics and Cognition in Normal and Impaired Hearing,* Advances in Experimental Medicine and Biology 894,
DOI 10.1007/978-3-319-25474-6_4

1 Introduction

Recent years have seen an increased interest in the role of individual differences in cognitive functioning in speech and language processing and their interaction with different types of listening tasks and conditions. The psychological construct that has received the most attention in the emerging field of cognitive hearing science is working memory (WM), possibly because it has been shown to be involved in a wide range of complex cognitive behaviours (e.g. reading comprehension, reasoning, complex learning). WM can be conceptualised as the cognitive system that is responsible for active maintenance of information in the face of ongoing processing and/or distraction. Its capacity (WMC) is generally assessed by so-called complex span tasks, requiring the temporary storage and simultaneous processing of information. For example, in one of the most widely used WM tasks, the Reading-Span test (Baddeley et al. 1985), visually presented sentences have to be read and their semantic correctness judged (processing component), while trying to remember parts of their content for recall after a variable number of sentences (storage component).

A growing body of evidence from studies using mainly older hearing-impaired (HI) listeners indeed confirms that higher WMC is related to better unaided and aided speech-in-noise (SiN) identification, with correlation coefficients frequently exceeding 0.50 (Lunner 2003; Foo et al. 2007; Lunner and Sundewall-Thorén 2007; Arehart et al. 2013). In addition, high-WMC listeners were less affected by signal distortion introduced by hearing-aid processing (e.g. frequency or dynamic-range compression).

Consistent with these results, models of speech/language processing have started incorporating active cognitive processes (Rönnberg et al. 2013; Heald and Nusbaum 2014). For example, according to the Ease of Language Understanding model (Rönnberg et al. 2013), any mismatch between the perceptual speech input and the phonological representations stored in long-term memory disrupts automatic lexical retrieval, resulting in the use of explicit, effortful processing mechanisms based on WM. Both internal distortions (i.e., related to the integrity of the auditory, linguistic, and cognitive systems) and external distortions (e.g. background noise) are purportedly susceptible to contribute to the mismatch. Consequently, it is assumed that WMC also plays a role when individuals with normal hearing (NH) have to process spoken language in acoustically adverse conditions.

However, Füllgrabe et al. (2015) recently failed to observe a link between Reading-Span scores and SiN identification in older listeners (≥60 years) with audiometrically NH (≤20 dB HL between 0.125 and 6 kHz), using a range of target speech (consonants and sentences), maskers (unmodulated and modulated noise, interfering babble), and signal-to-noise ratios (SNRs).

2 Study Survey

To assess the claim that individual variability in WMC accounts for differences in SiN identification even in the absence of peripheral hearing loss, we surveyed published and unpublished studies administering the Reading-Span test and a measure of SiN identification to participants with audiometrically NH. To ensure consistency with experimental conditions in investigations of HI participants, only studies presenting sentence material "traditionally" used in hearing research (i.e., ASL, Hagerman, HINT, IEEE, QuickSIN, or Versfeld sentences) against co-located background maskers were considered. In addition, we only examined studies in which the effect of age was controlled for (either by statistically partialling it out or by restricting the analysis to a "narrow" age range), in order to avoid inflated estimates of the correlation between WMC and SiN tasks caused by the tendency for performance in both kinds of tasks to worsen with age. Figure 1 summarizes the results of this survey.

Correlation coefficients in the surveyed studies are broadly distributed, spanning almost half of the possible range of r values (i.e., from −0.29 to 0.58). Confidence intervals (CIs) are generally large and include the null hypothesis in 21/25 and 24/25 cases for CIs of 95 and 99%, respectively, suggesting that these studies are not appropriately powered. For the relatively small number of studies included in this survey, there is no consistent trend for stronger correlations in more complex and/or informationally masking backgrounds or at lower SNRs, presumably corresponding to more adverse listening conditions.

Across studies restricting their sample to young (18–40 years) participants, the weighted average r value is 0.12, less than 2% of the variance in SiN identification. According to a power calculation, it would require 543 participants to have an 80% chance of detecting such a small effect with $p = 0.05$ (one-tailed)!

3 Analysis of Cohort Data for Audiometrically Normal-Hearing Participants

Given the mixed results from previous studies based on relatively small sample sizes, we re-analysed data from a subset of a large cohort of NH listeners taking part in another study.

3.1 Method

Participants were 132 native-English-speaking adults, sampled continuously from across the adult lifespan (range = 18–91 years). Older (≥60 years) participants were screened using the Mini Mental State Examination to confirm the absence of cognitive impairment. All participants had individual audiometric hearing thresholds of ≤20 dB HL at octave frequencies between 0.125 and 4 kHz, as well as at 3 kHz, in the test ear. Despite clinically "normal" audibility, the pure-tone average (PTA) for the tested frequency range declined as a function of age ($r = 0.65$, $p \leq 0.001$, one-tailed). Since changes in sensitivity even in the normal audiometric range can affect SiN identification (Dubno and Ahlstrom 1997), PTA is treated as a possible confounding variable in analyses involving the entire age group.

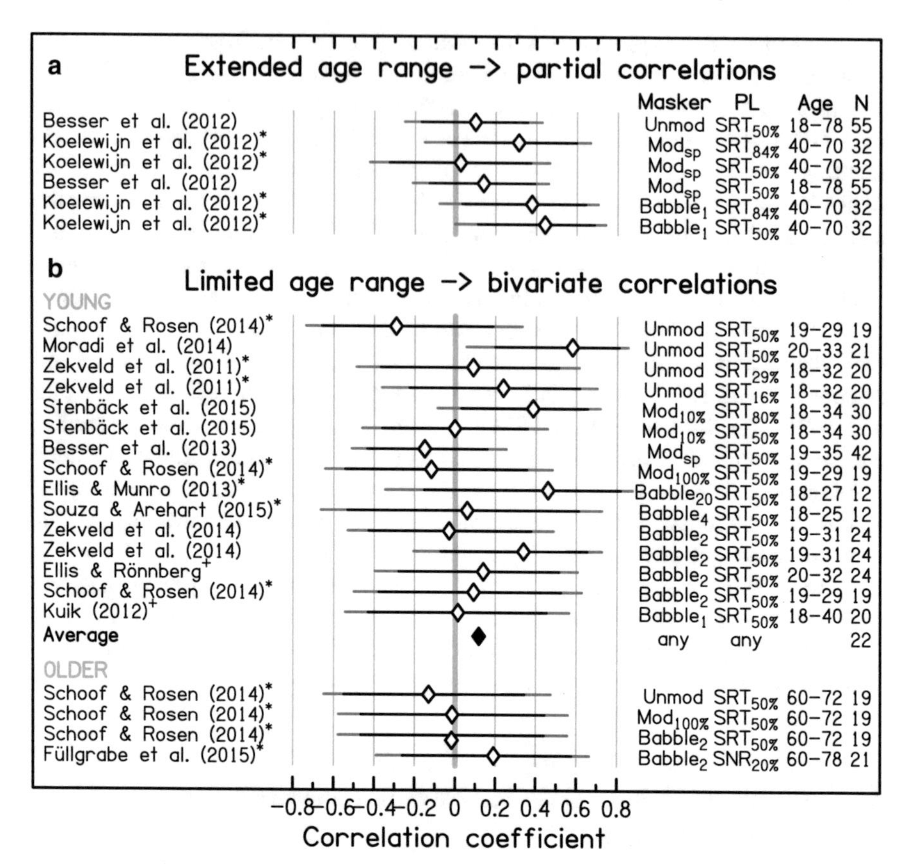

Fig. 1 Comparison of Pearson correlation coefficients (diamonds) and associated 95 % (*black*) and 99 % (*red*) confidence intervals for studies investigating the association between WMC and speech-in-"noise" identification in NH participants after controlling for the effect of age by **a** computing partial correlations, or **b** using a limited age range. When necessary, the sign of the correlation was changed so that a positive correlation represents good performance on the two tasks. A weighted average for correlations based only on young NH listeners is provided (multiple r values for the same study sample are entered as their average). Source references (* indicates re-analysed published data; + indicates unpublished data, personal communication) and experimental (type of masker (*Masker*); performance level (*PL*)) and participant (age range (*Age*); number of participants (*N*)) details are given in the figure. Masker: *Unmod* unmodulated noise, $Mod_{X\%}$ or $_{sp}$ noise modulated by an X % sinusoidal amplitude modulation or a speech envelope, $Babble_X$ X-talker babble. PL: $SRT_{X\%}$ adaptive procedure tracking the speech reception threshold corresponding to X %-correct identification, $SNR_{X\%}$ fixed SNR levels yielding, on average, X %-correct identification

WMC was assessed by means of the computerized version of the Reading-Span test (Rönnberg et al. 1989). Individual sentences were presented in three parts on a computer screen to be read aloud and judged as plausible or implausible. After three to six sentences, either the first or last word of each of the sentences had to be recalled. WMC corresponded to the number of correctly recalled words in any order.

SiN identification was assessed using the English version of the Matrix sentence test (Vlaming et al. 2011). Each target sentence, presented monaurally at 70 dB SPL, followed a fixed syntactic structure (proper noun—verb—numeral—adjective—noun) but had low semantic redundancy. The noise maskers had the same long-term spectrum as the target sentences and were either unmodulated or 100% sinusoidally amplitude modulated at 8 or 80 Hz. Target and masker were mixed together at SNRs ranging from −3 to −15 dB, and the mixture was lowpass-filtered at 4 kHz.

3.2 Results and Discussion

Identification scores were transformed into rationalized arcsine units (RAUs) and averaged across masker types and SNRs to reduce the effect of errors of measurement and to yield a composite intelligibility score representative of a range of test conditions.

Confirming previous results for audiometrically NH listeners (Füllgrabe et al. 2015), Reading-Span and SIN identification scores showed a significant decline with age, with Pearson's $r=-0.59$ and -0.68 (both $p\leq 0.001$, one-tailed), respectively. The scatterplot in Fig. 2 shows that, considering all ages, performances on the tasks were significantly related to each other ($r=0.64$, $p\leq 0.001$, one-tailed). This association remained significant after partialling out the effects of age and PTA ($r=0.39$, $p\leq 0.001$, one-tailed), contrasting with the results of Besser et al. (2012), using a cohort including only a few ($N=8$) older (≥ 60 years) participants, but being roughly consistent with those reported by Koelewijn et al. (2012) for a cohort comprised of middle-aged and older (≥ 40 years) participants (see Fig. 1a).

To further investigate the age dependency of the association between WMC and SiN identification, participants were divided into four age groups: "Young" (range =18–39 years, mean =28 years; $N=32$), "Middle-Aged" (range =40–59 years, mean =49 years; $N=26$), "Young-Old" (range =60–69 years, mean =65 years; $N=40$), and "Old-Old" (range =70–91 years, mean =77 years; $N=34$). Separate correlational analyses for each age group revealed that the strength of the association differed across groups (see Fig. 2). Consistent with the overall trend seen in Fig. 1, the correlation was weak and non-significant in the group of young participants ($r=0.18$, $p=0.162$, one-tailed). In contrast, the correlations were moderately strong and significant in the three older groups (all $r\geq 0.44$, all $p\leq 0.011$, one-tailed). Comparing the different correlation coefficients, after applying Fisher's r-to-z transformation, revealed a significant difference between the Young and Old-Old group ($z=-1.75$, $p=0.040$, one-tailed). There was no evidence for a difference in variance between these groups (Levene's test, $F_{(1,64)}<1$, $p=0.365$).

The age-related modulation of the strength of the correlation between WMC and SiN perception could be due to the different performance levels at which the age groups operated in this study (mean identification was 68, 60, 57, and 48 RAUs for the Young, Middle-Aged, Young-Old, and Old-Old group, respectively). However,

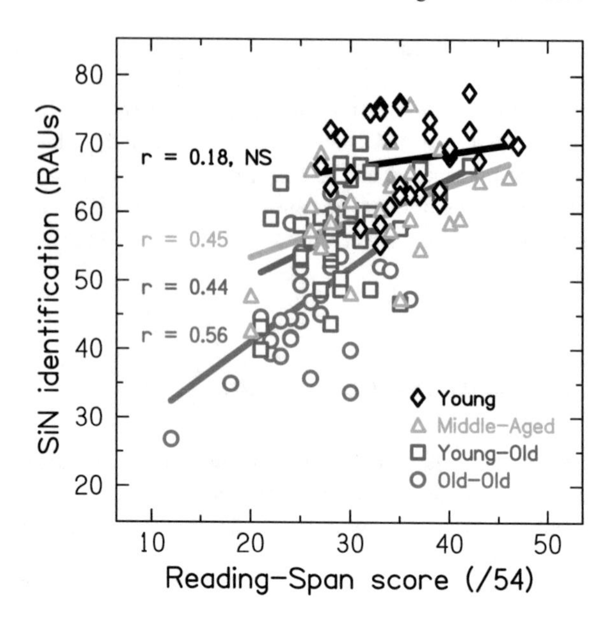

Fig. 2 Scatterplot relating SiN identification averaged across background noises and SNRs to Reading-Span scores for the four age groups. The best linear fit to the data (*thick lines*) and associated bivariate Pearson correlation coefficients for each age group are given in the figure

when performance only for the two lowest SNRs (corresponding to 46 RAUs) was considered, WMC was still not associated with SiN identification in the young participants (r=0.04, p=0.405, one-tailed).

4 Conclusions

Taken together, the reported results fail to provide evidence that, in *acoustically* adverse listening situations, WMC (as measured by the Reading-Span test) is a reliable and strong predictor of SiN *intelligibility* in young listeners with normal hearing. The new data presented here suggest that WMC becomes more important with age, especially in the oldest participants. One possible explanation for this increasing cognitive involvement with age could be the accumulation of age-related deficits in liminary but also supraliminary auditory processing (e.g. sensitivity to temporal-fine-structure and temporal-envelope cues; Füllgrabe 2013; Füllgrabe et al. 2015), resulting in under-defined and degraded internal representations of the speech signal, calling for WM-based compensatory mechanisms to aid identification and comprehension.

Our findings do not detract from the practical importance of cognitive assessments in the prediction of SiN identification performance in older HI listeners and the possible interaction between cognitive abilities and hearing-aid processing. Nor do they argue against the involvement of cognition in speech and language processing in young NH listeners *per se*. First, individual differences in WMC have been shown to explain some of the variability in performance in more linguistically complex task (such as in the comprehension of dynamic conversations; Keidser et al.

2015), presumably requiring memory or attentional/inhibitory processes associated with WMC (Conway et al. 2001; Kjellberg et al. 2008). Second, different cognitive measures, probing the hypothesized sub-processes of WM (e.g. inhibition, shifting, updating) or other domain-general cognitive primitives (e.g. processing speed) might prove to be better predictors of SiN processing abilities than the Reading-Span test.

In conclusion, and consistent with recent efforts to establish if and under which conditions cognitive abilities influence the processing of spoken language (e.g. Fedorenko 2014; Heinrich and Knight, this volume), the current results caution against the assumption that WM necessarily supports SiN identification independently of the age and hearing status of the listener.

Acknowledgments We would like to thank our colleagues who shared and reanalysed their data, and Dr. Oliver Zobay for his statistical advice. The MRC Institute of Hearing Research is supported by the Medical Research Council (grant number U135097130). This work was also supported by the Oticon Foundation (Denmark). CF is indebted to Prof. Brian Moore for granting access to the test equipment of his laboratory.

References

Arehart KH, Souza P, Baca R, Kates JM (2013) Working memory, age, and hearing loss: susceptibility to hearing aid distortion. Ear Hear 34(3):251–260

Baddeley A, Logie R, Nimmo-Smith I, Brereton N (1985) Components of fluent reading. J Mem Lang 24(1):119–131

Besser J, Zekveld AA, Kramer SE, Rönnberg J, Festen JM (2012) New measures of masked text recognition in relation to speech-in-noise perception and their associations with age and cognitive abilities. J Speech Lang Hear Res 55(1):194–209

Besser J, Koelewijn T, Zekveld AA, Kramer SE, Festen JM (2013) How linguistic closure and verbal working memory relate to speech recognition in noise–a review. Trends Amplif 17(2):75–93

Conway ARA, Cowan N, Bunting MF (2001) The cocktail party phenomenon revisited: the importance of working memory capacity. Psychon Bull Rev 8(2):331–335

Dubno JR, Ahlstrom JB (1997) Additivity of multiple maskers of speech. In: Jesteadt W (ed) Modeling sensorineural hearing loss. Lawrence Erlbaum Associates, Hillsdale, pp 253–272

Ellis RJ, Munro KJ (2013) Does cognitive function predict frequency compressed speech recognition in listeners with normal hearing and normal cognition? Int J Audiol 52(1):14–22

Fedorenko E (2014) The role of domain-general cognitive control in language comprehension. Front Psychol 5:335

Foo C, Rudner M, Rönnberg J, Lunner T (2007) Recognition of speech in noise with new hearing instrument compression release settings requires explicit cognitive storage and processing capacity. J Am Acad Audiol 18(7):618–631

Füllgrabe C (2013) Age-dependent changes in temporal-fine-structure processing in the absence of peripheral hearing loss. Am J Audiol 22(2):313–315

Füllgrabe C, Moore BC, Stone MA (2015) Age-group differences in speech identification despite matched audiometrically normal hearing: contributions from auditory temporal processing and cognition. Front Aging Neurosci 6:347

Heald SL, Nusbaum HC (2014) Speech perception as an active cognitive process. Front Syst Neurosci 8:35

Keidser G, Best V, Freeston K, Boyce A (2015) Cognitive spare capacity: evaluation data and its association with comprehension of dynamic conversations. Front Psychol 6:597

Kjellberg A, Ljung R, Hallman D (2008) Recall of words heard in noise. Appl Cognit Psychol 22(8):1088–1098

Koelewijn T, Zekveld AA, Festen JM, Rönnberg J, Kramer SE (2012) Processing load induced by informational masking is related to linguistic abilities. Int J Otolaryngol 2012:865731. 65731.

Kuik AM (2012) Speech reception in noise: on auditory and cognitive aspects, gender differences and normative data for the normal-hearing population under the age of 40. Bachelor's thesis. Vrije Universiteit Amsterdam, Amsterdam

Lunner T (2003) Cognitive function in relation to hearing aid use. Int J Audiol 42(Suppl 1):49–58

Lunner T, Sundewall-Thorén E (2007) Interactions between cognition, compression, and listening conditions: effects on speech-in-noise performance in a two-channel hearing aid. J Am Acad Audiol 18(7):604–617

Moradi S, Lidestam B, Saremi A, Rönnberg J (2014) Gated auditory speech perception: effects of listening conditions and cognitive capacity. Front Psychol 5:531

Rönnberg J, Arlinger S, Lyxell B, Kinnefors C (1989) Visual evoked potentials: relation to adult speechreading and cognitive function. J Speech Hear Res 32(4):725–735

Rönnberg J, Lunner T, Zekveld A, Sörqvist P, Danielsson H, Lyxell B, Dahlstrom O, Signoret C, Stenfelt S, Pichora-Fuller MK, Rudner M (2013) The Ease of Language Understanding (ELU) model: theoretical, empirical, and clinical advances. Front Syst Neurosci 7:31

Souza P, Arehart K (2015) Robust relationship between reading span and speech recognition in noise. Int J Audiol 54: 705–713

Stenbäck V, Hällgren M, Lyxell B, Larsby B (2015) The Swedish Hayling task, and its relation to working memory, verbal ability, and speech-recognition-in-noise. Scand J Psychol 56(3):264–272

Vlaming MSMG, Kollmeier B, Dreschler WA, Martin R, Wouters J, Grover B, Mohammadh Y, Houtgast T (2011) HearCom: hearing in the communication society. Acta Acust United Acust 97(2):175–192

Zekveld AA, Rudner M, Johnsrude IS, Festen JM, van Beek JH, Rönnberg J (2011) The influence of semantically related and unrelated text cues on the intelligibility of sentences in noise. Ear Hear 32(6):e16–e25

Zekveld AA, Rudner M, Kramer SE, Lyzenga J, Rönnberg J (2014) Cognitive processing load during listening is reduced more by decreasing voice similarity than by increasing spatial separation between target and masker speech. Front Neurosci 8:88

The Contribution of Auditory and Cognitive Factors to Intelligibility of Words and Sentences in Noise

Antje Heinrich and Sarah Knight

Abstract Understanding the causes for speech-in-noise (SiN) perception difficulties is complex, and is made even more difficult by the fact that listening situations can vary widely in target and background sounds. While there is general agreement that both auditory and cognitive factors are important, their exact relationship to SiN perception across various listening situations remains unclear. This study manipulated the characteristics of the listening situation in two ways: first, target stimuli were either isolated words, or words heard in the context of low- (LP) and high-predictability (HP) sentences; second, the background sound, speech-modulated noise, was presented at two signal-to-noise ratios. Speech intelligibility was measured for 30 older listeners (aged 62–84) with age-normal hearing and related to individual differences in cognition (working memory, inhibition and linguistic skills) and hearing ($PTA_{0.25-8\ kHz}$ and temporal processing). The results showed that while the effect of hearing thresholds on intelligibility was rather uniform, the influence of cognitive abilities was more specific to a certain listening situation. By revealing a complex picture of relationships between intelligibility and cognition, these results may help us understand some of the inconsistencies in the literature as regards cognitive contributions to speech perception.

Keywords Speech-in-noise perception · Listening situations · Cognitive and auditory variables

1 Introduction

Speech-in-noise (SiN) perception is something that many listener groups, including older adults, find difficult. Previous research has shown that hearing sensitivity cannot account for all speech perception difficulties, particularly in noise (Schneider

A. Heinrich (✉) · S. Knight
MRC Institute of Hearing Research, Nottingham NG7 2RD, UK
e-mail: antje.heinrich@ihr.mrc.ac.uk

S. Knight
e-mail: sarah.knight@ihr.mrc.ac.uk

P. van Dijk et al. (eds.), *Physiology, Psychoacoustics and Cognition in Normal and Impaired Hearing,* Advances in Experimental Medicine and Biology 894,
DOI 10.1007/978-3-319-25474-6_5

and Pichora-Fuller 2000; Wingfield and Tun 2007). Consequently, cognition has emerged as another key factor. While there is general agreement that a relationship between cognition and speech perception exists, its nature and extent remain unclear. No single cognitive component has emerged as being important for all listening contexts, although working memory, as tested by reading span, often appears to be important.

Working memory (WM) has no universally-accepted definition. One characterisation posits that WM refers to the ability to simultaneously store and process task-relevant information (Daneman and Carpenter 1980). WM tasks may emphasize either storage or processing. Storage-heavy tasks, such as Digit Span and Letter-Number Sequencing (Wechsler 1997), require participants to repeat back material either unchanged or slightly changed. Processing-heavy tasks, such as the Reading Span task (Daneman and Carpenter 1980), require a response that differs considerably from the original material and is only achieved by substantial mental manipulation.

The correlation between WM and speech perception, particularly in noise, tends to be larger when the WM task is complex i.e. processing-heavy (Akeroyd 2008). However, this is only a general trend: not all studies show the expected correlation (Koelewijn et al. 2012), and some studies show significant correlations between WM and SiN perception even though the WM measure was storage-, not processing-heavy (Humes et al. 2006; Rudner et al. 2008). Why these inconsistencies occurred remains to be understood.

WM can also be conceptualised in terms of inhibition of irrelevant information (Engle and Kane 2003), which has again been linked to speech perception. For instance, poor inhibition appears to increase susceptibility to background noise during SiN tasks (Janse 2012). Finally, general linguistic competence—specifically vocabulary size and reading comprehension—has also been shown to aid SiN perception in some situations (Avivi-Reich et al. 2014).

Some of the inconsistencies in the relationship between SiN perception and cognition are most likely caused by varying combinations of speech perception and cognitive tasks. Like cognitive tasks, speech perception tasks can be conceptualised in different ways. Speech tasks may vary along several dimensions including the complexity of the target (e.g. phonemes vs. sentences), of the background signal (e.g. silence vs. steady-state noise vs. babble), and/or the overall difficulty (e.g. low vs. high signal-to-noise ratio) of the listening situation. It is difficult to know if and to what extent variations along these dimensions affect the relationship to cognition, and whether these variations may explain, at least in part, inconsistencies between studies. For instance, could differences in target speech help explain why some studies (Desjardins and Doherty 2013; Moradi et al. 2014), using a complex sentence perception test, found a significant correlation between reading span and intelligibility, while another study (Kempe et al. 2012), using syllables and the same cognitive test, did not? The goal of this study is to systematically vary the complexity of the listening situation and to investigate how this variation affects the relationship between intelligibility and assorted cognitive measures.

Finally, it is important to note that a focus on cognitive contributions does not imply that auditory contributions to SiN perception are unimportant. Besides overall hearing sensitivity we also obtained a suprathreshold measure of temporal processing by measuring the sensitivity to change in interaural correlation. This task is assumed to estimate loss of neural synchrony in the auditory system (Wang et al. 2011), which in turn has been suggested to affect SiN perception.

2 Methods

2.1 Listeners

Listeners were 30 adults aged over 60 (mean: 70.2 years, SD: 6.7, range: 62–84) with age-normal hearing. Exclusion criteria were hearing aid use and non-native English language status.

2.2 Tasks

2.2.1 Speech Tasks

Sentences Stimuli were 112 sentences from a recently developed sentence pairs test (Heinrich et al. 2014). This test, based on the SPIN-R test (Bilger et al. 1984), comprises sentence pairs with identical sentence-final monosyllabic words, which are more or less predictable from the preceding context (e.g. 'We'll never get there at this rate' versus 'He's always had it at this rate'). High and low predictability (HP/LP) sentence pairs were matched for duration, stress pattern, and semantic complexity, and were spoken by a male Standard British English speaker.

Words A total of 200 words comprising the 112 final words from the sentence task and an additional 88 monosyllables were re-recorded using a different male Standard British English speaker.

Noise All speech stimuli were presented in speech-modulated noise (SMN) derived from the input spectrum of the sentences themselves. Words were presented at signal-to-noise ratios (SNRs) of + 1 and − 2 dB, sentences at − 4 and − 7 dB. SNR levels were chosen to vary the overall difficulty of the task between 20 and 80 % accuracy. Intelligibility for each of six conditions (words and LP/HP sentences at low and high SNRs) was measured.

2.2.2 Auditory Task

Temporal Processing Task (TPT) Duration thresholds were obtained for detecting a change in interaural correlation (from 0 to 1) in the initial portion of a 1-s broadband (0–10 kHz) noise presented simultaneously to both ears. A three-down, one-up 2AFC procedure with 12 reversals, of which the last eight were used to estimate the threshold, was used. Estimates were based on the geometric mean of all collected thresholds (minimum 3, maximum 5).

2.2.3 Cognitive Tasks

Letter-Number Sequencing (LNS) The LNS (Wechsler 1997) measures mainly the storage component of WM although some manipulation is required. Participants heard a combination of numbers and letters and were asked to recall the numbers in ascending order, then the letters in alphabetical order. The number of items per trial increased by one every three trials; the task stopped when all three trials of a given length were repeated incorrectly. The outcome measure was the number of correct trials.

Reading Span Task (RST) The RST places greater emphasis on manipulation (Daneman and Carpenter 1980). In each trial participants read aloud unconnected complex sentences of variable length and recalled the final word of each sentence at the end of a trial. The number of sentences per trial increased by one every five trials, from two to five sentences. All participants started with trials of two sentences and completed all 25 trials. The outcome measure was the overall number of correctly recalled words.

Visual Stroop In a variant on the original Stroop colour/word interference task (Stroop 1935) participants were presented with grids of six rows of eight coloured blocks. In two grids (control), each of the 48 blocks contained "XXXX" printed in 20pt font at the centre of the block; in another two grids (experimental), the blocks contained a mismatched colour word (e.g. "RED" on a green background). In both cases, participants were asked to name the colour of each background block as quickly and accurately as possible. Interference was calculated by subtracting the time taken to name the colours on the control grids from the time taken to name the colours in the mismatched experimental grids.

Mill Hill Vocabulary Scale (MH) The Mill Hill (Raven et al. 1982) measures acquired verbal knowledge in a 20-word multiple-choice test. For each word participants selected the correct synonym from a list of six alternatives. A summary score of all correct answers was used.

Nelson-Denny Reading Test (ND) The Nelson-Denny (Brown et al. 1981) is a reading comprehension test containing eight short passages. Participants were given 20 min to read the passages and answer 36 multiple-choice questions. The outcome measure was the number of correctly answered questions.

2.3 Procedure

Testing was carried out in a sound-attenuated chamber using Sennheiser HD280 headphones. With the exception of the TPT all testing was in the left ear only. Testing took place over the course of two sessions around a week apart. Average pure-tone air-conduction thresholds (left ear) are shown in Fig. 1a. The pure-tone average (PTA) of all measured frequencies was used as an individual measure of hearing sensitivity. In order to determine the presentation level for all auditory stimuli including cognitive tasks, speech reception thresholds were obtained using 30 sentences from the Adaptive Sentence List (MacLeod and Summerfield 1990). The speech level was adaptively varied starting at 60 dB SPL and the average presentation level of the last two reversals of a three-down, one-up paradigm with a 2 dB step size was used. Presenting all stimuli at 30 dB above that level was expected to account for differences in intelligibility in quiet.

TPT and cognitive tasks were split between sessions without a strict order. The word and sentence tasks were always tested in different sessions, with the order of word and sentence tasks counterbalanced across participants. Each sentence-final word was only heard once, either in the context of an HP or an LP sentence, and half the sentences of each type were heard with high or low SNR. Across all listeners, each sentence-final word was heard an equal number of times in all four conditions (sentence type × SNR). After hearing each sentence or word participants repeated as much as they could. Testing was self-paced. The testing set-up for the cognitive tasks was similar but adapted to task requirements.

3 Results

Figure 1b presents mean intelligibility for stimulus type and SNR A 3 stimulus type (words, LP sentences, HP sentences) by two SNR (high, low) repeated-measures ANOVA showed main effects of type ($F(2, 50)=192.55$ $p<0.001$, LP < words < HP) and SNR ($F(1, 25)=103.43$, $p<0.001$) but no interaction ($F(2, 50)=1.78$, $p=0.18$), and suggested that a 3-dB decrease in SNR reduced intelligibility for all three stimuli types by a similar amount (12%). There was also a significant difference in intelligibility between LP and HP sentences.

The effect of auditory and cognitive factors on intelligibility was examined in a series of separate linear mixed models (LMM with SSTYPE1). Each model included one auditory or cognitive variable, both as main effect and in its interactions with stimulus type (words, LP, HP) and SNR (L, H). The previously confirmed main effects for stimulus type and SNR were modelled but are not separately reported. Participants were included with random intercepts. Table 1 A displays the *p*-values for all significant effects. Non-significant results are not reported. Table 1 B displays bivariate correlations between each variable and the scores in each listening situation to aid interpretation. Note however that bivariate correlations do not need to be significant by themselves to drive a significant effect in an LMM.

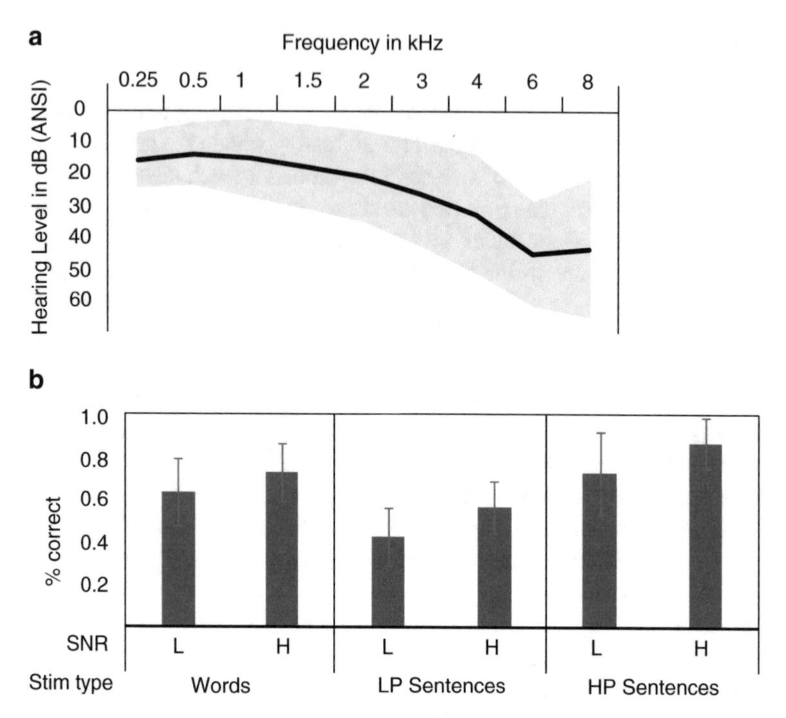

Fig. 1 **a** Audiometric thresholds (mean ± 1 SD) as a function of frequency for the left ear. **b** Intelligibility (mean + 1SD) for three stimuli types (words, LP/HP sentences) presented at high (*H*) and low (*L*) SNR

Table 1 A: Significant *p*-values in linear mixed models estimating the effect of each auditory and cognitive variable on listening situations varying in type of target speech (words, LP/HP sentences) and SNR. *PTA* pure-tone average 0.25–8 kHz (left ear), *LNS* letter-number sequencing, *RST* reading span task, *ND* Nelson-Denny reading comprehension, *ME* main effect. 1B: Pearson product-moment correlations between each auditory/cognitive variable and intelligibility in each of six listening situations. *Words* monosyllables, *LP* low-predictability, *HP* high-predictability. Significant correlations are in italics

		A. p-values			B. Pearson product-moment correlations					
					Words		LP sentences		HP sentences	
	var	ME	SNR *var	Type *SNR *var	Low SNR	High SNR	Low SNR	High SNR	Low SNR	High SNR
Auditory	PTA	<0.001	0.007	0.05	*−0.57*	*−0.60*	*−0.66*	*−0.53*	*−0.69*	*−0.61*
	TPT			0.04	−0.18	*−0.38*	−0.33	−0.20	−0.29	−0.26
Storage	LNS	0.003			0.34	*0.43*	*0.54*	*0.55*	0.38	*0.55*
Processing	RST		0.05		0.16	0.20	0.33	0.03	0.36	0.27
Inhibition	Stroop				0.01	0.06	−0.08	0.16	0.10	0.12
Vocab	MH				−0.03	−0.06	0.07	0.03	−0.02	0.22
Compr	ND				0.06	0.10	0.30	0.02	0.17	0.26

The main effect of PTA reflected the fact that average audiometric threshold was negatively correlated with intelligibility in all listening situations. The interaction with SNR occurred because the correlations tended to be greater for low SNRs than high SNRs. This was particularly true for the sentence stimuli leading to the three-way interaction (Type * SNR * PTA). The Type*SNR*TPT interaction occurred because TPT thresholds showed a significant negative correlation with intelligibility for word stimuli at high SNRs. The main effect of LNS occurred because a better storage-based WM score was beneficial for intelligibility in all tested listening situations. Examining the bivariate correlations to understand the interaction between WM processing-heavy RST and SNR suggests that a good score in the RST was most beneficial for listening situations with low SNR, although this relationship did not reach significance in any case. Intelligibility was not affected by inhibition, vocabulary size or reading comprehension abilities.

4 Discussion

In this study we assessed individual differences in auditory and cognitive abilities and investigated their predictive value for SiN perception across a range of tasks. By systematically varying the characteristics of the listening situation we hoped to resolve inconsistencies in the cognitive speech literature regarding correlations between cognitive and speech tasks. By assessing multiple relevant abilities we also aimed to understand if and how the contribution of a particular ability varied across listening situation.

The results suggest that the importance of a particular variable often, but not always, depends on the listening situation. Hearing sensitivity (PTA) and a basic WM task correlated with intelligibility in all tested situations. The results for PTA are somewhat surprising given that all speech testing was done at sensitivity-adjusted levels, which might have been expected to equate for PTA differences. The PTA measure may therefore capture some aspect of hearing that is not well represented by SRT.

The importance of WM for intelligibility is less surprising. However, the rather basic, storage-based WM task appeared to capture a more general benefit, at least in the tested listening situations, than the often-used RST. While the RST did predict intelligibility, the effect was stronger for acoustically difficult listening situations (low SNR). The data provide support for the notion that the relationship between RST and speech perception depends on the listening situations and that inconsistent results in the literature may have occurred because not all speech tasks engage WM processes enough to lead to a reliable correlation. Lastly, we showed an effect of individual differences in temporal processing on intelligibility, but this effect was limited to easily perceptible (high SNR) single words. Possibly, temporal information is most useful when the stimulus is clearly audible and no other semantic information is available.

These results suggest that different variables modulate listening in different situations, and that listeners may vary not only in their overall level of performance but also in how well they perceive speech in a particular situation depending on which auditory/cognitive abilities underpin listening in that situation and how successful the listener is at employing them.

Acknowledgments This research was supported by BBSRC grant BB/K021508/1.

References

Akeroyd MA (2008) Are individual differences in speech reception related to individual differences in cognitive ability? A survey of twenty experimental studies with normal and hearing-impaired adults. Int J Audiol 47(2):53–71

Avivi-Reich M, Daneman M, Schneider BA (2014) How age and linguistic competence alter the interplay of perceptual and cognitive factors when listening to conversations in a noisy environment. Front Syst Neurosci. doi:10.3389/fnsys.2014.00021

Bilger RC, Nuetzel JM, Rabinowitz WM, Rzeczkowski C (1984) Standardization of a test of speech perception in noise. J Speech Hear Res 27:32–48

Brown JI, Bennett JM, Hanna G (1981) The Nelson-Denny reading test. Riverside, Chicago

Daneman M, Carpenter PA (1980) Individual differences in working memory and reading. J Verbal Learning Verbal Behav 19:450–466

Desjardins JL, Doherty KA (2013) Age-related changes in listening effort for various types of masker noises. Ear Hear 34(3):261–272

Engle RW, Kane MJ (2003) Executive attention, working memory capacity, and a two-factor theory of cognitive control. In: Ross B (ed) Psychology of learning and motivation, vol 44. Elsevier, San Diego, pp 145–199

Heinrich A, Knight S, Young M, Moorhouse R, Barry J (2014) Assessing the effects of semantic context and background noise for speech perception with a new British English sentences set test (BESST). Paper presented at the British Society of Audiology Annual Conference, Keele University, UK

Humes LE, Lee JH, Coughlin MP (2006) Auditory measures of selective and divided attention in young and older adults using single-talker competition. J Acoust Soc Am 120(5):2926–2937. doi:10.1121/1.2354070

Janse E (2012) A non-auditory measure of interference predicts distraction by competing speech in older adults. Aging Neuropsychol Cogn 19:741–758. doi:10.1080/13825585.2011.652590

Kempe V, Thoresen JC, Kirk NW, Schaeffler F, Brooks PJ (2012) Individual differences in the discrimination of novel speech sounds: effects of sex, temporal processing, musical and cognitive abilities. PloS ONE 7(11):e48623. doi:10.1371/journal.pone.0048623

Koelewijn T, Zekveld AA, Festen JM, Ronnberg J, Kramer SE (2012) Processing load induced by informational masking is related to linguistic abilities. Int J Otolaryngol 1–11. doi:10.1155/2012/865731
MacLeod A, Summerfield Q (1990) A procedure for measuring auditory and audio-visual speech-reception thresholds for sentences in noise: rationale, evaluation, and recommendations for use. Br J Audiol 24:29–43
Moradi S, Lidestam B, Saremi A, Ronnberg J (2014) Gated auditory speech perception: effects of listening conditions and cognitive capacity. Front Psychol 5:531. http://doi.org/10.3389/fpsyg.2014.00531
Raven JC, Raven J, Court JH (1982) Mill Hill vocabulary scale. Oxford Psychologists Press, Oxford
Rudner M, Foo C, Sundewall-Thorén E, Lunner T, Rönnberg J (2008) Phonological mismatch and explicit cognitive processing in a sample of 102 hearing-aid users. Int J Audiol 47(2):91–98. doi:10.1080/14992020802304393
Schneider BA, Pichora-Fuller MK (2000) Implications of perceptual deterioration for cognitive aging research. In: Craik FIM, Salthouse TA (eds) The handbook of aging and cognition. Erlbaum, Mahwah, pp 155–219
Stroop JR (1935) Studies of interference in serial verbal reactions. J Exp Psychol 18:643–662
Wang M, Wu X, Li L, Schneider BA (2011) The effects of age and interaural delay on detecting a change in interaural correlation: the role of temporal jitter. Hear Res 275:139–149
Wechsler D (1997) Wechsler adult intelligence scale-3rd edition (WAIS-3®). Harcourt Assessment, San Antonio
Wingfield A, Tun PA (2007) Cognitive supports and cognitive constraints on comprehension of spoken language. J Am Acad Audiol 18:548–558

Do Hearing Aids Improve Affect Perception?

Juliane Schmidt, Diana Herzog, Odette Scharenborg and Esther Janse

Abstract Normal-hearing listeners use acoustic cues in speech to interpret a speaker's emotional state. This study investigates the effect of hearing aids on the perception of the emotion dimensions arousal (aroused/calm) and valence (positive/negative attitude) in older adults with hearing loss. More specifically, we investigate whether wearing a hearing aid improves the correlation between affect ratings and affect-related acoustic parameters. To that end, affect ratings by 23 hearing-aid users were compared for aided and unaided listening. Moreover, these ratings were compared to the ratings by an age-matched group of 22 participants with age-normal hearing.

For arousal, hearing-aid users rated utterances as generally more aroused in the aided than in the unaided condition. Intensity differences were the strongest indictor of degree of arousal. Among the hearing-aid users, those with poorer hearing used additional prosodic cues (i.e., tempo and pitch) for their arousal ratings, compared to those with relatively good hearing. For valence, pitch was the only acoustic cue that was associated with valence. Neither listening condition nor hearing loss severity (differences among the hearing-aid users) influenced affect ratings or the use

J. Schmidt (✉) · O. Scharenborg
Centre for Language Studies, Radboud University, PO Box 9103, 6500 HD Nijmegen, The Netherlands
e-mail: j.schmidt@let.ru.nl

O. Scharenborg
Donders Institute for Brain, Cognition and Behavior, P.O. Box 9010 // 066, 6500 GL Nijmegen, The Netherlands
e-mail: o.scharenborg@let.ru.nl

D. Herzog
Phonak AG, Laubisrütistrasse 28, 8712 Staefa, Switzerland
e-mail: Diana.Herzog@sonova.com

E. Janse
Centre for Language Studies, Radboud University, PO Box 9103, 6500 HD, Nijmegen, The Netherlands

Donders Institute for Brain, Cognition and Behavior, P.O. Box 9010, 066 6500 GL Nijmegen, The Netherlands

Max Planck Institute for Psycholinguistics, PO Box 310, 6500 AH Nijmegen, The Netherlands

P. van Dijk et al. (eds.), *Physiology, Psychoacoustics and Cognition in Normal and Impaired Hearing,* Advances in Experimental Medicine and Biology 894,
DOI 10.1007/978-3-319-25474-6_6

of affect-related acoustic parameters. Compared to the normal-hearing reference group, ratings of hearing-aid users in the aided condition did not generally differ in both emotion dimensions. However, hearing-aid users were more sensitive to intensity differences in their arousal ratings than the normal-hearing participants.

We conclude that the use of hearing aids is important for the rehabilitation of affect perception and particularly influences the interpretation of arousal.

Keywords Emotion perception · Arousal · Valence · Affective prosody · Acoustic parameters · Natural speech · Older adults · Hearing loss · Hearing aids · Mean F0 · Mean intensity

1 Introduction

People use several information sources to perceive and interpret emotions. Visual information, such as facial expressions, is most informative, but auditory, prosodic cues in the speech signal also provide important cues for emotion perception. For instance, prosodic cues may alter the meaning of a spoken message, as in the case of irony: the meaning of an utterance like "I like roses" can be interpreted as positive (I do like roses) or negative (I do not like roses), depending on the applied prosody. Prosodic cues, then, are acoustic parameters in speech, such as pitch, intensity, and tempo, from which a normal-hearing listener may perceive emotion in the speech signal (Banse and Scherer 1996; Scherer 2003; Coutinho and Dibben 2013). In an ideal communicative setting both visual and auditory information is available. Everyday communication settings, however, may frequently deprive the listener of visual information, (e.g., during a telephone conversation) so that listeners have to rely on auditory information only.

As hearing loss impairs the perception of auditory information perception of prosodic information may also suffer. Although hearing aids clearly improve speech intelligibility, it is unclear to what extent hearing aids sufficiently restore information needed for emotion perception in speech. Several studies with severely hearing-impaired children and adolescents indicate that aided hearing-impaired listeners perform poorly compared to their normal-hearing peers when rating affective prosody in speech (Most et al. 1993; Most and Michaelis 2012). Moreover, they found that affect perception in hearing-impaired participants was independent of their individual hearing loss. These findings, however, cannot be directly transferred to older hearing aid wearing adults, as younger and older adults differ in the perception of affective prosody, even if both groups have normal hearing (e.g., Paulmann et al. 2008). Moreover, older adults were normal-hearing when they acquired language, and will have learned to interpret the acoustic cues associated with affect, in contrast to hearing-impaired children, who never have had a normal development of hearing and perception. Finally, the two age groups may differ in the type of hearing loss, which also complicates the comparison.

To our knowledge, in older adults only the effect of mild hearing loss has been investigated so far. Findings concerning the link between individual hearing loss and affect perception have been inconsistent. Orbelo and colleagues (Orbelo et al. 2005) found no effect of hearing sensitivity on affect perception, while Rigo and Lieberman (Rigo and Lieberman 1989) found that low-frequency hearing loss ($PTA_{low\ (0.25,\ 0.5,\ 1\ kHz)}$ > 25 dB HL) impacted affect perception. Note that both these studies used acted speech. The lack of a global effect of hearing sensitivity on affect perception in these experiments could be due to the more prototypical prosodic expression of affect in acted compared to natural speech (Scherer 1986; Wilting et al. 2006). More extreme expressions of affect may be relatively easy to perceive, even for people with hearing loss (Grant 1987) thus obscuring a possible influence of hearing sensitivity on affect perception in natural communicative settings.

The current study investigates whether hearing aids restore affect perception, and how hearing loss in older adults influences affect perception. In particular, this study focuses on the question to what extent hearing aid use and hearing loss influence listeners' sensitivity to the acoustic parameters cueing affect. To that end, older (bilateral) hearing aid users are tested while wearing their hearing aid (aided condition) and without it (unaided condition). The relation between the acoustic parameters and the affect ratings are then evaluated for the two listening conditions. Moreover, the performance in the aided condition is compared to a control group of age-matched normal-hearing listeners. Participants will be tested on natural conversational speech stimuli in order to mimic realistic listening conditions.

2 Experimental Set-up

2.1 Participants

Two groups of older adults aged between 65 and 82 were tested. All participants were Swiss German native speakers and were financially compensated for their participation. The group of 23 older hearing aid users with bilaterally symmetric sensorineural hearing loss (M_{Age} = 73.5 years, SD_{Age} = 4.5; 17 men, 6 women) was recruited via the Phonak AG participant database. Participants have worn hearing aids bilaterally for at least 2 years. The group of 22 normal-hearing adults (M_{Age} = 70.8 years, SD_{Age} = 5.2; 10 men, 12 women) was recruited via the Phonak human resource department and a local senior club in Staefa, Switzerland.

Participants' hearing ability was tested by means of pure-tone audiometry (air conduction thresholds). The mean unaided pure-tone average (PTA) across 0.5, 1, 2, and 4 kHz for the hearing-impaired group was 49.8 dB HL (SD = 8.7, range: 32.5–68.8). The normal-hearing participants had age-normal thresholds (as defined in the ISO 7029:2000 standards for this age group). Thresholds below the ISO's maximum pure-tone average threshold (across 0.5, 1, 2, and 4 kHz) at the age of 70 for men (PTA = 33.5 dB HL) and women (PTA = 26.0 dB HL) were considered as normal hearing. Additionally, participants underwent a brief cognitive screening test to scan for mild cognitive impairment. We used the German version of the Montreal Cognitive Assessment Test (MOCA, Nasreddine et al. 2005) using a cutoff criterion of 67 % accuracy (cf. Waldron-Perrine and Axelrod 2012). The test was adjusted for hearing-impaired participants (Dupuis et al. 2015) by leaving out tasks in which auditorily presented items had to be memorized. All participants passed the test.

2.2 Task and Procedure

Affect perception was tested using the dimensional approach, in which participants indicate the level of the emotion dimensions arousal (calm vs. aroused) and valence (positive vs. negative attitude), separately on a rating scale (rather than labeling emotion categories such as "angry" or "sad").

Stimuli were short audio-only utterances from an authentic and affectively-colored German conversational speech corpus (Grimm et al. 2008). Emotion inferences from speech correlate across languages, particularly for similar languages (cf. Scherer et al. 2001). Given the close relationship between German and Swiss German, the way affect is encoded in Swiss German is not expected to differ considerably from that in German as spoken in Germany. The corpus comes with mean reference values for the degree of arousal and valence for each utterance. These reference values had been collected with a 5-step pictorial rating tool (Bradley and Lang 1994), ranging from −1 (calm/negative) to +1 (aroused/positive). The same rating tool was used to collect affective ratings in the current study. From the corpus, 24 utterances were selected for the arousal task (reference value range: −0.66 to 0.94) and 18 were selected for the valence task (reference value range: −0.80 to 0.77). All stimuli in our experiment were neutral regarding the content of what was said (e.g. '*Was hast du getan*?' 'What have you done?') to minimize semantic interference, were shorter than 3 s and were produced by multiple speakers. From these two stimuli sets two randomized lists were created differing in the order in which the stimuli were presented for each emotion dimension.

Participants were comfortably seated in a sound-treated room and were tested in the free field. The pictorial rating tool was displayed on a computer screen and stimuli were presented via a single loudspeaker which was placed at head level in front of the participant (0° azimuth) at a distance of 1 m. Participants received written and oral instructions and performed four practice trials before proceeding to the test stimuli of either rating task. Both rating tasks were completed at the participant's own pace. Utterances were rated one at a time and could be replayed if needed.

All participants performed the rating tasks in two conditions. For the hearing aid users, these two conditions were with (aided) and without their hearing aids (unaided). The normal-hearing participants completed the tasks in a normal listening condition and in a condition with simulated hearing loss (data of the latter condition are not reported here). In each listening condition, participants rated all stimulus utterances, so each participant rated each utterance twice. The order of the arousal/ratings rating tasks and listening conditions were counterbalanced across participants. Two different lists were used to present listeners with a different order of the stimuli in the two listening conditions. There was a short break between each of the four blocks (i.e., between the two listening conditions and between the two rating tasks).

2.3 *Acoustic Parameters*

Affect ratings provided by the participants in our study were related to four acoustic parameters which are traditionally related to affective prosody: mean F0 (e.g., Hammerschmidt and Jürgens 2007), mean intensity (e.g., Aubergé and Cathiard 2003), global temporal aspects (Mozziconacci and Hermes 2000), and spectral measures, which are related to vocal effort (e.g., Tamarit et al. 2008). In the current study, mean F0 and mean intensity were calculated for each utterance by averaging over the utterance using Praat (Boersma and Weenink 2013). As a measure of tempo, articulation rate was calculated by dividing the number of syllables in the canonical transcription of the utterance by the file length, excluding pauses longer than 100 ms. Spectral slope is reflected in the spectral information described by the Hammarberg Index (Hammarberg et al. 1980), which is defined as the intensity difference between the maximum intensity in a lower frequency band [0–2000 Hz] versus that in a higher frequency band [2000–5000 Hz]. In this study, the Hammarberg Index energy distribution measure was averaged across the entire utterance.

3 Results

The data were analyzed using R statistical software (R Development Core Team 2008). To investigate (a) whether hearing loss severity modulates affect ratings and (b) whether wearing a hearing aid makes listeners more sensitive to subtle differences in acoustic parameters, we compared affect ratings (the dependent variable) of the hearing-impaired listeners in the aided and unaided conditions using linear mixed-effects regression analyses with random intercepts for stimulus and participant. The initial models (one for arousal and one for valence) allowed for three-way interactions between listening condition (aided, unaided), individual hearing loss, and each of the acoustic parameters (mean F0, mean intensity, articulation rate, Hammarberg Index). Interactions and predictors that did not improve model fit (according to the Akaike Information Criterion) were removed using a stepwise exclusion procedure. Interactions were removed before simple effects, and those with the highest non-significant p-values were excluded first.

To investigate whether the use of a hearing aid restores affect perception to the level of normal-hearing older adults, we compared hearing aid users' performance in the aided condition to that of the normal-hearing listeners. The method and model-stripping procedure were identical to that of the first analysis. The initial models (for arousal and valence, respectively) allowed for two-way interactions between group (hearing aid users aided, normal hearing) and each of the four acoustic parameters.

3.1 *Aided Versus Unaided Listening*

For arousal, mean intensity was found to be a strong cue for arousal rating ($\beta = 6.606 \times 10^{-2}$, SE $= 1.528 \times 10^{-2}$, $p < 0.001$): higher intensity was associated with higher ratings of arousal in the aided and unaided conditions. Moreover, arousal ratings were generally higher in the aided condition than in the unaided condition (mapped on the intercept) ($\beta = 7.156 \times 10^{-2}$, SE $= 2.089 \times 10^{-2}$, $p < 0.001$). Significant interactions between listening condition and articulation rate ($\beta = 3.012 \times 10^{-2}$, SE $= 1.421 \times 10^{-2}$, $p < 0.05$) and listening condition and vocal effort ($\beta = 1.459 \times 10^{-2}$, SE $= 3.949 \times 10^{-3}$, $p < 0.001$) were observed: while vocal effort and articulation rate did not influence ratings in the unaided condition, their effects were larger in the aided condition. In the unaided condition, those with poorer hearing had lower ratings ($\beta = -9.772 \times 10^{-3}$, SE $= 4.093 \times 10^{-3}$, $p < 0.05$) than those with better hearing, but this was less the case in the aided condition ($\beta = 7.063 \times 10^{-3}$, SE $= 2.459 \times 10^{-3}$, $p < 0.01$). This suggests that wearing the hearing aid made the rating patterns of poorer and better-hearing participants more alike. Furthermore, those with poorer hearing associated increases in F0 ($\beta = 6.093 \times 10^{-5}$, SE $= 2.094 \times 10^{-5}$, $p < 0.01$) and in articulation rate ($\beta = 1.833 \times 10^{-3}$, SE $= 8.952 \times 10^{-4}$, $p < 0.05$) more with higher arousal than those with better hearing across listening conditions. This suggests that, among the hearing aid users, those with poorer hearing used additional prosodic cues compared to those with relatively good hearing.

For valence, a significant simple effect of mean F0 ($\beta = -4.813 \times 10^{-3}$, SE $= 8.856 \times 10^{-4}$, $p < 0.001$) was found: higher pitch was associated with lower ratings, i.e., more negative ratings. None of the other acoustic parameters was predictive of the valence ratings. Moreover, importantly, no effects for listening condition and hearing loss were observed: valence ratings were independent of whether the participants wore their hearing aids or not and were independent of individual hearing loss.

3.2 *Aided Listening Versus Normal-Hearing Controls*

Similar to the previous arousal analysis, a significant simple effect of mean intensity ($\beta = 0.071$, SE $= 0.014$, $p = 0.001$) was found: higher mean intensity was associated with higher arousal ratings. Although ratings of the hearing aid users did not differ significantly from the normal-hearing participants (mapped onto the intercept, $\beta = -0.030$, SE $= 0.053$, $p = 0.57$), use of mean intensity differed between the two listener groups: hearing aid users responded more strongly to differences in intensity than participants with age-normal hearing ($\beta = 0.009$, SE $= 0.004$, $p < 0.05$).

For valence, similar to the previous analysis, mean F0 was associated with lower valence ratings ($\beta = -4.602 \times 10^{-3}$, SE $= 1.168 \times 10^{-3}$, $p < 0.01$). No other acoustic parameters were predictive of the valence ratings. There was no effect of group, nor any interactions between group and the acoustic parameters.

4 Discussion

This study aimed to investigate whether the use of a hearing aid restores affect perception to the level of older adults with age-normal hearing. More specifically, our study investigated to what extent hearing aids and individual hearing loss modify sensitivity to the acoustic parameters cueing affect in older hearing aid users.

The study showed that the hearing aid restored affect perception in the sense that the use of the hearing aid makes rating patterns of hearing aid users with severe hearing loss more similar to those with less severe hearing loss. Secondly, the study showed that the use of a hearing aid changed the pattern of acoustic parameters that were used for arousal perception. Importantly, across the aided and unaided conditions, hearing loss modulated the extent to which listeners used alternative cues to interpret arousal (i.e., other cues than intensity): hearing-impaired listeners with more severe degrees of hearing loss made more use of articulation rate and mean F0. In other words, gradually acquired hearing loss causes listeners to rely on different cues for their interpretation of arousal, but restoring their hearing by means of a hearing aid will also change which cues they rely on for their interpretation of arousal. Older adults may only start using additional cues (such as articulation rate) for their interpretation of arousal with more severe hearing loss. In a related study (Schmidt et al., submitted), older adults with mild hearing loss were tested who were not wearing hearing aids. For this group with mild hearing impairment, intensity emerged as the only significant predictor of arousal. Note, however, that this reliance on multiple cues rather than on a single cue does not hold for valence, where F0 is the only prosodic cue listeners irrespective of their hearing sensitivity are using.

Hearing aid users wearing their hearing aid generally showed the same pattern of affect ratings as participants with age-normal hearing, especially for the valence dimension. However, for arousal ratings, those wearing a hearing aid were actually more sensitive to intensity differences than participants in the reference group. This may be because hearing in the reference group was normal for their age, but still implied elevated high-frequency thresholds. Consequently, older adults in the reference group were less sensitive, at least to some acoustic differences, than the hearing aid users.

In sum, the current study shows that older hearing aid users do not generally differ from their normal-hearing peers in their perception of arousal and valence, which underlines the importance of hearing aids in the rehabilitation of affect perception. While the perception of valence seems to be independent of listening condition and individual hearing loss, wearing hearing aids matters for the interpretation of rating prosodic information related to arousal. Due to this difference between emotion dimensions, future studies on affect perception in hearing aid users should treat perception of arousal and valence separately.

Acknowledgments This project has received funding from the European Union's Seventh Framework Programme for research, technological development and demonstration under grant agreement no FP7-PEOPLE-2011-290000. The Netherlands Organization for Scientific Research

(NWO) supported the research by Esther Janse (grant number 276-75-009) and Odette Scharenborg (grant number 276-89-003).

References

Aubergé V, Cathiard M (2003) Can we hear the prosody of smile? Speech Comm 40:87–97

Banse R, Scherer KR (1996) Acoustic profiles in vocal emotion expression. J Pers Soc Psychol 70(3):614–636

Boersma P, Weenink D (2013). Praat: doing phonetics by computer. www.praat.org/. Accessed 10 April 2013

Bradley MM, Lang PJ (1994) Measuring emotion: the self-assessment manikin and the semantic differential. J Behav Ther Exp Psychiatry 25(1):49–59

Coutinho E, Dibben N (2013) Psychoacoustic cues to emotion in speech prosody and music. Cogn Emot 27(4):658–684

Dupuis K, Pichora-Fuller MK, Chasteen AL, Marchuk V, Singh G, Smith SL (2015) Effects of hearing and vision impairments on the Montreal cognitive assessment. Aging Neuropsychol Cognit: J Norm Dysfunct Dev 22(4):413–437

Grant KW (1987) Identification of intonation contours by normally hearing and profoundly hearing-impaired listeners performance. J Acoust Soc Am 82(4):1172–1178

Grimm M, Kroschel K, Narayanan S (2008). The Vera am Mittag German audio-visual emotional speech database. In Proceedings of the IEEE International Conference on Multimedia and Expo (ICME) (pp 865–868). Hannover, Germany

Hammarberg B, Fritzell B, Gauffin J, Sundberg J, Wedin L (1980) Perceptual and acoustic correlates of abnormal voice qualities. Acta Otolaryngol 90:441–451

Hammerschmidt K, Jürgens U (2007) Acoustical correlates of affective prosody. J Voice: Off J Voice Found 21(5):531–540

International Organization for Standardization (2000) Acoustics—statistical distribution of hearing thresholds as a function of age. ISO 7029. ISO, Geneva

Most T, Michaelis H (2012) Auditory, visual, and auditory–visual perceptions of emotions by young children with hearing loss versus children with normal hearing. J Speech Lang Hear Res 55(4):1148–1162

Most T, Weisel A, Zaychik A (1993) Auditory, visual and auditory–visual identification of emotions by hearing and hearing-impaired adolescents. Br J Audiol 27:247–253

Mozziconacci SJL, Hermes DJ (2000) Expression of emotion and attitude through temporal speech variations. In Sixth International Conference on Spoken Language Processing (pp 373–378). Beijing, China

Nasreddine ZS, Phillips NA, Bedirian V, Charbonneau S, Whitehead V, Collin I, Cummings JL, Chertkow H (2005) http://www.mocatest.org/. Accessed 28 Nov 2013

Orbelo DM, Grim MA, Talbott RE, Ross ED (2005) Impaired comprehension of affective prosody in elderly subjects is not predicted by age-related hearing loss or age-related cognitive decline. J Geriatr Psychiatry Neurol 18(1):25–32

Paulmann S, Pell MD, Kotz SA (2008) How aging affects the recognition of emotional speech. Brain Lang 104(3):262–269

R Development Core Team (2008) R: a language and environment for statistical computing. R Foundation for Statistical Computing, Vienna. http://www.R-project.org/. Accessed 4 March 2013

Rigo TG, Lieberman DA (1989) Nonverbal sensitivity of normal-hearing and hearing-impaired older adults. Ear Hear 10(3):184–189

Scherer KR (1986) Vocal affect expression: a review and a model for future research. Psychol Bull 99(2):143–165

Scherer KR (2003) Vocal communication of emotion: a review of research paradigms. Speech Comm 40(1–2):227–256

Scherer KR, Banse R, Wallbott HG (2001) Emotion inferences from vocal expression correlate across languages and cultures. J Cross Cult Psychol 32(1):76–92

Tamarit L, Goudbeek M, Scherer K (2008). Spectral slope measurements in emotionally expressive speech. In Proceedings for ISCA ITRW Speech Analysis and Processing for Knowledge Discovery. Aalborg, Denmark

Waldron-Perrine B, Axelrod BN (2012) Determining an appropriate cutting score for indication of impairment on the Montreal Cognitive Assessment. Int J Geriatr Psychiatry 27(11):1189–1194

Wilting J, Krahmer E, Swerts M (2006) Real vs. acted emotional speech. In Proceedings of the 9th International Conference on Spoken Language Processing (Interspeech) (pp 805–808). Pittsburgh, Pennsylvania

Suitability of the Binaural Interaction Component for Interaural Electrode Pairing of Bilateral Cochlear Implants

Hongmei Hu, Birger Kollmeier and Mathias Dietz

Abstract Although bilateral cochlear implants (BiCIs) have succeeded in improving the spatial hearing performance of bilateral CI users, the overall performance is still not comparable with normal hearing listeners. Limited success can be partially caused by an interaural mismatch of the place-of-stimulation in each cochlea. Pairing matched interaural CI electrodes and stimulating them with the same frequency band is expected to facilitate binaural functions such as binaural fusion, localization, or spatial release from masking. It has been shown in animal experiments that the magnitude of the binaural interaction component (BIC) derived from the wave-eV decreases for increasing interaural place of stimulation mismatch. This motivated the investigation of the suitability of an electroencephalography-based objective electrode-frequency fitting procedure based on the BIC for BiCI users. A 61 channel monaural and binaural electrically evoked auditory brainstem response (eABR) recording was performed in 7 MED-EL BiCI subjects so far. These BiCI subjects were directly stimulated at 60% dynamic range with 19.9 pulses per second via a research platform provided by the University of Innsbruck (RIB II). The BIC was derived for several interaural electrode pairs by subtracting the response from binaural stimulation from their summed monaural responses. The BIC based pairing results are compared with two psychoacoustic pairing methods: interaural pulse time difference sensitivity and interaural pitch matching. The results for all three methods analyzed as a function of probe electrode allow for determining a matched pair in more than half of the subjects, with a typical accuracy of ± 1 electrode. This includes evidence for statistically significant tuning of the BIC as a function of probe electrode in human subjects. However, results across the three conditions were sometimes not consistent. These discrepancies will be discussed in the light of pitch plasticity versus less plastic brainstem processing.

Keywords Binaural interaction component · Electrically evoked auditory brainstem response · Bilateral cochlear implant · Pitch matching · Interaural pulse time difference · Interaural electrode pairing

H. Hu (✉) · B. Kollmeier · M. Dietz
Medizinische Physik, Universität Oldenburg, Cluster of Excellence "Hearing4all",
26111 Oldenburg, Germany
e-mail: hongmei.hu@uni-oldenburg.de

P. van Dijk et al. (eds.), *Physiology, Psychoacoustics and Cognition in Normal and Impaired Hearing,* Advances in Experimental Medicine and Biology 894,
DOI 10.1007/978-3-319-25474-6_7

1 Introduction

Bilateral cochlear implantation seeks to restore the advantages of binaural hearing to the profound deaf by providing binaural cues that are important for binaural perception, such as binaural fusion, sound localization and better detection of signals in noise. Most bilateral cochlear implant (BiCI) users have shown improvements compared to their ability when only one CI was used. However, compared to normal hearing (NH) individuals, the average performance of BiCI users is still worse and has a large variability in performance amongst them (Majdak et al. 2011; Litovsky et al. 2012; Goupell et al. 2013; Kan et al. 2013). One likely reason for the worse performance of BiCI users is the interaural electrodes mismatch between two CIs because of different surgery insertion depth or different implant length. Since the inputs to the NH binaural system from the two ears can be assumed to be well matched and binaural brainstem neurons are comparing only by interaurally place matched inputs, it is very important to determine interaural electrode pairs for frequency matching the electrode arrays in the two ears to compensate for any differences in the implanted cochleae. This interaural electrode pairing (IEP) is expected to become even more relevant in the future, to better exploit recent time information preserving coding strategies and with the advent of truly binaural cochlear implants, which will preserve, enhance, and/or optimize interaural cues.

It has been suggested that physiological measures of binaural interactions (e.g., evoked potentials) will likely be required to accomplish best-matched interaural electrode pairs (Pelizzone et al. 1990). More recently, the binaural interaction component (BIC) of the electrically evoked auditory brainstem responses (eABRs) have been obtained in both animals (Smith and Delgutte 2007) and humans (He et al. 2010; Gordon et al. 2012). Smith and Delgutte (2007) proposed a potential way by using evoked potentials to match interaural electrode pairs for bilaterally implanted cats. Their study shows that the interaural electrode pairings that produced the best aligned IC activation patterns were also those that yielded the maximum BIC amplitude. More recently, He et al. (2010) observed some evidence of a BIC/electrode-offset interaction at low current levels. In another follow up study, they used the same electroencephalography (EEG) procedure to examine whether the BIC amplitude evoked from different electrode pairings correlated with an interaural pitch comparison task (He et al. 2012). Their results show that there is no significant correlation between results of BIC measures and interaural pitch comparisons on either the individual or group levels. Gordon et al. (2012) demonstrated that binaural processing in the brainstem of children using bilateral CI occurs regardless of bilateral or unilateral deafness; it is disrupted by a large but not by a small mismatch in place of stimulation. All these studies suggest that BIC could be a potentially approaches for electrode pairing, especially for pediatrics, however, it is not accurate enough in all existing studies of human subjects.

In order to tackle the issue of accuracy, a 61-channel monaural and binaural eABR recording setup was developed together with a multi-step offline post-processing strategy specifically for eABR signals. Further, to address the question of method validity the BIC based pairing results are compared with two psychoacoustic pairing methods: interaural pulse time difference sensitivity and interaural pairwise pitch comparison.

2 Methods

Seven BiCIs (three male and four female; mean age = 53 year) participated in this study. All of them were post-lingual at their onset of bilateral severe-to-profound sensorineural hearing loss and were using MED-EL implant systems. The voluntary informed written consent was obtained with the approval of the Ethics Committee of the University of Oldenburg. The intra-cochlear electrodes of the MED-EL CI are numbered from 1 to 12 in an apical to basal direction. Here the electrode on the right or the left implant is named Rx or Lx. For example, L4 refers to the 4th electrode on the left implant.

A research platform was developed for psychoacoustic testing and eABR recording (Hu et al. 2014). The electrical stimuli were presented via a research interface box (RIB II, manufactured at University of Innsbruck, Austria) and a National Instruments I/O card. The monaural or bilaterally synchronized electrical pulses were applied to the CI, bypassing the behind-the-ear unit.

The experiment consisted of three psychophysical pretests, accompanied by three IEP methods. Pretest 1 was to determine the participant's maximum comfortable level (MCL) and the hearing threshold (HL) of the reference electrode (For more details about the stimuli, please refer to the last paragraph of section 2). From these two values the dynamic range (DR) was obtained. In the second pretest the interaural loudness-balanced level for each electrode pair was determined. The presentation level of the reference CI electrode was fixed at 60 % of DR, while the presentation level of the contralateral probe CI electrode was adapted according to the participant's response. The stimulation levels judged by the participant to be equally loud in both sides were saved and used in the pairwise pitch comparison and EEG recording procedures. In the interaural pairwise pitch comparison task interaural electrode pairs were stimulated sequentially. The participants were asked to indicate in which interval the higher pitch was perceived. This experiment was repeated 50 times per electrode pair. As a third pre-test, a lateralization task was performed on each electrode pair to ensure that a single, fused auditory image was perceived in the center of the head (Kan et al. 2013). The current level and electrode pairs that allowed for a fused and central image were again stored and employed for the second IEP method: IPTD discrimination. Two-interval trials, randomly located with left-leading (IPTD = −T/2) and right-leading stimuli (IPTD = T/2), were presented for the electrode pair using a constant stimulus procedure, for a fixed IPTD = T. The participant was required to indicate whether the stimulus in the second interval was perceived either to the left or the right of the first interval.

In the EEG measurement session, eABRs were differentially recorded from Ag/AgCl electrodes of a customized equidistant 63-channel braincap (Easycap) with an electrode at FPz serving as the ground, and the midline cephalic location (Cz) as the physical reference (Hu et al. 2015). The 63 scalp electrodes were connected to the SynAmps RT amplifier system (Neuroscan). Channel 49 (sub-Inion) and channel 59 (Inion) were primary interest in this study. Electrode impedances were kept below 10 KΩ. The sweeps were filtered by an analog antialiasing-lowpass with a corner frequency of 8 kHz, digitized with 20 kHz sampling rate via a 24 bit A/D convertor. The artifact rejection was turned off during the recording, since filtering, artifact analysis and averaging were done offline. 2100–3000 single sweeps for all electrode conditions were recorded in random order on a sweep-by-sweep basis. During the recording, the participants seated in a recliner and watched silent subtitled movies within an electrically shielded sound-attenuating booth.

The stimulus was a train of charge-balanced biphasic pulses presented at a rate of 19.9 pulses per second (pps) (He et al. 2010), with 50 or 60 μs phase duration, and 2.1 μs interphase gap presented repeatedly via monopolar stimulation mode. A phase duration of 60 μs was used only when the subject could not reach the MCL with 50 μs. In psychophysical sessions 1–4, a 10-pulse chain about 500 ms time duration was used. In the IPTD test, each interval was a 500 ms stimulus and there was 300 ms pause between the two intervals. For the EEG recoding, a continuous electrical pulse train was presented with a 5 ms trigger sent 25 ms before each the CI stimulation onset.

3 Results

The BIC is defined as the difference between the eABR with binaural stimulation (B) and the sum of the eABRs obtained with monaural stimulation (L + R): BIC = B − (L + R). For both eABR and BIC, the latency was defined as the time position of the respective peaks. The amplitude was defined as the difference between the positive peak and the following trough amplitude.

Figure 1a shows the morphology of the eABR and the BIC of subject S5 with reference electrode L4 and probe electrode R1, where the electric artifact was already removed (Hu et al. 2015). In general, wave eI was not observed due to the stimulation pulse artifact from the implant. The amplitude of wave eV was larger than an acoustically evoked ABR amplitude, and latency which occurs approximately 3.6 ms after the onset of the stimulus is shorter because the electrical stimulus directly activates the neural pathway (Starr and Brackmann 1979; Pelizzone et al. 1989). Figure 1b shows the results of the BIC across 12 electrode pairs (R1-R12), for S5 with reference electrode of L4. The latency of the BIC is about 4 ms, and the BIC amplitude is increasing first and then decreasing after electrode 6 (R6). Figure 1c shows the IEP results based on the best IPTD performance and the pitch comparison. The open circles on the dash-dotted line indicate the per-

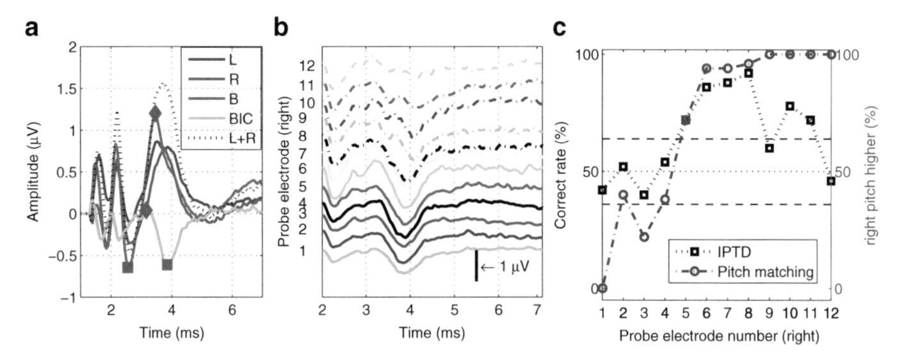

Fig. 1 EABRs and the BICs of subject S5 and the reference electrode is L4: **a** EABRs of *left* CI stimulated only, *right* CI stimulated only, both CI stimulated simultaneously, L + R and the BIC. The probe electrode is R1. The y-axis is their amplitude values in µV. The electric artifact was lined out, wave eV and BIC are visible at approximately 3.6 and 4.1 ms, respectively. **b** The results of the BIC across 12 electrode pairs. **c** The IEP results based on the best IPTD performance and pitch comparison. The x-axis is the number of the probe electrodes in the *right* implant

centage of how often the probe electrode (right side) resulted in a higher pitch percept than the reference electrode (left side) in the pitch matching experiment. The squares on the dashed line are the correct rates of lateralization judgment in the IPTD experiment. The x-axis indicates the probe electrode number in the right implant. The matched electrodes are R4.4 (pitch comparison), R6-R8 (best IPTD performance), and R6 (largest BIC amplitude). In contrast to the common assumption that pitch-matched pairs maximize sensitivity to binaural cues, for this subject the pairs with best IPTD sensitivity are 1.6–3.6 electrodes away from (interpolated) electrode identical pitch perception. The BIC based method indicates electrode R6 to be matched with L4, which is closer to the IPTD based match for this subject. For S5, there are three pairs that have a high IPTD sensitivity (R6 -R8). They are not significantly different from each other, rendering the assignment of a unique matched pair difficult. This is consistent with the previous finds that ITD JNDs do not change much until there is a 3–6 mm mismatch (van Hoesel and Clark 1997). Note that the pitch matched electrode pair does not even yield a significantly above chance IPTD sensitivity.

The results from the other methods and of the other subjects are not shown because of the space limitations. The general trends were: (1) Pairwise pitch comparison results highly depend on the participant's clinical frequency-place fitting map. The matched pair indicates a similar frequency range in the map, except for one subject which was implanted only 9 months before the experiment. (2) Both BIC and IPTD based pairing results indicate a large 2–3 electrode mismatch for some subjects. There is clear correlation between the mismatches indicated by these two methods. (3) In line with Poon et al. (2009) there is no correlation between the pitch matched pairs and the IPTD matched pair and in line with He et al. (2012) there is no correlation between the pitch matched pairs and the BIC matched pair.

4 Discussion

With the increasing number of bilateral CI users and new product development towards binaural CIs, it becomes increasingly important to find ways of matching the electrode arrays in the two ears to compensate for the differences between the two implanted cochleae because of insertion depth, electrode lengths, or neural survival. Current clinical cochlear implant fitting strategies generally treat each implant monaurally and allocate a generic frequency range to each electrode based on the electrode number often without concern about the actual position and the mismatching between the two implants. In a very few positive exceptions interaural pitch comparison techniques are employed. This study investigated the suitability of three IEP methods which have previously been tested in research studies but mostly only one method in isolation. In general, clear monaural and binaural eABRs and BICs were obtained in most of the subjects. All but one subject showed nicely tuned pitch matching, IPTD sensitivity, and BIC as a function of electrode number. However, the pairing results were not always consistent across methods. The IPTD paring results of some subjects are spanned across several electrodes, which is consistent with the previous finds that ITD JNDs do not changed much until there is a 3–6 mm mismatch compared to the best matched place condition (van Hoesel and Clark 1997).

Interaural pitch comparison shows some predictive value in selecting interaural electrode pairs, but it may not fully compensate for any underlying differences between the two implanted cochleae. In line with previous studies (van Hoesel and Clark 1997; Long et al. 2003; Poon et al. 2009) the pitch matched pair does not guarantee best IPTD sensitivity. All the CI subjects in this study have had at least 9 months bilateral CI experience prior to the experiment so acclimatization to the electrode-frequency-map or brain plasticity may explain the good pitch tuning curve. There is no correlation between the BIC amplitudes and results of interaural pitch matching as well, which is consistent with (He et al. 2012).

For the interpretation of the BIC, the straightest assumption is that if the reference electrode and the probe electrode are cochlear place matched, they stimulate similarly "tuned" auditory-nerve fibers that eventually result in binaural interaction in the brain stem and thus in a larger BIC than the other pairs. However, in case of a different neural survival across the ears or within one ear, also non-place matched electrodes may result in the maximum BIC. Further studies with more reference electrodes are necessary to test this assumption. This may also partially explain the similar ITD sensitivity of the neighboring electrodes around the best matched electrode.

In summary, all three methods can obtain reasonable interaural electrode pairing results. There is substantial variability in all the measurements both within and between individual CI users. There are pros and cons for all the three methods and there is no gold standard to judge which one is better. The acclimatization to the frequency to place map of the patient's CI seems unignorably in the pitch perception. Pitch matching data may be 'misleadingly good' because of acclimatization or brain plasticity which does not affect the IPTD and the BIC based matching. Knowledge about the location of the implant inside the cochlea maybe helpful in judging the pairing results. Longitudinal measurements are necessary in the future to investigate the plasticity of the binaural perception in the BiCI users. The BIC is a promising candidate for electrode pairing of BiCI subjects, especially for pediatric fitting.

Acknowledgments This work was funded by European Union under the Advancing Binaural Cochlear Implant Technology (ABCIT) grant agreement (No. 304912). We thank our ABCIT project partners at the UCL Ear Institute David McAlpine and Torsten Marquardt for inspirational discussions when planning this study. The authors are grateful to Stefan Strahl (MED-EL) for RIB II support and Stephan Ewert and Tom Campbell for their help during method development. We thank Karsten Plotz, Katrin Bomke (Klinik für Phoniatrie und Pädaudiologie, Ev. Krankenhaus Oldenburg), and Stefan Dazert (CI-Zentrum Ruhrgebiet, Universitätsklinik Bochum) for recruiting CI participants. The authors also thank all CI participants for joining the study and for giving us valuable feedback.

References

Gordon KA, Salloum C, Toor GS, van Hoesel R, Papsin BC (2012) Binaural interactions develop in the auditory brainstem of children who are deaf: effects of place and level of bilateral electrical stimulation. J Neurosci 32(12):4212–4223

Goupell MJ, Stoelb C, Kan A, Litovsky RY (2013) Effect of mismatched place-of-stimulation on the salience of binaural cues in conditions that simulate bilateral cochlear-implant listening. J Acoust Soc Am 133(4):2272–2287

He S, Brown CJ, Abbas PJ (2010) Effects of stimulation level and electrode pairing on the binaural interaction component of the electrically evoked auditory brain stem response. Ear Hear 31(4):457–470

He S, Brown CJ, Abbas PJ (2012) Preliminary results of the relationship between the binaural interaction component of the electrically evoked auditory brainstem response and interaural pitch comparisons in bilateral cochlear implant recipients. Ear Hear 33(1):57–68

Hu H, Ewert S, Campbell T, Kollmeier B, Dietz M (2014). An interaural electrode pairing clinical research system for bilateral cochlear implants. Paper presented at the the 2nd IEEE China Summit and International Conference on Signal and Information Processing (ChinaSIP' 14), Xi'an, China

Hu H, Kollmeier B, Dietz M (2015 Reduction of stimulation coherent artifacts in electrically evoked auditory brainstem responses. Biomed Signal Process Control xx(in Press):xx–xx

Kan A, Stoelb C, Litovsky RY, Goupell MJ (2013) Effect of mismatched place-of-stimulation on binaural fusion and lateralization in bilateral cochlear-implant users. J Acoust Soc Am 134(4):2923–2936

Litovsky RY, Goupell MJ, Godar S, Grieco-Calub T, Jones GL, Garadat SN, Agrawal S, Kan A, Todd A, Hess C, Misurelli S (2012) Studies on bilateral cochlear implants at the University of Wisconsin's Binaural Hearing and Speech Laboratory. J Am Acad Audiol 23(6):476–494

Long CJ, Eddington DK, Colburn HS, Rabinowitz WM (2003) Binaural sensitivity as a function of interaural electrode position with a bilateral cochlear implant user. J Acoust Soc Am 114(3):1565–1574

Majdak P, Goupell MJ, Laback B (2011) Two-dimensional localization of virtual sound sources in cochlear-implant listeners. Ear Hear 32(2):198–208

Pelizzone M, Kasper A, Montandon P (1989) Electrically evoked responses in Cochlear implant patients. Audiology 28(4):230–238

Pelizzone M, Kasper A, Montandon P (1990) Binaural interaction in a cochlear implant patient. Hear Res 48(3):287–290

Poon BB, Eddington DK, Noel V, Colburn HS (2009) Sensitivity to interaural time difference with bilateral cochlear implants: development over time and effect of interaural electrode spacing. J Acoust Soc Am 126(2):806–815

Smith ZM, Delgutte B (2007) Using evoked potentials to match interaural electrode pairs with bilateral cochlear implants. J Assoc Res Otolaryngol 8(1):134–151

Starr A, Brackmann DE (1979) Brain stem potentials evoked by electrical stimulation of the cochlea in human subjects. Ann Otol, Rhinol Laryngol 88(4 Pt 1):550–556

van Hoesel RJ, Clark GM (1997) Psychophysical studies with two binaural cochlear implant subjects. J Acoust Soc Am 102(1):495–507

Binaural Loudness Constancy

John F. Culling and Helen Dare

Abstract In binaural loudness summation, diotic presentation of a sound usually produces greater loudness than monaural presentation. However, experiments using loudspeaker presentation with and without earplugs find that magnitude estimates of loudness are little altered by the earplug, suggesting a form of loudness constancy. We explored the significance of controlling stimulation of the second ear using meatal occlusion as opposed to the deactivation of one earphone. We measured the point of subjective loudness equality (PSLE) for monaural vs. binaural presentation using an adaptive technique for both speech and noise. These stimuli were presented in a reverberant room over a loudspeaker to the right of the listener, or over lightweight headphones. Using the headphones, stimuli were either presented dry, or matched to those of the loudspeaker by convolution with impulse responses measured from the loudspeaker to the listener position, using an acoustic manikin. The headphone response was also compensated. Using the loudspeaker, monaural presentation was achieved by instructing the listener to block the left ear with a finger. Near perfect binaural loudness constancy was observed using loudspeaker presentation, while there was a summation effect of 3–6 dB for both headphone conditions. However, only partial constancy was observed when meatal occlusion was simulated. These results suggest that there may be contributions to binaural loudness constancy from residual low frequencies at the occluded ear as well as a cognitive element, which is activated by the knowledge that one ear is occluded.

Keywords Loudness summation · Perceptual constancy · Meatal occlusion · Monaural · Virtual acoustics

1 Introduction

When the same sound is presented to both ears, it is perceived to be louder than when it is presented to one ear only (Fletcher and Munson 1933). As a result, the loudness of diotically and monaurally presented stimuli, usually tones or noise, has

---

J. F. Culling (✉) · H. Dare
School of Psychology, Cardiff University, Tower Building, Park Place, Cardiff, CF10 3AT, UK
e-mail: CullingJ@cf.ac.uk

P. van Dijk et al. (eds.), *Physiology, Psychoacoustics and Cognition in Normal and Impaired Hearing,* Advances in Experimental Medicine and Biology 894,
DOI 10.1007/978-3-319-25474-6_8

been found to be equivalent when the monaural stimulus is between 3 and 10 dB more intense (Fletcher and Munson 1933; Reynolds and Stevens 1960; Zwicker and Zwicker 1991; Sivonen and Ellermeier 2006; Whilby et al. 2006; Edmonds and Culling 2009). Models of loudness therefore incorporate this phenomenon in various ways (Zwicker and Scharf 1965; Moore and Glasberg 1996a, b; Moore and Glasberg 2007). These models can be used in hearing-aid fitting (e.g. Moore and Glasberg 2004), so hearing aids fitted using these models will have relatively reduced gain when fitted bilaterally in order to achieve a comfortable maximum loudness level. However, some recent studies have cast doubt on the ecological validity of the loudness summation literature.

Cox and Gray (2001) collected loudness ratings (7 categories from "very soft" to "uncomfortable") for speech at different sound levels when listening monaurally and binaurally. These two modes of listening were compared by using either one or two earphones and, using a loudspeaker, by occluding one ear with a plug and ear muff. They found that results with one or two earphones produced a conventional loudness summation effect, whereby the mean rating was substantially higher at each sound level for binaural presentation. However, when listening to an external source in the environment (the loudspeaker) there was much less summation effect: occlusion of one ear had little effect on the loudness ratings compared to listening binaurally. This experiment showed for the first time that the loudness of external sounds may display constancy across monaural and binaural listening modes. However, the methods used were clinically oriented and difficult to compare with conventional psychophysical measurements.

Epstein and Florentine (2009, 2012) conducted similar tests, but using standard loudness estimation procedures and speech (spondees) either with or without accompanying video of the speaker's face. They also observed loudness constancy, but only when using the audiovisual presentation. Their tentative conclusion was that loudness constancy may only occur using stimuli of relatively high ecological validity. Ecological validity may be enhanced when an external source is used, when that source is speech, particularly connected speech rather than isolated words, and when accompanied by coherent visual cues. Since all these conditions are fulfilled when listening to someone in real life, the phenomenon of loudness constancy can be compellingly, if informally, demonstrated by simply occluding one ear with a finger when listening to someone talk; most people report that no change in loudness is apparent. The present study was inspired by this simple demonstration technique.

A limitation of previous studies is that the use of ear plugs and muffs means that different listening modes cannot be directly compared. It is commonplace in psychophysical testing to match the loudness of different stimuli by listening to them in alternation, but it is impractical to insert/remove an ear plug between one presentation interval and another. As a result comparisons are made over quite long time intervals. In contrast, a finger can be applied to the meatus in less than a second, so it is possible to perform a 2-interval, forced-choice procedure provided that the monaural interval is always the same one so that the listener can learn the routine of blocking one ear for the same interval of each trial. The present experiment adopted this technique and also explored the idea that ecological validity plays a key role by using either speech or noise as a sound source and by creating a close physical match between stimuli presented from a loudspeaker and those presented using headphones.

2 Methods

2.1 *Stimuli*

There were six conditions, comprised of two stimulus types (speech/noise) and three presentation techniques. The speech stimuli were IEEE sentences (Rothauser et al. 1969) and the noise stimuli were unmodulated noises filtered to have the same long-term excitation pattern (Moore and Glasberg 1987) as the speech. These two stimulus types were presented (1) dry over headphones, (2) from a loudspeaker to the listeners' right or (3) virtually through lightweight open-backed headphones (Sennheiser HD414). Monaural presentation was achieved in (1) and (3) by deactivating the left earphone for the second interval of each trial and in (2) by asking the listener to block their left ear with a finger for the second interval of each trial. For virtual presentation, the stimuli were convolved with binaural room impulse responses (BRIRs) recorded from a manikin (B&K HATS 4100) sitting in the listeners' position and wearing the lightweight headphones with the loudspeaker 1 m to the right in a standard office room with minimal furnishing (reverberation time, RT60 = 650 ms). Impulse responses were also recorded from the headphones and used to derive inverse filters to compensate for the headphone-to-microphone frequency response. Sound levels for the different stimuli were equalised at the right ear of the manikin and, for the reference stimuli, were equivalent to 57 dB(A) as measured at the head position with a hand-held sound-level meter.

2.2 *Procedure*

Twelve undergraduate-students with no known hearing impairments took part in a single 1 h session. Initially, five practice trials were presented, for which the listeners were simply required to learn the routine of blocking their left ear during the one-second inter-stimulus interval between two bursts of speech-shaped noise presented from a loudspeaker. During the experiment, the ordering of six conditions was randomly selected for each participant.

For each condition, listeners completed two loudness discrimination threshold tasks. These adaptive thresholds served to bracket the point of subjective loudness equality (PSLE): one was the 71 % threshold for identifying when a monaural stimulus was louder than a binaural one, and the other was the 71 % threshold for identifying when the monaural stimulus was quieter than a binaural one. For a given

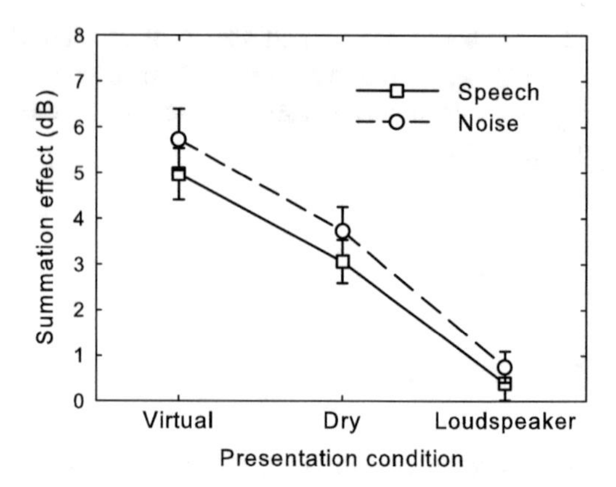

Fig. 1 The mean summation effect in Expt. 1 in each of the three presentation conditions and for speech and for noise sources. Error bars are one standard error

condition, the two adaptive tracks started with the adapted (monaural) sound level 15 dB above and below the level of the reference (binaural) stimulus, respectively. In each task, the listeners were required to identify the louder of two stimuli in a 2-down, 1-up adaptive procedure with six reversals. The two thresholds were averaged to yield a PSLE for that condition.

Within each trial the first interval was always the reference stimulus, while the adapted stimulus was presented in the second interval. The adapted stimulus was thus always the monaural one.

3 Results

Figure 1 shows that there was a strong effect of the presentation condition ($F(2,22)=22.1$, $p<0.001$). The summation effect is less than 1 dB in the Loudspeaker condition, in which monaural presentation was achieved by occluding the meatus with one finger, but is 3–6 dB in the Virtual and Dry conditions, in which monaural presentation was achieved by deactivating one earphone. Summation was also significantly smaller using speech than using noise as a source signal ($F(1,11)=5.6$, $p<0.05$).

4 Discussion

The lack of summation in the Loudspeaker condition supports the observations made by Cox and Gray (2001) and by Epstein and Florentine (2009, 2012), that, when monaural presentation is achieved by occluding one ear in a sound field, binaural summation can be dramatically reduced. The present results demonstrate this phenomenon for the first time using direct comparisons between successive stimuli. Consistent with suggestions that ecological validity is important to the effect, it was a little larger for speech than for noise, but this effect was rather small (0.8 dB).

In order to match the headphone and loudspeaker conditions physically, virtual presentation was used in a reverberant room. This approach resulted in a compelling impression of sound source externalisation for the headphone condition, which may have been accentuated by the presence of a loudspeaker at the appropriate position. A post-experiment test showed that listeners perceived the virtual binaural stimuli as emanating from the loudspeaker rather than the headphones. A consequence of this procedure was that when the left headphone was deactivated for monaural presentation, the externalisation of the sound collapsed, so that listeners found themselves comparing one sound that appeared to be from the loudspeaker with another that, although still reverberant and from the same side, appeared to come from the headphones. It is unclear whether this effect may have caused the apparent enhancement of the loudness summation effect in the Virtual condition.

In each of the studies in which binaural summation has been largely abolished, the second ear has been occluded by some means. The lack of summation could have two logical causes. First, there may be some residual sound energy at the occluded ear that is sufficient to maintain its contribution to binaural loudness. Second, there may be some cognitive element which generates loudness constancy through the knowledge that one ear is occluded. It is important, therefore to determine the amount of residual sound and the role it might play. Cox and Gray used a combination of ear plug and muff and reported that this combination produced a threshold shift of 23 dB at 250 Hz and 52 dB at 4 kHz, indicating that the residual sound was substantially attenuated. Epstein and Florentine used only an ear plug and reported the resulting attenuation caused a threshold shift of 20–24 dB at 1 kHz. In the present study, a finger was used. In order to determine the likely degree of attenuation achieved, we recorded binaural room impulse responses BRIRs with and without an experimenter's finger over the meatus. This was not possible with the B&K HATS 4100 because it has the microphone at the meatus, so KEMAR was used. Several repeated measurements found a consistent attenuation of ~30 dB across most of the frequency spectrum, but progressively declining at low frequencies to negligible levels at 200 Hz and below. Assuming this measurement gives a reasonably accurate reflection of the change in cochlear excitation when a human listener blocks the meatus, the question arises whether this residual sound might be sufficient to support summation.

First, we conducted an analysis of the virtual stimuli by applying the Moore and Glasberg (1996a, b) loudness model. Since similar effects were observed with noise as with speech, it was thought unnecessay to employ the model for time-varying sounds (Glasberg and Moore 2002). Despite presentation of stimuli from one side, the difference in predicted loudness at each unoccluded ear differed by only 4%. The interaural level difference was small due to the room reverberation. When the left ear was occluded, the predicted loudness at this ear fell by 52%. Older models assume that binaural loudness is the sum of that for each ear, but more recent models predict less binaural summation by invoking an inhibitory mechanism (Moore and Glasberg 2007). In either case, attenuation caused by meatal occlusion leading to a 52% drop at one ear seems more than adequate to observe a strong summation effect.

While modelling can be persuasive, we wanted to be sure that the relatively unattenuated low frequencies were not responsible for a disproportionate contribution to binaural loudness. A second experiment examined this possibility by simulating occlusion.

5 Methods

Eight listeners took part in a similar experiment, but using a new presentation condition that simulated the use of a finger to block the meatus. One of the BRIRs collected with a finger over the ear of a KEMAR was convolved with the source signals to form the monaural stimulus of a Virtual Finger condition. This condition was contrasted with a replication of the previous Virtual condition in which monaural presentation was achieved by silencing the left ear. This condition is now labelled Silence, so that the condition labels reflect the method of attenuating the left ear. For consistency, all BRIRs for the second experiment were collected using KEMAR. The binaural stimulus for each condition was identical. The same speech and noise sources were used.

6 Results

Figure 2 shows the mean summation effect from each of the four conditions of the second experiment. The Silence condition produces a comparable summation effect to the similarly constructed Virtual condition of Expt. 1. However, while the Virtual Finger condition does not result in an abolition of this summation effect, it does result in a substantial reduction in its size ($F(1,7)=102, p<0.001$). In contrast to Expt. 1, the noise stimulus produces a smaller summation effect than the speech stimulus ($F(1,7)=22, p<0.005$). The interaction fell short of significance.

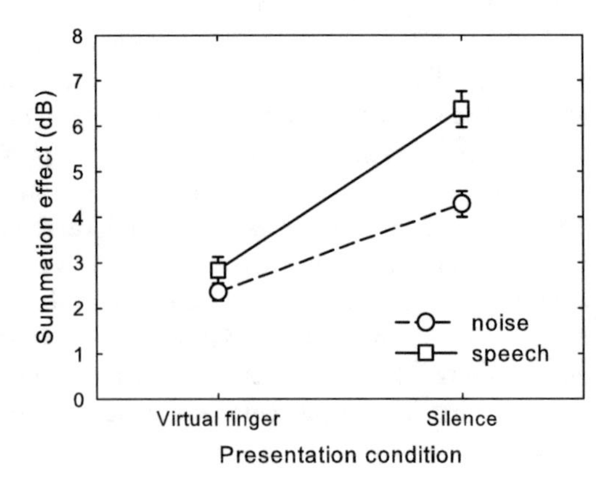

Fig. 2 The mean summation effect in Experiment 2 in each of the presentation conditions and for speech and for noise sources. Error bars are one standard error

7 Discussion

Since the Silence and Virtual-Finger conditions of experiment 2 produced significantly different summation effects, it seems that a portion of the loudness constancy effect observed in experiment 1 may have been mediated by the residual sound at the occluded ear. Both the Silence and the Virtual-Finger conditions produced a collapse in externalisation for the monaural case, so it seems unlikely that the difference in effect can be attributed to this cause.

On the other hand, 2.5–3 dB of summation was still observed in the Virtual-Finger condition, whereas less that 1 dB was observed in the Loudspeaker condition of experiment 1. It appears, therefore, that the listeners' awareness of meatal occlusion, which would only be present in the first case, may still play a role. A major caveat to these conclusions is that KEMAR was not designed to realistically simulate the effects of bone-conducted sound, which likely plays a major role in listening in a sound field with an occluded ear. While probe microphones in the ear canal might more accurately record the stimulation at the tympanic membrane, they would still not capture the stimulation occurring at the cochlea.

The small effect of stimulus type observed in experiment 1, was reversed in experiment 2. Taking these small and inconsistent effects together, we found no evidence that binaural loudness constancy is greater for more ecologically valid stimuli such as connected speech. Indeed, for both speech and noise, occlusion of the ear led to little reduction in loudness (a small summation effect), suggesting that loudness constancy occurs independently of stimulus type. It should be noted, however, that quite prominent reverberation was present in our experiment, which reliably cued the existence of an external sound source.

8 Conclusions

Experiment 1 demonstrated for the first time that binaural loudness constancy can be observed in a loudness matching task using direct comparisons of loudness between an occluded and unoccluded second ear. Experiment 2 showed that this effect could be partially mediated by the residual sound at the occluded ear, but the remaining effect would seem attributable to the listeners' awareness of the occlusion.

References

Cox RM, Gray GA (2001) Verifying loudness perception after hearing aid fitting. Am J Audiol 10:91–98

Edmonds BA, Culling JF (2009) Interaural correlation and the binaural summation of loudness. J Acoust Soc Am 125:3865–3870

Epstein M, Florentine M (2009) Binaural loudness summation for speech and tones presented via earphones and loudspeakers. Ear Hear 30:234–237

Epstein M, Florentine M (2012) Binaural loudness summation for speech and tones presented via earphones and loudspeakers. J Acoust Soc Am 131:3981–3988

Fletcher H, Munson WA (1933) Loudness, its definition, measurement and calculation. J Acoust Soc Am 5:82–108

Glasberg BR, Moore BCJ (2002) Model of loudness applicable to time-varying sounds. J. Audio Eng Society 53:331–342

Moore BCJ, Glasberg BR (1987) Formulae describing frequency selectivity as a function of frequency and level, and their use in calculating excitation patterns. Hear Res 28:209–225

Moore BCJ, Glasberg BR (1996a) A revision of Zwicker's loudness model. Acustica 82:335–345

Moore BCJ, Glasberg BR (1996b) A model for the prediction of thresholds, loudness, and partial loudness. J Audio Eng Society 45:224–240

Moore BCJ, Glasberg BR (2004) A revised model of loudness perception applied to cochlear hearing loss. Hear Res 188:70–88

Moore BCJ, Glasberg BR (2007) Modelling binaural loudness. J Acoust Soc Am 121:1604–1612

Reynolds GS, Stevens SS (1960) Binaural summation of loudness. J Acoust Soc Am 32:1337–1344

Rothauser EH, Chapman WD, Guttman N, Nordby KS, Silbiger HR, Urbanek GE, Weinstock M (1969) I.E.E.E. recommended practice for speech quality measurements. IEEE Trans Audio Electroacoust 17:227–246

Sivonen VP, Ellermeier W (2006) Directional loudness in an anechoic sound field, head-related transfer functions, and binaural summation. J Acoust Soc Am 119:2965–2980

Whilby S, Florentine M, Wagner E, Marozeau J (2006) Monaural and binaural loudness of 5- and 200-ms tones in normal and impaired hearing. J Acoust Soc Am 119:3931–3939

Zwicker E, Scharf B (1965) A model of loudness summation. Psychol Rev 72:3–26

Zwicker E, Zwicker UT (1991) Dependence of binaural loudness summation on interaural level differences, spectral distribution and temporal distribution. J Acoust Soc Am 89:756–7645

Intelligibility for Binaural Speech with Discarded Low-SNR Speech Components

Esther Schoenmaker and Steven van de Par

Abstract Speech intelligibility in multitalker settings improves when the target speaker is spatially separated from the interfering speakers. A factor that may contribute to this improvement is the improved detectability of target-speech components due to binaural interaction in analogy to the Binaural Masking Level Difference (BMLD). This would allow listeners to hear target speech components within specific time-frequency intervals that have a negative SNR, similar to the improvement in the detectability of a tone in noise when these contain disparate interaural difference cues. To investigate whether these negative-SNR target-speech components indeed contribute to speech intelligibility, a stimulus manipulation was performed where all target components were removed when local SNRs were smaller than a certain criterion value. It can be expected that for sufficiently high criterion values target speech components will be removed that do contribute to speech intelligibility. For spatially separated speakers, assuming that a BMLD-like detection advantage contributes to intelligibility, degradation in intelligibility is expected already at criterion values below 0 dB SNR. However, for collocated speakers it is expected that higher criterion values can be applied without impairing speech intelligibility. Results show that degradation of intelligibility for separated speakers is only seen for criterion values of 0 dB and above, indicating a negligible contribution of a BMLD-like detection advantage in multitalker settings. These results show that the spatial benefit is related to a spatial separation of speech components at positive local SNRs rather than to a BMLD-like detection improvement for speech components at negative local SNRs.

Keywords Speech intelligibility · Speech interferers · Multitalker situation · Binaural masking level differences · Binaural listening · Binaural detection · Masking · Masking release · Spatial unmasking · Speech segregation

E. Schoenmaker (✉) · S. van de Par
Acoustics Group, Cluster of Excellence "Hearing4All", Carl von Ossietzky University, Carl von Ossietzkystraße 9-11, D-26129 Oldenburg, Germany
e-mail: esther.schoenmaker@uni-oldenburg.de

S. van de Par
e-mail: steven.van.de.par@uni-oldenburg.de

P. van Dijk et al. (eds.), *Physiology, Psychoacoustics and Cognition in Normal and Impaired Hearing,* Advances in Experimental Medicine and Biology 894,
DOI 10.1007/978-3-319-25474-6_9

1 Introduction

When listening to speech in a noisy background, higher speech intelligibility is measured for spatially separated speech and interfering sources, as compared to collocated sources. The benefit of this spatial separation, known as spatial release from masking, has been attributed to binaural processing enabling a better signal detection, as can be measured by binaural masking level differences (BMLDs). These BMLDs describe the difference in thresholds for tone-in-noise detection when both the target tone and masking noise are presented interaurally in phase, in contrast to a reversed phase of the tone in one ear. In the latter case thresholds are significantly lower. Similar interaural relations of the stimuli can be achieved by presenting the noise from a spatial position directly in front of the listener and the signal from a different spatial location. Therefore a connection was hypothesized between the spatial benefit in speech intelligibility and the tone detection advantage that had been measured as BMLD.

Levitt and Rabiner (1967) predicted the spatial improvement in speech intelligibility using this hypothesis. The underlying assumption of their model is that a BMLD effect acts within each frequency band, leading to a within-band detection advantage. This was modeled as a reduction of the level of the noise floor equal in size to the BMLD for that particular frequency, resulting in a more advantageous signal-to-noise ratio (SNR). This would make previously undetectable speech elements audible and enable them to contribute to speech intelligibility. More recent models of speech intelligibility also make use of the concept of within-band improvement of SNR to model the spatial release from masking (e.g., Beutelmann et al. 2010; Lavandier et al. 2012; Wan et al. 2014).

Whereas Levitt and Rabiner (1967) proposed their model for speech in stationary noise, we are interested in speech interferers. Since speech shows strong modulations both in time and frequency, the level of the interferer will vary. We will therefore consider BMLD effects at the level of spectro-temporal regions rather than frequency channels. Thus we extend the original hypothesis and postulate that a BMLD effect might take place at the level of individual spectro-temporal units and that this detection advantage would be responsible for the spatially improved speech intelligibility.

The aim of this study is to investigate whether a BMLD-like effect, that would lead to improved detection of speech elements in a spatially separated source configuration, indeed is responsible for improved speech intelligibility in the presence of interfering speech.

We will test the contribution of a BMLD-like effect by deleting spectro-temporal regions of the target signal below a specific SNR criterion and measuring speech intelligibility. We assume that, if certain speech elements contribute to speech intelligibility, the performance on a speech intelligibility task will be lower after deleting them. At low values of the SNR criterion this manipulation will lead to deletion of spectro-temporal regions at which either little target energy was present or the target signal was strongly masked. At such values not much impact on speech intelligibility is expected. Deletion of target speech components at increasing SNRs, however, will start to affect speech intelligibility at some point.

The question of interest is whether the effect of eliminating spectro-temporal regions will follow a different course for collocated and spatially separated source configurations. Should a BMLD-like effect indeed act at the level of small spectro-temporal regions, then deletion of target speech from spectro-temporal regions with local SNR values between −15 and 0 dB (i.e. between dichotic and diotic masked thresholds, Hirsh (1948)), should degrade speech intelligibility for separated, but not for collocated sources, since those speech components would be inaudible in a collocated configuration anyway. This implies that over this same range of SNR criteria the spatial release from masking, as measured by the difference in performance between separated and collocated source configurations, should decrease.

2 Methods

2.1 Stimuli

German sentences taken from the Oldenburg Sentence Test (OLSA, Wagener et al. 1999) and spoken by a male speaker served as target speech. Each sentence consisted of five words with the syntactic structure *name—verb—numeral—adjective—object*. For each word type ten alternatives were available, the random combination of which yields syntactically correct but semantically unpredictable sentences.

In order to have a spatially complex masker, the interfering speech consisted of two streams of ongoing speech spoken by two different female talkers, and was obtained from two audio books in the German language. Any silent intervals exceeding 100 ms were cut from the signals to ensure a natural-sounding continuous speech stream without pauses. All target and masker signals were available at a sampling rate of 44.1 kHz.

A total of 200 masker samples for each interfering talker were created by randomly cutting ongoing speech segments of 3.5-s duration from the preprocessed audio book signals. This length ensured that all target sentences were entirely masked by interfering speech. The onsets and offsets of the segments were shaped by 200-ms raised-cosine ramps.

The target sentences were padded with leading and trailing zeros to match the length of the interfering signals. Some time roving was applied to the target speech by randomly varying the number of leading zeros between 25 and 75 % of the total number of zeros.

The relative levels of the three signals were set based on the mean RMS of the two stereo channels after spatialization. The two interfering signals were equalized to the same level, whereas the target speech (before application of the binary mask, see next section) was presented at a level of −8 dB relative to each single interferer. This resulted in a global SNR of approximately −11 dB. The sound level was set to 62 dB SPL for a single interferer. After signal manipulation as will be described below, the three signals were digitally added which resulted in a total sound level of approximately 65 dB SPL.

All speech signals were presented from virtual locations in the frontal horizontal plane. The spatial positions were rendered with the help of head-related transfer functions (HRTF), that had been recorded according to Brinkmann et al. (2013), using a Cortex MK2 head-and-torso simulator at a source-to-head distance of 1.7 m in an anechoic chamber.

Two different speaker configurations were used in this experiment, one with spatially separated speakers and one with collocated speakers. The target speech was presented from 0° azimuth, i.e. directly in front of the listener, in both configurations. In the collocated configuration the two sources of masking speech were presented from this same location. In the spatially separated configuration one interfering speaker was presented from the location 60° to the left and the other speaker from 60° to the right of the target.

2.2 *Target Signal Manipulation*

In order to investigate the contributions of target speech at various levels of local SNR, the target speech needed to be manipulated.

The first step in the stimulus manipulation consisted of the calculation of the local, i.e. spectro-temporal, SNR of the three-speaker stimulus. Spectrograms were calculated from 1024-point fast Fourier transforms on 1024-samples long, 50%-overlapping, square root Hann-windowed segments of the signals. This was performed for each of the two stereo channels of a spatialized speech signal to accommodate interaural level differences. This resulted in a pair of left and right spectrograms for each signal, from which one estimate of the spectro-temporal power distribution needed to be calculated. This was achieved by computing the squared mean of the absolute values of each set of two (i.e. left and right) corresponding Fourier components and repeating this for the complete spectro-temporal plane (Fig. 1, left column).

Subsequently, the resulting spectral power distributions were transformed to the equivalent rectangular bandwidth (ERB) scale. This was achieved by grouping power components within the frequency range of one ERB. The time-frequency (T-F) units thus created covered a fixed time length of 23 ms (i.e. 1024 samples) and a frequency band of 1 ERB width.

In the final step to calculate the spectro-temporal SNR, the spectro-temporal power of each target signal was compared to the summed spectro-temporal power of the two masker signals with which it was combined in a single stimulus. An SNR value was calculated for each T-F unit of the combined three-speaker signal, resulting in a 2-D matrix of local spectro-temporal SNR values (Fig. 1, center).

The spectro-temporal SNR representation was used to decide which components of the target speech signal would be passed on to the final signal. Any local SNR

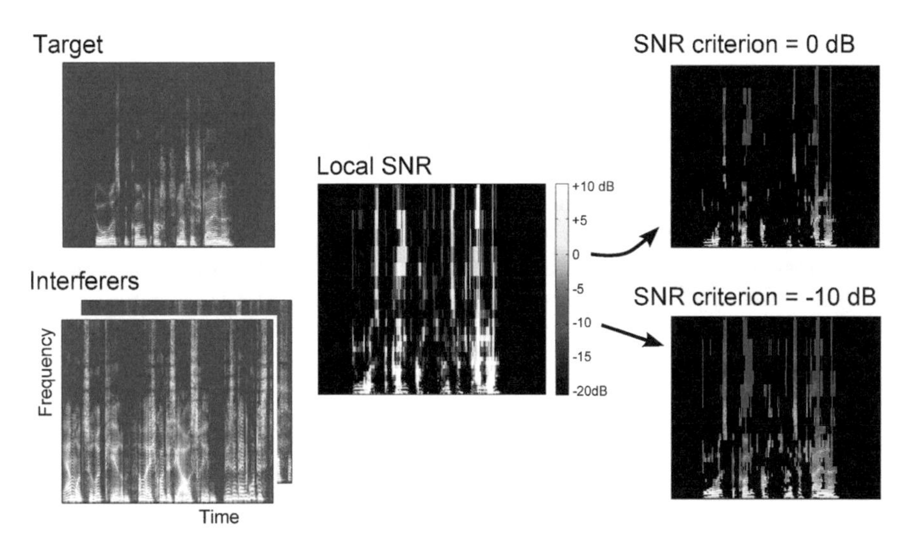

Fig. 1 Overview of the stimulus manipulation. *Left column*: Mean spectrograms of original target and interfering speech. *Center*: Local SNRs of T-F units. *Right column*: Spectrograms of target speech after discarding low-SNR T-F units. Examples are shown for SNR criterion values of 0 and −10 dB

values exceeding the selection criterion resulted in a value of 1 of a binary mask, and thus in selection of the corresponding spectro-temporal components of the target signal in both ears. Any local SNR values that did not pass the criterion resulted in a value of zero and deselection of the corresponding components of the target (Fig. 1, right column).

A total of 8 different local SNR criteria were included in the experiment based on pilot testing: −10,000, −4, 0, 2, 4, 6, 8 and 10 dB. The criterion of −10,000 dB was chosen to simulate an SNR of $-\infty$, and served as a control condition of unfiltered target speech in which the entire target signal was passed on to the final stimulus.

After addition of the selected spectro-temporal components of the target signal to the complete spectrograms of the two interferer signals, the final signal was reconstructed by calculation of the inverse fast Fourier transform, followed by windowing with square-root Hann windows and overlap-add.

2.3 Procedure

Six normal-hearing participants (4 females, 2 males, aged 22–29 years) participated in the experiment. All were native German speakers.

Stimuli were presented over Sennheiser HD650 headphones in a double-walled soundproof booth. The OLSA speech sentence test was run as a closed test and the 5×10 answer alternatives were shown on a PC screen. The participants were instructed to ignore the female speakers, and to mark the perceived words from the male speaker at the frontal position in a graphical user interface. They were forced to guess upon missed words. The listeners participated in a prior training session in which the target sentences had not been manipulated. The training session comprised 160 trials at a global SNR that started from 0 dB and was gradually decreased to the experimental SNR of −8 dB.

During the two experimental sessions that consisted of 160 trials each, all 16 conditions (8 SNR criteria ×2 spatial configurations) were tested interleaved in random order. One test list of 20 sentences, resulting in 100 test words, was used for each condition. Performance was measured as the percentage correct words.

3 Results

The top rows of Fig. 2 show the percentage of correctly recognized words versus SNR criterion for the six individual listeners. The graphs show a clear spatial release from masking for all listeners, reflected by the consistently higher scores for spatially separated speech. From Fig. 2 it can also be seen that the SNR criterion at which speech intelligibility starts to decrease below the performance in the

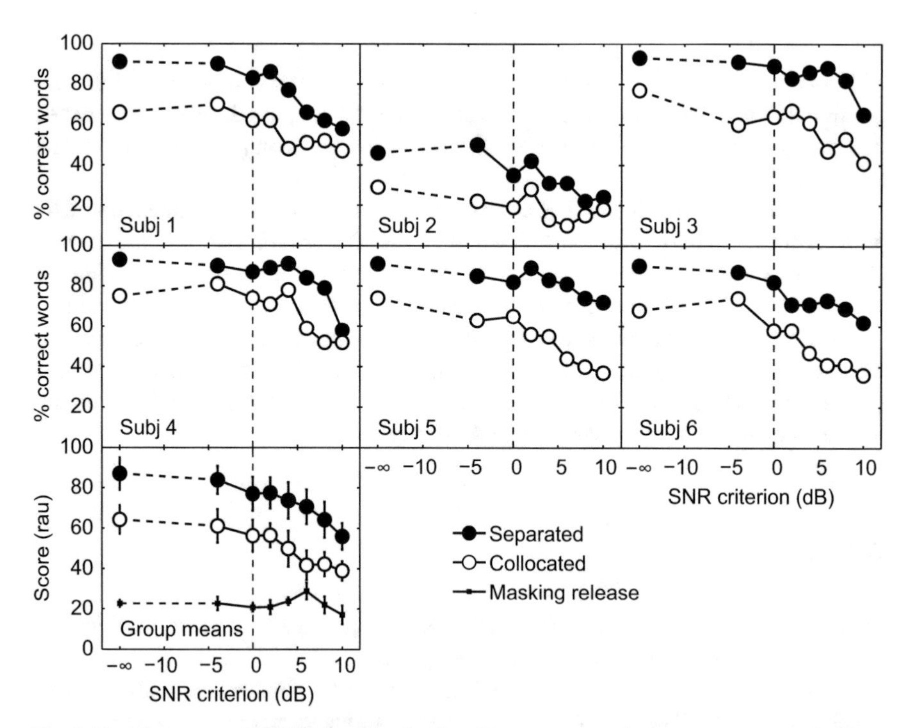

Fig. 2 *Upper two rows*: Individual results for six subjects, expressed as percentage correctly identified words versus SNR criterion. *Bottom row*: Mean results and spatial release from masking expressed in rau. The error bars represent standard errors

unfiltered condition (i.e. at SNR $=-\infty$) differs among listeners between about -4 and 5 dB.

The mean results are shown in Fig. 2 (bottom), together with the spatial release from masking which equals the difference between the spatially separated and collocated data at each SNR criterion. From this graph it becomes clear that the spatial release from masking remains more or less constant over the complete range of SNR criteria tested. The curves for the separated and collocated curves apparently do not converge.

In order to perform statistical analyses, the percentage values were first transformed into rationalized arcsine units (Studebaker 1985). A two-way repeated-measures ANOVA with factors of spatial condition (collocated or separated) and SNR criterion shows a significant effect of spatial condition [$F(1,5)=125.68$, $p<0.001$], a significant effect of SNR criterion [$F(7,35)=34.45$, $p<0.001$], and no significant interaction of spatial condition and SNR criterion [$F(7,35)=1.36$, $p=0.253$]. The absence of interaction between spatial condition and SNR criterion confirms the constant spatial release from masking that could be observed in Fig. 2, bottom.

Simple contrasts comparing scores at each test SNR criterion to the reference criterion at $-\infty$ dB, show that all SNR criteria including and exceeding 0 dB resulted in scores different from the reference criterion at the level of $p<0.05$ (one-sided Dunnett's test). Only the SNR criterion of -4 dB shows no significant difference from the reference criterion.

4 Discussion

The aim of this study was to gain more insight into the mechanism leading to spatial release from masking in a multispeaker situation. Specifically, we were interested in assessing whether a BMLD-like detection advantage at the spectro-temporal level could be responsible for access to more target speech elements in a configuration with spatially separated sources, as this could potentially contribute to a higher speech intelligibility.

An experiment in which spectro-temporal regions from the target speech signal were removed according to a variable local SNR criterion gave no indication for any BMLD-like detection advantage. The results of the experiment showed a decrease in speech intelligibility when elements with a local SNR of 0 dB and larger were removed from the target speech, suggesting that speech elements at local SNRs above -4 dB did contribute to speech intelligibility. Interestingly, this was the case for both the collocated and spatially separated speaker configurations. The spatial release from masking appeared to be independent of the extent to which lower-SNR components were removed from the target signal.

Our method of applying an SNR-dependent binary mask to the target speech resembles the technique of ideal time-frequency segregation (ITFS) that is known from computational auditory scene analysis studies (e.g., Wang 2005; Brungart 2006; Kjems et al. 2009). Although these studies used diotic signals and applied the masks to both the targets and interferers, the results are comparable. The ITFS studies found a decrease in target speech intelligibility for SNR criteria exceeding the global SNR of the mixture by 5–10 dB, similar to the local SNR value of about 0 dB (i.e. around 10 dB above global SNR) in our study.

An analysis of the local SNRs of all T-F units in ten unfiltered stimuli (see Fig. 3) was performed to determine the proportion of T-F units exceeding a certain SNR level. The analysis reveals that on average only about 20 % of the total target T-F units exceeded a local SNR value of –4 dB and about 7 % exceeded a value of 10 dB SNR. This corresponds to the proportion of target speech that was retained in the final signal at the SNR criteria mentioned. Note that at the criterion of −4 dB no decrease in speech intelligibility was observed as compared to the condition with intact target speech. This analysis thus demonstrates the sparse character of speech and the fact that only little target speech was needed to achieve a reasonable performance on this experimental task with a restricted speech vocabulary.

The stable spatial release from masking over the range of SNR criteria tested is one of the most important outcomes of this study. It suggests that recovery of sub-threshold speech information by binaural cues does not play a role in the spatial release from masking in a multispeaker situation. Instead, masking release appeared to be sustained even when only high-SNR components of target speech were left in the signal. These supra-threshold components should be detectable monaurally in both the collocated and spatially separated conditions and thus their detection should not rely on a binaural unmasking mechanism as is the case for BMLDs.

An explanation for this spatial release from masking can be provided by directional cues of the stimuli that support a better allocation of speech components to the source of interest. The fact that speech sources originate from different spatial positions can eliminate some confusion that arises from the simultaneous presence

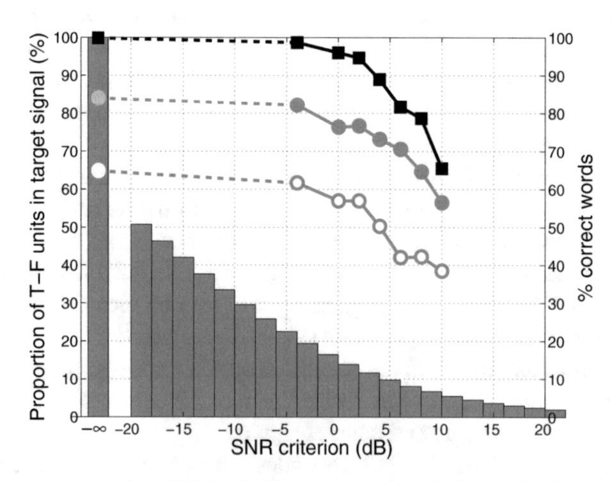

Fig. 3 The average proportion of T-F units that were retained in the manipulated target signal for a given SNR criterion are shown by bars (*left axis*). *Black squares* show the mean intelligibility of this target speech presented in absence of interferers (*right axis*). The mean data from Fig. 2 are shown in *grey* for reference

of multiple speech sources (Durlach et al. 2003). A further improvement can be seen for target speech presented in isolation, but otherwise manipulated identically as before (black squares in Fig. 3). These data show the inherent intelligibility of the sparse target speech, ruling out confusion from interfering speech.

Acknowledgments This work was supported by the DFG (SFB/TRR31 "The Active Auditory System").

References

Beutelmann R, Brand T, Kollmeier B (2010) Revision, extension, and evaluation of a binaural speech intelligibility model. J Acoust Soc Am 127(4):2479–2497

Brinkmann F, Lindau A, Weinzierl S, Geissler G, van de Par S (2013) A high resolution head-related transfer function database including different orientations of head above the torso. Fortschritte der Akustik. AIA-DAGA 2013, Merano, Italy (pp 596–599): DEGA e.V. Berlin

Brungart DS, Chang P S, Simpson BD, Wang D (2006) Isolating the energetic component of speech-on-speech masking with ideal time-frequency segregation. J Acoust Soc Am 120(6):4007–4018

Durlach NI, Mason CR, Gerald Kidd J, Arbogast TL, Colburn HS, Shinn-Cunningham BG (2003) Note on informational masking (L). J Acoust Soc Am 113(6):2984–2987

Hirsh IJ (1948) The influence of interaural phase on interaural summation and inhibition. J Acoust Soc Am 20(4):536–544

Kjems U, Boldt JB, Pedersen MS, Lunner T, Wang D (2009) Role of mask pattern in intelligibility of ideal binary-masked noisy speech. J Acoust Soc Am 126(3):1415–1426

Lavandier M, Jelfs S, Culling JF, Watkins AJ, Raimond AP, Makin SJ (2012). Binaural prediction of speech intelligibility in reverberant rooms with multiple noise sources. J Acoust Soc Am 131(1):218–231

Levitt H, Rabiner LR (1967) Predicting binaural gain in intelligibility and release from masking for speech. J Acoust Soc Am 42(4):820–829

Studebaker GA (1985) A "rationalized" arcsine transform. J Speech, Lang Hear Res 28(3):455–462

Wagener K, Brand T, Kollmeier B (1999) Entwicklung und Evaluation eines Satztests für die deutsche Sprache I: Design des Oldenburger Satztests. Z Audiol 38:4–15

Wan R, Durlach NI, Colburn HS (2014) Application of a short-time version of the equalization-cancellation model to speech intelligibility experiments with speech maskers. J Acoust Soc Am 136(2):768–776

Wang D (2005) On ideal binary mask as the computational goal of auditory scene analysis. In: P Divenyi (ed) Speech separation by humans and machines. Kluwer Academic, Norwell, pp 181–197

On the Contribution of Target Audibility to Performance in Spatialized Speech Mixtures

Virginia Best, Christine R. Mason, Jayaganesh Swaminathan, Gerald Kidd, Kasey M. Jakien, Sean D. Kampel, Frederick J. Gallun, Jörg M. Buchholz and Helen Glyde

Abstract Hearing loss has been shown to reduce speech understanding in spatialized multitalker listening situations, leading to the common belief that spatial processing is disrupted by hearing loss. This paper describes related studies from three laboratories that explored the contribution of reduced target audibility to this deficit. All studies used a stimulus configuration in which a speech target presented from the front was masked by speech maskers presented symmetrically from the sides. Together these studies highlight the importance of adequate stimulus audibility for optimal performance in spatialized speech mixtures and suggest that reduced access to target speech information might explain a substantial portion of the "spatial" deficit observed in listeners with hearing loss.

Keywords Speech intelligibility · Spatial release from masking · Hearing loss · Amplification · Glimpsing

1 Introduction

In the context of speech communication, spatial release from masking (SRM) refers to an improvement in intelligibility when competing sounds are spatially separated from the talker of interest. This improvement can arise as a result of acoustic benefits (such as the "head-shadow" advantage) or by effective increases in signal-to-noise

V. Best (✉) · C. R. Mason · J. Swaminathan · G. Kidd
Department of Speech, Language and Hearing Sciences, Boston University, Boston, MA, USA
e-mail: ginbest@bu.edu

C. R. Mason
e-mail: cmason@bu.edu

J. Swaminathan
e-mail: jswamy@bu.edu

G. Kidd
e-mail: gkidd@bu.edu

P. van Dijk et al. (eds.), *Physiology, Psychoacoustics and Cognition in Normal and Impaired Hearing,* Advances in Experimental Medicine and Biology 894,
DOI 10.1007/978-3-319-25474-6_10

ratio resulting from neural processing of binaural cues (so called "masking level differences"). In other cases it appears that the perceived separation of sources drives the advantage by enabling attention to be directed selectively.

In many situations, listeners with sensorineural hearing impairment (HI) demonstrate reduced SRM compared to listeners with normal hearing (NH). This observation commonly leads to the conclusion that spatial processing is disrupted by hearing loss. However, convergent evidence from other kinds of spatial tasks is somewhat lacking. For example, studies that have measured fine discrimination of binaural cues have noted that individual variability is high, and some HI listeners perform as well as NH listeners (e.g. Colburn 1982; Spencer 2013). Free-field localization is not strongly affected by hearing loss unless it is highly asymmetric or very severe at low frequencies (e.g. Noble et al. 1994). Other studies have tried to relate SRM in multitalker environments to localization ability (Noble et al. 1997; Hawley et al. 1999) or to binaural sensitivity (Strelcyk and Dau 2009; Spencer 2013) with mixed results. Finally, it has been observed that SRM is often inversely related to the severity of hearing loss (e.g. Marrone et al. 2008). This raises the question of whether in some cases apparent spatial deficits might be related to reduced audibility in spatialized mixtures.

A popular stimulus paradigm that has been used in recent years consists of a frontally located speech target, and competing speech maskers presented symmetrically from the sides. This configuration was originally implemented to minimize the contribution of long-term head-shadow benefits to SRM (Noble et al. 1997; Marrone et al. 2008) but has since been adopted as a striking case in which the difference between NH and HI listeners is large. This paper describes related studies from three different laboratories that used the "symmetric masker" configuration to explore the interaction between target audibility and performance under these conditions.

K. M. Jakien · S. D. Kampel · F. J. Gallun
National Center for Rehabilitative Auditory Research, VA Portland Health Care System, Portland, OR, USA
e-mail: kasey.jakien@va.gov

S. D. Kampel
e-mail: sean.kampel@va.gov

F. J. Gallun
e-mail: frederick.gallun@va.gov

J. M. Buchholz · H. Glyde
National Acoustic Laboratories, Macquarie University, Sydney, NSW, Australia
e-mail: Jorg.Buchholz@nal.gov.au

H. Glyde
e-mail: helen.glyde@nal.gov.au

2 Part 1

2.1 Motivation

Gallun et al. (2013) found that the effect of hearing loss on separated thresholds was stronger when one target level was used for all listeners (50 dB SPL) compared to when a sensation level (SL) of 40 dB was used (equivalent to a range of 47–72 dB SPL). They speculated that a broadband increase in gain was not sufficient to combat the non-flat hearing losses of their subjects. Thus in their most recent study (Jakien et al., under revision), they performed two experiments. In the first, they directly examined the effect of SL on SRM, while in a second experiment they carefully compensated for loss of audibility within frequency bands for each listener.

2.2 Methods

Target and masker stimuli were three male talkers taken from the Coordinate Response Measure corpus. Head-related transfer functions (HRTFs) were used to position the target sentences at 0° azimuth and the maskers either colocated with the target or at ±45° (Gallun et al. 2013; Xie 2013). In the first experiment the target sentences were fixed at either 19.5 dB SL (low SL condition) or 39.5 dB SL (high SL condition) above each participant's speech reception threshold (SRT) in quiet. To estimate masked thresholds (target-to-masker ratio, TMR, giving 50 % correct), the levels of the two masking sentences were adjusted relative to the level of the target sentences using a one-up/one-down adaptive tracking algorithm. In the second experiment the spectrum of the target sentences was adjusted on a frequency band-by-band basis to account for differences in the audiogram across participants. The initial level was set to that of the high SL condition of the first experiment for a listener with 0 dB HL. Target and masking sentences were then filtered into six component waveforms using two-octave-wide bandpass filters with center frequencies of 250, 500, 1000, 2000, 4000, and 8000 Hz. The level of each component was adjusted based on the difference between the audiogram of the listener being tested and the audiogram of a comparison listener with 0 dB HL thresholds at each of the six octave frequencies, and then the six waveforms were summed. To estimate thresholds, the levels of the two masking sentences were adjusted relative to the level of the target sentence according to a progressive tracking algorithm which has been shown to be comparable to and more efficient than adaptive tracking (Gallun et al. 2013).

Thirty-six listeners participated in both experiments, and an additional 35 participated in just the second experiment. All 71 participants had four frequency (500, 1000, 2000, 4000 Hz) average hearing losses (4FAHL) below 37 dB HL (mean 12.1 dB ± 8.2 dB) and all had fairly symmetrical hearing at 2000 Hz and below. Ages of the listeners were between 18 and 77 years (mean of 43.1 years) and there was a

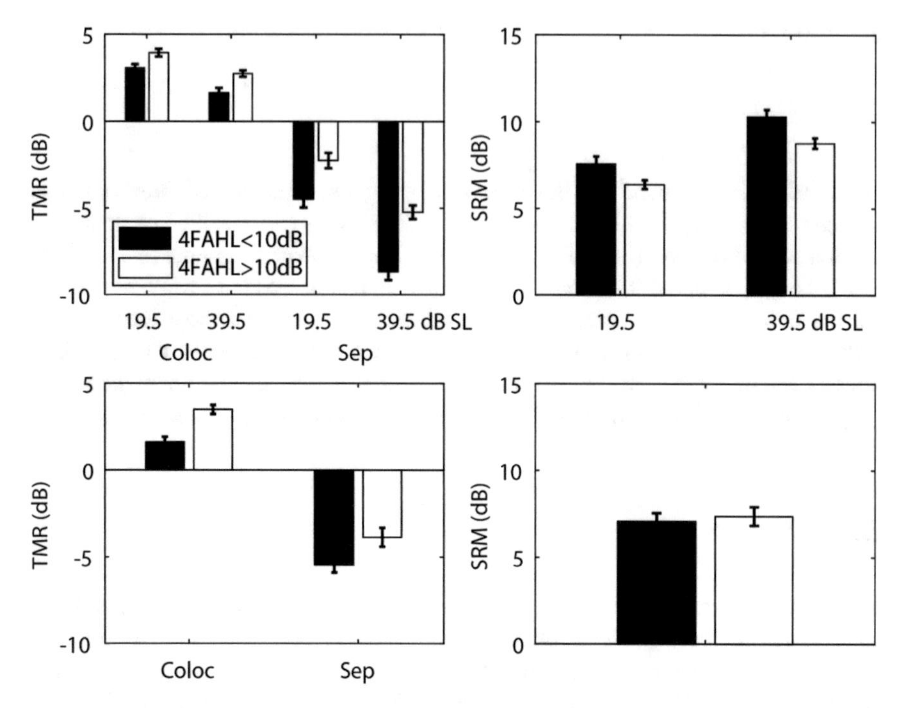

Fig. 1 Group-mean TMRs at threshold in the colocated and separated configurations (*top left panel*) and SRM (*top right panel*) as a function of SL. Group mean TMRs at threshold for the colocated and separated configurations (*bottom left panel*) and SRM (*bottom right panel*) in the equal-audibility condition. Error bars show standard errors

significant correlation ($r=0.59, p<0.001$) between 4FAHL and age. For simplicity, the listeners have been divided into those with 4FAHLs below 10 dB HL ($n=22$ in the first experiment; $n=35$ in the second experiment) and those with 4FAHLs above 10 dB HL ($n=14$ in the first experiment; $n=36$ in the second experiment).

2.3 Results

In the first experiment (top row of Fig. 1), those with lower 4FAHLs had better thresholds, and thresholds and SRM in both 4FAHL groups improved with an increase in SL. In the second experiment (bottom row of Fig. 1), despite equating audibility across listeners, there was a group difference in both the colocated and separated thresholds, but SRM was equivalent between groups.

For the 36 listeners who participated in both experiments, correlations between SRM and 4FAHL were examined. In the first experiment, 4FAHL was negatively correlated with SRM in both the low SL ($r=-0.33, p=0.05$) and high SL ($r=-0.39, p=0.02$) conditions. In the second experiment, 4FAHL was not significantly correlated with SRM ($r=-0.10, p=0.57$).

In summary, increasing SL improved performance and increased SRM for all listeners. Furthermore, careful equalization of audibility across listeners reduced the effects of hearing loss on SRM. On the other hand, no manipulation was able to guarantee equal performance across listeners with various degrees of hearing loss. This suggests that while audibility is clearly an important factor, other factors may impact speech-in-speech intelligibility (e.g. aging, auditory filter width, or comorbidities associated with cognition, working memory and attention).

3 Part 2

3.1 Motivation

Glyde et al. (2013) showed that even with frequency-specific gain applied according to the individual audiogram using the NAL-RP hearing aid prescription, a strong relationship between SRM and hearing status persisted. The authors noted that with the relatively low presentation levels used in their experiment (55 dB SPL masker), the NAL-RP prescription may not have provided sufficient gain especially in the high frequency region. Thus in a follow-up experiment (Glyde et al., 2015), they examined the effect of providing systematically more high-frequency gain than that provided by NAL-RP. They tested HI subjects as well as NH subjects with a simulated hearing loss.

3.2 Methods

The data is compiled from different studies but each group contained at least 12 NH (mean age 28.8–33.6 years) and 16 older HI (mean age 68.8–73.1 years). The HI listeners had a moderate, bilaterally symmetric, sloping sensorineural hearing loss with a 4FAHL of 48 ± 5dB.

Subjects were assessed with a Matlab version of the LiSN-S test (Glyde et al. 2013), in which short, meaningful sentences (e.g., "The brother carried her bag") were presented in an ongoing two-talker background. Target and distractors were spoken by the same female talker and target sentences were preceded by a brief tone burst. Using HRTFs, target sentences were presented from 0° azimuth and the distractors from either 0° azimuth (colocated condition) or ± 90° azimuth (spatially separated condition). The combined distractor level was fixed at 55 dB SPL and the target level was adapted to determine the TMR at which 50 % of the target words were correctly understood. Subjects were seated in an audiometric booth and repeated the target sentences to a conductor.

Stimuli were presented over equalized headphones and for HI listeners had different levels of (linear) amplification applied to systematically vary audibility: amplification according to NAL-RP, NAL-RP plus 25 % of extra gain (i.e. on top of NAL-RP), and NAL-RP plus 50 % of extra gain. An extra gain of 100 % would have restored normal audibility, but was impossible to achieve due to loudness discomfort. Given the sloping hearing loss, an increase in amplification mainly resulted in an increased high-frequency gain and thus in an increase in audible bandwidth. NH subjects were tested at the same audibility levels. This was realized by first applying attenuation filters that mimicked the average audiogram of the HI subjects and then applying the same gains as described above. No other aspects of hearing loss were considered. Details of the processing can be found in Glyde et al. (2015). The NH subjects were also tested with no filtering.

3.3 Results

Thresholds for the colocated conditions (Fig. 2 left panel) were basically independent of amplification level, and for the NH subjects were about 2.5 dB lower than for the HI subjects. Thresholds for the spatially separated condition (middle panel) clearly improved with increasing amplification for both the NH and HI subjects and were maximal for "normal" audibility. However, thresholds for the NH subjects were on average 4.8 dB lower than for the HI subjects. The corresponding SRM, i.e., the difference in threshold between the colocated and separated conditions (right panel), increased with increasing gain similarly to the spatially separated thresholds, but the overall difference between NH and HI subjects was reduced to about 2.5 dB. It appears that under these conditions, a large proportion of the SRM deficit in the HI (and simulated HI) group could be attributed to reduced audibility.

4 Part 3

4.1 Motivation

Accounting for audibility effects in speech mixtures is not straightforward. While it is common to measure performance in quiet for the target stimuli used, this does not incorporate the fact that portions of the target are completely masked, which greatly reduces redundancy in the speech signal. Thus a new measure of "masked target audibility" was introduced.

Simple energy-based analyses have been used to quantify the available target in monaural speech mixtures (e.g. the ideal binary mask described by Wang 2005; ideal time-frequency segregation as explored by Brungart et al. 2006; and the glimpsing model of Cooke 2006). The basic approach is to identify regions in the time-frequency plane where the target energy exceeds the masker energy. The number of these glimpses is reduced as the SNR decreases, or as more masker talkers are added to the mixture. To define the available glimpses in symmetric binaural mix-

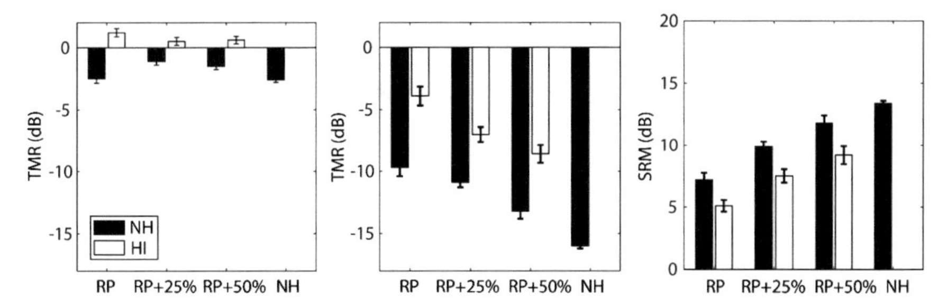

Fig. 2 Group-mean TMRs at threshold for the colocated (*left panel*) and separated (*middle panel*) configurations and the corresponding SRM (*right panel*). Error bars show standard errors

tures, we simply applied a monaural glimpsing model separately to the two ears. For symmetric listening situations, the glimpses can occur in either ear, and often occur in both ears for a particular time-frequency tile. However the glimpses may not all be above threshold, particularly for listeners with hearing loss. Thus we conducted an experiment in which we presented "glimpsed" stimuli to NH and HI listeners to measure their ability to use the available target information. Performance was compared to natural binaural performance to determine to what extent target audibility/availability can explain performance.

4.2 Methods

Six NH (mean age 23 years) and six HI (mean age 26 years) listeners participated. The HI listeners had a moderate, bilaterally symmetric, sloping sensorineural hearing loss with a 4FAHL of 49 ± 14dB.

Speech materials were taken from a corpus of monosyllabic words (Kidd et al. 2008), in which five-word sentences are assembled by selecting one word from each of five categories (e.g., "Sue bought two red toys"). Using HRTFs, one target sentence was presented at 0° azimuth, and two or four different masker sentences were presented at ±90° azimuth, or ±45°/±90° azimuth. All talkers were female, and the target was identified by its first word "Sue". Each masker was fixed in level at 60 dB SPL, and the target was varied in level to set the TMR at one of five values (from −25 to −5 dB in the NH group; from −20 to 0 dB in the HI group). Stimuli for HI listeners had individualized NAL-RP gain applied.

To generate the glimpsed stimuli, an ideal binary mask was applied separately to the two ears of the binaural stimuli using the methods of Wang (2005) and Brungart et al. (2006). In short, the signals were analyzed using 128 frequency channels between 80 and 8000 Hz, and 20-ms time windows with 50% overlap. Tiles with target energy exceeding masker energy were assigned a mask value of one and the remaining tiles were assigned a value of zero. The binary mask was then applied to the appropriate ear of the binaural stimulus before resynthesis. As a control condition, the mask was also applied to the target alone.

Stimuli were presented over headphones to the listener who was seated in an audiometric booth fitted with a monitor, keyboard and mouse. Responses were given by selecting five words from a grid presented on the monitor.

4.3 Results

Figure 3 shows 50 % thresholds extracted from logistic fits to the data. Performance was better overall with two maskers as compared to four maskers, and for NH than HI listeners. With two maskers, the difference in thresholds between groups was 11 dB in the natural condition, and 9 dB in the glimpsed mixture condition. For four maskers these deficits were 7 and 9 dB, respectively. In other words, the group differences present in the natural binaural condition were similar when listeners were presented with the good time-frequency glimpses only. In the control condition where the glimpses contained only target energy, the group differences were even larger (12.6 and 10.8 dB). This suggests that the HI deficit is related to the ability to access or use the available target information and not to difficulties with spatial processing or segregation. Individual performance for natural stimuli was strongly correlated with performance for glimpsed stimuli ($r = 0.92$), again suggesting a common limit on performance in the two conditions.

5 Conclusions

These studies demonstrate how audibility can affect measures of SRM using the symmetric masker paradigm, and suggest that reduced access to target speech information might in some cases contribute to the "spatial" deficit observed in listeners with hearing loss. This highlights the importance of adequate stimulus audibility for optimal performance in spatialized speech mixtures, although this is not always feasible due to loudness discomfort in HI listeners, and technical limitations of hearing aids.

Acknowledgments This work was supported by NIH-NIDCD awards DC04545, DC013286, DC04663, DC00100 and AFOSR award FA9550-12-1-0171 (VB, CRM, JS, GK), DC011828 and the VA RR&D NCRAR (KMJ, SDK, FJG), the Australian Government through the Department of Health, and the HEARing CRC, established and supported under the Cooperative Research Centres Program—an initiative of the Australian Government (VB, JB, HG).

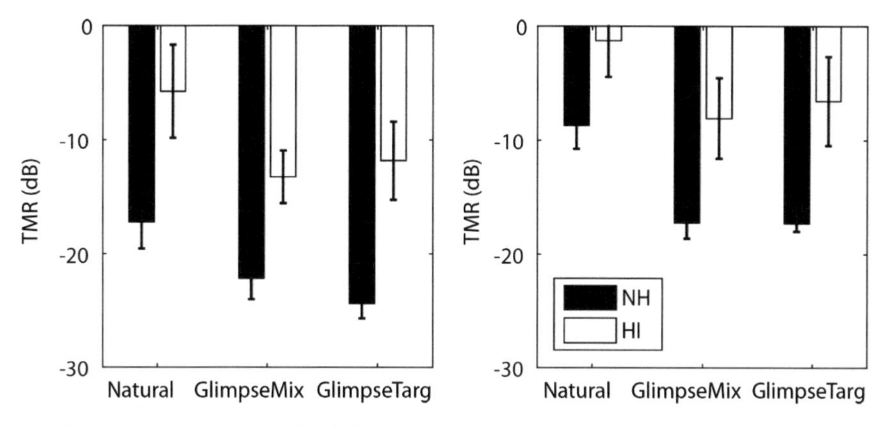

Fig. 3 Group-mean TMRs at threshold for the two-masker (*left panel*) and four-masker (*right panel*) conditions. Error bars show standard errors

References

Brungart DS, Chang PS, Simpson BD, Wang DL (2006) Isolating the energetic component of speech-on-speech masking with ideal time-frequency segregation. J Acoust Soc Am 120:4007–4018

Colburn HS (1982) Binaural interaction and localization with various hearing impairments. Scand Audiol Suppl 15:27–45

Cooke M (2006) A glimpsing model of speech perception in noise. J Acoust Soc Am 119:1562–1573

Gallun FJ, Kampel SD, Diedesch AC, Jakien KM (2013) Independent impacts of age and hearing loss on spatial release in a complex auditory environment. Front Neurosci 252(7):1–11

Glyde H, Cameron S, Dillon H, Hickson L, Seeto M (2013) The effects of hearing impairment and aging on spatial processing. Ear Hear 34(1):15–28

Glyde H, Buchholz JM, Nielsen L, Best V, Dillon H, Cameron S, Hickson L (2015) Effect of audibility on spatial release from speech-on-speech masking. J Acoust Soc Am 138:3311–3319

Hawley ML, Litovsky RY, Colburn HS (1999) Speech intelligibility and localization in a multi-source environment. J Acoust Soc Am 105:3436–3448

Kidd Jr G, Best V, Mason CR (2008) Listening to every other word: examining the strength of linkage variables in forming streams of speech. J Acoust Soc Am 124:3793–3802

Marrone N, Mason CR, Kidd Jr G (2008) The effects of hearing loss and age on the benefit of spatial separation between multiple talkers in reverberant rooms. J Acoust Soc Am 124:3064–3075

Noble W, Byrne D, Lepage B (1994) Effects on sound localization of configuration and type of hearing impairment. J Acoust Soc Am 95:992–1005

Noble W, Byrne D, Ter-Host K (1997) Auditory localization, detection of spatial separateness, and speech hearing in noise by hearing impaired listeners. J Acoust Soc Am 102:2343–2352

Spencer N (2013). Binaural benefit in speech intelligibility with spatially separated speech maskers. PhD Dissertation, Boston University

Strelcyk O, Dau T (2009) Relations between frequency selectivity, temporal fine-structure processing, and speech reception in impaired hearing. J Acoust Soc Am 125:3328–3345

Wang DL (2005) On ideal binary mask as the computational goal of auditory scene analysis. In: Divenyi P (ed) Speech separation by humans and machines. Kluwer Academic, Norwell, pp 181–197

Xie B (2013). Head-related transfer function and virtual auditory display, 2nd edn. J. Ross Publishing, Plantation

Optimization of a Spectral Contrast Enhancement Algorithm for Cochlear Implants Based on a Vowel Identification Model

Waldo Nogueira, Thilo Rode and Andreas Büchner

Abstract Speech intelligibility achieved with cochlear implants (CIs) shows large variability across different users. One reason that can explain this variability is the CI user's individual electrode nerve interface which can impact the spectral resolution they can achieve. Spectral resolution has been reported to be related to vowel and consonant recognition in CI listeners. One measure of spectral resolution is the spectral modulation threshold (SMT), which is defined as the smallest detectable spectral contrast in a stimulus. In this study we hypothesize that an algorithm that improves SMT may improve vowel identification, and consequently produce an improvement in speech understanding for CIs. With this purpose we implemented an algorithm, termed spectral contrast enhancement (SCE) that emphasizes peaks with respect to valleys in the audio spectrum. This algorithm can be configured with a single parameter: the amount of spectral contrast enhancement entitled "SCE factor". We would like to investigate whether the "SCE factor" can be individualized to each CI user. With this purpose we used a vowel identification model to predict the performance produced by the SCE algorithm with different "SCE factors" in a vowel identification task.

In five CI users the new algorithm has been evaluated using a SMT task and a vowel identification task. The tasks were performed for SCE factors of 0 (no enhancement), 2 and 4. In general it seems that increasing the SCE factor produces a decrease in performance in both the SMT threshold and vowel identification.

W. Nogueira (✉) · A. Büchner
Dept. of Otolaryngology and Hearing4all, Medical University Hannover, Hannover, Germany
e-mail: nogueiravazquez.waldo@mh-hannover.de

A. Büchner
e-mail: buechner@hoerzentrum-hannover.de

T. Rode
HörSys GmbH, Hannover, Germany
e-mail: rode.thilo@hzh-gmbh.de

P. van Dijk et al. (eds.), *Physiology, Psychoacoustics and Cognition in Normal and Impaired Hearing,* Advances in Experimental Medicine and Biology 894,
DOI 10.1007/978-3-319-25474-6_11

Keywords Cochlear implant · Model · Vowel identification · Spectral contrast enhancement

1 Introduction

Cochlear implants (CIs) can restore the sense of hearing in profound deafened adults and children. CI signal processing strategies have been developed for speech understanding in quiet, such that many post-lingually deafened adults with CIs can recognize 60–80 % of sentences presented in quiet (Friesen et al. 2001). However, speech intelligibility in noise and music perception, although very variable, remain generally poor for CI listeners.

For example it is still challenging for many CI users to discriminate vowels and phonemes in a closed set identification task without background noise (Sagi et al. 2010; Svirsky et al. 2011). These difficulties might be produced by the limited spectral resolution delivered by CI devices. Spectral resolution may be degraded by the broad electrical fields created in the cochlea when the electrodes are stimulated.

In a recent study, the identification of spectrally smeared vowels and consonants was improved by spectral contrast enhancement (SCE) in a group of 166 normal hearing listeners (Alexander et al. 2011). Spectral contrast is defined as the level difference between peaks and valleys in the spectrum. In CIs, spectral contrast is degraded because of the limited number of stimulation electrodes and overlapping electric fields activating the nervous system through the bony structure of the cochlea. This might reduce the differences in amplitudes between peaks and valleys in the input making it more difficult to locate spectral dominance (i.e., formants) which provide crucial cues to speech intelligibility and instrument identification. Loizou and Poroy 2001 showed that CI users need a higher spectral contrast than normal hearing listeners in vowel identification tasks.

In this study we propose a new sound coding strategy that uses SCE for CI users. The working principle of the coding strategy can affect speech intelligibility. For example "NofM" strategies such as ACE were developed in the 1990s to separate speech signals into M sub-bands and derive envelope information from each band signal. N bands with the largest amplitude are then selected for stimulation (N out of M). One of the consequences here is that the spectral contrast of the spectrum is enhanced, as only the N maxima are retained for stimulation. In this work, we want to investigate whether additional spectral enhancement can provide with improved speech intelligibility.

When designing speech coding strategies, the large variability in speech intelligibility outcomes has to be considered. For example, two sound coding strategies can produce opposite effects in speech performance, even when the CI users are post-locutive adults and have enough experience with their CIs. One possible reason that might explain this variability is the electrode nerve interface of each individual which can impact the spectral resolution they can achieve. Spectral resolution has been reported to be closely related to vowel and consonant recognition in cochlear implant (CI) listeners (Litvak et al. 2007). One measure of spectral resolution is the spectral modulation threshold (SMT), which is defined as the smallest detectable spectral contrast in the spectral ripple stimulus (Litvak et al. 2007). In this study we hypothesize that an SCE algorithm may be able to improve SMT and therefore may also be able to improve vowel recognition.

Recently a relatively simple model of vowel identification has been used to predict confusion matrices of CI users. Models of sound perception are not only beneficial in the development of sound coding strategies to prototype the strategy and create hypotheses, but also to give more robustness to the results obtained from an evaluation in CI users. Evaluations with CI users are time consuming and results typically show large variability. In this study we use the same model developed by (Sagi et al. 2010) and (Svirsky et al. 2011) to show the potential benefits of SCE in "NofM" strategies for CIs.

2 Methods

2.1 The Signal Processing Method: SCE in NofM Strategies for CIs

The baseline or reference speech coding strategy is the advanced combinational encoder (ACE, a description of this strategy can be found in Nogueira et al. 2005). The ACE strategy can be summarized in five signal processing blocks: (1) The Fast Fourier Transform (FFT); (2) The envelope detector; (3) The NofM band selection; (4) The loudness growth function (LGF) compression and (5) The channel mapping. The new SCE strategy incorporates a new processing stage just before the NofM band selection. The goal of this stage is to enhance spectral contrast by attenuating spectral valleys while keeping spectral peaks constant. The amount of spectral contrast enhancement can be controlled by a single parameter termed SCE factor. A more detailed description of the algorithm will be published elsewhere (Nogueira et al. 2016).

2.2 Hardware Implementation

All the stimuli were computed in Matlab© using the ACE and the SCE strategies. The stimuli were output from a standard PC to the Nucleus Cochlear© implant using the Nucleus Interface Communicator (NIC). The Matlab toolbox was used to process the acoustic signals and compute the electrical stimuli delivered to the CI. For each study participant we used their clinical map, i.e., their clinical stimulation rate, comfort and threshold levels, number of maxima and frequency place allocation table. For the experiments presented in the report we used three different configurations of the SCE strategy which only differed in the amount of spectral contrast enhancement applied. The three strategies are denoted by SCE0, SCE2 and SCE4. SCE0 means no spectral enhancement and is exactly the same strategy as the clinical ACE strategy.

2.3 Experiments in Cochlear Implant Users

2.3.1 Participants

Five CI users of the Freedom/System5 system participated in this study. The relevant details for all subjects are presented in Table 1. All the test subjects used the ACE strategy in daily life and all had a good speech performance in quiet.

Each CI user participated in a study to measure SMTs and vowel idenitification performance.

2.3.2 Spectral Modulation Threshold

We used the spectral ripple test presented by (Litvak et al. 2007) to estimate the spectral modulation thresholds of each CI user. This task uses a cued two interval, two-alternative, forced choice procedure. In the first interval the standard stimulus was always presented. The standard stimulus had a flat spectrum with bandwidth extending from 350 to 5600 Hz. The signal and the second standard were randomly presented in the other two intervals. Both signals were generated in the frequency domain assuming a sampling rate of 44,100 Hz. The spectral shape of the standard and the signal were generated using the equation:

$$|F(f)| = \begin{Bmatrix} 10^{\frac{c}{2}\sin(2\pi(log_2(\frac{f}{350})f_c+\theta_0))}, \text{ for } 350 < f < 5600 \\ 0, \text{ otherwise} \end{Bmatrix}$$

where F is the amplitude of a bin with center frequency f (in Hertz), f_c is the spectral modulation frequency (in cycles/octave), and θ_0 is the starting phase. Next, noise

Table 1 Patients details participating in the study

Id.	Age	Side	Cause of deafness	Implant experience in years
1	48	Right	Sudden Hearing Loss	1
2	38	Left	Antibiotics	1.8
3	46	Left	Unknown	7.5
4	25	Right	Ototoxika	7
5	22	Left	Unkwown	3.2

was added to the phase of each bin prior to computing the inverse Fourier transform. The standard was generated using a spectral contrast c equal to 0. The amplitude of each stimulus was adjusted to an overall level of 60 dB sound pressure level (SPL). Independent noise stimuli were presented on each observation interval. The stimulus duration was 400 ms. A 400 ms pause was used between the stimuli.

Thresholds were estimated using an adaptive psychophysical procedure employing 60 trials. The signal contrast level was reduced after three consecutive correct responses and increased after a single incorrect response. Initially the contrast was varied in a step size of 2 dB, which was reduced to 0.5 dB after three reversals in the adaptive track (Levitt 1971). Threshold for the run was computed as the average modulation depth corresponding to the last even number of reversals, excluding the first three. Using the above procedure, modulation detection thresholds were obtained for the modulation frequency of 0.5 cycles/octave which is the one that correlates best with vowel identification (Litvak et al. 2007).

2.3.3 Vowel Identification Task

Speech understanding was assessed using a vowel identification task. Vowel stimuli consisted of eight long vowels 'baat', 'baeaet', 'beet', 'biit', 'boeoet', 'boot', 'bueuet', 'buut'. All vowels had a very similar duration of around 180 ms. The stimuli were uttered by a woman. An 8-alternative forced choice task procedure 8-AFC was created where 2 and 4 repetitions of each vowel were used for training and testing respectively. The vowels were presented at the same 60 dB SPL level as the spectral ripples with a loudness roving of+/- 1.5 dB.

2.3.4 The standard Multidimensional Phoneme Identification Model

We used a model of vowel identification to select the amount of spectral contrast enhancement SCE factor. The model is based on the multidimensional phoneme identification (MPI) model (Sagi et al. 2010; Svirsky et al. 2011). A basic block diagram of the model is presented in Fig. 1.

The model estimates relevant features from electrodograms generated by the CI sound processor. Because we are modelling a vowel identification task it makes sense to extract features related to formant frequencies. Moreover, because we are analyzing the effect of enhancing spectral contrast it seems logical to use spectral contrast features between formants. In this study the number of features was limited to two formants and therefore the MPI model is two dimensional. Next, the MPI model adds noise to the features. The variance of the noise is set based on the individual abilities of a CI user to perceive the features extracted from the electrodogram. In our implementation we used the results of the SMT task to set the noise variance. The obtained jnd from the SMT task was scaled between 0.001 and 0.5. This number was applied as the variance of the Gaussian noise applied in the MPI model and for this reason is termed jnd noise in Fig. 1.

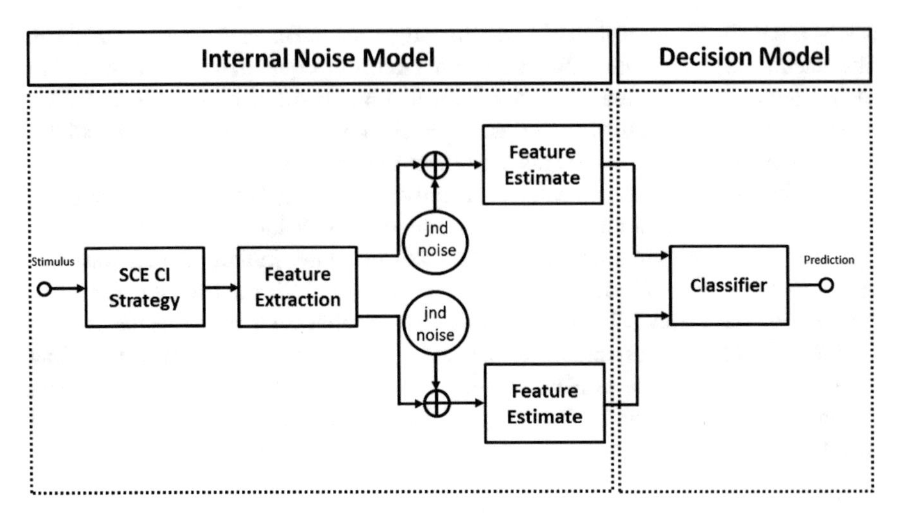

Fig. 1 Two dimensional implementation of the MPI model from Sagi et al. 2010 and Svirsky et al. 2011

3 Results

3.1 Results from the MPI model

The MPI model has been used to model the performance of the SCE strategy for five virtual CI users. All virtual CI user differs from each other in their most comfortable and threshold levels used in the map. That means that the speech processor of each virtual CI user will generate different electrodograms for the same vowels. Next, formant features were extracted from the electrodograms based on the spectral contrast between formants 1 and 2. Noise was added to the spectral contrast features (jnd noise). Three amounts of noise were added 0.01, 10 and 50% of the magnitude of the formants extracted. Figure 2 presents the results predicted by the model in percentage of correct vowels identified for five different SCE factors (0, 0.5, 1, 2 and 4).

From the modelling results it can be observed that maximum performance is achieved for SCE factors 2 and 4. The other interesting aspect is that there is no difference in performance across the five different virtual CI users for "Noise Factor" 0.001 (Fig. 2a). That means that the feature extraction (i.e. the extraction of the formants) is robust to the different electrodograms of each virtual CI user. Differences in performance between the five virtual CI users can only be observed for "Noise Factor" 0.1 and 0.5, meaning that the "jnd noise" is the only parameter explaining the differences.

From the MPI modelling results we decided to use SCE factors 2 and 4 (equivalent to increasing the original spectral contrast of the spectrum by 3 and by 5 in a dB scale) to be investigated in CI users.

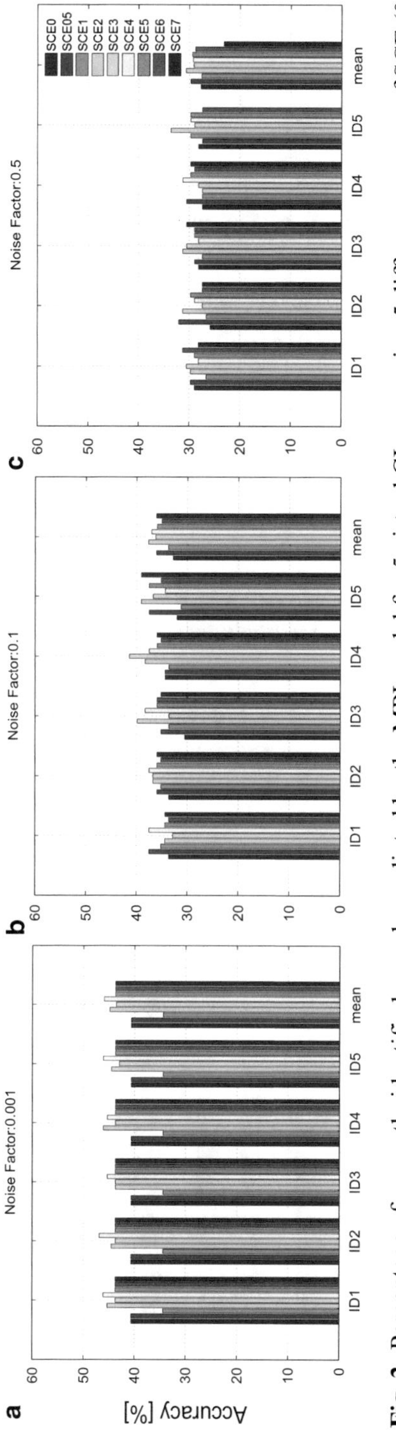

Fig. 2 Percentage of correctly identified vowels predicted by the MPI model for 5 virtual CI users using 5 different amounts of SCE (0, 0.5, 1, 2 and 4) and for different amounts of internal noise. (Noise Factors 0.001, 0.1 and 0.5)

3.2 Results Vowel Identification in CI users

Figure 3 presents the averaged results of the vowel identification task for the five subjects participating in this study.

3.3 Results Spectral Modulation Threshold in CI users

Figure 4 presents the individual and averaged results for the spectral ripple task.

Unexpectedly, additional spectral contrast, which in turn increases the spectral modulation depth, could no produce an improvement in jnd SMT.

3.4 Correlation Between Spectral Modulation Threshold and Vowel Identification

An important question for the analysis was whether the results obtained from the spectral ripple task could be used to predict the outcome of the vowel identification

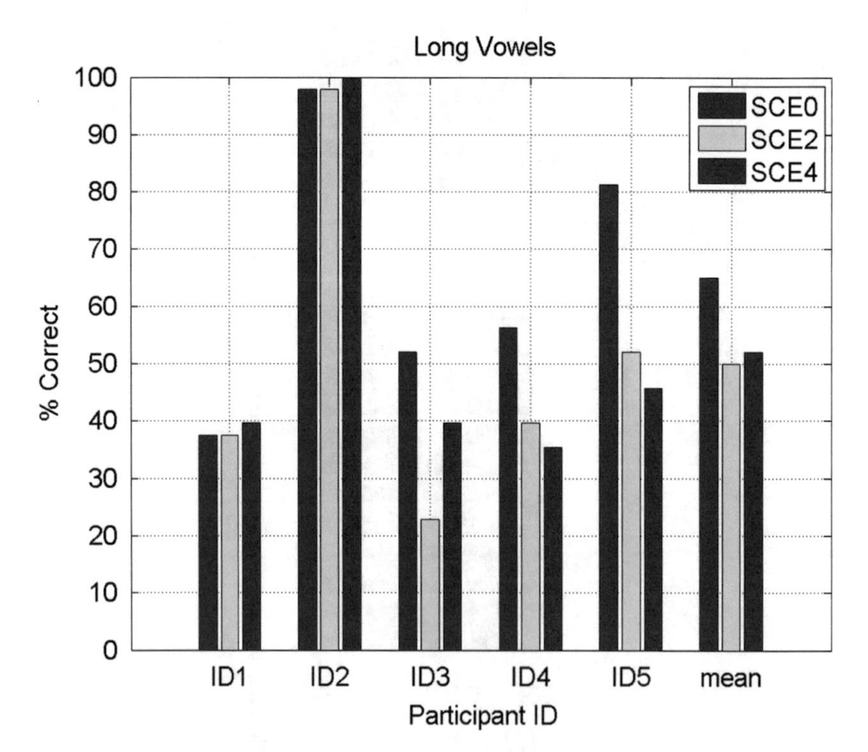

Fig. 3 Results of the vowel identification task for the three strategies (SCE0, SCE2 and SCE4) for 5 CI users and averaged as % correct

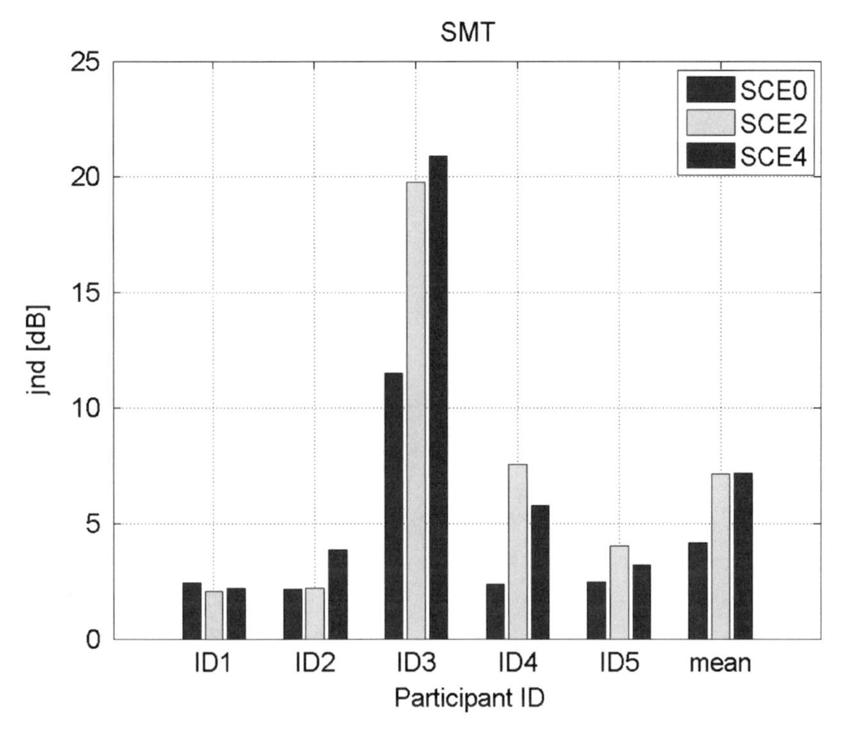

Fig. 4 Results of the spectral modulation threshold task for the three strategies (SCE0, SCE2 and SCE4) given as just noticeable differences (jnd SMT) in decibels. The lower is the jnd the better is the result

task. This can be seen in the left plot of Fig. 5 using an SCE factor of 0. Probably because of the low number of participants only a relatively weak correlation between the two measures was observed.

In the same manner, the middle and right plots in Fig. 5 show the relationship between the improvements of the two tasks comparing the results using an SCE factor 0 to those using SCE factors 2 and 4 respectively. Again, the correlation observed is weak but still a trend for the relationship can be seen. It seems that for the SCE factors used the decline in performance in the SMT is somewhat connected to a decline in performance in the vowel identification task. It remains unclear if an increase of the number of participants would confirm this trend.

4 Discussion

A new sound coding strategy that enhances spectral contrast has been designed. The amount of spectral contrast enhancement can be controlled by a single parameter termed SCE factor.

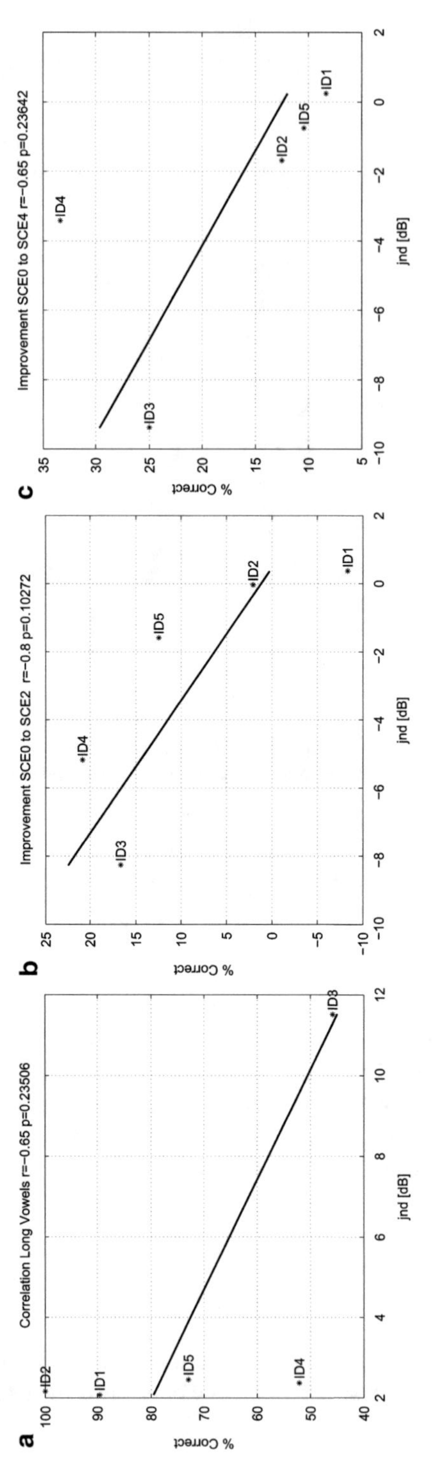

Fig. 5 **a** Correlation between vowel identification performance and jnd SMT in dB. **b** Correlation between the difference (SCE0–SCE2) in vowel identification performance and the difference in jnd SMT (SCE0–SCE2). **c** Correlation between the difference (SCE0–SCE4) in vowel identification performance and the difference in jnd SMT (SCE0–SCE4)

A model of vowel identification has been used to investigate the effect of SCE on vowel identification. The model predicts that increasing the amount of SCE increases vowel identification accuracy. Based on these results we decided to use SCE factors 2 and 4 (equivalent to increasing the original spectral contrast of the spectrum by 3 and by 5 in a dB scale).

The new SCE strategy has been evaluated in CI users. Results from a vowel identification task and a SMT task in five CI users show differences in vowel identification scores for different SCE factors. In general, it seems that SCE produces a detrimental effect in spectral modulation detection and vowel identification in CI users. These results are contrary to the model predictions. Previous studies in the literature give reasons to believe that spectral contrast enhancement would result in a benefit for the chosen tasks. It is possible that spectral valleys are attenuated too much and relevant information required by the CI users to understand speech is lost. These effects are not taken into account by the MPI model, and this could explain the contradictory results between experiments and CI users and modelling results. Still, it is possible that the SCE factors selected where too high, for this reason we think that a follow-up study should investigate whether lower amounts SCE can provide improvements in CI users.

Acknowledgments The authors would like to thank the subjects who have participated in the experiments and the two anonymous reviewers for their comments on different versions of this manuscript. This work was supported by the DFG Cluster of Excellence EXC 1077/1 "Hearing4all" and Cochlear.

References

Alexander JM, Jenison RL, Kluender KR (2011) Real-time contrast enhancement to improve speech recognition. PLoS One 6(9):e24630. http://doi.org/10.1371/journal.pone.0024630

Friesen LM, Shannon RV, Baskent D, Wang X (2001) Speech recognition in noise as a function of the number of spectral channels: comparison of acoustic hearing and cochlear implants. J Acoust Soc Am 110(2):1150. http://doi.org/10.1121/1.1381538

Levitt H (1971) Transformed up-down methods in psychoacoustics. J Acoust Soc Am 49(2B):467–477. http://doi.org/ http://dx.doi.org/10.1121/1.1912375

Litvak LM, Spahr AJ, Saoji AA, Fridman GY (2007) Relationship between perception of spectral ripple and speech recognition in cochlear implant and vocoder listeners. J Acoust Soc Am 122(2):982–991. http://doi.org/ http://dx.doi.org/10.1121/1.2749413

Loizou PC, Poroy O (2001) Minimum spectral contrast needed for vowel identification by normal hearing and cochlear implant listeners. J Acoust Soc Am 110(3):1619–1627. http://doi.org/ http://dx.doi.org/10.1121/1.1388004
Nogueira W, Rode T, Büchner A (2016) Spectral contrast enhancement improves speech intelligibility in 1 noise in NofM strategies for cochlear implants. J Acoust Soc Am
Nogueira W, Büchner A, Lenarz T, Edler B (2005) A psychoacoustic "NofM"-type speech coding strategy for cochlear implants. Eurasip J Appl Signal Process 2005(18):3044–3059
Sagi E, Meyer TA, Kaiser AR, Teoh SW, Svirsky MA (2010) A mathematical model of vowel identification by users of cochlear implants. J Acoust Soc Am 127(2):1069–1083. http://doi.org/ http://dx.doi.org/10.1121/1.3277215
Svirsky MA, Sagi E, Meyer TA, Kaiser AR, Teoh SW (2011) A mathematical model of medial consonant identification by cochlear implant users. J Acoust Soc Am 129(4):2191–2200. http://doi.org/ http://dx.doi.org/10.1121/1.3531806

Roles of the Contralateral Efferent Reflex in Hearing Demonstrated with Cochlear Implants

Enrique A. Lopez-Poveda, Almudena Eustaquio-Martín, Joshua S. Stohl, Robert D. Wolford, Reinhold Schatzer and Blake S. Wilson

Abstract Our two ears do not function as fixed and independent sound receptors; their functioning is coupled and dynamically adjusted via the contralateral medial olivocochlear efferent reflex (MOCR). The MOCR possibly facilitates speech recognition in noisy environments. Such a role, however, is yet to be demonstrated because selective deactivation of the reflex during natural acoustic listening has not been possible for human subjects up until now. Here, we propose that this and other roles of the MOCR may be elucidated using the unique stimulus controls provided by cochlear implants (CIs). Pairs of sound processors were constructed to mimic or not mimic the effects of the contralateral MOCR with CIs. For the non-mimicking condition (STD strategy), the two processors in a pair functioned independently of each other. When configured to mimic the effects of the MOCR (MOC strategy), however, the two processors communicated with each other and the amount of compression in a given frequency channel of each processor in the pair decreased

E. A. Lopez-Poveda (✉)
INCYL, IBSAL, Dpto. Cirugía, Facultad de Medicina, Universidad de Salamanca, C/Pintor Fernando Gallego n 1, 37007 Salamanca, Spain
e-mail: ealopezpoveda@usal.es

A. Eustaquio-Martín
INCYL, IBSAL, Universidad de Salamanca, C/Pintor Fernando Gallego n 1, 37007 Salamanca, Spain
e-mail: aeustaquio@usal.es

J. S. Stohl
MED-EL Corporation, Durham, NC 27713, USA
e-mail: josh.stohl@medel.com

R. D. Wolford
MED-EL Corporation, Durham, NC 27713, USA
e-mail: bob.wolford@medel.com

R. Schatzer
Institute of Mechatronics, University of Innsbruck, Innsbruck, Austria
e-mail: reinhold.schatzer@medel.com

B. S. Wilson
Duke University, Durham, NC, USA
e-mail: blake.wilson@duke.edu

P. van Dijk et al. (eds.), *Physiology, Psychoacoustics and Cognition in Normal and Impaired Hearing,* Advances in Experimental Medicine and Biology 894,
DOI 10.1007/978-3-319-25474-6_12

with increases in the output energy from the contralateral processor. The analysis of output signals from the STD and MOC strategies suggests that in natural binaural listening, the MOCR possibly causes a small reduction of audibility but enhances frequency-specific inter-aural level differences and the segregation of spatially non-overlapping sound sources. The proposed MOC strategy could improve the performance of CI and hearing-aid users.

Keywords Olivocochlear reflex · Speech intelligibility · Spatial release from masking · Unmasking · Dynamic compression · Auditory implants · Hearing aids · Speech encoding · Sound processor

1 Introduction

The central nervous system can control sound coding in the cochlea via medial olivocochlear (MOC) efferents (Guinan 2006). MOC efferents act upon outer hair cells, inhibiting the mechanical gain in the cochlea for low- and moderate-intensity sounds (Cooper and Guinan 2006) and may be activated involuntarily (i.e., in a reflexive manner) by ipsilateral and/or contralateral sounds (Guinan 2006). This suggests that auditory sensitivity and dynamic range could be dynamically and adaptively varying during natural, binaural listening depending on the state of activation of MOC efferents.

The roles of the MOC reflex (MOCR) in hearing remain controversial (Guinan 2006, 2010). The MOCR may be essential for normal development of cochlear active mechanical processes (Walsh et al. 1998), and/or to minimize deleterious effects of noise exposure on cochlear function (Maison et al. 2013). A more widely accepted role is that the MOCR possibly facilitates understanding speech in noisy environments. This idea is based on the fact that MOC efferent activation restores the dynamic range of auditory nerve fibre responses in noisy backgrounds to values observed in quiet (Fig. 5 in Guinan 2006), something that probably improves the neural coding of speech embedded in noise (Brown et al. 2010; Chintanpalli et al. 2012; Clark et al. 2012). The evidence in support for this unmasking role of the MOCR during natural listening is, however, still indirect (Kim et al. 2006). A direct demonstration would require being able to selectively deactivate the MOCR while keeping afferent auditory nerve fibres functional. To our knowledge, such an experiment has not been possible for human subjects up until now. Here, we argue that the experiment can be done in an approximate manner using cochlear implants (CIs).

Cochlear implants can enable useful hearing for severely to profoundly deaf persons by bypassing the cochlea via direct electrical stimulation of the auditory nerve. The sound processor in a CI (Fig. 1) typically includes an instantaneous back-end compressor in each frequency channel of processing to map the wide dynamic range

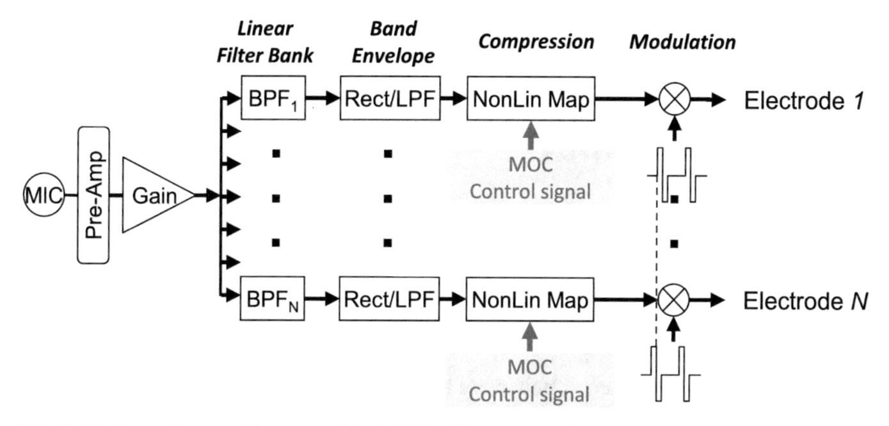

Fig. 1 Envisaged way of incorporating MOC efferent control to a standard pulsatile CI sound processor. The elements of the standard sound processor are depicted in *black* (Wilson et al. 1991) and the MOC efferent control in *green*. See the main text for details

of sounds in the acoustic environment into the relatively narrow dynamic range of electrically evoked hearing (Wilson et al. 1991). The compressor in a CI serves the same function as the cochlear mechanical compression in a healthy inner ear (Robles and Ruggero 2001). In the fitting of modern CIs, the amount and endpoints of the compression are adjusted for each frequency channel and its associated intracochlear electrode(s). The compression once set is fixed. Cochlear implant users thus lack MOCR effects, but these effects can be reinstated, and thus the role(s) of the MOCR in hearing can be assessed by using dynamic rather than fixed compression functions. For example, contralateral MOCR effects could be reinstated by making compression in a given frequency channel vary dynamically depending upon an appropriate control signal from a contralateral processor (Fig. 1). Ipsilateral MOCR effects could be reinstated similarly, except that the control signal would originate from the ipsilateral processor. Using this approach, the CI offers the possibility to mimic (with dynamic compression) or not mimic (with fixed compression) the effects of the MOCR. Such activation or deactivation of MOCR effects is not possible for subjects who have normal hearing.

Our long-term goal is to mimic MOCR effects with CIs. This would be useful not only to explore the roles of the MOCR in acoustic hearing but to also reinstate the potential benefits of the MOCR to CI users. The present study focuses on (1) describing a bilateral CI sound processing strategy inspired by the *contralateral* MOCR; (2) demonstrating, using CI processor output-signal simulations, some possible benefits of the contralateral MOCR in acoustic hearing.

2 Methods

2.1 A Bilateral CI Sound Processor Inspired by the Contralateral MOCR

To assess the potential effects of the contralateral MOCR in hearing using CIs, pairs of sound processors (one per ear) were implemented where the amount of compression in a given frequency channel of each processor in the pair was either constant (STD strategy) or varied dynamically in time in a manner inspired by the contralateral MOCR (MOC strategy).

The two processors in each pair were as shown in Fig. 1. They included a high-pass pre-emphasis filter (first-order Butterworth filter with a 3-dB cutoff frequency of 1.2 kHz); a bank of twelve, sixth-order Butterworth band-pass filters whose 3-dB cut-off frequencies followed a modified logarithmic distribution between 100 and 8500 Hz; envelope extraction via full-wave rectification (Rect.) and low-pass filtering (LPF) (fourth-order Butterworth low-pass filter with a 3-dB cut-off frequency of 400 Hz); and a back-end compression function.

The back-end compression function was as follows (Boyd 2006):

$$y = \log(1 + c \cdot x) / \log(1 + c) \qquad (1)$$

where x and y are the input and output amplitudes to/from the compressor, respectively, both of them assumed to be within the interval [0, 1]; and c is a parameter that determines the amount of compression. In the STD strategy, c was constant ($c = c_{max}$) and identical at the two ears. In the MOC strategy, by contrast, c varied dynamically in time within the range $[c_{max}, c_{min}]$, where c_{max} and c_{min} produce the most and least amount of compression, respectively. For the natural MOCR, the more intense the contralateral stimulus, the greater the amount of efferent inhibition and the more linear the cochlear mechanical response (Hood et al. 1996). On the other hand, it seems reasonable that the amount of MOCR inhibition depends on the magnitude of the output from the cochlea rather than on the level of the acoustic stimulus. Inspired by this, we assumed that the instantaneous value of c for any given frequency channel in an MOC processor was inversely related to the output energy, E, from the corresponding channel in the contralateral MOC processor. In other words, the greater the output energy from a given frequency channel in the left-ear processor, the less the compression in the corresponding frequency channel of the right-ear processor, and vice versa (contralateral 'on-frequency' inhibition). We assumed a relationship between c and E such that in absence of contralateral energy, the two MOC processors in the pair behaved as a STD ones (i.e., $c = c_{max}$ for $E = 0$).

Inspired by the exponential time courses of activation and deactivation of the contralateral MOCR (Backus and Guinan 2006), the instantaneous output energy from the contralateral processor, E, was calculated as the root-mean-square (RMS) output amplitude integrated over a preceding exponentially decaying time window with two time constants.

2.2 Evaluation

The compressed output envelopes from the STD and MOC strategies were compared for different signals in different types and amounts of noise or interferers (see below). Stimulus levels were expressed as RMS amplitudes in full-scale decibels (dBFS), i.e., dB re the RMS amplitude of a 1-kHz sinusoid with a peak amplitude of 1 (i.e., dB re 0.71).

The effects of contralateral inhibition of compression in the MOC strategy were expected to be different depending upon the relative stimulus levels and frequency spectra at each ear. For this reason, and to assess performance in binaural, free-field scenarios, all stimuli were filtered through KEMAR head-related transfer functions (HRTFs) (Gardner and Martin 1995) prior to their processing through the MOC or STD strategies. HRTFs were applied as 512-point finite impulse response filters. STD and MOC performance was compared assuming sound sources at eye level (0° of elevation).

For the present evaluations, c_{max} was set to 1000, a value typically used in clinical CI sound processor, and c_{min} to 1. The energy integration time constants were 4 and 29 ms. These parameter values were selected *ad hoc* to facilitate visualising the effects of contralateral inhibition of compression in the MOC strategy and need not match natural MOCR characteristics or improve CI-user performance (see the Discussion).

All processors were implemented digitally in the time domain, in Matlab™ and evaluated using a sampling frequency of 20 kHz.

3 Results

3.1 The MOC Processor Enhances Within-channel Inter-aural Level Differences

The compressed envelopes at the output from the STD and MOC processors were first compared for a 1-kHz pure tone presented binaurally from a sound source in front of the left ear. The tone had a duration of 50 ms (including 10-ms cosine squared onset and offset ramps) and its level was −20 dBFS. Figure 2a shows the output envelopes for channel #6. Given that the stimulus was binaural and filtered through appropriate HRTFs, there was an acoustic inter-aural level difference (ILD), hence the greater amplitude for the left (blue traces) than for the right ear (red traces). Strikingly, however, the inter-aural output difference (IOD, black arrows in Fig. 2a) was greater for the MOC than for the STD strategy. This is because the gain (or compression) in the each of the two MOC processors was inhibited in direct proportion to the output energy from the corresponding frequency channel in the contralateral processor. Because the stimulus level was higher on the left ear, the output amplitude was also higher for the left- than for the right-ear MOC processor.

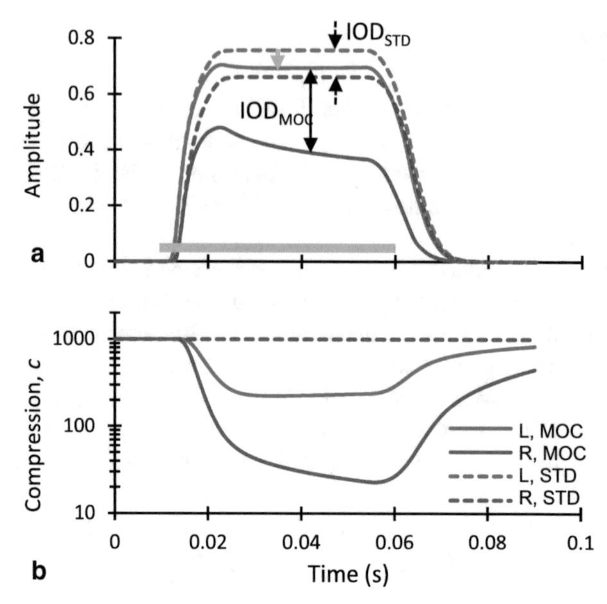

Fig. 2 Compressed output envelopes (**a**) and compression parameter (**b**) for frequency channel #6 of the STD and MOC processors in response to a 1-kHz pure tone presented binaurally from a sound source in front of the left ear. *Red and blue traces* depict signals at the left (*L*) and right (*R*) ears, respectively. The *thick horizontal line* near the abscissa depicts the time period when the stimulus was on. Note the overlap between the *dashed red and blue traces* in the *bottom panel*

Hence, the left-ear MOC processor inhibited the right-ear MOC processor more than the other way around.

Of course, the gains of the left-ear MOC processor were also inhibited by the right-ear processor to some extent. As a result, the output amplitude in the left-ear processor was slightly lower for the MOC than for the STD strategy (green arrow in Fig. 2a). The amount of this inhibition was, however, smaller than for the right-ear MOC processor because (1) the stimulus level was lower on the right ear, and (2) the gain in the right-ear processor was further inhibited by the higher left-ear processor output. In other words, the asymmetry in the amount of contralateral inhibition of gain (or compression) was enhanced because it depended on the processors' *output* rather on their input amplitudes.

Figure 2b shows that the compression parameter, c, in channel #6 of the MOC processors was dynamically varying, hence the IOD was relatively smaller at the stimulus onset and increased gradually in time until it reached a plateau effect. The figure also illustrates the time course of this effect. Note that at time = 0 s, compression was identical for the right-ear and the left-ear MOC processors, and equal to the compression in the STD processors ($c = 1000$). Then, as the output amplitude from the left-ear processor increased, compression in the right-ear processor decreased (i.e., Eq. (1) became more linear). This decreased the output amplitude from the right-ear processor even further, which in turn reduced the inhibitory effect of the right-ear processor on the left-ear processor.

Unlike the MOC processor, whose output amplitudes and compression varied dynamically depending on the contralateral output energy, compression in the two STD processors was fixed (i.e., *c* was constant as depicted by the *dashed lines* in Fig. 2b). As a result, IOD_{STD} was a compressed version of the acoustic ILD and smaller than IOD_{MOC}. In summary, the MOC processor enhanced the within-channel ILDs and presumably conveyed a better lateralized signal compared to the STD processor.

3.2 The MOC Processor Enhances the Spatial Segregation of Simultaneous Sounds

Spatial location is a powerful cue for auditory streaming (Bregman 2001). This section is devoted to showing that the MOC processor can enhance the spatial segregation of speech sounds in a multi-talker situation.

Figure 3 shows example 'electrodograms' (i.e., graphical representations of processors' output amplitudes as a function of time and frequency channel number) for the STD and MOC strategies for two disyllabic Spanish words uttered simultaneously in the free field by speakers located on either side of the head: the word '*diga*' was uttered on the left side, while the word '*sastre*' was uttered on the right side of head. In this example, the two words were actually uttered by the same female speaker and were presented at the same level (−20 dBFS). Approximate spectrograms for the two words are shown in Fig. 3a and b as the output from the proces-

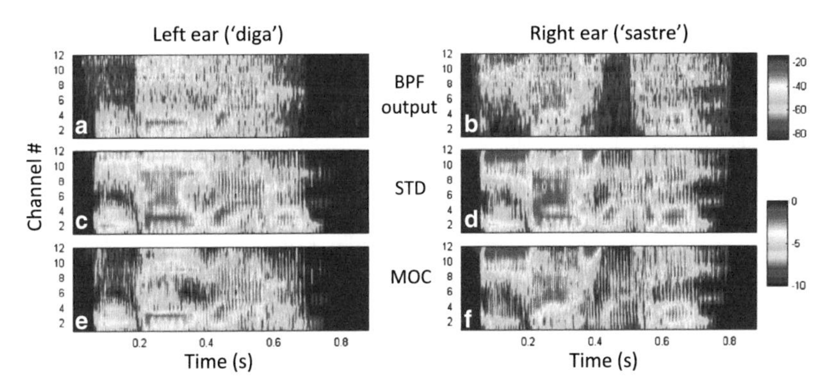

Fig. 3 A comparison of left- and right-ear 'electrodograms' for the *STD* and *MOC* strategies. The stimulus mimicked a free-field multi-talker condition where a speaker in front of the *left* ear uttered the Spanish word 'diga' while a second speaker in front of the *right* ear uttered the word 'sastre'. **a, b** Output from the processors linear filter banks (*BPF* in Fig. 1) for each separate word. **c, d**. Electrodograms for the left- and right-ear *STD* processors. **e, f**. Electrodograms for the left- and right-ear *MOC* processors. *Colour* depicts output amplitudes in dBFS

sors' filter banks (BPF in Fig. 1). Note that up to the filterbank stage, processing was identical and linear for the STD and MOC strategies.

The overlap between the words' most significant features was less for the MOC (Fig. 3e–f) than for the STD strategy (Fig. 3c–d). For the STD strategy, significant features of the two words were present in the outputs of the left- and right-ear processors. By contrast, one can almost 'see' that the word *diga*, which was uttered on the left side of the head, was encoded mostly in the output of the left MOC processor, and the word *sastre*, which was uttered on the right side of the head, was encoded mostly in the right MOC processor output, with little interference between the two words at either ear. This suggests that the MOC processor may enhance the lateralization of speech, and possibly spatial segregation, in situations with multiple spatially non-overlapping speakers. Although not shown, something similar occurred for speech presented in competition with noise. This suggests that the MOC processor may (1) enhance speech intelligibility in noise or in 'cocktail party' situations, and (2) reduce listening effort.

4 Discussion and Conclusions

Our main motivation was to gain insight into the roles of the MOCR in acoustic hearing by using the unique stimulus controls provided by CIs. As a first step, we have proposed a bilateral CI sound coding strategy (the MOC strategy) inspired by the contralateral MOCR and shown that, compared to using functionally independent sound processors on each ear, our strategy can enhance ILDs on a within-channel basis (Fig. 2a) and the segregation of spatially distant sound sources (Fig. 3). An extended analysis (Lopez-Poveda 2015) revealed that it can also enhance within-channel amplitude modulations (a natural consequence from using overall less compression) and improve the speech-to-noise ratio in the speech ear for spatially non-overlapping speech and noise sources. The proposed MOC strategy might thus improve the intelligibility of speech in noise, and the spatial segregation of concurrent sound sources, while reducing listening effort. These effects and their possible benefits have been inferred from analyses of MOC-processor output envelopes for a limited number of spatial conditions. Though promising, further research is necessary to confirm that they hold for other spatial conditions and for behavioural hearing tasks.

The proposed MOC strategy is inspired by the contralateral MOCR but is not an accurate model of it. So far, we have disregarded that contralateral MOCR inhibition is probably stronger for apical than for basal cochlear regions (Lilaonitkul and Guinan 2009a; Aguilar et al. 2013) and that there is a half-octave shift in frequency between the inhibited cochlear region and the spectrum of contralateral MOCR elicitor sound (Lilaonitkul and Guinan 2009b). In addition, the present MOC-strategy parameters were selected *ad hoc* to facilitate visualization of the reported effects rather than to realistically account for the time course (Backus and Guinan 2006) or the magnitude (Cooper and Guinan 2006) of contralateral MOCR gain inhibition.

Our next steps include assessing the potential benefits of more accurate MOC-strategy implementations and parameters.

Despite these inaccuracies, it is tempting to conjecture that the natural contralateral MOCR can produce effects and benefits for acoustic hearing similar to those reported here for the MOC processor. Indeed, the results in Fig. 3 are consistent with the previously conjectured role of the contralateral MOCR as a 'cocktail party' processor (Kim et al. 2006). The present analysis shows that this role manifests during 'free-field' listening and results from a combination of head-shadow frequency-specific ILDs with frequency-specific contralateral MOCR inhibition of cochlear gain. This and the other possible roles of the MOCR described here would complement its more widely accepted role of improving the neural encoding of speech in noisy environments by restoring the effective dynamic range of auditory nerve fibre responses (Guinan 2006). The latter effect is neural and cannot show up in the MOC-processor output signals used in the present analyses.

The potential benefits attributed here to the contralateral MOCR are unavailable to current CI users and this might explain part of their impaired performance understanding speech in challenging environments. The present results suggest that the MOC processor, or a version of it, might improve CI performance. The MOC strategy may also be adapted for use with bilateral hearing aids.

Acknowledgments We thank Enzo L. Aguilar, Peter T. Johannesen, Peter Nopp, Patricia Pérez-González, and Peter Schleich for comments. This work was supported by the Spanish Ministry of Economy and Competitiveness (grants BFU2009-07909 and BFU2012-39544-C02-01) to EAL-P and by MED-EL GmbH.

References

Aguilar E, Eustaquio-Martin A, Lopez-Poveda EA (2013) Contralateral efferent reflex effects on threshold and suprathreshold psychoacoustical tuning curves at low and high frequencies. J Assoc Res Otolaryngol 14(3):341–357

Backus BC, Guinan JJ (2006) Time-course of the human medial olivocochlear reflex. J Acoust Soc Am 119(5):2889–2904

Boyd PJ (2006) Effects of programming threshold and maplaw settings on acoustic thresholds and speech discrimination with the MED-EL COMBI 40+ cochlear implant. Ear Hear 27(6):608–618

Bregman AS (2001) Auditory scene analysis: the perceptual organization of sound. Second MIT Press, Massachusetts Institute of Technology, Cambridge (Massachusetts)

Brown GJ, Ferry RT, Meddis R (2010) A computer model of auditory efferent suppression: implications for the recognition of speech in noise. J Acoust Soc Am 127(2):943–954

Chintanpalli A, Jennings SG, Heinz MG, Strickland EA (2012) Modeling the anti-masking effects of the olivocochlear reflex in auditory nerve responses to tones in sustained noise. J Assoc Res Otolaryngol 13(2):219–235
Clark NR, Brown GJ, Jurgens T, Meddis R (2012) A frequency-selective feedback model of auditory efferent suppression and its implications for the recognition of speech in noise. J Acoust Soc Am 132(3):1535–1541
Cooper NP, Guinan JJ (2006) Efferent-mediated control of basilar membrane motion. J Physiol 576(Pt 1):49–54
Gardner WG, Martin KD (1995) HRTF measurements of a KEMAR. J Acoust Soc Am 97(6):3907–3908
Guinan JJ (2006) Olivocochlear efferents: anatomy, physiology, function, and the measurement of efferent effects in humans. Ear Hear 27(6):589–607
Guinan JJ (2010) Cochlear efferent innervation and function. Curr Opin Otolaryngol Head Neck Surg 18(5):447–453
Hood LJ, Berlin CI, Hurley A, Cecola RP, Bell B (1996) Contralateral suppression of transient-evoked otoacoustic emissions in humans: intensity effects. Hear Res 101(1–2):113–118
Kim SH, Frisina RD, Frisina DR (2006) Effects of age on speech understanding in normal hearing listeners: relationship between the auditory efferent system and speech intelligibility in noise. Speech Commun 48(7):855–862
Lilaonitkul W, Guinan JJ (2009a) Human medial olivocochlear reflex: effects as functions of contralateral, ipsilateral, and bilateral elicitor bandwidths. J Assoc Res Otolaryngol 10(3):459–470
Lilaonitkul W, Guinan JJ (2009b) Reflex control of the human inner ear: a half-octave offset in medial efferent feedback that is consistent with an efferent role in the control of masking. J Neurophysiol 101(3):1394–1406
Lopez-Poveda EA (2015) Sound enhancement for cochlear implants. Patent WO2015169649 A1.
Maison SF, Usubuchi H, Liberman MC (2013) Efferent feedback minimizes cochlear neuropathy from moderate noise exposure. J Neurosci 33(13):5542–5552
Robles L, Ruggero MA (2001) Mechanics of the mammalian cochlea. Physiol Rev 81(3):1305–1352
Walsh EJ, McGee J, McFadden SL, Liberman MC (1998) Long-term effects of sectioning the olivocochlear bundle in neonatal cats. J Neurosci 18(10):3859–3869
Wilson BS, Finley CC, Lawson DT, Wolford RD, Eddington DK, Rabinowitz WM (1991) Better speech recognition with cochlear implants. Nature 352(6332):236–238

Deactivating Cochlear Implant Electrodes Based on Pitch Information for Users of the ACE Strategy

Deborah Vickers, Aneeka Degun, Angela Canas, Thomas Stainsby and Filiep Vanpoucke

Abstract There is a wide range in performance for cochlear implant (CI) users and there is some evidence to suggest that implant fitting can be modified to improve performance if electrodes that do not provide distinct pitch information are de-activated. However, improvements in performance may not be the same for users of all CI devices; in particular for those with Cochlear devices using n-of-m strategies (ACE or SPEAK).

The goal of this research was to determine for users of Cochlear devices (CP810 or CP900 series processors) if speech perception could be improved when indiscriminable electrodes were de-activated and this was also compared to when the same number of discriminable electrodes were de-activated.

A cross-over study was conducted with 13 adult CI users who received experimental maps with de-activated channels for a minimum of 2 months and these were compared to optimised clinical maps.

The findings showed that there were no significant benefits of electrode de-activation on speech perception and that there was a significant deterioration in spectro-temporal ripple perception when electrodes were switched off. There were no significant differences between de-activation of discriminable or indiscriminable electrodes.

These findings suggest that electrode de-activation with n-of-m strategies may not be beneficial.

D. Vickers (✉) · A. Degun · A. Canas
UCL Ear Institute, 332 Grays Inn Road, London WC1X 8EE, UK
e-mail: d.vickers@ucl.ac.uk

A. Degun
e-mail: aneeka.degun.10@ucl.ac.uk

A. Canas
e-mail: a.canas@ucl.ac.uk

T. Stainsby · F. Vanpoucke
Cochlear Technology Centre, Schaliënhoevedreef 20, building I, 2800 Mechelen, Belgium

P. van Dijk et al. (eds.), *Physiology, Psychoacoustics and Cognition in Normal and Impaired Hearing*, Advances in Experimental Medicine and Biology 894,
DOI 10.1007/978-3-319-25474-6_13

Keywords Cochlear implant · Electrode discrimination · Electrode deactivation · Fitting · Mapping · Electrode-neurone interface · Speech perception · Pitch ranking · ACE strategy · n-of-m

1 Introduction

Various approaches have been explored to look at the effects of re-mapping cochlear implants (CIs) to overcome the problems associated with neural dead regions or poor electrode placement (electrode-neuron interface (ENI)). The mapping interventions have typically explored ways to identify electrodes for de-activation using Computerized Tomography (CT), threshold profiling with focussed stimulation, detection of modulation and pitch-based electrode discrimination. Electrodes indicated as being in a region with a poor ENI have typically been deactivated with the intention of improving spectral resolution by reducing overlapping electrode stimulation patterns.

If a poor ENI arises due to an underlying neural dead region, the transmission of information in that frequency channel could be impeded; if a poor ENI arises due to the presence of broad electrical fields, this will lead to corruption of channel specific information. For both cases it could be beneficial to deactivate electrodes, however when doing so frequency information is re-allocated to other electrodes making analysis bands broader and distorting frequency-to-electrode mapping. These effects are small for individual electrode deactivation but can be substantial for deactivation of multiple electrodes, potentially resulting in poorer precision of information transmission; a completely opposite effect to that intended.

Some n-of-m (n channels with highest input amplitudes stimulated out of a possible m channels; e.g. ACE (8-of-22) or SPEAK) strategies may not benefit from deactivation of electrodes due to the simplification of the information delivery already inherent in the strategy.

The findings from electrode deactivation studies are mixed. Noble et al. (2014) used a CT-guided technique in which pre- and post-operative scans were used to determine the electrode positioning with respect to the modiolus. Poorly positioned electrodes were deactivated and additional mapping adjustments made. Assessments were conducted at baseline and following a 3–6 week exposure period with the adjusted map. Learning effects were not accounted for. Thirty-six out of 72 (50 %) participants demonstrated a benefit on one outcome measure (CNC words, AzBio sentences (quiet, +10 dB, +5 dB signal-to-noise ratio (SNR)) and BKBSin), 17 had a significant degradation in performance (24 %), three cochlear users had mixed results (4 %) and the final 16 participants did not demonstrate any effect of the de-activation (22 %). When broken down by device, the results showed significant benefit on one measure for 16 out of 26 AB and MEDEL users (62 %; seven showed significant deterioration) and 20 out of 46 Cochlear users (44 %; ten showed significant deterioration and three a mixture of significant deterioration and significant improvement). Overall the results demonstrated significant improvements in speech understanding in quiet and noise for re-mapped participants, with no particular test exhibiting the greatest improvements. Cochlear users were less likely to benefit from electrode deactivation than users of the other devices. As speculated earlier it could be that the n-of-m strategies used in the Cochlear device may be less sensitive to channel overlap effects because not all channels are stimulated in each cycle.

Zwolan et al. (1997) assessed users of an older generation Cochlear device (Mini-22) stimulated in bipolar mode using an MPEAK sound processing strategy. They demonstrated positive results when deactivating electrodes based on pitch discrimination. Testing was acute with care to randomise between clinical and experimental maps in the assessment, but the experimental map exposure was far briefer than that of the clinical map. They found significant improvements in speech perception in at least one speech test for seven out of nine participants; while two participants showed a significant decrease on at least one speech test. This study demonstrated some benefit from channel deactivation for Cochlear users; however the MPEAK strategy was based on feature extraction and not n-of-m.

Work by Saleh (Saleh et al. 2013; Saleh 2013) using listeners with more up-to-date sound processing strategies also demonstrated positive findings when deactivating electrodes. They based channel selection on acoustic pitch ranking. Electrodes were identified for deactivation if they did not provide distinct or tonotopic pitch information. Twenty-five participants were enrolled. They received take-home experimental maps for a 1-month time period before being assessed. Assessments were conducted with the clinical map at baseline and again following experience with the experimental map. They found statistically significant improvements for electrode deactivation on speech perception tests for 16 of the 25 participants (67%). For the Cochlear users 56% (five out of nine) gained benefit compared to 69% for the AB and MED-EL users (11 out of 16). Participant numbers were small but there was a trend for fewer Cochlear users to benefit compared to other devices; similar to that observed in Noble et al. (2014).

Garadat et al. (2013) deactivated electrodes based on modulation detection in a group of ACE users. Up to five electrodes were switched off and they were selected such that only individual non-adjacent contacts were deactivated to avoid large regions being switched off. Following deactivation they demonstrated significant benefit for consonant perception and CUNY sentences in speech-shaped noise (8 out of 12 showed benefit) but demonstrated a deterioration in performance for vowel perception (7 out of 12 showed degradation). Testing was acute so learning effects were not controlled for and listeners were not given time to adapt to the experimental maps; with further experience and acclimatisation users may have performed better on the vowel task, particularly because it is known that vowel perception is sensitive to filtering adjustments (Friesen et al. 1999). These findings are potentially positive for ACE users but without a control condition they are not definitive proof.

Some researchers have not observed benefit from electrode deactivation and have even demonstrated potentially deleterious effects for Cochlear device users. Henshall and McKay (2001) used a multidimensional scaling technique to identify and deactivate electrodes having non-tonotopic percepts. They did not find an improvement in speech perception when channels were switched off and hypothesized that this could be due to the shift in frequency-to-electrode allocation produced by clinical software when electrodes are deactivated; particularly in the apical region.

The findings from different deactivation studies are inconclusive, partly due to the difficulties of comparing experimental to clinical maps that the CI users had adapted to and partly due to the impact that deactivating channels can have on other factors (e.g. rate of stimulation; filter-frequency allocation) which differ between devices and sound processing strategies. There is a trend suggesting that deactivating electrodes may be less effective for ACE or SPEAK users than for the CIS based sound processing strategies incorporated in AB or MEDEL devices.

The goal of this research was to explore the impact of electrode deactivation for adult users of the ACE strategy ensuring that participants received sufficient adaptation time with experimental maps and that learning effects were controlled for. Additionally participants' clinical maps were optimised to ensure comparisons were made with an optimal control condition. A pitch ranking task was used to identify indiscriminable electrodes for deactivation and an additional control condition was included in which discriminable electrodes were deactivated to determine if any benefits observed were due to switching off electrodes with a poor ENI or just due to having fewer channels.

The research objectives were:

1. To determine if deactivating electrodes that are indiscriminable based on pitch ranking leads to significant changes in speech perception for adult ACE users when compared to an optimised clinical map.
2. To determine if deactivating the same number of discriminable electrodes leads to significantly different performance to the deactivation of indiscriminable electrodes.

2 Method

2.1 Ethics Approval

Research was ethically approved by National Research Evaluation Service (PRSC 23/07/2014).

2.2 Participants

Thirteen post-lingually deafened adult ACE users with Cochlear CP810 or CP900 series devices were recruited. Individuals with ossification or fibrosis were excluded and participants had to have a minimum of 12 months CI experience and primarily speak English. Median age was 67 years (range 36–85 years); the median length of CI use was 2 years (mean 4 years). Median duration of severe-to-profound deafness prior to surgery was 4 years (mean 2 years).

2.3 Study Design

A crossover single-blinded (participant unaware of mapping interventions) randomised control trial was conducted. Prior to starting each participant had their clinical map optimised by adjusting individual channel levels and they received a minimum of 1 month acclimatisation to this optimised map. The crossover study used an A-B-B-A/B-A-A-B design in which map A had indiscriminable electrodes deactivated and map B had an equal amount of discriminable electrodes deactivated.

2.4 Equipment

For the majority of the assessments the sound processor was placed inside an Otocube; a sound proof box designed for testing CI users to simulate listening over a loudspeaker in a sound treated room. Stimulus delivery and collection of responses was controlled using a line in from a laptop to the Otocube and sounds were presented over the Otocube loudspeaker positioned at 0 ° azimuth to the sound processer.

The direct stimulation pitch-ranking task was conducted using the Nucleus Implant Communicator (NIC) to send stimuli directly to individual electrodes.

2.5 Test Materials

CHEAR Auditory Perception Test (CAPT)
The CAPT was used because it is sensitive to spectral differences in hearing aid fitting algorithms (Marriage et al. 2011) and was known to be highly repeatable (inter-class correlation of 0.70 for test and re-test conditions; Vickers et al. 2013).

The CAPT is a four-alternative-forced-choice monosyllabic word-discrimination test spoken by a female British English speaker. It contained ten sets of four minimally-contrastive real words; eight sets with a contrastive consonant, e.g. fat, bat, cat, mat and two sets with a contrastive vowel, e.g. cat, cot, cut, cart. The participant selected from four pictures on a computer screen. The d' score was calculated.

Stimuli were presented at 50 dBA in quiet; a level selected in pilot work because it resulted in performance falling on the slope of the psychometric function.

Children's Coordinate Response Measure (CCRM)
The adaptive CCRM was based on the test developed by Bolia et al. (2000) and Brungart (2001). It was used because of low contextual cues, and ease of task.

Each stimulus sentence took the form:

'Show the dog where the (colour) (number) is?': e.g. 'show the dog where the green three is?'

There were six colour options (blue, black, green, pink, red and white) and eight possible numbers (1–9, excluding 7). Stimuli were spoken by a British female speaker.

Two different maskers were used: 20-talker babble (created by modulating a speech-shaped noise with the amplitude envelope of male sentences) or speech-shaped noise (average long-term spectrum of the sentences). Random sections of the noise were selected on each trial.

The sentences were presented at 65 dBA and the noise adjusted adaptively (2-down/1-up) on the basis of whether or not both the colour and number were identified correctly. Initial step size was 9 dB, and decreased after two reversals to 3 dB. A further four reversals were run and averaged to obtain the speech reception threshold (SRT), but no test was longer than 26 trials.

Spectral-Temporally Modulated Ripple Task (SMRT)
The SMRT was used to assess spectral resolution. The SMRT test was chosen over other spectral ripple tasks, because it avoided potential confounds such as loudness and edge-frequency cues whilst still being sensitive to spectral resolution (Aronoff and Landsberger 2013).

Stimuli were 500 ms long with 100 ms ramps, nonharmonic tone complexes with 202 equal amplitude pure-tone frequency components, spaced every 1/33.3 octave from 100 to 6400 Hz. The amplitudes of the pure tones were modulated by a sine wave with a 33 % modulation depth. A three-interval, two-alternative forced choice task was used, with a reference stimulus of 20 ripples per octave (RPO) presented at 65 dBA. The target stimulus initially had 0.5 RPO and the number of ripples was modified using a 1-up/1-down adaptive procedure with a step size of 0.2 RPO. The test was completed after ten reversals, the last six reversals were averaged to calculate the threshold.

Pitch Ranking Approaches
The determination of discriminable and indiscriminable electrodes was based on two approaches, acoustic tones presented at the centre frequency of a filter via the System for Testing Auditory Responses (STAR; Saleh et al. 2013) and direct stimulation with biphasic pulses delivered using the NIC.

The pitch ranking task was the same for both approaches. A two-interval two-alternative forced choice paradigm was used. A single presentation consisted of two intervals in which the higher tone was randomly assigned to either the first or second interval, and the listener indicated which interval sounded higher in pitch. The 2 tones/pulses were presented sequentially with duration of 1000 ms and were 500 ms apart. Five pairs of tones/pulses were initially presented and responses recorded. If the participant got all five correct the pair was considered discriminable. If they scored less than five a further five presentations were given and scored. If the participant got eight out of ten presentations correct the pair passed. This was based on binomial significance at the $p<0.05$ level (Skellam 1948). A run consisted of presentation of all possible electrode pairs presented either five or ten times, the duration of the test varied depending on accuracy of response and number of electrodes activated, but was typically 25 min long for the best performers with a full electrode array but up to 50 min for poorer performers. For the STAR delivery the stimuli were presented at 65 dBA and the NIC presentation was at 85 % of the upper comfort level (C level). Level roving was applied to each presentation and this was +/−3 dB for STAR and +/−3 clinical units (CU) for NIC.

The responses from the STAR and the NIC test were highly correlated ($rho = 0.82$, $N = 13$, $p = 0.001$) so the results from both tests were combined to create a composite score for each electrode by adding up the passes and fails across both tests thus increasing the calculation power. After determination of the "indiscriminable" electrode set, a "discriminable" electrode set of the same size was selected, using electrodes that were as near as possible to the indiscriminable set and had not failed on either task.

Fitting Procedure for De-activation
Each electrode in the entire A/B set was deactivated. The overall level of the experimental map was checked and if necessary all C levels adjusted to be approximately equally loud to the optimised clinical map. All other fitting parameters remained at the default settings (900 pulses-per-second, 25 μs pulse width, eight maxima and MP1+2 stimulation mode).

3 Results

Non-parametric statistics (Wilcoxon) were used due to the small number of participants ($N = 13$) and the CCRM and SMRT data exceeded acceptable limits for skewness and kurtosis. Each mapping condition was tested on two occasions and the highest score was used in the analysis.

Fig. 1 shows group results, comparing the optimised clinical map to the "indiscriminable" and "discriminable" deactivation conditions.

For CAPT, CCRM in speech-shaped noise and babble there were no significant differences between any of the mapping options at the $p < 0.05$ level. However, the SMRT was significantly poorer for the indiscriminable electrodes deactivated than the clinical map ($z = -2.9$, $p = 0.004$) and also significantly poorer when the discriminable electrodes were de-activated ($z = -2.62$, $p = 0.009$). There was no significant difference between the de-activation of the indiscriminable or discriminable electrodes.

4 Discussion

Results indicated that there was no measurable benefit for deactivating channels based on pitch ranking for ACE users. Effects were negligible regardless of whether the deactivated channels were indiscriminable or discriminable based on pitch ranking. Care was taken to balance the design and ensure that all maps were used for sufficient time to obtain benefit. Prior to experimental map adjustments, clinical maps were optimised and each participant given an acclimatisation period prior to starting the study.

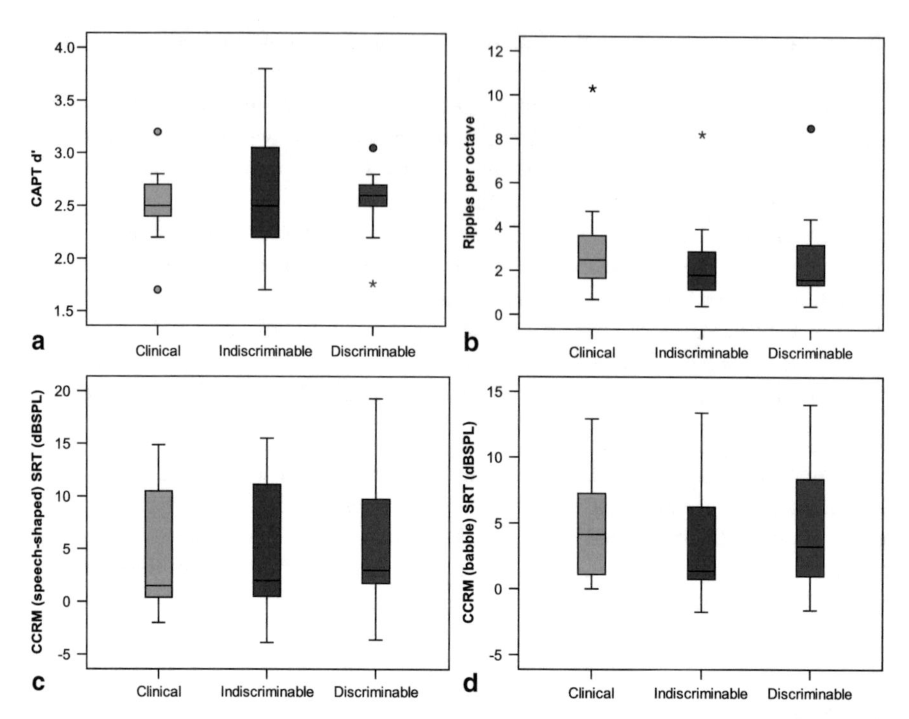

Fig. 1 Boxplots for group results for CAPT d' (panel **a**), SMRT ripples per octave (panel **b**), CCRM in speech-shaped noise (panel **c**), CCRM in babble (panel **d**). *Boxes* represent inter-quartile range, the *line* in the box shows the median and the *whiskers* indicate range and *asterisks* and *circles* show outliers

There was a range in the number of channels selected for deactivation (1–6; median = four channels), which, when combined with the electrodes already switched off, resulted in a de-activation range of 1–10 (median = six channels). For an n-of-m strategy this number of deactivated electrodes may have been too small to demonstrate differences in performance; the selection of channels to deactivate was based purely on electrodes that were indiscriminable from one another in pitch ranking and for many users the number of indiscriminable electrode pairs was low so only small numbers were deactivated.

The only statistically significant finding indicated that spectral resolution was poorer when electrodes were de-activated. When the channels were de-activated the rate of stimulation never changed, however the default filter allocation was used, which may have affected the spectral information (Henshall and McKay 2001). For the deactivation of discriminable channels, the selected channels were as near in site to the deactivated indiscriminable channels in an attempt to avoid dramatic differences between the two conditions for filter allocations. However, for many people the changes to the frequency-to-electrode allocation would have been large when compared to the clinical map.

There should be further exploration of the fitting parameter space for users of n-of-m strategies. The limits or extent of effective use for channel de-activation should be determined and predictive factors to aid the fitting process defined. It is also essential to determine the most critical factors to modify for those with poorer performance levels.

Acknowledgments This project was part-funded by Cochlear Research and Development Ltd. Thanks to Stuart Rosen for assistance with speech-in-noise testing.

References

Aronoff JM, Landsberger DM (2013) The development of a modified spectral ripple test. J Acoust Soc Am 134(2):EL217–EL222

Bolia RS, Nelson WT, Ericson MA, Simpson BD (2000) A speech corpus for multitalker communications research. J Acoust Soc Am 107(2):1065–1066

Brungart DS (2001) Informational and energetic masking effects in the perception of two simultaneous talkers. J Acoust Soc Am 109(3):1101–1109

Friesen LM, Shannon RV, Slattery WH (1999) The effect of frequency allocation on phoneme recognition with the Nucleus-22 cochlear implant. Am J Otol 20:729–734

Garadat SN, Zwolan TA, Pfingst BE (2013) Using temporal modulation sensitivity to select stimulation sites for processor MAPs in cochlear implant listeners. Audiol Neurootol 18(4):247–260

Henshall KR, McKay CM (2001) Optimizing electrode and filter selection in cochlear implant speech processor maps. J Am Acad Audiol 12(9):478–489

Marriage JE, Vickers DA, Baer T, Moore BCJ (2011). Using speech perception measures to guide the choice of amplification. In Seewald RC, Bamford JM (eds) A sound foundation through early amplification. Phonak, Staefa, pp 273–279

Noble JH, Gifford RH, Hedley-Williams AJ, Dawant BM, Labadie RF (2014) Clinical evaluation of an image-guided cochlear implant programming strategy. Audiol Neurootol 19(6):400–411

Saleh SM (2013) Fitting cochlear implants based on pitch perception. Doctoral Dissertation. University College London

Saleh SM, Saeed SR, Meerton L, Moore DR, Vickers DA (2013) Clinical use of electrode differentiation to enhance programming of cochlear implants. Cochlear Implants Int 14:16–18

Skellam JG (1948) A probability distribution Derived from the binomial distribution by regarding the probability of success as variable between the sets of trials. J R Stat Soc Ser B Stat Methodol 10(2):257–261

Vickers D, Backus B, Macdonald N, Rostamzadeh N, Marriage J, Mahon M (2013) Using personal response systems to assess speech perception within the classroom: an approach to determine the efficacy of sound field re-enforcement in primary school classrooms. Ear Hearing 34(4):491–502

Zwolan TA, Collins LM, Wakefield GH (1997) Electrode discrimination and speech recognition in postlingually deafened adult cochlear implant subjects. J Acoust Soc Am 102(6):3673–3685

Speech Masking in Normal and Impaired Hearing: Interactions Between Frequency Selectivity and Inherent Temporal Fluctuations in Noise

Andrew J. Oxenham and Heather A. Kreft

Abstract Recent studies in normal-hearing listeners have used envelope-vocoded stimuli to show that the masking of speech by noise is dominated by the temporal-envelope fluctuations inherent in noise, rather than just overall power. Because these studies were based on vocoding, it was expected that cochlear-implant (CI) users would demonstrate a similar sensitivity to inherent fluctuations. In contrast, it was found that CI users showed no difference in speech intelligibility between maskers with and without inherent envelope fluctuations. Here, these initial findings in CI users were extended to listeners with cochlear hearing loss and the results were compared with those from normal-hearing listeners at either equal sensation level or equal sound pressure level. The results from hearing-impaired listeners (and in normal-hearing listeners at high sound levels) are consistent with a relative reduction in low-frequency inherent noise fluctuations due to broader cochlear filtering. The reduced effect of inherent temporal fluctuations in noise, due to either current spread (in CI users) or broader cochlear filters (in hearing-impaired listeners), provides a new way to explain the loss of masking release experienced in CI users and hearing-impaired listeners when additional amplitude fluctuations are introduced in noise maskers.

Keywords Cochlear hearing loss · Hearing in noise · Speech perception

---

A. J. Oxenham (✉)
Departments of Psychology and Otolaryngology, University of Minnesota—Twin Cities, Minneapolis, MN 55455, USA
e-mail: oxenham@umn.edu

H. A. Kreft
Department of Otolaryngology, University of Minnesota—Twin Cities, Minneapolis, MN 55455, USA

P. van Dijk et al. (eds.), *Physiology, Psychoacoustics and Cognition in Normal and Impaired Hearing,* Advances in Experimental Medicine and Biology 894,
DOI 10.1007/978-3-319-25474-6_14

1 Introduction

Speech perception is a major communication challenge for people with hearing loss and with cochlear implants (CIs), particularly when the speech is embedded in background noise (e.g., Humes et al. 2002; Zeng 2004). Recent work has suggested that it is not so much the overall noise energy that limits speech perception in noise, as suggested by earlier work (French and Steinberg 1947; Kryter 1962; George et al. 2008), but rather the energy in the inherent temporal-envelope modulations in noise (Dubbelboer and Houtgast 2008; Jorgensen and Dau 2011; Stone et al. 2011, 2012; Jorgensen et al. 2013; Stone and Moore 2014). In a recent study (Oxenham and Kreft 2014) we examined the effects of inherent noise fluctuations in CI users. In contrast to the results from normal-hearing (NH) listeners, we found that CI users exhibited *no* benefit of maskers without inherent fluctuations. Further experiments suggested that the effective inherent noise envelope fluctuations were reduced in the CI users, due to the effects of current spread, or interactions between adjacent electrodes, leading to smoother temporal envelopes.

One remaining question is whether HI listeners exhibit the same loss of sensitivity to inherent noise fluctuations as CI users. If so, the finding may go some way to explaining why both HI and CI populations exhibit less masking release than NH listeners when additional slow fluctuations are imposed on noise maskers (e.g., Festen and Plomp 1990; Nelson and Jin 2004; Stickney et al. 2004; Gregan et al. 2013). On one hand, cochlear hearing loss is generally accompanied by a loss of frequency selectivity, due to loss of function of the outer hair cells; on the other hand this "smearing" of the spectrum occurs before extraction of the temporal envelope, rather than afterwards, as is the case with CI processing. Because the smearing is due to wider filters, rather than envelope summation, the resultant envelopes from broader filters still have Rayleigh-distributed envelopes with the same overall relative modulation power as the envelopes from narrower filters (i.e., the same area under the modulation power spectrum, normalized to DC; see Fig. 1). In contrast, with CIs, the envelopes derived from summing the envelope currents from adjacent electrodes are no longer Rayleigh distributed and have lower modulation power, relative to the overall (DC) power in the envelope (Hu and Beaulieu 2005).

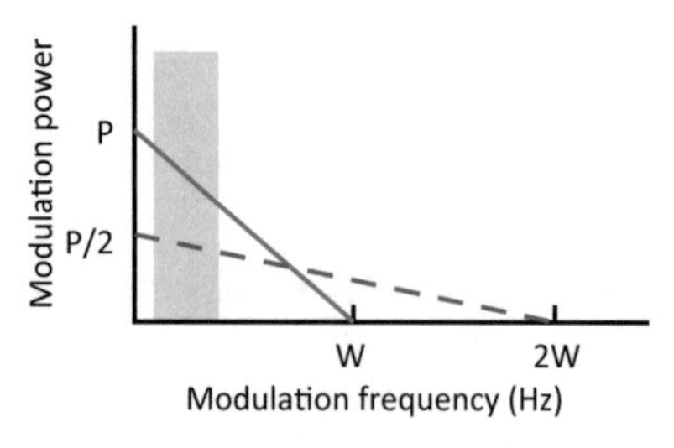

Fig. 1 Schematic diagram of the effect of broadening the filter from a bandwidth of W to a bandwidth of 2 W on the modulation spectrum of filtered Gaussian noise. The relative modulation power (area under the *line*) remains constant, but the area under the lines within the speech-relevant range (*shaded rectangle*) is reduced

One factor suggesting that HI listeners may also experience less influence of inherent noise fluctuations is that the modulation spectrum is altered by broadening the filters: for an ideal rectangular filter, the modulation power of Gaussian noise after filtering has a triangular distribution, reaching a minimum of no power at a frequency equal to the bandwidth of the filter (Lawson and Uhlenbeck 1950). Although widening the filter does not alter the area under the modulation spectrum, it results in relatively less power at lower modulation frequencies (see Fig. 1). Given that low modulation frequencies are most important for speech, the relative reduction in modulation power at low modulation frequencies may reduce the influence of the inherent fluctuations for listeners with broader filters, due to hearing loss.

The aim of this experiment was to test the resulting prediction that hearing loss leads to less effect of inherent noise fluctuations on speech masking. We compared the results of listeners with cochlear hearing loss with the performance of young NH listeners and age-matched NH listeners. Performance was compared for roughly equal sensation levels (SL), and for equal sound pressure levels (SPL) to test for the effect of overall level on performance in NH listeners.

2 Methods

2.1 Listeners

Nine listeners with mild-to-moderate sensorineural hearing loss (4 male and 5 female; mean age 61.2 years) took part in this experiment. Their four-frequency pure-tone average thresholds (4F-PTA from 500, 1000, 2000, and 4000 Hz) ranged from about 25 to 65 dB HL (mean ~40 dB HL). Nine listeners with clinically normal hearing (defined as 20 dB HL or less at octave frequencies between 250 and 4000 Hz; mean 4F-PTA 7.6 dB HL), who were matched for age (mean age 62.2 years) and gender with the HI listeners, were run as the primary comparison group. In addition, a group of four young (mean age 20.5 years; mean 4F-PTA 2.8 dB HL) NH listeners were tested. All experimental protocols were approved by the Institutional Review Board of the University of Minnesota, and all listeners provided informed written consent prior to participation.

2.2 Stimuli

Listeners were presented with sentences taken from the AZBio speech corpus (Spahr et al. 2012). The sentences were presented to the HI listeners at an overall rms level of 85 dB SPL. The sentences were presented to the NH listeners at two different levels: in the equal-SL condition, the sentences were presented at

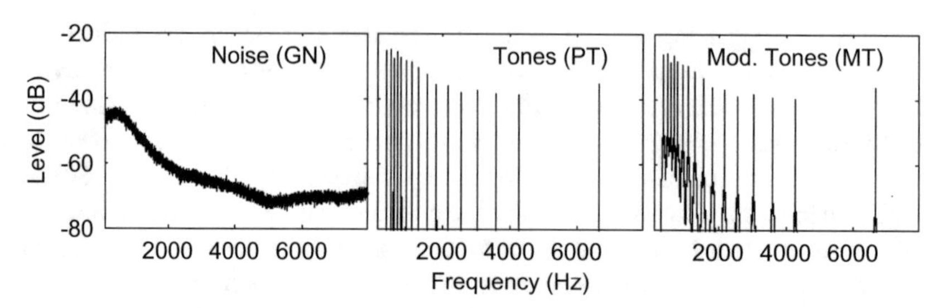

Fig. 2 Representation of the three masker types used in the experiment. The three panels provide spectral representations of the noise (*top*), tone (*middle*), and modulated tone (*bottom*) maskers

40 dB above the detection threshold of the speech for the NH listeners, similar to the sensation level (SL) of the 85-dB SPL speech for the HI listeners; in the equal-SPL condition, the speech was presented at 85 dB SPL. The sentences were presented in three different types of masker, as in Oxenham and Kreft (2014); see Fig. 2. The first was a Gaussian noise, spectrally shaped to match the long-term spectrum of the target speech. The second was comprised of 16 pure tones, approximately evenly spaced on a logarithmic frequency scale from 333 to 6665 Hz, corresponding to the center frequencies on a standard CI map for Advanced Bionics. The amplitudes of the tones were selected to produce the same long-term output of a 16-channel vocoder as the target speech (or Gaussian noise masker), assuming the same center frequencies. The third was comprised of the same 16 tones, but each tone was modulated independently with the temporal envelope of a noise masker, bandpass filtered using a vocoder filter with the same center frequency as the tone carrier. These three maskers produced equal amounts of speech masking in the CI users tested by Oxenham and Kreft (2014). The masker was gated on 1 s before the beginning of each sentence, and was gated off 1 s after the end of each sentence. The masker in each trial was a sample of a longer 25-s sound file, cut randomly from within that longer waveform. The speech and masker were mixed before presentation and the signal-to-masker ratios were selected in advance, based on pilot data, to span a range of performance between 0 and 100 % word recognition.

The speech and the masker were mixed and low-pass filtered at 4000-Hz, and were either presented unprocessed or were passed through a tone-excited envelope vocoder that simulates certain aspects of CI processing (Dorman et al. 1998; Whitmal et al. 2007). The stimulus was divided into 16 frequency subbands, with the same center frequencies as the 16 tone maskers. The temporal envelope from each subband was extracted using a Hilbert transform, and then the resulting envelope was lowpass filtered with a 4th-order Butterworth filter and a cutoff frequency of 50 Hz. This cutoff frequency was chosen to reduce possible voicing periodicity cues, and to reduce the possibility that the vocoding produced spectrally resolved components via the amplitude modulation. Each temporal envelope was then used to modulate a pure tone at the center frequency of the respective subband.

2.3 Procedure

The stimuli were generated digitally, converted via a 24-bit digital-to-analog converter, and presented via headphones. The stimuli were presented to one ear (the better ear in the HI listeners), and the speech-shaped noise was presented in the opposite ear at a level 30 dB below the level of the speech. The listeners were seated individually in a double-walled sound-attenuating booth, and responded to sentences by typing what they heard via a computer keyboard. Sentences were scored for words correct as a proportion of the total number of keywords presented. One sentence list (of 20 sentences) was completed for each masker type and masker level. Presentation was blocked by condition (natural and vocoded, and speech level), and the order of presentation was counterbalanced across listeners. The test order of signal-to-masker ratios was random within each block.

3 Results

The results from the conditions where the speech was presented at roughly equal sensation level to all participants (85 dB SPL for the HI group, and 40 dB SL for the NH groups) are shown in Fig. 3. The upper row shows results without vocoding; the lower row shows results with vocoding. The results from the young NH, age-matched NH, and HI listeners are shown in the left, middle and right columns, respectively.

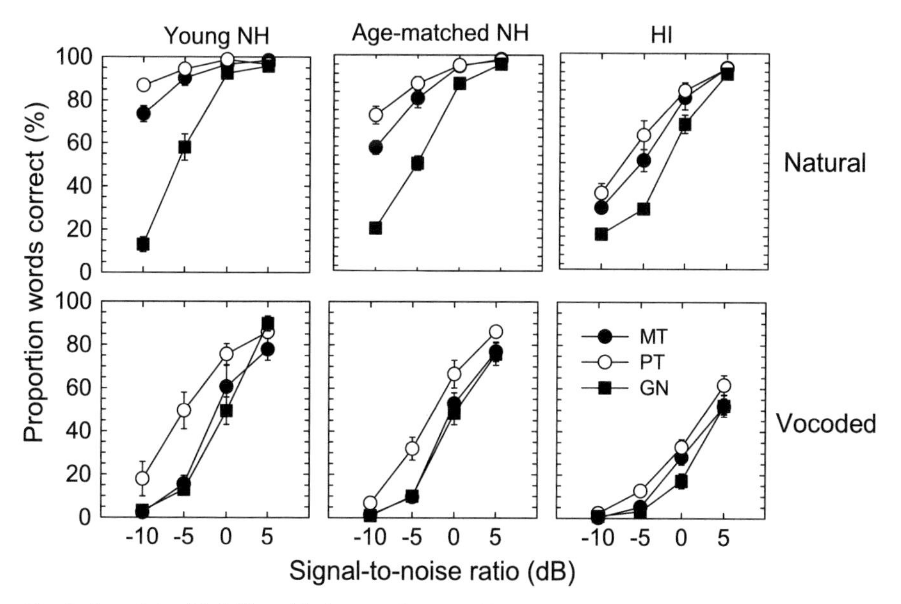

Fig. 3 Speech intelligibility with the speech at roughly equal sensation level *(SL)* across the three groups. Sentence recognition was tested in speech-shaped Gaussian noise *(GN)*, pure tones *(PT)*, and modulated tones *(MT)*

Consider first the results from the non-vocoded conditions (upper row of Fig. 3). Performance in the presence of Gaussian noise (GN; filled squares) is quite similar across the three groups but, as expected, performance is slightly poorer in the HI group. When the noise masker is replaced by 16 pure tones (PT; open circles), performance is markedly better—all three groups show a large improvement in performance, although there appears to be a systematic decrease in performance from young NH to age-matched NH to the HI group. Imposing noise envelope fluctuations to the pure-tone maskers (MT; filled circles) results in a small reduction in performance for all three groups. It appears that both the spectral sparsity of the tonal maskers (PT and MT), as well as the lack of fluctuations (PT), contributes to the release from masking relative to the GN condition.

Consider next the results from the vocoded conditions (lower row of Fig. 3). Here, the benefits of frequency selectivity have been reduced by limiting spectral resolution to the 16 vocoder channels. As expected, the MT masker produces very similar results to the GN masker, as both produce very similar outputs from the tone vocoder. The young NH and the age-matched NH groups seem able to take similar advantage of the lack of inherent masker fluctuations in the PT condition. In contrast, the differences between the PT and MT seem less pronounced in the HI listeners; although some differences remain (in contrast to CI users), they are smaller than in the NH listeners.

The results using 85-dB SPL speech for all listeners are shown in Fig. 4. Only three speech-to-masker ratios (−5, 0, and 5 dB) were tested in this part of the experiment. The relevant data from the HI listeners are simply replotted from Fig. 3. Performance was generally worse overall for both the young and age-matched NH

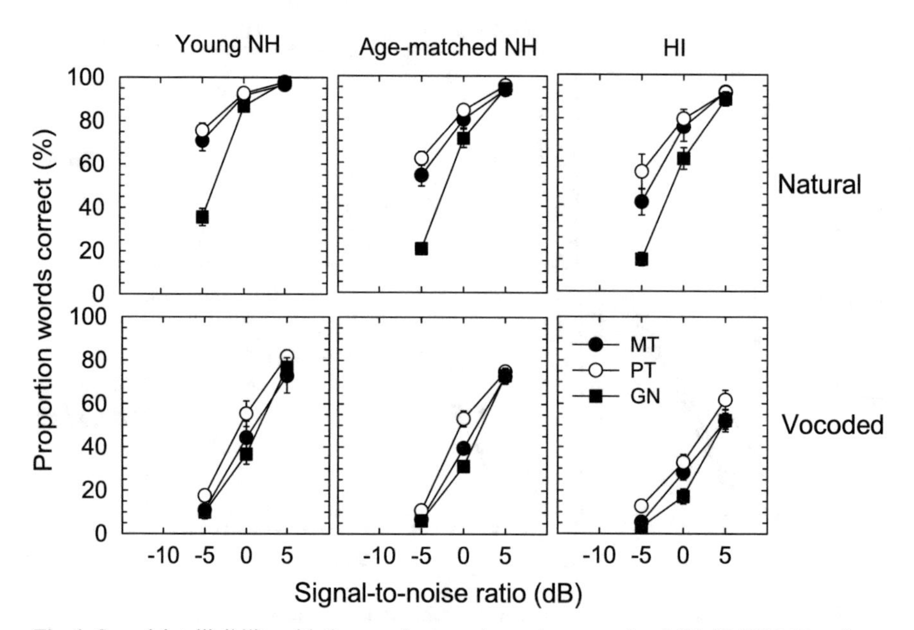

Fig. 4 Speech intelligibility with the speech at equal sound pressure level (85 dB SPL). Data from the HI listeners are replotted from Fig. 3

listeners, compared to results at the lower sound level. In addition, the difference between the PT and MT conditions was reduced. This reduction was observed even in the non-vocoded conditions, but was particularly apparent in the vocoded conditions. In fact, in the vocoded conditions, the differences between the NH groups and the HI group were relatively small.

4 Discussion

Overall, the HI group showed smaller-than-normal differences between maskers with and without inherent fluctuations. This loss of sensitivity to inherent fluctuations was particularly apparent in the vocoded conditions (Fig. 3; lower right panel). However, in contrast to earlier results from CI users (Oxenham and Kreft 2014), some differences remained between conditions with and without inherent fluctuations, suggesting that the effects of poorer frequency selectivity are not as profound as for CI users. This could be for two reasons: First, frequency selectivity in HI listeners with mild-to-moderate hearing loss is not as poor as that in CI users. Second, the difference between the interaction occurring before and after envelope extraction may affect outcomes to some extent, in that the reduction in masker modulation power may be greater for CI users than for HI listeners, even with a similar loss of frequency selectivity.

One interesting outcome was that the differences between the NH groups and the HI group were not very pronounced when the speech was presented to the groups at the same high SPL. It is well known that frequency selectivity becomes poorer in NH listeners at high levels (e.g., Nelson and Freyman 1984; Glasberg and Moore 2000). Apparently the filter broadening with level in NH listeners is sufficient to reduce the effects of inherent masker fluctuations.

Overall, the results show that the importance of inherent masker fluctuations in determining speech intelligibility in noise depends to some extent on the conditions and the listening population. It cannot be claimed that inherent masker fluctuations always limit speech perception, as the effect of the fluctuations is non-existent in CI users (Oxenham and Kreft 2014), and is greatly reduced in HI listeners and in NH listeners at high sound levels. Finally, the reduced effect of inherent fluctuations may also provide a reason for why HI listeners (as well as CI users) exhibit less masking release in the presence of maskers with imposed additional fluctuations.

Acknowledgments This work was supported by NIH grant R01 DC012262.

References

Dorman MF, Loizou PC, Fitzke J, Tu Z (1998) The recognition of sentences in noise by normal-hearing listeners using simulations of cochlear-implant signal processors with 6–20 channels. J Acoust Soc Am 104(6):3583–3585

Dubbelboer F, Houtgast T (2008) The concept of signal-to-noise ratio in the modulation domain and speech intelligibility. J Acoust Soc Am 124(6):3937–3946. doi:10.1121/1.3001713

Festen JM, Plomp R (1990) Effects of fluctuating noise and interfering speech on the speech-reception threshold for impaired and normal hearing. J Acoust Soc Am 88(4):1725–1736

French NR, Steinberg JC (1947) Factors governing the intelligibility of speech sounds. J Acoust Soc Am 19:90–119

George EL, Festen JM, Houtgast T (2008) The combined effects of reverberation and nonstationary noise on sentence intelligibility. J Acoust Soc Am 124(2):1269–1277. doi:10.1121/1.2945153

Glasberg BR, Moore BCJ (2000) Frequency selectivity as a function of level and frequency measured with uniformly exciting notched noise. J Acoust Soc Am 108:2318–2328

Gregan MJ, Nelson PB, Oxenham AJ (2013) Behavioral measures of cochlear compression and temporal resolution as predictors of speech masking release in hearing-impaired listeners. J Acoust Soc Am 134(4):2895–2912. doi:10.1121/1.4818773

Hu J, Beaulieu NC (2005) Accurate simple closed-form approximations to Rayleigh sum distributions and densities. IEEE Commun Lett 9:109–111. doi:10.1109/LCOMM.2005.02003

Humes LE, Wilson DL, Barlow NN, Garner C (2002) Changes in hearing-aid benefit following 1 or 2 years of hearing-aid use by older adults. J Speech Lang Hear Res 45(4):772–782

Jorgensen S, Dau T (2011) Predicting speech intelligibility based on the signal-to-noise envelope power ratio after modulation-frequency selective processing. J Acoust Soc Am 130(3):1475–1487. doi:10.1121/1.3621502

Jorgensen S, Ewert SD, Dau T (2013) A multi-resolution envelope-power based model for speech intelligibility. J Acoust Soc Am 134(1):436–446. doi:10.1121/1.4807563

Kryter KD (1962) Methods for the calculation and use of the articulation index. J Acoust Soc Am 34:467–477

Lawson JL, Uhlenbeck GE (1950) Threshold signals, vol 24. McGraw Hill, New York

Nelson DA, Freyman RL (1984) Broadened forward-masked tuning curves from intense masking tones: delay-time and probe level manipulations. J Acoust Soc Am 75:1570–1577

Nelson PB, Jin SH (2004). Factors affecting speech understanding in gated interference: Cochlear implant users and normal-hearing listeners. J Acoust Soc Am 115(5 Pt 1):2286–2294

Oxenham AJ, Kreft HA (2014). Speech perception in tones and noise via cochlear implants reveals influence of spectral resolution on temporal processing. Trends Hear 18. doi:10.1177/2331216514553783

Spahr AJ, Dorman MF, Litvak LM, Van Wie S, Gifford RH, Loizou PC, Loiselle LM, Oakes T, Cook S (2012) Development and validation of the AzBio sentence lists. Ear Hear 33(1):112–117. doi:10.1097/AUD.0b013e31822c2549

Stickney GS, Zeng FG, Litovsky R, Assmann P (2004) Cochlear implant speech recognition with speech maskers. J Acoust Soc Am 116(2):1081–1091

Stone MA, Fullgrabe C, Mackinnon RC, Moore BCJ (2011) The importance for speech intelligibility of random fluctuations in "steady" background noise. J Acoust Soc Am 130(5):2874–2881. doi:10.1121/1.3641371

Stone MA, Fullgrabe C, Moore BCJ (2012) Notionally steady background noise acts primarily as a modulation masker of speech. J Acoust Soc Am 132(1):317–326. doi:10.1121/1.4725766

Stone MA, Moore BCJ (2014) On the near non-existence of "pure" energetic masking release for speech. J Acoust Soc Am 135(4):1967–1977

Whitmal NA, Poissant SF, Freyman RL, Helfer KS (2007) Speech intelligibility in cochlear implant simulations: effects of carrier type, interfering noise, and subject experience. J Acoust Soc Am 122:2376–2388

Zeng FG (2004) Trends in cochlear implants. Trends Amplif 8(1):1–34

Effects of Pulse Shape and Polarity on Sensitivity to Cochlear Implant Stimulation: A Chronic Study in Guinea Pigs

Olivier Macherey and Yves Cazals

Abstract Most cochlear implants (CIs) stimulate the auditory nerve with trains of symmetric biphasic pulses consisting of two phases of opposite polarity. Animal and human studies have shown that both polarities can elicit neural responses. In human CI listeners, studies have shown that at suprathreshold levels, the anodic phase is more effective than the cathodic phase. In contrast, animal studies usually show the opposite trend. Although the reason for this discrepancy remains unclear, computational modelling results have proposed that the degeneration of the peripheral processes of the neurons could lead to a higher efficiency of anodic stimulation. We tested this hypothesis in ten guinea pigs who were deafened with an injection of sysomycin and implanted with a single ball electrode inserted in the first turn of the cochlea. Animals were tested at regular intervals between 1 week after deafening and up to 1 year for some of them. Our hypothesis was that if the effect of polarity is determined by the presence or absence of peripheral processes, the difference in polarity efficiency should change over time because of a progressive neural degeneration. Stimuli consisted of charge-balanced symmetric and asymmetric pulses allowing us to observe the response to each polarity individually. For all stimuli, the inferior colliculus evoked potential was measured. Results show that the cathodic phase was more effective than the anodic phase and that this remained so even several months after deafening. This suggests that neural degeneration cannot entirely account for the higher efficiency of anodic stimulation observed in human CI listeners.

Keywords Inferior colliculus · Electrical stimulation · Asymmetric pulses · Auditory nerve

---

The original version of this chapter was revised. An erratum to this chapter can be found at https://doi.org/10.1007/978-3-319-25474-6_51

O. Macherey (✉)
LMA-CNRS, UPR 7051, Aix-Marseille Univ., Centrale Marseille,
4 Impasse Nikola Tesla, 13013 Marseille, France
e-mail: macherey@lma.cnrs-mrs.fr

Y. Cazals
LNIA-CNRS, UMR 7260, Aix-Marseille Univ., 3 place Victor Hugo, 13331 Marseille, France
e-mail: yves.cazals@univ-amu.fr

P. van Dijk et al. (eds.), *Physiology, Psychoacoustics and Cognition in Normal and Impaired Hearing*, Advances in Experimental Medicine and Biology 894,
DOI 10.1007/978-3-319-25474-6_15

1 Introduction

Most contemporary cochlear implants (CIs) stimulate the auditory nerve with trains of symmetric biphasic pulses consisting of two phases of opposite polarity presented in short succession. Although both polarities can elicit neural responses (Miller et al. 1998, 1999; Klop et al. 2004; Undurraga et al. 2013), they are not equally efficient. Animal recordings performed at the level of the auditory nerve usually show longer latencies and lower thresholds for cathodic than for anodic stimulation (e.g., Miller et al. 1999). This observation is consistent with theoretical models of extracellular electrical stimulation of the nerve showing that a cathodic current produces a larger depolarization and elicits action potentials more peripherally than an anodic current of the same level (Rattay 1989).

In past years, studies performed with human CI listeners have shown different results. While at hearing threshold, no consistent difference between polarities has been observed, a large number of studies have reported that anodic stimulation is more effective than cathodic stimulation at suprathreshold levels (Macherey et al. 2006, 2008; Undurraga et al. 2013; Carlyon et al. 2013). This has been shown psychophysically in loudness balancing, pitch ranking and masking experiments and electrophysiologically in measures of electrically-evoked compound action potentials and auditory brainstem responses. Although this result contrasts with the conclusions of most animal reports, an exception is the study by Miller et al. (1998) who found a similar trend in their guinea pigs. Because the results of this specific study are at odds with other measurements performed in the guinea pig (Klop et al. 2004) and with all measurements performed in cats, the first aim of the present study was to clarify the effect of polarity in guinea pigs. This was done by measuring the inferior colliculus evoked potential in response to intracochlear electrical stimulation using various pulse shapes differing in polarity and in current level.

Although the reason for the higher sensitivity of human CI users to anodic stimulation remains unknown, predictions from two different -among the few- computational models of the implanted cochlea offer possible explanations (Macherey et al. 2006). First, the modeling results of Rattay et al. (2001) suggest that polarity sensitivity may strongly depend on the state of degeneration and demyelination of the peripheral processes of the neurons. Specifically, the model predicts lower anodic thresholds for degenerated peripheral processes. Although post-mortem studies have shown that such degeneration can occur in CI listeners, there is no currently available method to evaluate the presence of peripheral processes in patients. The second aim of this study was to investigate polarity sensitivity in guinea pigs as a function of time after deafening. Our reasoning was based on the observation that spiral ganglion neurons degenerate over time in deafened animals and that this degeneration first affects the peripheral processes (Koitchev et al. 1982; Cazals et al. 1983). If this degeneration is responsible for the higher anodic sensitivity, we hypothesized that the effect of polarity would change over time.

An alternative explanation for the higher anodic sensitivity at high current levels comes from a model of the implanted guinea-pig cochlea (Frijns et al. 1996). This model predicts that cathodic stimulation should elicit action potentials at the periphery of the neurons but that above a certain current level, the propagation of these action potentials may be blocked by the hyperpolarization produced more centrally on the fibers. This blocking was not observed in computer simulations using anodic stimuli because action potentials were in this case elicited at a more central locus on the fibers. A third aim of the present study was to investigate if this behavior could be observed at the level of the inferior colliculus of the guinea pigs. If, at some level in the dynamic range, action potentials are being blocked for cathodic stimulation, this should be reflected by a slower rate of increase of the neural response for cathodic than for anodic stimulation and/or potentially by a non-monotonic growth of response amplitude as a function of current level.

Finally, most animal studies on polarity sensitivity have been performed using monophasic pulses which cannot safely be used in humans. This is why studies in CI subjects use charge-balanced asymmetric waveforms which are hypothesized to make one polarity dominate over the other. Some of these waveforms have the advantage of requiring lower current levels than the commonly-used symmetric waveform to elicit auditory sensations in implant listeners (Macherey et al. 2006). The fourth aim of the present study was to evaluate the efficiency of these different pulse shapes and to compare it to human psychophysical data obtained previously.

2 Methods

2.1 Animal Preparation

Ten Albino guinea pigs (G1–G10) of the Dunkin-Hartley strain were used in these studies. The experimental procedures were carried out in accordance with French national legislation (JO 87–848, European Communities Council Directive (2010/63/EU,74) and local ethics committee "Direction Départementale de la Protection des Populations", with permit number 00002.01.

In a first step, using stereotaxic procedures, the animals were chronically implanted with an electrode in the right inferior colliculus (IC). This technique allows easy and stable recordings over months as indicated in several previous experiments (e.g., Popelár et al. 1994). Using a micromanipulator, a Teflon-coated platinum electrode (127 mm in diameter obtained from A-M systems) was lowered into the IC. Then the electrode was fixed to the skull with acrylic cement and soldered to a small connector to which two screws in the frontal bones were also connected and served as reference and ground, respectively. The ensemble was finally fixed to the surface of the skull with acrylic cement.

After a recovery period of about 1 week, the animal was sedated with xylazine (5 mg/kg), a small loudspeaker (MF1 from TDT) was placed at 1 cm from its left ear pinna and the animal's head connector was linked to the input of an amplifier (A-M systems model 1700). Tone pips with 2 ms linear rise/fall times at octave frequencies between 2 to 32 kHz were presented at levels of 90–0 dB SPL in 10-dB steps. In response to sound stimuli, electrophysiological responses were amplified 10,000 times and filtered between 100 and 5 kHz, then they were averaged over 100–300 times (CED 1401 A/D converter).

About a week later, the animal was anesthetized and the left ear bulla was approached from behind the pinna. The round window was gently pierced and a mixture of sisomycin at a concentration of 50 mg/ml was injected into the round window niche until the bulla was completely filled with the mixture. The animal was left so for 2 h and the sisomycin was then aspirated from the bulla. This sisomycin treatment is known from previous experiments (Cazals et al. 1983) to completely destroy cochlear hair cells and lead to progressive spiral ganglion cells degeneration over months. Then a teflon-coated platinum wire (200 mm in diameter) with a tip bare over about 0.5 mm was introduced through the round window over a length of about 2 mm (about the 16–32 kHz tonotopy). Because of deficiencies in handling and care in animals' quarter during the months of the experiment, for some animals, the head connector came off the skull, and the animal was operated again, sometimes replacing the cochlear implant electrode.

2.2 *Stimuli*

During the experiment, electrical stimulation was delivered using a stimulus isolator current generator (AM systems model 2200), the output of which was connected through the head connector to the cochlear implant electrode (active) and to one screw in the frontal bone (reference). This generator delivered a current proportional to an arbitrary input voltage waveform. The voltage waveform was the output of an Advanced Bionics HiRes90k test implant connected to a purely resistive load. This allowed us to use pulse shapes similar to those used in previous human CI studies. The stimuli were designed and controlled using the BEDCS software (Advanced Bionics).

Stimuli were trains of electrical pulses delivered at a rate of 8 Hz. The four pulse shapes illustrated in Fig. 1 were tested, including symmetric biphasic (SYM), reversed pseudomonophasic (RPS), reversed pseudomonophasic with an interphase gap of 6.5 ms (RPS-IPG) and triphasic (TRI) pulses. The top pulse shapes are assumed to favor cathodic stimulation, hence the "C" at the end of their acronyms while the bottom pulse shapes are assumed to favor anodic stimulation, hence the "A". The recordings were synchronized to the onset of the cathodic phase for the top pulse shapes and to the onset of the anodic phase for the bottom pulse shapes.

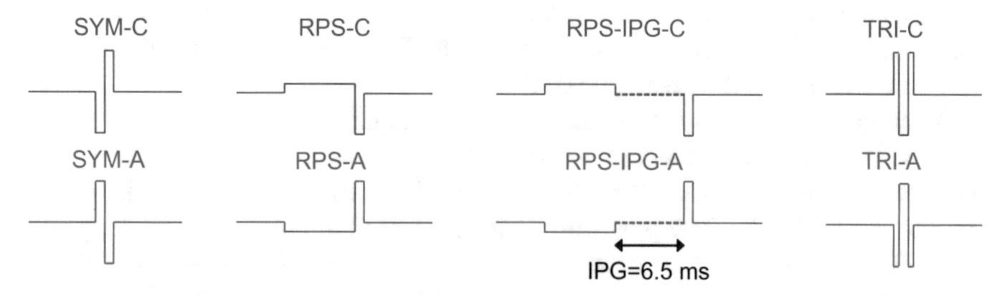

Fig. 1 Schematic representation of the electrical pulse shapes used for stimulation, including symmetric biphasic cathodic-first (SYM-C) and anodic-first (SYM-A), Reversed pseudomonophasic with a short, high-amplitude cathodic (RPS-C) or anodic (RPS-A) phase, Reversed pseudomonophasic with a 6.5-ms interphase gap (RPS-IPG-C and RPS-IPG-A) and triphasic with a cathodic (TRI-C) or anodic (TRI-A) central phase

The duration of the phases of SYM, of the short, high-amplitude phases of RPS and RPS-IPG and of the central phase of TRI was 86 μs. The long, low-amplitude phases of RPS and RPS-IPG were 8 times longer and lower in amplitude to maintain the total charge balanced. The first and third phases of TRI had a duration of 43 μs and the same amplitude as the central phase.

2.3 Sessions

Over months, several sessions of electrical stimulation from the cochlear implant and recording from the right IC were performed. Each session lasted about 1 h during which the animals were sedated. In each session and for each pulse shape, a threshold level was first defined as the level producing a detectable IC evoked potential after averaging; the maximum level of stimulation was defined as the level triggering a detectable motor response of the animal's left cheek. Between these two values, a series of current levels was defined in 1.5-dB steps. For each pulse shape and each level tested, the recordings alternated between polarities. Between 50 and 100 averages were obtained for each combination of pulse shape, polarity and level and the corresponding IC evoked potentials were recorded.

3 Results

3.1 Morphology and Response Amplitude of the IC Evoked Potential

The bottom traces in Figs. 2a and b show examples of IC evoked potentials obtained for subject G2 2 weeks after deafening in response to RPS-IPG-C and RPS-IPG-A, respectively. For comparison, the top traces in each panel show recordings obtained from the same subject in response to 16-kHz tone pips *before deafening*.

The morphology of the response was similar for acoustic and electric stimulation and consisted of a first positive peak (P1) followed by a negative peak (N1) and a second positive peak (P2). The amplitude of the response was arbitrarily defined as the difference in voltage between the negative peak and the second positive peak. The response amplitude varied from about 20 to 600 μV depending on subjects and pulse shapes. The dynamic range, as defined by the difference between the maximum stimulation level and the minimum level that produced a visually detectable response ranged from about 3 to 18 dB, which compare well with usual values of dynamic range for individual CI electrodes.

Figure 2c shows the response amplitude as a function of current level for the same subject and the same pulse shape. It can be seen that at all current levels, the response was larger for the cathodic (RPS-IPG-C) than for the anodic (RPS-IPG-A) pulse. As depicted in the next section, this trend was observed in most of our subjects. Because the RPS-IPG stimulus had a large interphase gap between the two phases, the responses illustrated here presumably represent the responses to

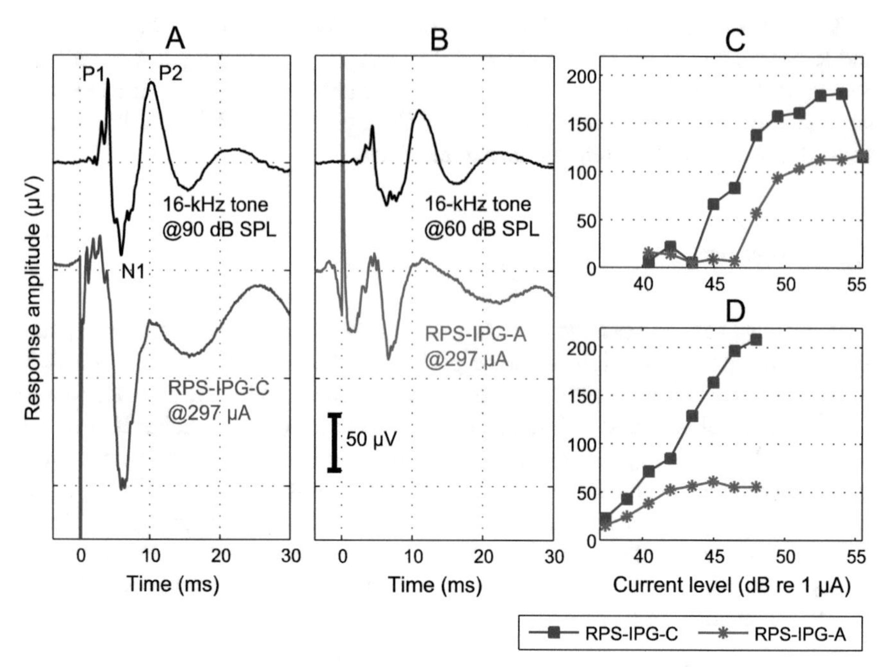

Fig. 2 **a** Inferior colliculus evoked potential obtained 2 weeks after deafening for G2 in response to RPS-IPG-C pulses. The top trace shows the response obtained before deafening for an acoustic tone pip presented at 90 dB SPL. **b** Same as a for RPS-IPG-A and for a 60-dB SPL tone pip. **c** Response amplitude (N1-P2) as a function of current level for G2 2 weeks after deafening. **d** Same as c 7 months after deafening

each individual polarity. Input/output growth functions from the same animal collected 7 months after deafening are shown in Fig. 2d. The effect of polarity was consistent with that obtained just after deafening. Matching the response amplitudes obtained for electrical stimulation to those measured acoustically before deafening, the difference in response amplitude between the two polarities was, for this animal, equivalent to an acoustic level difference of about 30 dB.

Figure 3 shows the amplitude of the response for cathodic stimulation as a function of that for anodic stimulation for the four pulse shapes SYM, RPS, RPS-IPG and TRI and all subjects and levels tested. A very large majority of the data points lie above the diagonal. This illustrates that so-called cathodic pulses usually produced a larger response than anodic pulses for all pulse shapes. For subjects G8-G10 who had several electrodes implanted in the IC, only the recordings from the electrode giving the largest responses are shown. However, the effects of pulse shape and polarity remained consistent across the different recording sites.

3.2 Effect of Polarity as a Function of Time

For a given pulse shape and session, the input/output growth functions obtained for the two polarities were analyzed to determine the current levels needed for each of them to reach an arbitrarily-defined IC response amplitude. This arbitrary amplitude was equal

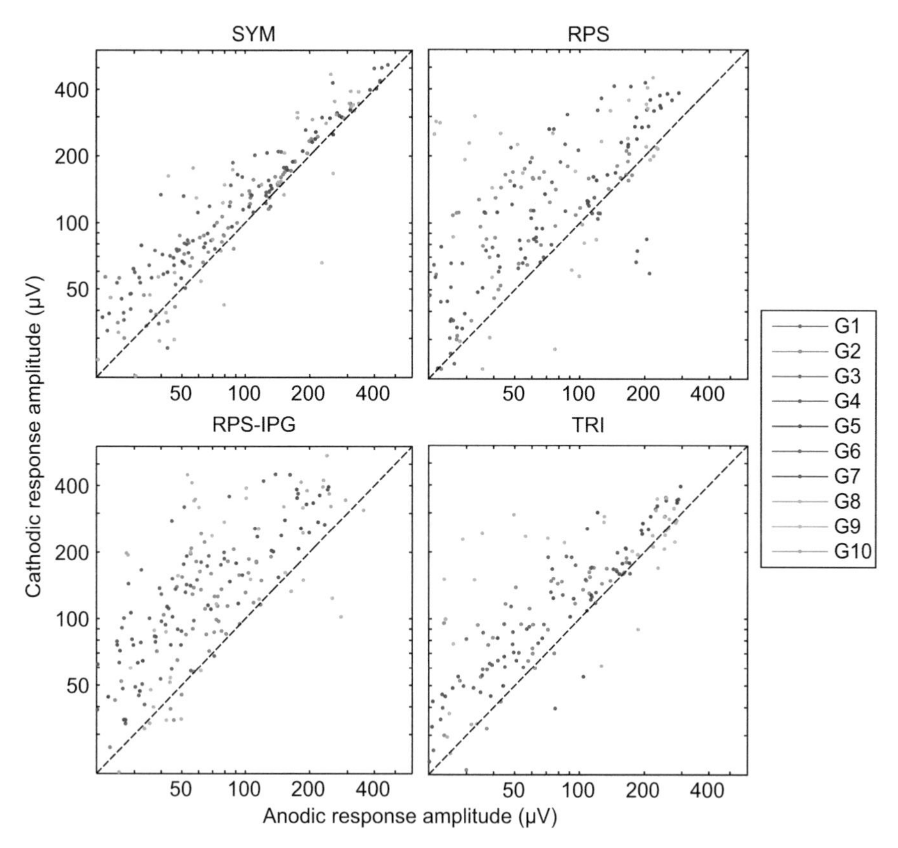

Fig. 3 Summary of cathodic and anodic response amplitudes obtained for all subjects, sessions and levels

to 50% of the maximum response amplitude obtained for this particular pulse shape, subject and session and could, therefore, differ from one session to the next.

Figure 4a shows the difference in current level between RPS-C and RPS-A when both produced the same response amplitude. This quantity referred to as the "polarity effect" is illustrated as a function of time after deafening for the different subjects. Although only the data for RPS are shown, similar results were obtained for RPS-IPG and TRI. If neural degeneration had an effect on polarity sensitivity, we would expect the polarity effect to increase over time and to reach positive values when fibers are degenerated. This does not seem to be the case as the level difference between polarities rather tends to decrease right after deafening for several subjects and then to remain roughly constant.

Given the absence of an effect of duration of deafness, the polarity effects obtained in all sessions were averaged for each animal and each pulse shape. As shown by the star symbols in Fig. 4b, the magnitude of the polarity effect differed across pulse shapes. The mean across subjects was −1.4, −3.1, −5.3 and −2.3 dB for SYM, RPS, RPS-IPG and TRI, respectively. Paired-sample t-tests with Bonferroni correction showed that this effect was larger for RPS than for SYM ($t(8)=3.6$,

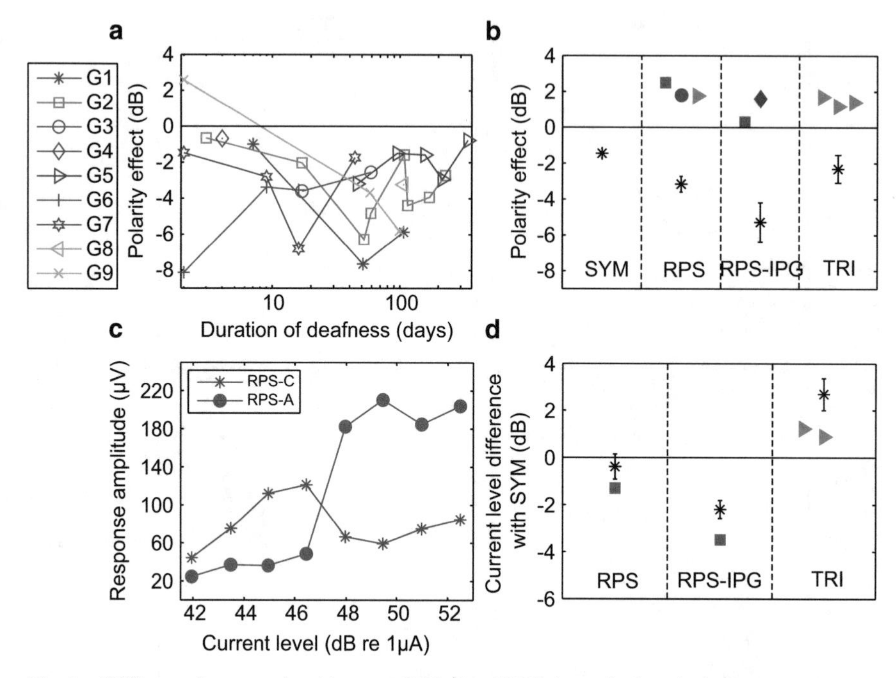

Fig. 4 **a** Difference in current level between RPS-C and RPS-A required to obtain the same response amplitude in the IC. **b** Star symbols: mean polarity effect across guinea-pigs subjects (+/−1 s.e.). Filled symbols: mean polarity effects obtained at most comfortable loudness for similar pulse shapes in previous human studies using electrodes in the middle of the array. *Squares*: Macherey et al. 2006; *circles*: Undurraga et al. 2013; *triangles*: Carlyon et al. 2013; *diamonds*: Macherey et al. 2008. Human data were obtained with non-reversed versions of RPS and RPS-IPG, meaning that the short, high-amplitude phase was presented first. The three data points for comparison with TRI correspond to the 'TP' condition of Carlyon et al. 2013 for Advanced Bionics (1st point) and Med-EL (2nd point) subjects and to their 'QP' condition for Cochlear subjects (3rd point). **c** Example of non-monotonic growth function obtained for RPS in G7. **d** Difference in current level between each cathodic asymmetric pulse shape and SYM-C needed to obtain the same response amplitude. Star symbols: results of the present study averaged across subjects (+/−1 s.e.). *Filled symbols* show corresponding human psychophysical data measured at most comfortable loudness for anodic pulse shapes. Triangles correspond to the TP condition for Advanced Bionics (1st point) and Med-EL (2nd point) subjects

$p<0.05$) and also for RPS-IPG than for SYM ($t(9)=3.7$, $p<0.05$). For comparison, filled symbols show the polarity effects obtained using loudness-balancing experiments in human CI listeners for similar pulse shapes (Macherey et al. 2006, 2008; Undurraga et al. 2013; Carlyon et al. 2013).

Although the polarity effect is opposite in sign for the two species, it is interesting to note that its magnitude is similar for RPS and for TRI. For RPS-IPG, the smaller effect obtained psychophysically may be due to the fact that the delayed long, low-amplitude phase contributed to perception whereas it did not contribute to the neural responses recorded here.

Closer inspection of the data revealed that the input/output growth functions sometimes had non-monotonic shapes. An example obtained for G7 1 month after deafening is illustrated in Fig. 4c for RPS-A and RPS-C. The function for RPS-C shows, at some point in the dynamic range, a decrease in response amplitude with increases in level. In this case, the cathodic pulse was more effective than the anodic pulse at low levels but less effective at high levels. Such non-monotonicities were observed in seven of the ten subjects but only in about 13 % of the cathodic and 4 % of the anodic growth functions. Furthermore, they did not always occur for the same pulse shape across sessions.

3.3 *Effect of Pulse Shape*

The difference in level between SYM-C and the various asymmetric pulse shapes RPS-C, RPS-IPG-C and TRI-C when they all produced the same response amplitude was determined. The response amplitude at which the pulse shapes were compared was equal to 50 % of the maximum amplitude of SYM-C. Given there was no effect of duration of deafness on the magnitude of these effects, the data were averaged across sessions for each animal and analyzed in a one-way repeated measures ANOVA. The star symbols in Fig. 4d show the mean *across* animal subjects. The effect of pulse shape was highly significant ($F(3,18)=22.76$, $p<0.001$). RPS-IPG required 2.2 dB less current than SYM ($p<0.05$) while TRI required 2.7 dB more. There was no significant difference between SYM and RPS. For comparison, the filled symbols show the effect of pulse shape as observed in human CI listeners. For these data, however, the level difference was obtained using loudness balancing between *anodic* pulse shapes. Here again, the magnitude of the effect of pulse shape is similar in both species.

4 Discussion

Most guinea-pigs tested in the present study showed higher sensitivity to cathodic stimulation. This effect of polarity was consistent across all levels tested, ranging from threshold to maximum level. This observation corroborates the results obtained in most previous animal experiments performed in cats and guinea pigs.

Duration of deafness did not influence the effect of polarity, suggesting that neural degeneration may not *entirely* account for the higher sensitivity to anodic stimulation observed in humans. However, the morphology of the nerve fibers and the cochlear geometry are different in the two species and may play an additional role on neural excitation (Rattay et al. 2001).

In few cases, input/output growth functions were non-monotonic, showing a decrease in response amplitude with increases in level. Similar non-monotonic functions have already been observed at the level of the auditory nerve (Miller et al. 1998). They may reflect the blocking of action potentials along the nerve fibers (Frijns et al. 1996) and, therefore, remain a possible candidate explanation for the higher efficiency of anodic stimulation at suprathreshold levels in humans.

Finally, the difference in efficiency between the various pulse shapes was consistent with results obtained in previous animal and human experiments (Miller et al. 2001; Carlyon et al. 2013), showing that, to elicit responses of equal amplitude, pseudomonophasic pulses with a gap between the two phases require a lower current level than symmetric pulses while triphasic pulses require a *higher* current level.

Acknowledgments The authors acknowledge funding from the Agence Nationale de la Recherche (grant numbers ANR-11-PDOC-0022 and ANR-TECSAN-040-STESU).

References

Carlyon RP, Deeks JM, Macherey O (2013) Polarity effects on place pitch and loudness for three cochlear-implant designs and at different cochlear sites. J Acoust Soc Am 134(1):503–509

Cazals Y, Aran JM, Charlet de Sauvage R (1983) Artificial activation and degeneration of the cochlear nerve in guinea pigs. Arch Otorhinolaryngol 238(1):1–8

Frijns JH, de Snoo SL, ten Kate JH (1996) Spatial selectivity in a rotationally symmetric model of the electrically stimulated cochlea. Hear Res 95(1–2):33–48

Klop WM, Hartlooper A, Briaire JJ, Frijns JH (2004) A new method for dealing with the stimulus artefact in electrically evoked compound action potential measurements. Acta Otolaryngol 124(2):137–143

Koitchev K, Guilhaume A, Cazals Y, Aran JM (1982) Spiral ganglion changes after massive aminoglycoside treatment in the guinea pig. Counts and ultrastructure. Acta Otolaryngol 94(5–6):431–438

Macherey O, Carlyon RP, van Wieringen A, Deeks JM, Wouters J (2008) Higher sensitivity of human auditory nerve fibers to positive electrical currents. J Assoc Res Otolaryngol 9(2):241–251

Macherey O, van Wieringen A, Carlyon RP, Deeks JM, Wouters J (2006) Asymmetric pulses in cochlear implants: effects of pulse shape, polarity, and rate. J Assoc Res Otolaryngol 7(3):254–266

Miller CA, Abbas PJ, Robinson BK, Rubinstein JT, Matsuoka AJ (1999) Electrically evoked single-fiber action potentials from cat: responses to monopolar, monophasic stimulation. Hear Res 130(1–2):197–218

Miller CA, Abbas PJ, Rubinstein JT, Robinson BK, Matsuoka AJ, Woodworth G (1998) Electrically evoked compound action potentials of guinea pig and cat: responses to monopolar, monophasic stimulation. Hear Res 119(1–2):142–154

Miller CA1, Robinson BK, Rubinstein JT, Abbas PJ, Runge-Samuelson CL (2001) Auditory nerve responses to monophasic and biphasic electric stimuli. Hear Res 151:79–94

Popelár J, Erre JP, Aran JM, Cazals Y (1994) Plastic changes in ipsi-contralateral differences of auditory cortex and inferior colliculus evoked potentials after injury to one ear in the adult guinea pig. Hear Res 72(1–2):125–134

Rattay F (1989) Analysis of models for extracellular fiber stimulation. IEEE Trans Biomed Eng 36(7):676–682

Rattay F, Lutter P, Felix H (2001) A model of the electrically excited human cochlear neuron. I. Contribution of neural substructures to the generation and propagation of spikes. Hear Res 153(1–2):43–63

Undurraga JA, Carlyon RP, Wouters J, van Wieringen A (2013) The polarity sensitivity of the electrically stimulated human auditory nerve measured at the level of the brainstem. J Assoc Res Otolaryngol 14(3):359–377

Assessing the Firing Properties of the Electrically Stimulated Auditory Nerve Using a Convolution Model

Stefan B. Strahl, Dyan Ramekers, Marjolijn M. B. Nagelkerke, Konrad E. Schwarz, Philipp Spitzer, Sjaak F. L. Klis, Wilko Grolman and Huib Versnel

Abstract The electrically evoked compound action potential (eCAP) is a routinely performed measure of the auditory nerve in cochlear implant users. Using a convolution model of the eCAP, additional information about the neural firing properties can be obtained, which may provide relevant information about the health of the auditory nerve. In this study, guinea pigs with various degrees of nerve degeneration were used to directly relate firing properties to nerve histology. The same convolution model was applied on human eCAPs to examine similarities and ultimately to examine its clinical applicability. For most eCAPs, the estimated nerve firing probability was bimodal and could be parameterised by two Gaussian distributions with an average latency difference of 0.4 ms. The ratio of the scaling factors of the late and early component increased with neural degeneration in the guinea pig. This ratio decreased with stimulation intensity in humans. The latency of the early component decreased with neural degeneration in the guinea pig. Indirectly, this was observed in humans as well, assuming that the cochlear base exhibits more neural degeneration than the apex. Differences between guinea pigs and humans were observed, among other parameters, in the width of the early component: very robust in guinea pig, and

S. B. Strahl (✉) · K. E. Schwarz · P. Spitzer
R&D MED-EL GmbH, Innsbruck, Austria
e-mail: stefan.strahl@medel.com

K. E. Schwarz
e-mail: konrad.schwarz@medel.com

P. Spitzer
e-mail: philipp.spitzer@medel.com

M. M. B. Nagelkerke
Department of Otorhinolaryngology and Head & Neck Surgery,
University Medical Center Utrecht, Utrecht, The Netherlands
e-mail: m.m.b.nagelkerke@gmail.com

D. Ramekers · S. F. L. Klis · W. Grolman · H. Versnel
Department of Otorhinolaryngology and Head & Neck Surgery, Brain Center Rudolf Magnus,
University Medical Center Utrecht, Utrecht, The Netherlands
e-mail: d.ramekers@umcutrecht.nl

P. van Dijk et al. (eds.), *Physiology, Psychoacoustics and Cognition in Normal and Impaired Hearing,* Advances in Experimental Medicine and Biology 894,
DOI 10.1007/978-3-319-25474-6_16

dependent on stimulation intensity and cochlear region in humans. We conclude that the deconvolution of the eCAP is a valuable addition to existing analyses, in particular as it reveals two separate firing components in the auditory nerve.

Keywords Auditory nerve · eCAP · Cochlear implant · Deconvolution · Firing probability · Neural health

1 Introduction

Most cochlear implant (CI) systems allow for the recording of the auditory nerve's response to an electric stimulus—the electrically evoked compound action potential (eCAP). This objective measure is an important clinical tool to assess the quality of the electrode-nerve interface of a CI recipient (Miller et al. 2008). Routinely, the lowest stimulation level that evokes an eCAP ('threshold') is determined. The morphology of suprathreshold eCAP waveforms is usually not evaluated.

1.1 Mathematical Model of the Compound Action Potential

The recorded compound action potential (*CAP*) is described as the convolution of the unit response (*UR*, i.e., the waveform resulting from a single action potential) with the compound discharge latency distribution (*CDLD*), the sum of spike events over time of all individual nerve fibres (Goldstein and Kiang 1958, see also Fig. 1).

$$\mathrm{CAP}(t) = \int_{-\infty}^{t} \mathrm{CDLD}(\tau)\mathrm{UR}(t-\tau)d\tau \quad (1)$$

This mathematical model was validated with simultaneous recordings of acoustically evoked CAPs from the round window and single-fibre responses from the auditory nerve in guinea pigs (Versnel et al. 1992b).

Most mathematical models of the CAP are concerned with solving the forward problem, i.e. predicting the CAP by modelling the activation of single nerve fibres and assuming a convolution with the unit response (Teas et al. 1962; de Boer 1975; Elberling 1976b; Kiang et al. 1976; Versnel et al. 1992b; Frijns et al. 1996; Miller et al. 1999; Briaire and Frijns 2005).

S F.L. Klis
e-mail: s.klis@umcutrecht.nl

W. Grolman
e-mail: w.grolman@umcutrecht.nl

H. Versnel
e-mail: h.versnel@umcutrecht.nl

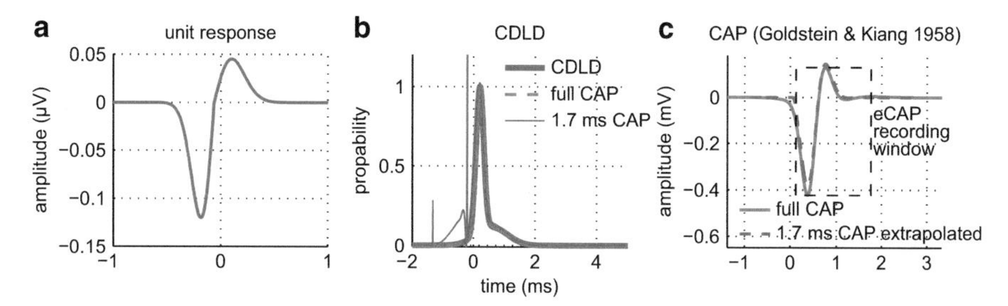

Fig. 1 The assumed elementary unit response of a single nerve fibre is shown in **a**. Its convolution with a CDLD (**b**, *thick gray line*) results in the CAP waveform shown in **c** (*red line*). The monotonic extrapolation of the time-limited eCAP recording is shown in **c** (*blue dashed line*). The derived CDLDs obtained by Eq. 4 are shown in **b** for the full eCAP (*red dashed line*) and a simulation of the time-limited and extrapolated eCAP (*thin blue line*)

A few studies addressed the inverse problem, i.e. predicting the firing probability by deconvolution of acoustically evoked CAPs with a unit response (Elberling 1976a; Charlet de Sauvage et al. 1987). To our knowledge no study investigated the inverse problem for eCAPs. In this study, we perform a deconvolution of eCAPs to examine CDLDs, presuming that they could provide information about the neural status of the auditory periphery. It has been shown that the eCAP in a guinea pig exhibits differences in its morphology depending on the neural health of the auditory nerve (Ramekers et al. 2014, 2015). Therefore, eCAPs from deafened guinea pigs are analysed to assess effects of the status of the auditory nerve on its firing probability. eCAPs from human CI users are evaluated for comparison to the effects observed in the animal model.

2 Methods

2.1 eCAP Recording System

eCAP measurements were performed with MED-EL PULSAR or SONATA CIs (for details see Neustetter et al. 2012). A charge-balanced biphasic pulse with an interphase gap of 2.1 µs and a phase duration of 30 µs was presented at an intra-cochlear stimulation electrode in a monopolar configuration. The stimulation amplitude was defined in current units (cu), where 1 cu corresponds to approximately 1 µA. The eCAP was recorded on a neighbouring electrode with a minimum delay of 125 µs from stimulus onset to reduce stimulation artefacts. The recording window was 1.7 ms. Single eCAP recordings were separated by at least 20 ms, assuming a relative refractory time below 10 ms (Brill et al. 2009; Ramekers et al. 2015). Any stimulation or recording artefact was removed using an alternating polarity and zero amplitude template paradigm, respectively (for details see Brill et al. 2009).

2.2 eCAP Data Sets

2.2.1 Guinea Pig

Data were acquired from 18 guinea pigs (Ramekers et al. 2014, 2015), which were divided into 3 groups of 6 animals: normal-hearing (NH), 2 weeks deaf (2WD) and 6 weeks deaf (6WD). Methods of deafening and eCAP recordings are described in detail in Ramekers et al. (2014).

Briefly, profound cochlear hair cell loss was induced by co-treatment of kanamycin and furosemide. It resulted in mild degeneration of the auditory nerve after 2 weeks (25 % loss of spiral ganglion cells, SGCs) and severe degeneration after 6 weeks (60 % loss of SGCs).

The eCAP recordings were performed in isoflurane-anaesthetized animals. eCAPs were evoked and recorded with two electrodes placed in the basal turn through a cochleostomy. In this study, three different stimulation intensities were analysed: just above threshold (first visible eCAP), halfway the input/output function ('intermediate'), and at maximum stimulation intensity typically corresponding to saturation (800 cu). Waveforms were averaged over 900 iterations.

2.2.2 Human

From a multicentre study (Senn et al. 2012) eCAPs recorded post-operatively with the clinical system software MAESTRO (MED-EL GmbH, Innsbruck) in 52 awake human subjects were selected. The selection criteria were the availability of a pre-operative audiogram and eCAPs of at least 100 μV amplitude. The mean age at onset of hearing loss was 19 years (range: 0–72). The mean age at implantation was 46 years (range: 15–79).

The recordings were manually analysed by three experts. Similar to the animal model, three different stimulation levels were selected: threshold, intermediate, and maximum. In contrast to the animal model, maximum typically corresponded to the loudest acceptable presentation level. Intermediate corresponded to 50 % eCAP amplitude compared to maximum.

The eCAPs were obtained from stimulation electrodes in the apical (contact 2), middle (contact 4 or 5) and basal cochlear region (contact 9 or 10). Waveforms were averaged over 25 iterations and 5 kHz low-pass filtered (fifth-order Butterworth) to remove any remaining recording noise.

2.3 Deconvolution of the eCAP

Following Eq. (1) and Versnel et al. (1992a) the unit response *UR* is assumed constant and was modelled by Eqs. 2 and 3 with $U_N = 0.12\ \mu V$, $\sigma_N = 0.12$ ms describing the negative and U_P=0.045 μV, σ_P=0.16 ms describing the positive part. The cross point is defined with t_0=−0.06 ms (Fig. 1a).

$$UR(t) = \frac{U_N}{\sigma_N}(t - t_0)e^{\frac{1}{2}-\frac{(t-t_0)^2}{2\sigma_N{}^2}}, t < t_0 \tag{2}$$

$$UR(t) = \frac{U_P}{\sigma_P}(t - t_0)e^{\frac{1}{2}-\frac{(t-t_0)^2}{2\sigma_P{}^2}}, t \geq t_0 \tag{3}$$

Having assumed the same unit response for all contributing fibres, the CDLD can be obtained directly from the recorded eCAP by deconvolution. The deconvolution was performed in the frequency domain with

$$CDLD(t) = \mathcal{F}^{-1}\left(\frac{\mathcal{F}(eCAP(t))}{\mathcal{F}(UR(t))}\right) \tag{4}$$

where $\mathcal{F}$ represents the Fourier transform and $\mathcal{F}^{-1}$ the inverse Fourier transform.

The baseline was estimated over the final 200 μs and, assuming that the eCAP decays monotonically to baseline before and after the recording window, the eCAPs were linearly extrapolated to baseline (Fig. 1c) before performing the deconvolution. The extrapolation only affected the CDLD before stimulus onset (see Fig. 1b). The deconvolution of the eCAP can be understood as applying an inverse low-pass filter, amplifying unwanted high-frequency noise. Therefore, the extrapolated eCAPs were low-pass filtered by a 50-point moving-average filter applied twice (once in forward and once in reverse direction) to achieve a zero-phase digital filtering. Any remaining high-frequency noise in the CDLD was removed by a 2.5 kHz fifth-order Butterworth low-pass filter.

2.4 *Parametrisation of the CDLD*

The CDLDs derived from eCAP responses from both data sets exhibited a skewed, quasi-bimodal, distribution (see Fig. 2). To perform further analysis, a parametrisation of the CDLD was performed by a two-component Gaussian mixture model (GMM).

$$CDLD = a_1\mathcal{N}(\mu_1, \sigma_1) + \mathrm{a}_2\mathcal{N}(\mu_2, \sigma_2) \tag{5}$$

The GMM was fitted to the CDLDs using a nonlinear least-squares regression (nlinfit, MATLAB; Mathworks, Natick, MA, USA) with initial start values being manually optimised if needed to achieve an adequate fit (R^2>0.95). As outcome parameters we considered μ_1 and μ_2 (corresponding to the peak latencies), σ_1 and σ_2 (reflecting peak width), and the ratio of the components a_2/a_1.

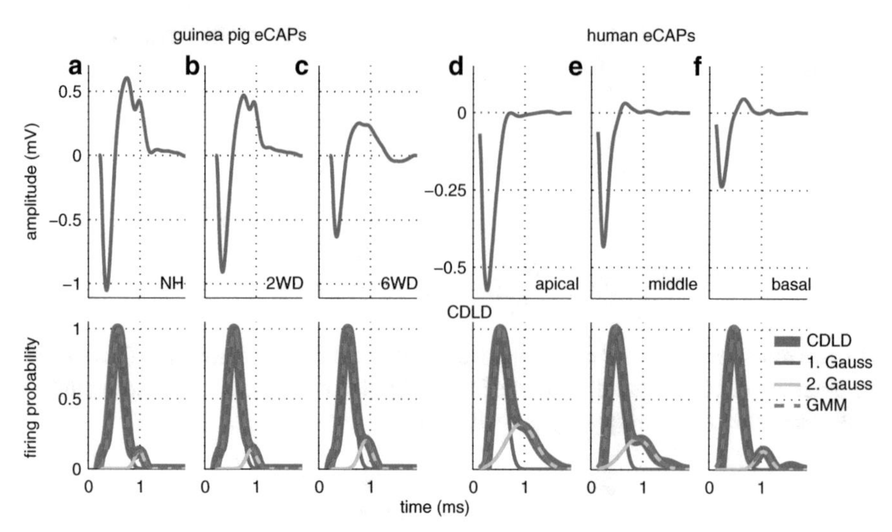

Fig. 2 Example eCAPs for maximum stimulation level (*upper row*) and the corresponding CDLDs and GMM fits (*bottom row*). The *left* three panels show examples of each of the three guinea pig groups; the *right* three panels are from each of the three cochlear regions in one human subject

3 Results

3.1 Guinea Pig Deafness Model eCAPs

Figure 2 shows eCAP and CDLD examples for NH (a), 2WD (b) and 6WD animals (c). In each animal the deconvolution of the recorded eCAPs resulted in CDLDs with a fast rising peak, typically narrow (<0.5 ms width), in most cases followed by a shoulder-shaped component or a second peak. In those cases the CDLD could be well fitted with the two-component GMM (Eq. 5), remaining single-peak cases (8 % of all CDLDs) could be well fitted with a single Gaussian. Figure 3 and Table 1 show group averages of CDLD parameters. Two variables significantly varied with group (i.e., with degeneration of the auditory nerve): μ_1 and a_2/a_1 ($p<0.05$; rmANOVA with stimulation intensity as within factor, and group as between factor). With more nerve degeneration (6WD vs. 2WD and NH) the peak latency μ_1 was shorter and a_2/a_1 was larger.

Table 1 CDLD parameters for guinea pig groups (upper section), and humans grouped by cochlear region (lower section), averaged over stimulation level

		μ_1 (ms)	μ_2–μ_1 (ms)	σ_1 (ms)	σ_2 (ms)	a_2/a_1
Guinea pig	*NH*	0.58	0.38	0.14	0.10	0.090
	2WD	0.60	0.36	0.16	0.10	0.068
	6WD	0.52	0.37	0.14	0.13	0.33
Human	*apex*	0.54	0.42	0.13	0.27	0.96
	middle	0.50	0.42	0.14	0.27	0.86
	base	0.51	0.46	0.14	0.26	0.85

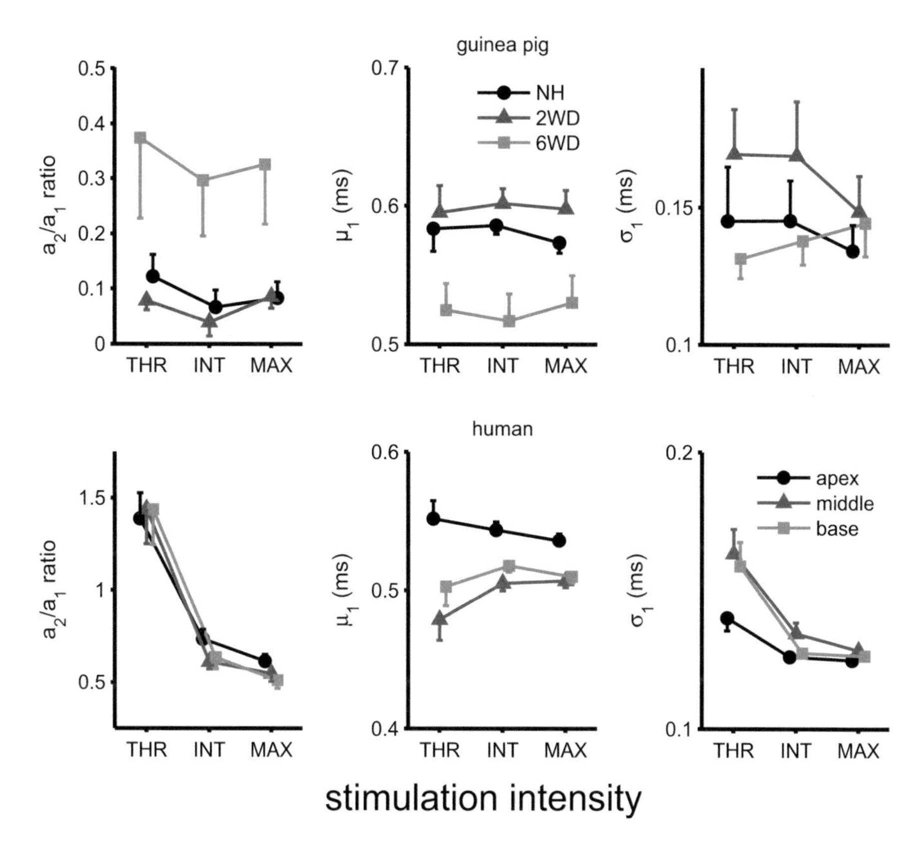

Fig. 3 Average CDLD parameters a_2/a_1, μ_1 and σ_1 shown for guinea pigs (*left*) and humans (*right*). *THR* threshold, *INT* intermediate, *MAX* maximum. Error bars are SEM

3.2 Human Clinical eCAPs

Figure 2 shows examples of eCAPs and corresponding CDLDs for an apical (d), middle (e) and basal (f) stimulation site in one human subject. Human CDLDs showed a morphology comparable to the guinea pig data, with 15 % of CDLDs being single-peaked. Figure 3 and Table 1 show averages of CDLD parameters for different cochlear regions. Assuming that any residual hearing correlates with the auditory nerve's neural health, subjects were divided into two groups having pre-operative 0.5–4 kHz pure tone averages ≤ 105 dB hearing loss (HL) (24 subjects) or > 105 dB HL (28 subjects). No significant between-subject group effects were found ($p = 0.054$; rmANOVA with cochlear region and stimulation intensity as within factors, and group as between factor). A significant decrease of a_2/a_1 with increasing stimulation intensity ($p < 0.001$) and distance from the apex (apex vs. middle and base, $p < 0.05$) was found. The latency μ_1 was significantly longer in the apical cochlear region compared to middle and base ($p<0.001$). The width of the first Gaussian σ_1 was significantly smaller in the apex vs. middle and base ($p<0.05$) and with increasing stimulation intensity ($p<0.001$).

4 Discussion

4.1 Comparing Deconvolution Results from Guinea Pigs and Humans

We observed several similarities between the obtained CDLDs of guinea pigs and humans. First, the CDLDs could be described in the large majority of cases by two Gaussian distributions (Fig. 2). Second, the peak latency of the early component decreased with increasing neural degeneration, assuming for the human data more degeneration at the cochlear base than near the apex (Fig. 3, μ_1). Notable differences were observed in particular with respect to the width of the early component, which was very robust in guinea pigs, and quite variable with stimulation intensity and cochlear region in humans.

4.2 Choice of UR Waveform

The main challenge of the eCAP deconvolution is the dependence of the CDLDs on the choice of UR. For instance, with a larger U_P a_2/a_1 increases since the P_2 component of eCAP can be reduced by increasing the N_2. We have chosen our parameters based on experimental estimates of the UR in guinea pigs (Prijs 1986; Versnel et al. 1992a). Therefore, we can be fairly confident about our choice for the eCAP deconvolutions in the guinea pig. The UR in humans is thought to be similar (Briaire and Frijns 2005; Whiten 2007). However, the cochlear morphology is quite different in humans. The UR may change with stimulation site in the human cochlea and explain in part the cochlear region dependent variation of the CDLD seen in the human dataset. Also our assumption of the same UR across the nerve population might be too coarse (Westen et al. 2011), in particular considering location-dependent effects of degeneration, such as demyelination and cell shrinkage. To further address this point, a sensitivity analysis of the convolution model is needed.

4.3 Parametrisation of the CDLD

Single-fibre responses to electric pulses (estimated by post-stimulus time histograms) showing a fast rise and a slow decline (van den Honert and Stypulkowski 1984; Sly et al. 2007) may be modelled best with a gamma function (Versnel et al. 1992b). However, the distribution of response latencies between fibres may be Gaussian. Therefore, we tested fits of CDLDs with gamma functions and with Gaussian functions, and both in guinea pigs and humans we observed superior fits with Gaussian functions. Due to the large population size and assuming that the discharge jitter is smaller within than between fibres, the resulting convolution of the gamma function with the Gaussian distribution could explain the Gaussian-like CDLD.

4.4 Interpretation of CDLD Parameters

In most cases a_2/a_1 is significant, in both guinea pigs and humans. This suggests either two subpopulations of fibres and/or neural elements (peripheral vs. central site of excitation) with different latencies and proportions, or repeated firings of the same neurons. The latter option is supported by the following arguments (see also in Ramekers et al. 2015). First, the interval between the two components, μ_2–μ_1, is about 0.5 ms, which is around the absolute refractory period estimated from masker-probe responses. Second, the N_2–N_1 interval of the eCAP waveform has been found to correlate with recovery measures (Ramekers et al. 2015).

Alternatively, a reduced relative contribution of the second component with increasing stimulation intensity and more basal position within the cochlea as observed in humans could be explained by the former option. The late component can be ascribed to excitation peripheral to the cell body (Stypulkowski and van den Honert 1984) which is thought to occur to a larger extent near threshold (Briaire and Frijns 2005).

5 Conclusion

Obviously, there are multiple differences in experimental settings and biology between the guinea pig and human, which may contribute to different CDLD outcomes. Among others, cause of deafness, cochlear anatomy, duration of deafness, and rate of neural degeneration differ, and there is awake vs. anaesthetized state, chronic vs. acute cochlear implantation, and long vs. short electrode array. Taking these differences into account, a few characteristics appeared to be quite robust across species and experimental setting. Importantly, the deconvolution of the eCAP revealed two separate firing components, which could not easily be detected from the eCAP waveforms themselves. The ratio of the components and the latency of the early component came out as potential markers to assess the condition of the auditory nerve.

References

Briaire JJ, Frijns JHM (2005) Unraveling the electrically evoked compound action potential. Hear Res 205(1–2):143–156

Brill S, Müller J, Hagen R, Möltner A, Brockmeier S-J, Stark T et al (2009). Site of cochlear stimulation and its effect on electrically evoked compound action potentials using the MED-EL standard electrode array. BioMed Eng OnLine 8(1):40

Charlet de Sauvage R, Aran JM, Erre JP (1987) Mathematical analysis of VIIIth nerve cap with a linearly-fitted experimental unit response. Hear Res 29(2–3):105–115

De Boer E (1975) Synthetic whole-nerve action potentials for the cat. J Acoust Soc Am 58(5):1030–1045

Elberling C (1976a) Deconvolution of action potentials recorded from the ear canal in man. In: Stevens SDG (ed) Disorders of auditory function II. Academic, London

Elberling C (1976b) Simulation of cochlear action potentials recorded from the ear canal in man. In: Rubin R, Elberling C, Salomon G (eds) Electrocochleography. University Park Press, Baltimore, pp 151–168

Frijns JHM, Schoonhoven R, Grote JJ (1996) The influence of stimulus intensity on spike timing and the compound action potential in the electrically stimulated cochlea: a model study. Proceedings of the 18th Annual International Conference of the IEEE engineering in medicine and biology society, 1996. Bridging Disciplines for Biomedicine, vol 1, pp 327–328

Goldstein JMH, Kiang NY-S (1958) Synchrony of neural activity in electric responses evoked by transient acoustic stimuli. J Acoust Soc Am 30(2):107–114

Kiang NYS, Moxon EC, Kahn AR (1976) The relationship of gross potentials recorded from the cochlea to single unit activity in the auditory nerve. In: Ruben RJ, Elberling C, Salomon G (eds) Electrocochleography. University Park Press, Baltimore, pp 95–115

Miller CA, Abbas PJ, Rubinstein JT (1999) An empirically based model of the electrically evoked compound action potential. Hear Res 135(1–2):1–18

Miller CA, Brown CJ, Abbas PJ, Chi S-L (2008) The clinical application of potentials evoked from the peripheral auditory system. Hear Res 242(1–2):184–197

Neustetter C, Zangerl M, Spitzer P, Zierhofer C (2012) In-vitro characterization of a cochlear implant system for recording of evoked compound action potentials. BioMed Eng OnLine 11(1):22

Prijs VF (1986) Single-unit response at the round window of the guinea pig. Hear Res 21(2):127–133

Ramekers D, Versnel H, Strahl SB, Klis SFL, Grolman W (2015) Recovery characteristics of the electrically stimulated auditory nerve in deafened guinea pigs: relation to neuronal status. Hear Res 321:12–24

Ramekers D, Versnel H, Strahl SB, Smeets EM, Klis SFL, Grolman W (2014) Auditory-nerve responses to varied inter-phase gap and phase duration of the electric pulse stimulus as predictors for neuronal degeneration. J Assoc Res Otolaryngol 15(2):187–202

Senn P, van de Heyning PL, Arauz S, Atlas M, Baumgartner W-D, Caversaccio M et al (2012) Electrically evoked compound action potentials in patients supplied with CI. Presented at the

12th International Conference on cochlear implants and other implantable auditory technologies, Baltimore
Sly DJ, Heffer LF, White MW, Shepherd RK, Birch MGJ, Minter RL et al (2007) Deafness alters auditory nerve fibre responses to cochlear implant stimulation. Eur J Neurosci 26(2):510–522
Stypulkowski PH, van den Honert C (1984) Physiological properties of the electrically stimulated auditory nerve. I. Compound action potential recordings. Hear Res 14(3):205–223
Teas DC, Eldridge DH, Davis H (1962) Cochlear responses to acoustic transients: an interpretation of whole-nerve action potentials. J Acoust Soc Am 34(9, Pt. II):1438–1459
van den Honert C, Stypulkowski PH (1984) Physiological properties of the electrically stimulated auditory nerve. II. Single fiber recordings. Hear Res 14(3):225–243
Versnel H, Prijs VF, Schoonhoven R (1992a) Round-window recorded potential of single-fibre discharge (unit response) in normal and noise-damaged cochleas. Hear Res 59(2):157–170
Versnel H, Schoonhoven R, Prijs VF (1992b) Single-fibre and whole-nerve responses to clicks as a function of sound intensity in the guinea pig. Hear Res 59(2):138–156
Westen AA, Dekker DMT, Briaire JJ, Frijns JHM (2011) Stimulus level effects on neural excitation and eCAP amplitude. Hear Res 280(1–2):166–176
Whiten DM (2007) Electro-anatomical models of the cochlear implant (Thesis). Massachusetts Institute of Technology. http://dspace.mit.edu/handle/1721.1/38518. Accessed 1 April 2015

Modeling the Individual Variability of Loudness Perception with a Multi-Category Psychometric Function

Andrea C. Trevino, Walt Jesteadt and Stephen T. Neely

Abstract Loudness is a suprathreshold percept that provides insight into the status of the entire auditory pathway. Individuals with matched thresholds can show individual variability in their loudness perception that is currently not well understood. As a means to analyze and model listener variability, we introduce the *multi-category psychometric function* (MCPF), a novel representation for categorical data that fully describes the probabilistic relationship between stimulus level and categorical-loudness perception. We present results based on categorical loudness scaling (CLS) data for adults with normal-hearing (NH) and hearing loss (HL). We show how the MCPF can be used to improve CLS estimates, by combining listener models with maximum-likelihood (ML) estimation. We also describe how the MCPF could be used in an entropy-based stimulus-selection technique. These techniques utilize the probabilistic nature of categorical perception, a novel usage of this dimension of loudness information, to improve the quality of loudness measurements.

Keywords Loudness · Categorical · Psychoacoustics · Normal hearing · Hearing loss · Suprathreshold · Perception · Maximum likelihood · Modeling · Probability

1 Introduction

Loudness is a manifestation of nonlinear suprathreshold perception that is altered when cochlear damage exists (Allen 2008). There are a number of techniques for measuring loudness perception (Florentine 2011); in this work we will focus on categorical loudness scaling (CLS). CLS is a task that has a well-studied relationship with hearing loss, uses category labels that are ecologically valid (e.g., "Loud", "Soft"), can be administered in a clinic relatively quickly (<5 min/frequency), and requires little training on the part of the tester/listener. For these reasons, this task

A. C. Trevino (✉) · W. Jesteadt · S. T. Neely
Boys Town National Research Hospital, Omaha, USA
e-mail: Andrea.Trevino@boystown.org

P. van Dijk et al. (eds.), *Physiology, Psychoacoustics and Cognition in Normal and Impaired Hearing,* Advances in Experimental Medicine and Biology 894,
DOI 10.1007/978-3-319-25474-6_17

has been used in a number of loudness studies (Allen et al. 1990; Al-Salim et al. 2010; Brand and Hohmann 2002; Elberling, 1999; Heeren et al. 2013).

A listener's CLS function, generally computed from the median stimulus level for each response category, is the measure typically used in analysis. This standard approach treats the variability in the response data as something to be removed. We propose to instead model the probabilistic nature of listener responses; these probabilistic representations can be used to improve loudness measurements and further our understanding of the mechanisms underlying this perception. At each stimulus level, there is a probability distribution across categorical loudness responses; when plotted as a function of stimulus level, these distributions form a multi-category psychometric function (MCPF), modeling the statistics of the listeners' categorical perception. This model can be applied to listener simulations or when incorporating probabilistic analysis techniques such as estimation, information, or detection theory into analysis and measurements.

We describe one such application (i.e., ML estimation) to improve the accuracy of the CLS function. The ISO recommendations (Kinkel 2007) for adaptive CLS testing are designed to constrain the test time to a duration that is clinically acceptable. Due to the relatively low number of experimental trials, the natural variability of responses can create an inaccurate CLS function. Our proposed ML estimation approach is inspired by the work of Green (1993), who developed a technique for using estimation theory to determine the psychometric function of a "yes-no" task. We modify and extend this concept to estimate the MCPF that best describes an individual listener's categorical loudness perception.

In this paper, we (1) introduce the concept of the MCPF representation, (2) describe a parameterization method for the representation, (3) use principal component analysis (PCA) to create a representative catalog, (4) show MCPFs for NH and HL listeners, and (5) demonstrate how the catalog can be paired with a ML procedure to estimate an individual's MCPF.

2 Methods

2.1 Multi-Category Psychometric Function

A psychometric function describes the probability of a particular response as a function of an experimental variable. We introduce the concept of a multi-category psychometric function, which represents the probability distribution across multiple response categories as a function of an experimental variable.

In a MCPF, a family of curves demarcates the probabilities of multiple categories as a function of stimulus level. As an example, we demonstrate the construction of a hypothetical CLS MCPF for a NH listener, for an 11-category CLS scale, in Fig. 1. Figure 1a shows the listener's probability distribution across categories at 60 dB SPL. One can see that the loudness judgments are not constrained to one cat-

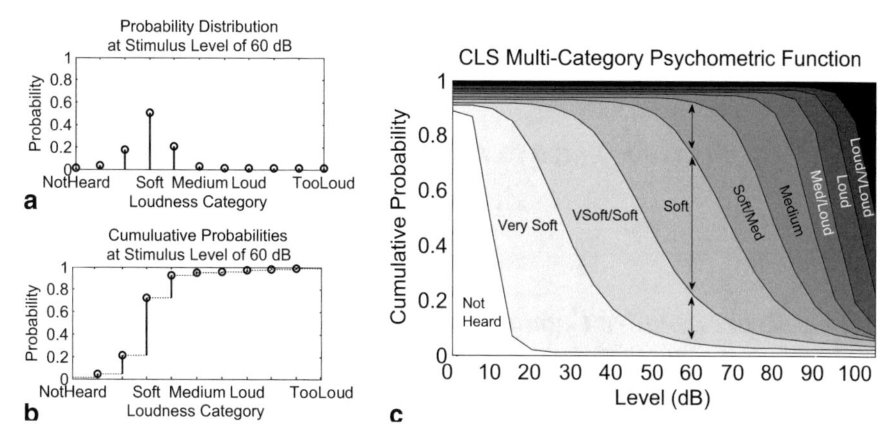

Fig. 1 Example MCPF, constructed from categorical probabilities. **a** Probability distribution across loudness categories at 60 dB SPL. **b** Cumulative probability function for loudness categories at 60 dB SPL; same data as **a.** **c** MCPF from 0 to 105 dB SPL. The vertical distance between curves represents the probability of each category. The 3 highest category probabilities at the 60 dB SPL stimulus level are marked by arrows. Maximum categories are not labeled

egory, but, instead, form a unimodal distribution. Figure 1b shows the cumulative probability density function for the data in (a). The MCPF (Fig. 1c) is constructed by plotting the cumulative distribution (marked by circles in (b)) as a function of stimulus level. In Fig. 1c, the probabilities of the top categorical response at 60 dB SPL are marked with arrows; the vertical distance between the curves matches the probabilities shown in Fig. 1a, 1b.

2.2 Parameterization

The MCPF curves are logistic functions that can be parameterized. A four-parameter logistic function $f(x \mid \theta)$, was selected to represent each of the curves of the MCPF. For our 11-category CLS scale, each MCPF, $F(x,i \mid \theta)$ consists of 10 category-boundary curves. Thus, the modeling implementation has 10 sets of parameters θ_i that define the 10 curves of each MCPF.

$$f(x \mid \theta) = D + \frac{A - D}{1 + (x \mid C)^B} \tag{1}$$

$$x = stimulus\ level, \theta = \{A, B, C, D\}$$

$$A = \textit{minimum asymptote}, B = \textit{slope}, C = \textit{inflection point}, D = \textit{maximum asymptote}$$

$$F(x,i \mid \theta) = f_i(x \mid \theta_i) = D_i + \frac{A_i - D_i}{1+(x/C_i)^{B_i}} \quad i = 1,\ldots,10 \tag{2}$$

2.3 A Representative Catalog

We have developed a "catalog" of MCPFs to be representative of both NH and HL listener CLS outcomes based on listener data. Each function in the catalog has the same form (Eq. 2) and is defined by a set of parameters, $(\theta_i, i = 1,\ldots,10)$. A PCA analysis decomposes the primary sources of variability in the parameters. Let the matrix of all listener parameters be $X = \{\theta^1,\ldots\theta^M\}$, where M is the number of listeners, then the set of listener weightings, $W = \{w^1,\ldots,W^M\}$, from the projections on the PCA eigenvectors v is $Xv = W$. Permutations of the sampled weightings were used to reconstruct the MCPFs that comprise the catalog. The superset of derived MCPF parameter sets that defines the catalog is denoted as Θ.

2.4 Maximum-Likelihood Estimation

The catalog can be used to compute a ML estimate of a listener's MCPF. This may be particularly useful when the number of experimental observations is relatively low, but an accurate estimate of a listener's loudness perception is needed, as is the case in most adaptive-level CLS methods that would be used in the clinic.

We denote a listener's raw CLS data as $(x_1,\ldots x_N)$, where N is the total number of experimental observations. The likelihood, $\mathcal{L}(\cdot)$, of these observations is computed for each catalog MCPF, $F(x \mid \theta)$, maximizing over all potential parameter sets (i.e., all $\theta \in \Theta$). The maximization is computed over the functionally-equivalent log-likelihood.

$$\operatorname{argmax}_{\theta\in\Theta} \ln \mathcal{L}(\theta, x_1,\ldots,x_N) = \operatorname{argmax}_{\theta\in\Theta} \sum_{i=1}^{N} \ln F(x_i \mid \theta) \tag{3}$$

Once the ML parameter set has been determined, it can be used to construct the listener's MCPF, $F(x \mid \theta)$, via Eqs. 1 and 2.

Table 1 NH and HL listener characteristics. N is the number of participants. Age, audiometric threshold, and loudness discomfort level (*LDL*) are reported. Standard deviations are shown in parentheses. NA: tallies listeners that did not report a LDL from 0 to 105 dB SPL

	N	Age (yr)	Threshold, 1 kHz	Threshold, 4 kHz	LDL, 1 kHz (dB SPL)	LDL, 4 kHz (dB SPL)
NH	15	38 (±10)	2 dB HL (±5)	5 dB HL (±5)	102 (±4) [6 NA]	102 (±4) [9 NA]
HL	22	53 (±16)	31 dB HL (±18)	51 dB HL (±12)	102 (±6) [3 NA]	100 (±4) [11 NA]

2.5 *Experiment*

2.5.1 Participants

Sixteen NH listeners and 25 listeners with HL participated. One ear was tested per participant. NH participants had audiometric octave thresholds ≤ 10 dB Hearing Level; participants with sensorineural HL had octave thresholds of 15–70 dB Hearing Level (Table 1).

2.5.2 Stimuli

Pure-tone stimuli (1000-ms duration, 25-ms onset/off cosine-squared ramps) were sampled at 44100 Hz. CLS was measured at 1 and 4 kHz. Stimuli were generated and presented using MATLAB.

2.5.3 Fixed-Level Procedure

The CLS experimental methods generally followed the ISO standard (Kinkel 2007). Stimuli were presented monaurally over Sennheiser Professional HD 25-II headphones in a sound booth. Eleven loudness categories were displayed on a computer monitor as horizontal bars increasing in length from bottom to top. Every other bar between the "Not Heard" and "Too Loud" categories had a descriptive label (i.e., "Soft", "Medium", etc.). Participants clicked on the bar that corresponded to their loudness perception.

The CLS test was divided into two phases: (1) estimation of the dynamic range, and (2) the main experiment. In the main experiment, for each frequency, a fixed set of stimuli was composed of 20 repetitions at each level, with the presentation levels spanning the listener's dynamic range in 5 dB steps. Stimulus order was pseudorandomized such that there were no consecutive identical stimuli and level differences between consecutive stimuli did not exceed 45 dB.

2.5.4 ISO Procedure for Testing ML Estimation

Five additional NH listeners, whose data were not used in the construction of the catalog, were recruited to complete an additional CLS task that used an adaptive stimulus-level selection technique, which conformed to the ISO standard. The adaptive technique calculated 10 levels that evenly spanned the dynamic range and presented these 3 times. For reference, for a listener with a 0–105 dB SPL dynamic range, the fixed-level procedure required 440 experimental trials while the ISO adaptive-level procedure required 30 trials.

3 Results

3.1 Individual Listener MCPFs

In each MCPF, the vertical distance between the upper and lower boundary curves for a category is the probability of that category. The steeper the slope of a curve, the more defined the distinction between categories, whereas shallower curves coincide with probability being more distributed across categories, over a range of levels. A wider probability distribution across categories indicates that the listener had more uncertainty (i.e. entropy) in their responses.

The NH and HL examples in Fig. 2 show some general patterns across both the 1 and 4 kHz stimuli. The CLS functions show the characteristic loudness growth that has been documented for NH and HL listeners in the literature. The most uncertainty (shallow slopes) is observed across the 5–25 CU categories. The most across-listener variability in category width is observed for 5 CU, which spans in width from a maximum of 50 dB to a minimum of 3 dB. A higher threshold and/or lower LDL results in a horizontally compressed MCPF; this compression narrows all intermediate categories.

The top two rows of Fig. 2 show data for two representative NH listeners, NH04 at 1 kHz (1st row) and NH01 at 4 kHz (2nd row). Listener NH04 had a low amount of uncertainty in their categorical response choices, with the most sharply-defined category boundaries at the lowest levels. Compared to NH04, listener NH01 has a wider range of levels that correspond to 5 CU ("Very Soft") and a lower LDL, leaving a smaller dynamic range for the remaining categories. This results in a horizontally-compressed MCPF, mirroring the listener's sharper growth of loudness in the CLS function. The bottom two rows of Fig. 2 show representative data for listeners with varying degrees of HL. Listener HL17 (3rd row) is an example of a listener with an elevated LDL. Listener HL11 (4th row) has LDL that is within the ranges of a NH listener; the higher threshold level causes the spacing between category boundary curves to be compressed horizontally. Despite this compression, this listener with HL has well-defined perceptual boundaries between categories (i.e., sharply-sloped boundary curves).

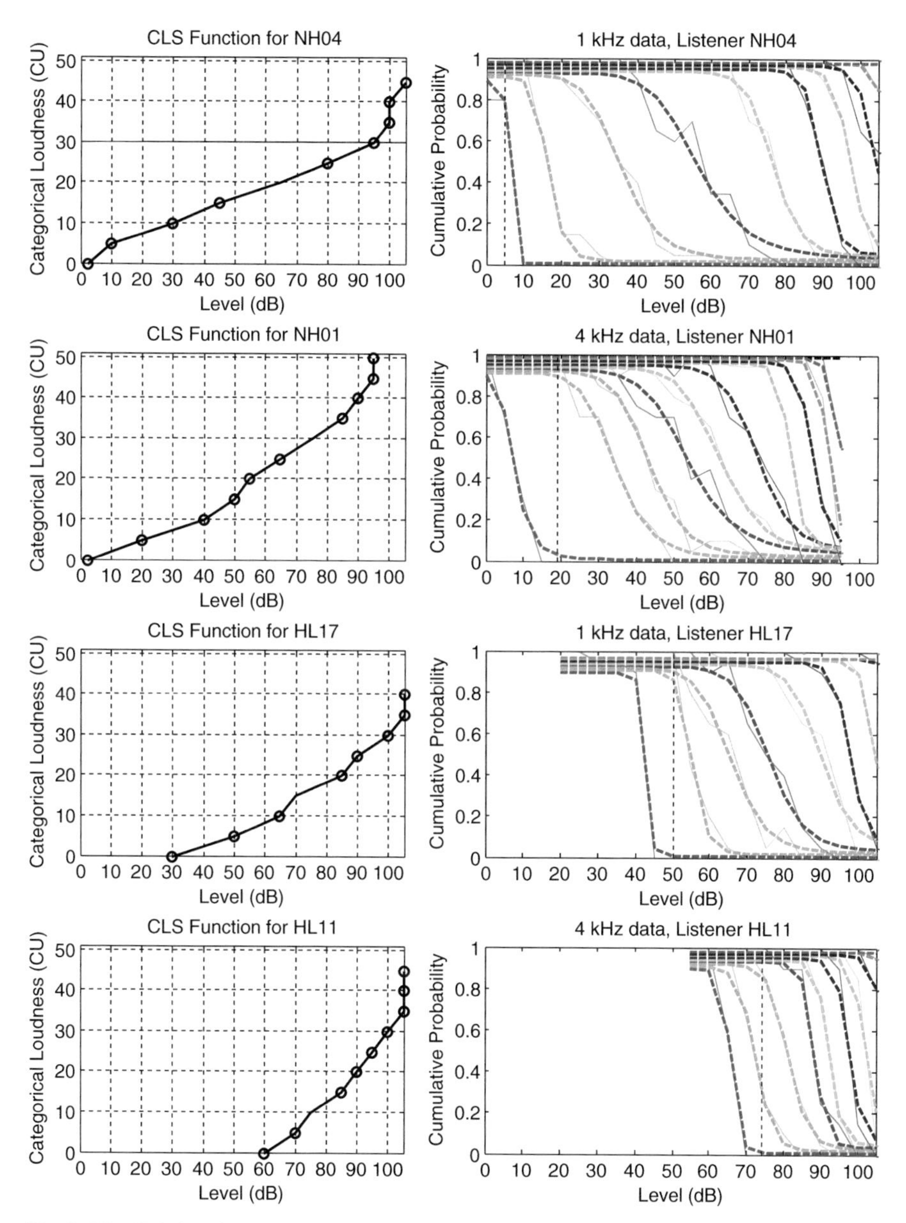

Fig. 2 The CLS function and corresponding MCPF for 4 listeners: (1st row) NH04, 1 kHz stimuli, (2nd row) NH01, 4 kHz stimuli, (3rd row) HL17, 1 kHz stimuli, (4th row) HL11, 4 kHz stimuli. The CLS function plots the median level for each CU. The MCPFs show the raw data as solid lines and the parameterized logistic fits as *dashed* lines. Each listener's audiometric threshold is marked with a vertical *black* line

3.2 Construction of the MCPF Catalog

A PCA of the combined NH and HL data revealed that the first two eigenvectors were sufficient for capturing at least 90 % of the variability in the listener parameters (θ). Vector weightings for creating the MCPF catalog were evenly sampled from the range (2 standard deviations) of the individual listener weightings. Permutations of 66 sampled weightings were combined to create the 1460 MCPFs that constitute the catalog.

3.3 Application to ML estimation

The MCPF catalog contains models of loudness perception for a wide array of listener types. Here, we demonstrate how a ML technique can be used with the catalog to estimate a novel listener's MCPF from a low number ($\leq$30) of CLS experimental trials. As this adaptive technique has a maximum of three presentations at each stimulus level, a histogram-estimation of the MCPF from this sparse data can be inaccurate. The 50 % intercepts of the ML MCPF category boundary curves may be used to estimate the CLS function (examples in Fig. 3). The average root mean squared error (RMSE) (fixed-level data used as baseline) for the median of the adaptive stimulus-level technique was 6.7 dB, and the average RMSE for the ML estimate, based on the same adaptive technique data, was 4.2 dB. The ML MCPF estimate improves the accuracy of the resulting CLS function and provides a probabilistic model of the listener's loudness perception.

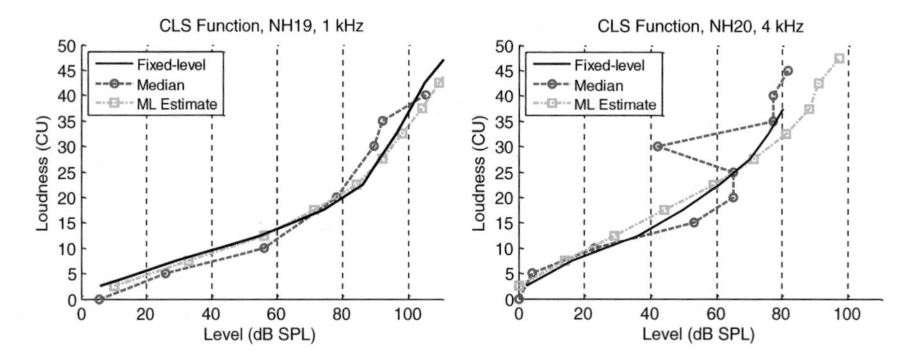

Fig. 3 Comparison of NH CLS results. The CLS function based on the fixed-level experiment is shown as a *black line*, representing the 'best' estimate of the listener's loudness perception. The ISO-adaptive median is shown with *red circles*. The ML-based estimate is shown with *green squares*

4 Discussion

The CLS MCPF describes how categorical perception changes with stimulus level. The results demonstrate that loudness perception of NH and HL listeners is a random variable at each stimulus level, i.e., each level has a statistical distribution across loudness categories. The development of a probabilistic model for listener loudness perception has a variety of advantages. The most common usage for probabilistic listener models is to simulate listener behavior for the development of experiments or listening devices. Probabilistic models also allow one to apply concepts from detection, information, and estimation theory to the analysis of results and the methodology of the experiment.

In this paper, we show how a catalog of MCPFs can be used to find the ML estimate of a listener's MCPF, when a relatively small number of experimental trials are available. The ISO adaptive recommendation for CLS testing results in a relatively low number of experimental trials ($\approx$15–30), in order to reduce the testing time and make the test more clinically realizible. Although this is an efficient approach, due to the low number of samples, the resulting CLS function can be inaccurate. The ML estimate of a listener's loudness perception based on this lower number of experimental trials is able to more accurately predict a listener's underlying CLS function, without removing outliers or using assumptions about the shape of the function to smooth the result. The MCPF catalog may be further exploited to develop optimal measurement methods; one such method would select experimental stimulus levels adaptively, such that each presentation maximizes the expected information. The ML MCPF estimate provides greater insight into the nature of the listener's categorical perception (Torgerson 1958), while still allowing for clinically-acceptable test times.

Acknowledgements This research was supported by grants from the NIH: T32 DC000013, R01 DC011806 (WJ), R01 DC008318 (STN), and P30 DC004662.

References

Allen JB (2008). Nonlinear cochlear signal processing and masking in speech perception. Springer handbook of speech processing. pp 27–60

Allen JB, Hall JL, Jeng PS (1990) Loudness growth in 1/2-octave bands (LGOB)—a procedure for the assessment of loudness. JASA 88(2):745–753

Al-Salim SC, Kopun JG, Neely ST, Jesteadt W, Stiegemann B, Gorga MP (2010) Reliability of categorical loudness scaling and its relation to threshold. Ear Hearing 31(4):567–578

Brand T, Hohmann V (2002) An adaptive procedure for categorical loudness scaling. JASA 112(4):1597–1604

Elberling C (1999) Loudness scaling revisited. JAAA 10(5):248–260

Florentine M (2011). Loudness. Springer New York, New York

Green DM (1993) A maximum-likelihood method for estimating thresholds in a yes–no task. JASA 93(4):2096–2105

Heeren W, Hohmann V, Appell JE, Verhey JL (2013) Relation between loudness in categorical units and loudness in phons and sones. JASA 133(4):EL314–EL319

Kinkel M (2007). The new ISO 16832 'Acoustics–loudness scaling by means of categories'. 8th EFAS Congress/10th Congress of the German Society of Audiology, Heidelberg

Torgerson W (1958) Theory and methods of scaling. Wiley, New York

Auditory fMRI of Sound Intensity and Loudness for Unilateral Stimulation

Oliver Behler and Stefan Uppenkamp

Abstract We report a systematic exploration of the interrelation of sound intensity, ear of entry, individual loudness judgments, and brain activity across hemispheres, using auditory functional magnetic resonance imaging (fMRI). The stimuli employed were 4 kHz-bandpass filtered noise stimuli, presented monaurally to each ear at levels from 37 to 97 dB SPL. One diotic condition and a silence condition were included as control conditions. Normal hearing listeners completed a categorical loudness scaling procedure with similar stimuli before auditory fMRI was performed. The relationship between brain activity, as inferred from blood oxygenation level dependent (BOLD) contrasts, and both sound intensity and loudness estimates were analyzed by means of linear mixed effects models for various anatomically defined regions of interest in the ascending auditory pathway and in the cortex. The results indicate distinct functional differences between midbrain and cortical areas as well as between specific regions within auditory cortex, suggesting a systematic hierarchy in terms of lateralization and the representation of sensory stimulation and perception.

Keywords Neural activation · Sound pressure level · Categorical loudness scaling · Auditory cortex · Auditory pathway · Monaural stimuli

1 Introduction

Loudness is mainly the perceptual correlate of sound intensity, which is usually expressed in dB SPL as sound pressure level on a logarithmic scale. However, loudness judgments are also affected by several other variables, including physical sound parameters like e.g. duration, spectral content and temporal modulation

O. Behler (✉)
Medizinische Physik, Carl von Ossietzky Universität Oldenburg, 26111 Oldenburg, Germany
e-mail: oliver.behler@uni-oldenburg.de

S. Uppenkamp
Medizinische Physik and Cluster of Excellence Hearing4All, Carl von Ossietzky Universität Oldenburg, 26111 Oldenburg, Germany
e-mail: stefan.uppenkamp@uni-oldenburg.de

P. van Dijk et al. (eds.), *Physiology, Psychoacoustics and Cognition in Normal and Impaired Hearing*, Advances in Experimental Medicine and Biology 894,
DOI 10.1007/978-3-319-25474-6_18

as well many more acoustic and non-acoustic factors, including the particular task or context and personal factors like the individual hearing status (Florentine et al. 2011). While the relationship between sound intensity and neural activity in the human central auditory pathway has been extensively studied by means of neuroimaging techniques, only a small number of studies have investigated the interrelation of sound intensity, loudness and the corresponding brain activity (for a review, see Uppenkamp and Röhl 2014). Some auditory functional magnetic resonance imaging (fMRI) studies suggest that neural activation, at least in auditory cortex (AC), might be more a representation of perceived loudness rather than of physical sound pressure level (Hall et al. 2001; Langers et al. 2007; Röhl and Uppenkamp 2012). The current literature still does not provide definite answers to the following questions: (1) At what stage or stages along the auditory pathway is sound intensity transformed into its perceptual correlate (i.e. loudness)? (2) What are the functional differences across regions within AC with respect to loudness-related activation? Promoting a better understanding of the neural basis of loudness might, in the long run, improve diagnostics and treatment of hearing disorders characterized by a distorted loudness perception, e.g. loudness recruitment.

The present study is aimed at extending the current literature by providing a detailed characterization of the neural representation of sound intensity and loudness, as reflected by functional MRI. In a group of normal hearing listeners, we systematically explored the interrelation of ear of entry, sound pressure level, individual loudness and brain activation, as inferred from blood oxygenation level dependent (BOLD) contrasts, in the ascending auditory pathway and within AC.

2 Methods

2.1 Participants and Procedure

Thirteen normal hearing volunteers (aged 34±8 years, 4 females) participated in this study. Each participant attended two experimental sessions. In the first session, standard audiometry and an adaptive categorical loudness scaling procedure (Brand and Hohmann 2002) were performed in a sound booth. In the second session, auditory fMRI was performed while subjects were doing a simple listening task in the MRI scanner.

2.2 Stimuli

All stimuli consisted of 1/3 octave band-pass low-noise noise (Pumplin 1985) bursts at 4 kHz center frequency and were delivered via MRI compatible insert earphones (Sensimetrics S14, Sensimetrics Corporation, Malden, MA). In the loudness scaling procedure, single noise bursts with a maximum intensity of 105 dB SPL were used under left monaural, right monaural and diotic stimulus conditions. In the MRI experiment, trains of noise bursts with a total duration of 4.75 s were presented left and right monaurally at 37, 52, 67, 82 and 97 dB SPL and diotically at 82 dB SPL.

2.3 MRI Data Acquisition

Functional and structural images were acquired with a 3-Tesla MRI system (Siemens MAGNETOM Verio). Functional images were obtained using T2*-weighted gradient echo planar imaging (EPI), with a sparse temporal sampling paradigm to reduce the influence of the acoustic noise created by the scanner (Hall et al. 1999). Stimuli were presented in pseudorandomized order during 5 s gaps of scanner silence in between two successive volume acquisitions. Stimuli of each of the eleven conditions plus a silence condition, which served as baseline, were presented 36 times over the course of the experiment. For the purpose of maintaining the participants' attention towards the acoustic stimuli, they were asked to count the number of occasionally presented deviants, characterized by a transient dip in sound level in one of the noise bursts.

2.4 Psychoacoustic Evaluation

Individual loudness judgments obtained in the scaling procedure were used to fit loudness functions for each participant by means of a recently suggested fitting method (Oetting et al. 2014). Loudness estimates for the stimulus conditions used in the MRI experiment were extracted from the individual loudness functions and were used for further analyses.

2.5 MRI Data Analysis

Standard preprocessing of the imaging data (including spatial smoothing with a 5 mm FWHM Gaussian kernel) and general linear model (GLM) estimation was done using SPM8 (Functional Imaging Laboratory, The Wellcome Department of Imaging Neuroscience, London, UK, http://www.fil.ion.ucl.ac.uk/spm/). A general linear model was set up to model the BOLD signal in every voxel as a function of ear of entry and sound level for each participant. The model included one separate regressor for each of the 11 stimulus conditions, while the silence condition was implicitly modeled as baseline. Region of interest (ROI) analyses were carried out to characterize the relationship between neural activation and sound intensity or loudness for left or right stimuli across subjects. For each participant, twelve auditory ROIs were defined based on anatomical landmarks in the individual structural images: Left and right inferior colliculus (IC), medial geniculate body (MGB),

Planum temporale (PT), posterior medial (HGpm), central (HGc) and anterolateral (HGal) parts of the first Heschl's gyrus. The average signal change from baseline of all voxels within spheres of 5 mm radius centered at individual ROI coordinates was calculated based on the regression coefficients for every stimulus condition and entered into linear mixed effects models (LMMs) with random intercepts and slopes. For each of the twelve ROIs, eight separate LMMs ($2 \times 2 \times 2$) were estimated modeling the ROI percent signal change as a linear or a quadratic function of sound intensity (expressed in dB SPL) or individual loudness (in categorical units from 0 to 50), for left or right stimuli. Model parameters were estimated by means of maximum-likelihood. Likelihood-ratio tests were conducted to assess significance of the models: First, linear models were tested against "null models" containing only the constant terms. Then, quadratic models were tested against the corresponding linear models. In both steps, models were considered significant at a level of $p<0.05$, Bonferroni-corrected. To provide measures of the models' goodness-of-fits in terms of explanatory power, marginal R^2 statistics (R^2_m), representing that part of variance in the model explained by the fixed effects, were calculated as suggested by (Johnson 2014).

3 Results

3.1 Categorical Loudness Scaling

Figure 1 shows the results of the categorical loudness scaling procedure for the group of 13 participants. Group averaged fitted loudness curves (Oetting et al. 2014) for left monaural, right monaural and binaural stimuli are shown, along with the interindividual standard deviations of loudness estimates for the stimulus intensities presented in the MRI experiment. All three curves are characterized by a nearly linear growth of categorical loudness with sound intensity between 20 and 80 dB SPL

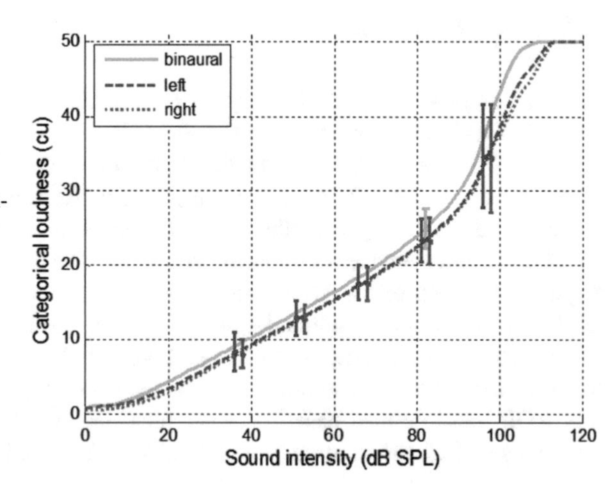

Fig. 1 Categorical loudness as a function of sound intensity and ear of entry. The three curves represent group averages of individual loudness fits. Error bars represent interindividual standard deviations of loudness estimates for the stimulus intensities presented in the MRI experiment

and an increase in the steepness of the slope around 90 dB SPL. There was virtually no difference in perceived loudness between left and right ear of entry. For diotic stimulation, the expected effect of binaural loudness summation is clearly visible.

3.2 Region-of-Interest Analysis

The results of the ROI analysis are illustrated in Fig. 2. R^2_m statistics, representing that part of variance explained by the fixed effects, display a largely symmetrical pattern across hemispheres. The highest values were found in the posterior medial parts of Heschl's gyri, whereas the subcortical ROIs and anterolateral HG were characterized by comparatively low R^2_m. Throughout all investigated regions, explanatory powers of sound levels or loudness estimates were higher for contralateral as compared to ipsilateral stimuli, albeit with varying degrees of lateralization across ROIs. In all regions, linear loudness models yielded at least slightly better goodness-of-fits as compared to linear sound intensity models, although still being outmatched by quadratic fits with sound intensity in the majority of cortical ROIs. The example plots of the different model fits in the left and right posterior medial HG (Fig. 2b) reveal that the significant quadratic component is attributable to a steeper increase of BOLD responses at the highest sound levels (in this regard, the examples are representative for all ROIs). Across ROIs, only 4 (out of the total 24) quadratic loudness models reached significance, indicating that the relationship between the responses and categorical loudness is sufficiently described by a linear model.

4 Discussion

We used auditory fMRI to measure responses to unilateral narrowband noise stimuli in the human auditory system of normal hearing subjects. Based on these measures, we investigated the interrelation of ear of entry, sound pressure level, individual loudness and the corresponding brain activity by means of linear mixed effects models for a large number of distinct regions of interest in the ascending auditory pathway and in the cortex. This approach allowed us to characterize the neural representation of sound intensity and loudness in a detailed way.

4.1 Response Characteristics in Relation to Sound Intensity

Throughout all investigated stages of the auditory system, except for ipsilateral stimuli in IC bilaterally and in left MGB, neural activation as reflected by the fMRI BOLD response was significantly and positively related to physical sound intensity. Specifically, at cortical level, changes in sound pressure level were reflected by nonlinear (quadratic) increases of activation magnitude, with steeper slopes at the

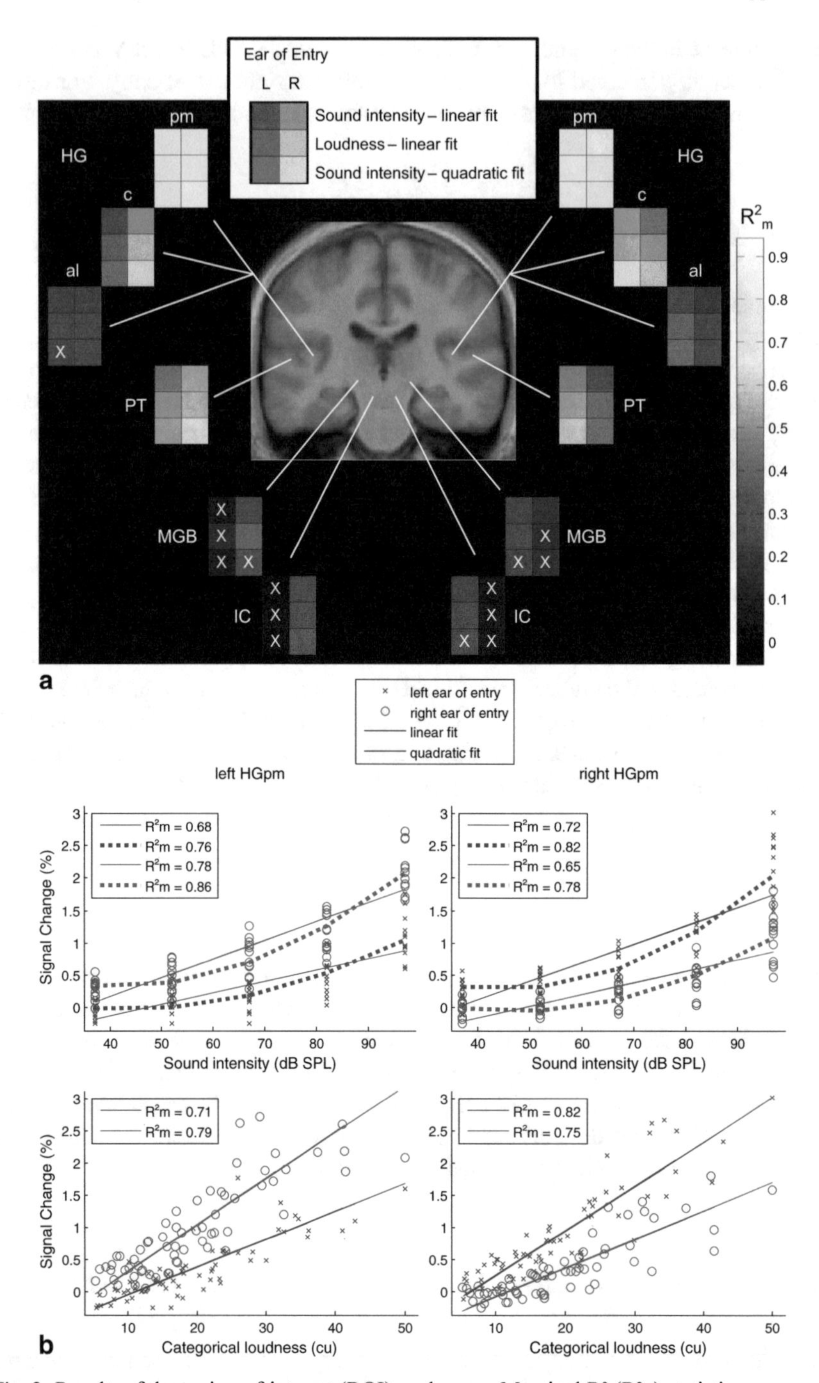

Fig. 2 Results of the region-of-interest (ROI) analyses. **a** Marginal R^2 (R^2_m) statistics, representing that part of variance explained by fixed effects, of linear and quadratic model fits with sound intensity as well as linear fits with loudness for *left* (L) and *right* (R) monaural stimuli corresponding to each auditory ROI. Non-significant models are marked with a white "X". **b** Percent signal change from baseline in the *left* and *right* HGpm plotted as a function of ear of entry and sound

highest sound levels. While this is in agreement with some previous findings (Hart et al. 2002, 2003), other studies reported a more or less linear increase of BOLD signal strength with sound intensity (Hall et al. 2001; Langers et al. 2007; Röhl and Uppenkamp 2012) and still others found indications of response saturation at the highest levels (Mohr et al. 1999). To what extent these differences can be explained in terms of the use of different types of stimuli, dynamic ranges, fMRI paradigms or other factors (e.g., perceived loudness) is an interesting topic by itself, but this is not the scope of the present study. As opposed to AC, activation in the investigated subcortical structures increased predominantly linearly with increasing sound pressure level, which is in line with previous findings (Röhl and Uppenkamp 2012).

Our findings further suggest considerable differences along the auditory pathway and across auditory regions in cortex with respect to the strength of the relationship between neural activity and sound intensity. The relationship was most pronounced in the posterior medial section of HG (which includes primary AC), a little less so in PT and comparatively weak in the anterolateral sections of HG and in subcortical regions. These findings largely conform to earlier studies reporting the most robust level dependence in primary AC, but less consistent results with regard to PT and a generally weak sensitivity to level changes in more lateral and anterior regions of HG (Hart et al. 2002; Brechmann et al. 2002; Ernst et al. 2008).

4.2 Interrelation With the Ear of Entry

Activity at all investigated stages of the auditory system appeared to be more closely related to changes in sound level for stimuli presented at the contralateral ear. This is in good agreement with the anatomy and with previous investigations suggesting that auditory information crosses to the contralateral hemisphere in the lower brain stem followed by ipsilateral signal propagation towards AC (e.g., Langers et al. 2005). However, in the present study, the degree of lateralization differed considerably between regions, being particularly pronounced in the IC of the brain stem and rather weak in the posteromedial and anterolateral sections of HG. In fact, in stark contrast to the IC, changes in sound intensity were highly predictive of signal changes in HGpm irrespective of the ear of entry, pointing towards strong interhemispheric connections in the upper stages of the auditory system.

4.3 Transformation of Sound Intensity into Perceived Loudness

As expected from the strong relationship between sound intensity and loudness, a very similar pattern across auditory regions was found for the relationship between

intensity or categorical loudness. Overlaid are the fitted curves as predicted by the fixed effects of the linear (*straight lines*) and quadratic (*dotted curves*) models with their corresponding R^2_m statistics. Quadratic fits with loudness are not shown, since only 4 of the corresponding models reached significance across ROIs

BOLD responses and individual loudness. Nonetheless, it was possible to identify distinct differences with regard to the representation of sensory stimulation and perception in the human auditory system. Specifically, while changes in sound pressure level were reflected by a quadratic growth of activation magnitude, especially in cortical areas, the relation between activation and categorical loudness can be described as predominantly linear in all investigated auditory regions. This finding, in conjunction with the similarities between the growth of categorical loudness (Fig. 1) and BOLD signal changes (Fig. 2b) with sound intensity, lends support to previous auditory fMRI studies suggesting that neural activation in AC might be more a direct linear reflection of perceived loudness rather than of physical sound pressure level (Langers et al. 2007; Röhl and Uppenkamp 2012). By contrast, in the previous study by Röhl and Uppenkamp (2012) no significant relationship between the BOLD signal strength and categorical loudness was found at the level of the IC or MGB. One interpretation in that study was that at the level of the IC, "neural activation might be more a reflection of physical sound intensity". The current results are different, given that activity in subcortical regions was—albeit only slightly—more closely related to individual loudness as compared to sound pressure. However, the interpretation of BOLD fMRI results from these regions appears difficult, given the poor signal-to-noise ratio caused by the small size of the structures and their susceptibility to cardiac-related, pulsatile motion effects. Taking measures to improve the signal quality in subcortical regions, such as cardiac gated image acquisition (Guimaraes et al. 1998), is therefore strongly advised for future investigations.

The largely linear relationship between categorical loudness and sound intensity over the range of presented levels, as well as the fact that our participants were from a rather homogenous group of normal hearing listeners with similar loudness perception, might have limited the ability to differentiate between the contributions of both measures to neural activation in the human auditory system. A promising next step would be the inclusion of hearing impaired participants characterized by an altered relationship between sound intensity and perceived loudness (e.g., loudness recruitment). Moreover, this approach could provide additional valuable information about the nature of hearing disorders associated with distorted loudness perceptions. This in the end might open a potential for the diagnostic use of auditory fMRI. Although it is yet to prove how well the present findings pertain to the hearing impaired listeners, our results so far point towards the activation of the posterior medial section of Heschl's gyrus as the most reliable indicator of individually perceived loudness.

4.4 Conclusion

Our findings support the notion that neural activation in auditory cortex as well as in certain stages of the ascending auditory pathway is more a direct linear reflection of perceived loudness rather than of physical sound intensity. There are, however, systematic differences between the investigated auditory regions in terms of the strength of this relationship and the degree of lateralization. We therefore suggest that functional differentiation, both between cortical and subcortical regions as well as between regions of auditory cortex, is an important issue to consider in the pursuit of a complete and comprehensive understanding of the physiological correlates of loudness perception.

Acknowledgments Oliver Behler is funded by the PhD programme 'Signals and Cognition' (Niedersächsisches Ministerium für Wissenschaft und Kultur).

References

Brand T, Hohmann V (2002) An adaptive procedure for categorical loudness scaling. J Acoust Soc Am 112:1597–1604

Brechmann A, Baumgart F, Scheich H (2002) Sound level-dependent representation of frequency modulations in human auditory cortex: a low-noise fMRI study. J Neurophysiol 87(1):423–433

Ernst SMA, Verhey JL, Uppenkamp S (2008) Spatial dissociation of changes of level and signal-to-noise ratio in auditory cortex for tones in noise. Neuroimage 43(2):321–328

Florentine M, Popper AN, Fay RR (Eds., 2011) Loudness. Springer handbook of auditory research, vol 37. Springer, New York

Guimaraes AR, Melcher JR, Baker JR, Ledden P, Rosen BR, Kiang NYS, Fullerton BC, Weisskopf RM (1998) Imaging subcortical auditory activity in humans. Hum Brain Mapp 6:33–41

Hall DA, Haggard MP, Summerfield AQ, Akeroyd MA, Palmer AR (2001) Functional magnetic resonance imaging of sound-level encoding in the absence of background scanner noise. J Acoust Soc Am 109:1559–1570

Hall DA, Summerfield AQ, Goncalves MS, Foster JR, Palmer AR, Bowtell RW (1999) "Sparse" temporal sampling in auditory fMRI. Hum Brain Mapp 7:213–223

Hart HC, Hall DA, Palmer AR (2003) The sound-level-dependent growth in the extent of fMRI activation in Heschl's gyrus is different for low- and high-frequency tones. Hear Res 179(1–2):104–112

Hart HC, Palmer AR, Hall DA (2002) Heschl's gyrus is more sensitive to tone level than non-primary auditory cortex. Hear Res 171(1–2):177–190

Johnson PCD (2014) Extension of Nakagawa & Schielzeth's R^2_{GLMM} to random slope models. Meth Ecol Evol 5(9):944–946

Langers DRM, van Dijk P, Backes WH (2005) Lateralization, connectivity and plasticity in the human central auditory system. Neuroimage 28(2):490–499

Langers DRM, van Dijk P, Schoenmaker ES, Backes WH (2007) fMRI activation in relation to sound intensity and loudness. Neuroimage 35(2):709–718

Mohr CM, King WM, Freeman AJ, Briggs RW, Leonard CM (1999) Influence of speech stimuli intensity on the activation of auditory cortex investigated with functional magnetic resonance imaging. J Acoust Soc Am 105:2738–2745

Oetting D, Brand T, Ewert SD (2014) Optimized loudness-function estimation for categorical loudness scaling data. Hear Res 316:16–27

Pumplin J (1985) Low-noise noise. J Acoust Soc Am 78:100–104

Röhl M, Uppenkamp S (2012) Neural coding of sound intensity and loudness in the human auditory system. J Assoc Res Otolaryngol 13(3):369–379

Uppenkamp S, Röhl M (2014) Human auditory imaging of sound intensity and loudness. Hear Res 307:65–73

Tinnitus- and Task-Related Differences in Resting-State Networks

Cris Lanting, Aron Woźniak, Pim van Dijk and Dave R. M. Langers

Abstract We investigated tinnitus-related differences in functional networks in adults with tinnitus by means of a functional connectivity study. Previously it was found that various networks show differences in connectivity in patients with tinnitus compared to controls. How this relates to patients' ongoing tinnitus and whether the ecological sensory environment modulates connectivity remains unknown. Twenty healthy controls and twenty patients suffering from chronic tinnitus were enrolled in this study. Except for the presence of tinnitus in the patient group, all subjects were selected to have normal or near-normal hearing. fMRI data were obtained in two different functional states. In one set of runs, subjects freely viewed emotionally salient movie fragments ("fixed-state") while in the other they were not performing any task ("resting-state"). After data pre-processing, Principal Component Analysis was performed to obtain 25 components for all datasets. These were fed into an Independent Component Analysis (ICA), concatenating the data across both groups and both datasets, to obtain group-level networks of neural origin, each consisting of spatial maps with their respective time-courses. Subject-specific maps and their time-course were obtained by back-projection (Dual Regression). For each of the components a mixed-effects linear model was composed with factors group (tinnitus vs. controls), task (fixed-state vs. resting state) and their interaction. The neural components comprised the visual, sensorimotor, auditory, and limbic systems, the default mode, dorsal attention, executive-control, and frontoparietal networks, and the cerebellum. Most notably, the default mode network (DMN) was less extensive and shows significantly less connectivity in tinnitus patients than in controls. This group difference existed in both paradigms. At the same time, the DMN was stronger during resting-state than during fixed-state in the controls but not the patients. We attribute this pattern to the unremitting engaging effect of the tinnitus percept.

C. Lanting (✉) · P. van Dijk · D. R. M. Langers · A. Woźniak
Department of Otorhinolaryngology / Head and Neck Surgery, University of Groningen, University Medical Center Groningen, Groningen, The Netherlands
e-mail: c.p.lanting@umcg.nl

Graduate School of Medical Sciences (Research School of Behavioural and Cognitive Neurosciences), University of Groningen, Groningen, The Netherlands

P. van Dijk et al. (eds.), *Physiology, Psychoacoustics and Cognition in Normal and Impaired Hearing,* Advances in Experimental Medicine and Biology 894,
DOI 10.1007/978-3-319-25474-6_19

Keywords Tinnitus · Resting-state fMRI · Independent component analysis · Functional connectivity

1 Introduction

Tinnitus is a percept of sound that is not related to an acoustic source outside the body. For many forms of tinnitus, mechanisms in the central nervous system are believed to play a role in the pathology. Despite its high prevalence of 5–10%, relatively little is known about the neural mechanisms and causes of tinnitus. Neuroimaging methods have been applied in the last decade to study potential mechanisms, and techniques such as electroencephalography (EEG), magnetoencephalography (MEG), positron emission tomography (PET), and functional magnetic resonance imaging (fMRI) have been extensively used to investigate neural correlates of tinnitus. Previously, neuroimaging studies on tinnitus relied mostly on task-related paradigms, such as presentation of sound stimuli that elicit sound-evoked responses, or manipulation of patients' tinnitus by somatic modulation (i.e., jaw protrusion or a change of gaze) to temporarily affect patients' tinnitus (Lanting et al. 2009). Alternatively, studies focused on anatomical differences in brain structure (Adjamian et al. 2014).

However, since a few years, new data-analysis techniques, such as independent component analysis (ICA), have emerged allowing for identification of coherent patterns of spatially independent, temporally correlated signal fluctuations in neuroimaging data during "resting-state", that is, without a predefined task. These patterns have since been termed "resting-state networks" (RSNs) (Greicius et al. 2004) and represent functionally relevant and connected networks or systems supporting core perceptual and cognitive processes. Resting-state networks are generally reported to show reliable and consistent patterns of functional connectivity (Zhang et al. 2008).

Moreover, advances in data-analysis currently allow for between-group analyses where individual networks can be compared across tasks and groups. Recently, resting-state fMRI has been used to investigate functional connectivity differences in individuals with tinnitus (Burton et al. 2012; Davies et al. 2014; Kim et al. 2012; Maudoux et al. 2012a, b). Results indicate differences in connectivity between healthy and affected subject groups. One result showed reduced connectivity between the left and right auditory cortex, indicative of disturbed excitation and inhibition across hemispheres (Kim et al. 2012). A different study found that in individuals with tinnitus connectivity was increased in numerous brain regions, including the cerebellum, parahippocampal gyrus, and sensorimotor areas. Decreased connectivity was found in the right primary auditory cortex, left prefrontal, left fusiform gyrus, and bilateral occipital regions (Maudoux et al. 2012a, b). Very recently, altered interhemispheric functional connectivity was shown and linked with specific tinnitus characteristics in chronic tinnitus patients (Chen et al. 2015a). In addition, enhanced functional connectivity between auditory cortex and the homologue frontal gyrus was observed, indicative of enhanced connectivity of auditory networks with attention networks (Chen et al. 2015b). All these results combined are thought to represent the influence tinnitus has on the networks encompassing memory, attention and emotion. Whether this is causal and which direction the causality goes remains unknown.

In this chapter we describe tinnitus-related changes in functional networks in adults with tinnitus compared to adults without tinnitus. We aim to study whether group-differences can be observed in functional connectivity at the group level and whether performance of an engaging task alters these connectivity patterns.

2 Methods

2.1 Participants

Twenty healthy controls and twenty patients suffering from chronic tinnitus were enrolled in this fMRI study after providing written informed consent, in accordance with the medical ethical committee at the University Medical Center Groningen, the Netherlands. Subjects were recruited from the hospital's tinnitus outpatient clinic as well as from advertisements in local media. None of them had a history of neurological or psychiatric disorders. Subjects' characteristics are listed in Table 1 (Langers et al. 2012). Except for the presence of tinnitus in the patient group, all subjects were selected to have normal or near-normal hearing up to 8 kHz. Thresholds were determined in a frequency range of 0.25–16 kHz by means of pure-tone audiometry. Subjects also performed the adaptive categorical loudness scaling (ACALOS) procedure (Brand and Hohmann 2002). To characterize the participants' self-reported complaints, all subjects filled out the 14-item Hyperacusis Questionnaire, relating to the attentional, social, and emotional aspects of auditory hypersensitivity (Khalfa et al. 2002).

The tinnitus patients also filled out Dutch translations of questionnaires related to their tinnitus, including the Tinnitus Handicap Inventory (THI) measuring tinnitus severity in daily life (Newman et al. 1996), the Tinnitus Reaction Questionnaire (TRQ) that assesses the psychological distress associated with tinnitus (Wilson et al. 1991), and the Tinnitus Coping Style Questionnaire (TCSQ) that quantifies effective as well as maladaptive coping strategies (Budd and Pugh 1996). Patients were asked about the laterality of their tinnitus and the character of the tinnitus sound. Using a modified tinnitus spectrum test, patients were asked to indicate the "likeness" of their tinnitus to narrow-band noises of varying center frequencies.

Table 1 Several group characteristics. Numbers indicate mean ± standard deviation (range). All questionnaire scales were expressed to range from 0 to 100 for ease of interpretation. The significance of group differences was based on Fisher's exact test (for gender and handedness) or Student's t-test (for all other comparisons), and classified as: ***$p<0.001$; **$p<0.01$; *$p<0.05$; #$p\geq0.05$. Numbers do not add up to equal the group size because for some patients not all tinnitus characteristics were obtained

Group		Healthy controls	Tinnitus patients	*p*
		(n=20)	(n=20)	
Demographics				
	Gender	16 female, 4 male	12 female, 8 male	#
	Handedness	17 right, 3 left	19 right, 1 left	#
	Age (years)	33±13 (21–60)	46±11 (26–60)	**
Audiometry				
	Average threshold (dB HL)	5±5 (−1–18)	8±5 (0–23)	#
	Loudness range (dB)	98±8 (84–113)	84±14 (56–105)	***
Self-reported symptoms				
	Hyperacusis (%)	25±14 (8–51)	59±16 (29–82)	***
	Depression (%)	8±9 (0–31)	27±27 (0–100)	**
	Dysthymia (%)	18±14 (0–44)	38±24 (0–75)	**
Tinnitus effects				
	Tinnitus Handicap (%)	–	43±22 (6–88)	–
	Tinnitus Reaction (%)	–	38±21 (0–88)	–
	Effective Coping (%)	–	54±14 (17–68)	–
	Maladaptive Coping (%)	–	29±15 (3–60)	–
Tinnitus percept				
	Lateralization	–	8 central, 4 right, 2 left	–
	Type	–	16 steady, 3 pulsatile	–
	Bandwidth	–	10 tone, 7 hiss, 2 ring	–
	Frequency	–	16 high, 2 other, 1 low	–

2.2 Imaging Paradigm

The equipment used for this study was a 3.0-T MR scanner (Philips Intera, Best, the Netherlands), supplied with an 8-channel phased-array (SENSE) transmit/receive head coil. The functional imaging sessions performed on subjects in head-first-supine position in the bore of the system included two runs for the resting-state and two runs for the fixed-state designs. Each resting-state run consisted of a dynamic series of 49 identical high-resolution T2*-sensitive gradient-echo echo-planar imaging (EPI) volume acquisitions (TR 15.0 s; TA 2.0 s; TE 22 ms; FA 90°; matrix 64 × 64 × 48; resolution 3 × 3 × 3 mm^3; interleaved slice order, no slice gap), whereas fixed-state runs consisted of the same acquisition parameters except the TR that for the fixed-state runs was equal to the TA of 2.0 s (continuous scanning) and an ascending slice ordering. The acquisition volume was oriented parallel to the Sylvian fissure, and centered approximately on the superior temporal sulci. In total were 361 volumes acquired.

A sparse, clustered-volume sequence was employed in the (unstimulated) resting-state runs to avoid interference from acoustic scanner noise, while continuous acquisitions were employed in the (stimulated) fixed-state runs to provide higher temporal resolution and power. The scanner coolant pump was turned off during to further reduce noise-levels. In order to control subjects' attentional state, they were asked to focus on the fixation cross throughout the imaging time for resting-state runs, or watch four "emotional" 3-min movie fragments (video and audio) for fixed-state runs, while being scanned simultaneously.

To ensure that the ambient noise did not eliminate the patients' tinnitus, they were asked to rate their tinnitus several times (before, between and after imaging runs) on a 0–10 scale, where 0 indicated no perception of tinnitus at all and 10 indicated the strongest tinnitus imaginable. Ratings did not decrease significantly throughout the session (conversely, some patient's indicated that it systematically increased with stress and fatigue) and never reached zero (2 was the lowest rating).

2.3 Stimuli

MR-compatible electrodynamic headphones (MR Confon GmbH, Magdeburg, Germany; (Baumgart et al. 1998)) were used during the fixed-state runs to deliver sound stimuli from a standard PC with soundcard to participants. Subjects were asked to wear foam earplugs underneath the headset to further diminish the acoustic noise created by the scanner. Visual stimuli were presented by means of a built-in MR-compatible LCD display.

2.4 Preprocessing

Nipype (Gorgolewski et al. 2011) was used to preprocess the fMRI data. Preprocessing consisted of realignment (FSL), artifact rejection (RapidArt) and spatiotemporal filtering to reduce physiological noise (Behzadi et al. 2007). Further preprocessing steps included band-pass filtering (0.1–1.0 Hz), registration to an individual high-resolution T1-weighted anatomical image and normalization into MNI stereotaxic space using ANTs (Klein et al. 2009). The data were resampled to a 3-mm isotropic resolution and moderately smoothed using an isotropic 5-mm FWHM Gaussian kernel to improve signal to noise ratio.

2.5 Group Analysis

Next, Principal Component Analysis (PCA) was performed to obtain 25 components for each dataset and each subject separately, explaining on average 79.8% of the total variance. These components were subsequently fed into a group Probabilistic Independent Component Analysis as implemented in FSL's MELODIC (v. 3.14) using multi-session temporal concatenation. The obtained networks consisted of spatial maps with their respective time-courses. Subject-specific maps and their time-courses were obtained by back-projection using FSL's Dual Regression. This method uses two stages of regression where in a first stage the individual time-courses are obtained that best fit the group maps. In a second stage, the raw data is regressed against these time-courses to obtain subject-specific component-maps corresponding to the overall group-map.

Importantly, the outputs of stage one of the dual regression —the individual time-courses— were variance normalized before the second stage of the dual regression, allowing for assessment of the shape and amplitude of the component maps. Manual inspection of the time-courses and maps as well as the (spatial) correlation of the maps with previously reported resting-state networks (Smith et al. 2009) were used to distinguish meaningful components from components containing non-neural signals.

For each of the components, a mixed-effects linear model was composed in AFNIs 3dLME (Chen et al. 2013) with factors group (tinnitus vs. controls) and task (fixed-state vs. resting state) and their interaction. The subjects' component maps were fed into this analysis and the significance of the factors was assessed for each component. Further, we assessed the power spectral density (PSD) of the corresponding time-courses.

3 Results

A total of ten networks were identified as neural in origin (Fig. 1). Based on a spatial correlation coefficient with previous networks (Smith et al. 2009) we identified clear sensorimotor networks: two visual networks [Vis1, Vis2], one auditory network [Aud1], a somatosensory/motor network [Sens.Motor] and the cerebellum [Cereb]. In addition, typical resting-state networks were found, such as the limbic system [Limbic], the dorsal attention network [DAN1, DAN2], the default mode network [DMN] and a frontoparietal network [Front.Par].

From these group-level networks, we obtained subject-specific component-maps by applying a dual-regression technique (Filippini et al. 2009). This method uses two stages of regression, similar to back-projection of individual data to group maps used in other studies. By applying dual-regression we allow for variation between subjects, while maintaining the correspondence of components across subjects.

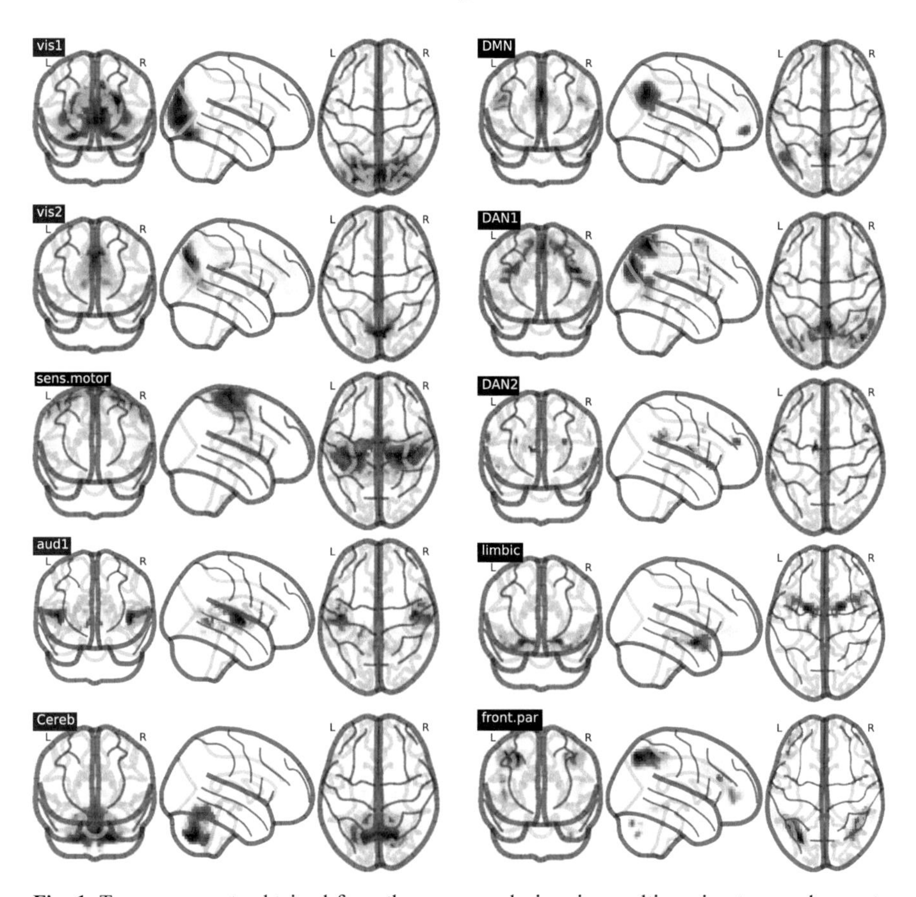

Fig. 1 Ten components obtained from the group analysis using multi-session temporal concatenation of all individual data. Each of the components is thresholded at $z > 3$. The maps in the *left* column are classical task-related sensory networks while those in the *right* column show typical resting-state networks

The subject-specific maps were subsequently fed into a linear mixed model (AFNI's 3dLME) with factors group and task. Importantly, the outputs of stage one of the dual regression—the individual time-courses—were variance normalized before the second stage of the dual regression. In this manner the linear-mixed model now tests for both differences between groups in the shape of the network as well as in the amplitude of the network. The results from these analyses indicated that several networks behaved differently in controls than in subjects with tinnitus. Most notably it identified the default mode network (DMN), roughly encompassed by two distinct but similar networks in the data

Figure 2 shows the two networks that partly overlap. Panel A indicates the DMN that also correlates well with previous demarcations of the DMN (spatial correlation = 0.8) (Smith et al. 2009). In addition, panel E shows a component consisting of a visual area and the posterior and anterior cingulate gyrus. In both the DMN

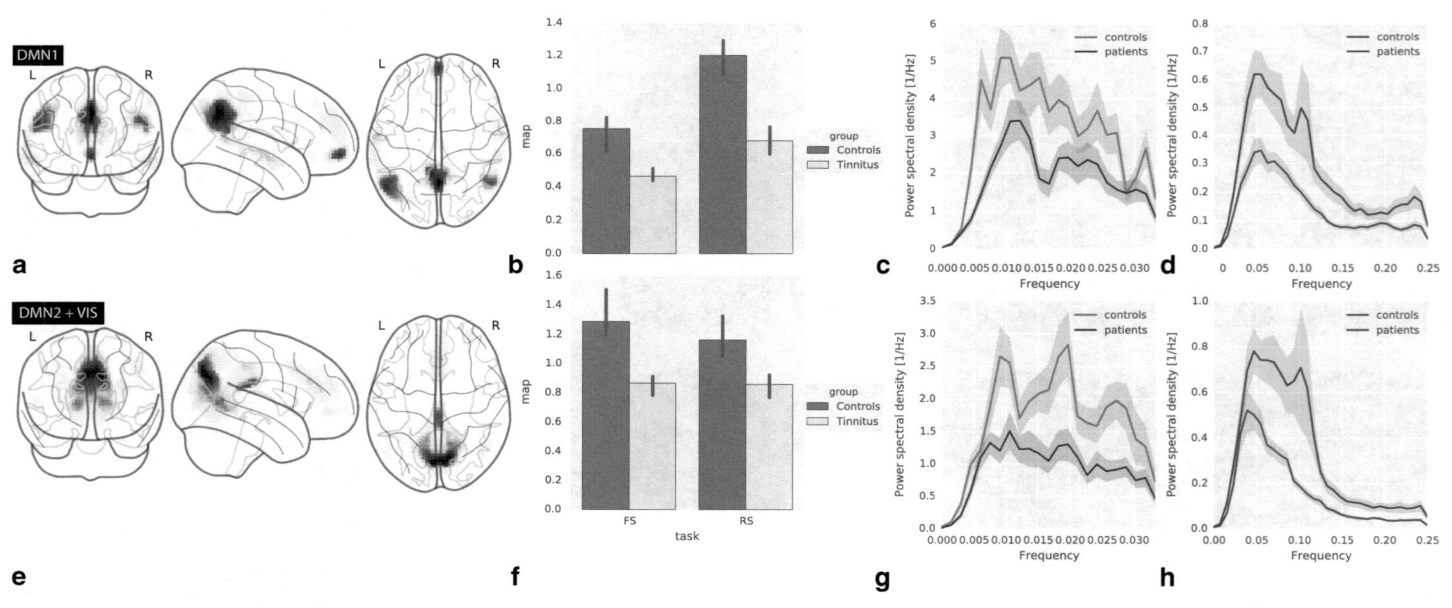

Fig. 2 Group differences within two neural networks. The *top* row shows the default mode network (DMN) whereas the *bottom* row shows the combination of a visual area and the anterior and posterior cingulate gyrus. Panels B and F show the differences in map strength between the groups and tasks. Panels C and G show the power spectra of the two groups for the fixed-state data whereas panels D and H show the same for the resting state data

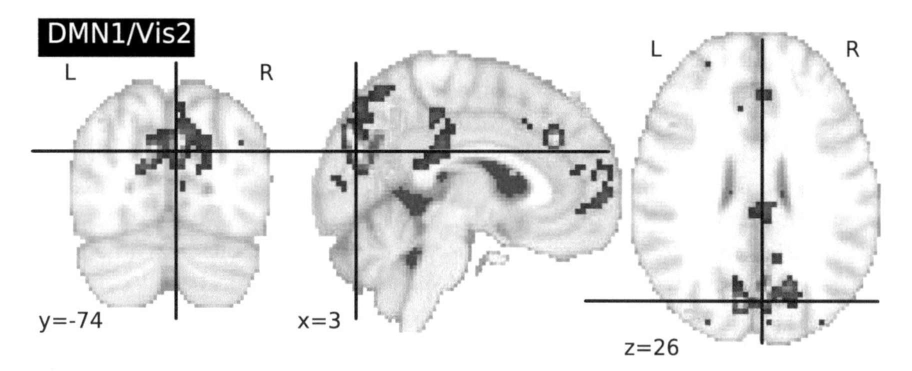

Fig. 3 Group differences within two neural networks. The *red-yellow* colours indicate the significant differences in the network representing a visual area and the cingulate gyrus (see Fig. 2e), and the *blue-purple* colours indicate the differences in the default-mode network (Fig. 2a). These areas are remarkably adjacent, corresponding to the *left* and *right* (pre)cuneus. This area shows significantly reduced functional connectivity in patients compared to controls in both networks

and the latter component, there are adjacent voxels that show a clear and significant group difference ($p < 0.001$ for each of the maps, corresponding to a F-value of 12.71). When looking at the map strength, a proxy for functional connectivity, patients consistently show decreased functional connectivity (panels B and F). Moreover, whereas controls show a clear task-effect, as would be expected for the DMN, this task-effect is much smaller in patients (panel B). For the network with the visual cortex and cingulate gyrus (panel E), this task-effect is less clear (panel F) and shows in controls a decrease during resting state compared to watching a movie, presumably due to the lack of visual stimulation throughout the acquisition of resting-state data. The pattern of reduced connectivity is also clearly visible as a decreased power in the time-courses of the same voxels. This is shown for the both the fixed-state data (panels C and G) and the resting-state data (panels D and F). In all cases, the power is significantly lower in the patient groups.

Figure 3 shows the areas where the subject groups differ significantly are almost if not completely adjacent, indicating the potential relevance of the two spatial networks in tinnitus.

4 Discussion

This study shows that by using resting-state fMRI in combination with independent component analysis we are able to detect differences in neural networks between substantially-sized subject groups. Resting-state fMRI is a valuable addition to traditional task-related fMRI, where a difference between conditions is obtained. This might not necessarily be the best paradigm to study tinnitus. The consequence of a task-related design is that, by using a subtraction approach, the tinnitus-related activity, present in both conditions, will not be present in the subtraction image unless the task affects the perception of tinnitus (Lanting et al. 2009).

Our data analysis is different from other studies that have used resting-state fMRI to study tinnitus. Often the decomposition of the data into independent component is done by performing two (or more) separate ICA analyses on two (or more) groups (see e.g. Schmidt et al. 2013). This leads to a difficult interpretation since it is then unclear whether group differences represent (true) differences between subject groups, or whether the data-decomposition was slightly different (biases) for the two groups.

In contrast, we adopted a strategy where the data is pooled over all participants and data-modalities (resting- and fixed-state data). In this way, the data-decomposition is not determined in a potentially different way for both groups, leaving out the necessity of post-ICA matching of corresponding component maps. This allows for a better group-estimate of the components. In addition, by using a dual regression technique we can obtain subject-specific maps and perform group-wise analyses on these maps (Filippini et al. 2009).

This study shows that by using resting-state fMRI in combination with independent component analysis we are able to detect differences in neural networks between substantially-sized subject groups. Resting-state fMRI is a valuable addition to traditional task-related fMRI, where a difference between conditions is obtained. This might not necessarily be the best paradigm to study tinnitus. The consequence of a task-related design is that, by using a subtraction approach, the tinnitus-related activity, present in both conditions, will not be present in the subtraction image unless the task affects the perception of tinnitus (Lanting et al. 2009)

Our data analysis is different from other studies that have used resting-state fMRI to study tinnitus. Often the decomposition of the data into independent component is done by performing two (or more) separate ICA analyses on two (or more) groups (see e.g. (Schmidt et al. 2013)). This lead to a difficult interpretation since it is then unclear whether group differences represent (true) differences between subject groups, or whether the data-decomposition was slightly different (biases) for the two groups.

In contrast, we adopted a strategy where the data is pooled over all participants and data-modalities (resting- and fixed-state data). In this way, the data-decomposition is not determined in a potentially different way for both groups, leaving out the necessity of post-ICA matching of corresponding component maps. This allows for a better group-estimate of the components. In addition, by using a dual regression technique we can obtain subject-specific maps and perform group-wise analyses on these maps (Filippini et al. 2009). Using this methodological approach we found a significant group effects for the DMN.

Our interpretation is that the activity of the DMN in patients with tinnitus, whether as the source or the consequence of tinnitus, is reduced in comparison to the activity in controls. The DMN is relatively better connected internally (i.e. larger spontaneous fluctuations, indicative of stronger connectivity) when the mind is at rest and allowed to wander or daydream. This network is believed to be key in how a person introspectively understands themselves and others, and forms beliefs, intentions, and desires through autobiographical memory. Recently it was shown that a dysfunctional DMN is related to autism (Washington et al. 2014). Perhaps the

DMN plays a similar role in tinnitus, where through the on-going tinnitus patients can't get into a state of mind-wandering and therefore show a reduced activity in the DMN compared to controls.

This DMN has been previously implicated in other diseases such as Alzheimer's disease. Lower functional connectivity scores in the DMN were found in an Alzheimer's disease group when compared to normal groups (Binnewijzend et al. 2012). In contrast, results of resting-state data-analysis in schizophrenia have been variable (Greicius 2008). Schizophrenia is regarded as a large cluster of profound neuropsychiatric symptoms. A subtype of patients (also) experience phantom perception of sounds. One interpretation is that, since the DMN is different between groups in previous work and ours, the DMN somehow plays a role in 'hearing' internally generated sound (whether it is meaningful like in patients with schizophrenia or meaningless like tinnitus).

Our study is novel in the sense that it specifically assesses the influence of performing a task on the scale of brain networks as a whole. Specifically in patients with tinnitus, the DMN does not clearly show a task-effect, which in contrast is clearly present in controls. This observation enhances the hypothesis that subjects with tinnitus cannot get the 'rest' that is necessary for a healthy life. In conclusion, we attribute the pattern of reduced connectivity in the DMN to the unremitting engaging effect of the tinnitus percept.

Acknowledgments DRML was funded by VENI research grant 016.096.011 from the Netherlands organization for scientific research (NWO) and the Netherlands organization for health research and development (ZonMw). Further financial support was provided by the Heinsius Houbolt Foundation.

References

Adjamian P, Hall DA, Palmer AR, Allan TW, Langers DRM (2014). Neuroanatomical abnormalities in chronic tinnitus in the human brain. Neurosci Biobehav Rev 45:119–133

Baumgart F, Kaulisch T, Tempelmann C, Gaschler-Markefski B, Tegeler C, Schindler F, Scheich H (1998) Electrodynamic headphones and woofers for application in magnetic resonance imaging scanners. Med Phys 25(10):2068–2070

Behzadi Y, Restom K, Liau J, Liu TT (2007) A component based noise correction method (CompCor) for BOLD and perfusion based fMRI. Neuroimage 37(1):90–101. http://doi.org/10.1016/j.neuroimage.2007.04.042

Binnewijzend MAA, Schoonheim MM, Sanz-Arigita E, Wink AM, van der Flier WM, Tolboom N, Barkhof F (2012) Resting-state fMRI changes in Alzheimer's disease and mild cogni-

tive impairment. Neurobiol Aging 33(9):2018–2028. http://doi.org/10.1016/j.neurobiolaging.2011.07.003
Brand T, Hohmann V (2002) An adaptive procedure for categorical loudness scaling. J Acoust Soc Am 112(4):1597–1604. http://doi.org/10.1121/1.1502902
Budd RJ, Pugh R (1996) Tinnitus coping style and its relationship to tinnitus severity and emotional distress. J Psychosom Res 41(4):327–335. http://doi.org/10.1016/S0022-3999(96)00171-7
Burton H, Wineland A, Bhattacharya M, Nicklaus J, Garcia KS, Piccirillo JF (2012) Altered networks in bothersome tinnitus: a functional connectivity study. BMC Neurosci 13(1):3. http://doi.org/10.1186/1471-2202-13-3
Chen G, Saad ZS, Britton JC, Pine DS, Cox RW (2013) Linear mixed-effects modeling approach to fMRI group analysis. Neuroimage 73:176–190. http://doi.org/10.1016/j.neuroimage.2013.01.047
Chen Y, Zhang J, Li X, Xia W, Feng X, Qian C, Teng G (2015a) Altered intra- and interregional synchronization in resting-state cerebral networks associated with chronic tinnitus. Neural Plast 2015:1–11. http://doi.org/10.1155/2015/475382
Chen Y, Zhang J, Li X, Xia W, Feng X, Qian C, Teng G (2015b) Altered intra- and interregional synchronization in resting-state cerebral networks associated with chronic tinnitus. Neural Plast 2015:1–11. http://doi.org/10.1155/2015/475382
Davies J, Gander PE, Andrews M, Hall Da (2014) Auditory network connectivity in tinnitus patients: a resting-state fMRI study. Int J Audiol 53(3):192–198. http://doi.org/10.3109/14992027.2013.846482
Filippini N, MacIntosh BJ, Hough MG, Goodwin GM, Frisoni GB, Smith SM, Mackay CE (2009) Distinct patterns of brain activity in young carriers of the APOE-epsilon4 allele. Proc Natl Acad Sci USA 106(17):7209–7214. http://doi.org/10.1073/pnas.0811879106
Gorgolewski K, Burns CD, Madison C, Clark D, Halchenko YO, Waskom ML, Ghosh SS (2011). Nipype: a flexible, lightweight and extensible neuroimaging data processing framework in python. Front Neuroinform 5(August):13. http://doi.org/10.3389/fninf.2011.00013
Greicius M (2008) Resting-state functional connectivity in neuropsychiatric disorders. Curr Opin Neurol 21(4):424–430. http://doi.org/10.1097/WCO.0b013e328306f2c5
Greicius MD, Srivastava G, Reiss AL, Menon V (2004) Default-mode network activity distinguishes Alzheimer's disease from healthy aging: evidence from functional MRI. Proc Natl Acad Sci USA 101(13):4637–4642. http://doi.org/10.1073/pnas.0308627101
Khalfa S, Dubal S, Veuillet E, Perez-Diaz F, Jouvent R, Collet L (2002) Psychometric normalization of a hyperacusis questionnaire. ORL 64:436–442. http://doi.org/10.1159/000067570
Kim J, Kim Y, Lee S, Seo J-H, Song H-J, Cho JH, Chang Y (2012). Alteration of functional connectivity in tinnitus brain revealed by resting-state fMRI?: a pilot study. Int J Audiol. http://doi.org/10.3109/14992027.2011.652677
Klein A, Andersson J, Ardekani B, Ashburner J, Avants B, Chiang M-C, Parsey RV (2009) Evaluation of 14 nonlinear deformation algorithms applied to human brain MRI registration. Neuroimage 46(3):786–802. http://doi.org/10.1016/j.neuroimage.2008.12.037
Langers DRM, de Kleine E, van Dijk P (2012). Tinnitus does not require macroscopic tonotopic map reorganization. Front Syst Neurosci 6:2. http://doi.org/10.3389/fnsys.2012.00002
Lanting CP, de Kleine E, van Dijk P (2009) Neural activity underlying tinnitus generation: results from PET and fMRI. Hear Res 255(1–2):1–13. http://doi.org/10.1016/j.heares.2009.06.009
Maudoux A, Lefebvre P, Cabay J-E, Demertzi A, Vanhaudenhuyse A, Laureys S, Soddu A (2012a) Auditory resting-state network connectivity in tinnitus: a functional MRI study. PloS One 7(5):e36222. http://doi.org/10.1371/journal.pone.0036222
Maudoux, A, Lefebvre P, Cabay J-E, Demertzi A, Vanhaudenhuyse A, Laureys S, Soddu A (2012b) Connectivity graph analysis of the auditory resting state network in tinnitus. Brain Res 1485:10–21. http://doi.org/10.1016/j.brainres.2012.05.006
Newman CW, Jacobson GP, Spitzer JB (1996). Development of the tinnitus handicap inventory. Arch Otolaryngol Head Neck Surg 122:143–148. http://doi.org/10.1001/archotol.1996.01890140029007

Schmidt SA, Akrofi K, Carpenter-Thompson JR, Husain FT (2013) Default mode, dorsal attention and auditory resting state networks exhibit differential functional connectivity in tinnitus and hearing loss. PloS One 8(10):e76488. http://doi.org/10.1371/journal.pone.0076488

Smith SM, Fox PT, Miller KL, Glahn DC, Fox PM, Mackay CE, Beckmann CF (2009) Correspondence of the brain's functional architecture during activation and rest. Proc Natl Acad Sci USA 106(31):13040–13045. http://doi.org/10.1073/pnas.0905267106

Washington SD, Gordon EM, Brar J, Warburton S, Sawyer AT, Wolfe A, Vanmeter JW (2014) Dysmaturation of the default mode network in autism. Hum Brain Mapp 35(4):1284–1296. http://doi.org/10.1002/hbm.22252

Wilson PH, Henry J, Bowen M, Haralambous G (1991) Tinnitus reaction questionnaire: psychometric properties of a measure of distress associated with tinnitus. J Speech Hear Res 34(1):197–201

Zhang D, Snyder AZ, Fox MD, Sansbury MW, Shimony JS, Raichle ME (2008) Intrinsic functional relations between human cerebral cortex and thalamus. J Neurophysiol 100(4):1740–1748. http://doi.org/10.1152/jn.90463.2008

The Role of Conduction Delay in Creating Sensitivity to Interaural Time Differences

Catherine Carr, Go Ashida, Hermann Wagner, Thomas McColgan and Richard Kempter

Abstract Axons from the nucleus magnocellularis (NM) and their targets in nucleus laminaris (NL) form the circuit responsible for encoding interaural time difference (ITD). In barn owls, NL receives bilateral inputs from NM, such that axons from the ipsilateral NM enter NL dorsally, while contralateral axons enter from the ventral side. These afferents act as delay lines to create maps of ITD in NL. Since delay-line inputs are characterized by a precise latency to auditory stimulation, but the postsynaptic coincidence detectors respond to ongoing phase difference, we asked whether the latencies of a local group of axons were identical, or varied by multiples of the inverse of the frequency they respond to, i.e., to multiples of 2π phase. Intracellular recordings from NM axons were used to measure delay-line latencies in NL. Systematic shifts in conduction delay within NL accounted for the maps of ITD, but recorded latencies of individual inputs at nearby locations could vary by 2π or 4π. Therefore microsecond precision is achieved through sensitivity to phase delays, rather than absolute latencies. We propose that the auditory system "coarsely" matches ipsilateral and contralateral latencies using physical delay lines,

C. Carr (✉)
Department of Biology, University of Maryland, College Park, MD, USA
e-mail: cecarr@umd.edu

G. Ashida
Cluster of Excellence "Hearing4all", University of Oldenburg, Oldenburg, Germany
e-mail: go.ashida@uni-oldenburg.de

H. Wagner
Institute for Biology II, RWTH Aachen, Aachen, Germany
e-mail: wagner@bio2.rwth-aachen.de

T. McColgan · R. Kempter
Institute for Theoretical Biology, Department of Biology,
Humboldt-Universität zu Berlin, Berlin, Germany
e-mail: thomas.mccolgan@gmail.com

R. Kempter
e-mail: r.kempter@biologie.hu-berlin.de

P. van Dijk et al. (eds.), *Physiology, Psychoacoustics and Cognition in Normal and Impaired Hearing,* Advances in Experimental Medicine and Biology 894,
DOI 10.1007/978-3-319-25474-6_20

so that inputs arrive at NL at about the same time, and then "finely" matches latency modulo 2π to achieve microsecond ITD precision.

Keywords Coding · Interaural time difference · Plasticity · Models · Rate

1 Introduction

A key feature of ITD sensitivity is how the observed microsecond delays may be created by neural elements that are both noisy and slow. There are two possibilities, illustrated by the example of a coincidence detector neuron that responds to a pure sine tone, presented to both ears at an ITD of 0 μs, or sound in front of the animal. One strategy requires equal conduction latencies from each ear to the coincidence detector, such that the inputs would arrive with equal latency and equal phase. The second strategy relies on relative delays and would simply require that inputs arrive at the same phase. Due to the periodicity of the sine wave, both scenarios would produce the same output. This coding strategy remains viable for non-periodic sounds because the tonotopic organization of the auditory system generates ringing properties of the inputs characteristic of the particular frequency (Carney et al. 1999; Recio et al. 1998; Wagner et al. 2005). Previously we had shown that the inputs from the cochlear nuclei act as delay line inputs to form maps of ITD in the barn owl (Carr and Konishi 1990), but could not determine whether the delay lines had identical conduction latencies or were matched for phase only.

To differentiate between the two hypotheses, we combined measurements of the extracellular potential or neurophonic with intracellular measures of click latency in NL. For mapping the ITD we used the neurophonic, because it reflects local ITD sensitivity (Sullivan and Konishi 1986; Kuokkanen et al. 2010; Mc Laughlin et al. 2010; Kuokkanen et al. 2013), and to measure delay we used clicks, because their temporal occurrence is precise to within about 20 μs (Wagner et al. 2005). We found the microsecond precision needed to construct maps of ITD is achieved through precisely regulated phase delays, rather than by regulation of absolute latency, consistent with the observation that nucleus laminaris neurons respond over a wide range of integer multiples of their preferred interaural phase difference (Christianson and Peña 2006). Thus latencies from the ear to the point of coincidence detection need only be adjusted within a stimulus period, creating a flexible yet precise system for detection of sound source sound location. This high precision approach should also benefit other binaural properties of the auditory system, such as resistance to background noise and reverberation.

2 Methods

The experiments were conducted at the University Maryland. Four barn owls (*Tyto alba*) were used to collect these data, and procedures conformed to NIH guidelines for Animal Research and were approved by the Animal Care and Use Committee of the Universities of Maryland. Anaesthesia was induced by intramuscular injections of 10–20 mg/kg ketamine hydrochloride and 3–4 mg/kg xylazine, with supplementary doses to maintain a suitable plane of anaesthesia. Recordings were made in a sound-attenuating chamber (IAC, New York). Intracellular recordings were amplified by an Axoclamp 2B, and then by a spike conditioner (PC1, Tucker-Davis Technologies (TDT), Gainesville, FL). Acoustic stimuli were digitally generated by custom-made software ("Xdphys" written in Dr. M. Konishi's lab at Caltech) driving a signal-processing board (DSP2 (TDT)). Acoustic signals were fed to earphones, and inserted into the owl's left and right ear canals, respectively. At a given recording site, we measured frequency tuning, then tuning to ITD, and responses to monaural clicks.

3 Results

Axons from NM form presynaptic maps of ITD in NL. These inputs generate a field potential that varies systematically with recording position and can be used to map ITDs. In the barn owl, the representation of best ITD shifts systematically in NL, forming multiple, largely overlapping maps of ITD (Fig. 1a). We had previously found that conduction delays could account for the shift in maps of ITD (Carr et al. 2013).

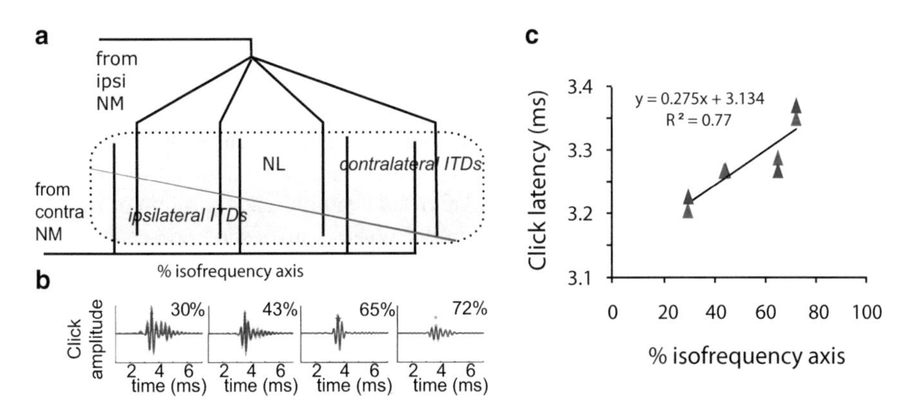

Fig. 1 **a** Schematic outline of an isofrequency slab in NL showing delay-line inputs from each side, with the "0 µs-ITD line" in *red*. **b** Click responses were recorded at best ITDs around 0 µs, at four different locations along a 3.6 kHz iso-frequency slab in NL. Superimposed ipsilateral (*blue*) and contralateral (*red*) averages of 128 click responses from recording sites at 30, 43, 65 and 72 % along the isofrequency axis, measured from 3-D reconstructions of NL with lesions at each recording location. **c** Click delays from b. showed a systematic increase in latency with mediolateral position in NL for both ipsi- and contralateral responses

3.1 Measuring Latency with Click Delays

Our mapping studies showed that the representation of frontal space (0 μs ITD) in NL shifted systematically to more ventral locations with the progression from medial to lateral along each tonotopic slab (red line, Fig. 1a, see Carr et al. 2013). To determine if changes in conduction velocity formed the basis of the systematic shift, we used click delay to measure latencies at different mediolateral positions. Measurements of click latency provided reliable measures of conduction delay (Köppl and Carr 2008; Wagner et al. 2005), because click stimuli were temporally precise. Nevertheless, click stimuli evoked an oscillatory response (Ruggero et al. 1986; Wagner et al. 2005; Wagner et al. 2009), in which typically several peaks and troughs could be distinguished (Fig. 1b).

Neurophonic responses to ipsi- and contralateral clicks at best ITDs at or near 0 μs generated similar ipsi- and contralateral click responses, which largely overlapped (Fig. 1b). Responses to both ipsilateral and contralateral stimulation showed a steady increase in latency along the iso-frequency axis in NL (Fig. 1c), i.e., systematic changes in conduction delay underlay the formation of the maps of ITD. To determine whether the delay lines had identical conduction latencies or were matched for phase only, we compared neurophonic multiunit click recordings with intracellular recordings from individual NM axons in NL.

3.2 Latencies of Adjacent Recordings Can Vary by Multiples of 2π

Intracellular recordings from NM axons in NL supported the hypothesis that oscillatory neurophonic click responses were composed of many NM click responses. Averaging intracellular click recordings from NM afferents revealed two to five PSTH peaks (Fig. 2a). PSTH peaks were separated in time by intervals equal to the inverse of the neuron's best frequency (BF), a 2π phase interval. Individual spike timing varied, consistent with phase locking (Anderson et al. 1970; Sachs et al. 1974; Joris et al. 1994; Köppl 1997). For each intracellular recording, we quantified the temporal distance between the PSTH peaks. All showed a similar 2π separation of peaks. Adjacent recordings responded at very similar phases, suggesting adjacent axons fire in synchrony or at multiples of 2π (Fig. 2b, note clustered adjacent symbols).

In the penetration through the dorsoventral dimension of 5.2 kHz NL shown in Fig. 2b and 2c, recordings between 205–270 μm depths yielded three contralateral units with similar phase, but a latency difference of 4π (Fig. 2c, black arrow). The first two recordings had latencies of 3.33 and 3.32 ms, the third 2.91 ms. Thus, the recording with the shorter latency occurred almost 4π or 0.38 ms earlier than the first two, given a stimulus period of 192 μs (3.33 ms − 2 * 0.192 ms = 2.94 ms). In other words, the phase delay of these peaks exhibited a narrow scatter, while the signal-front delay showed a large scatter, as had also been observed for the neurophonic (Wagner et al. 2005). A similar measurement at 865 μm had a latency of

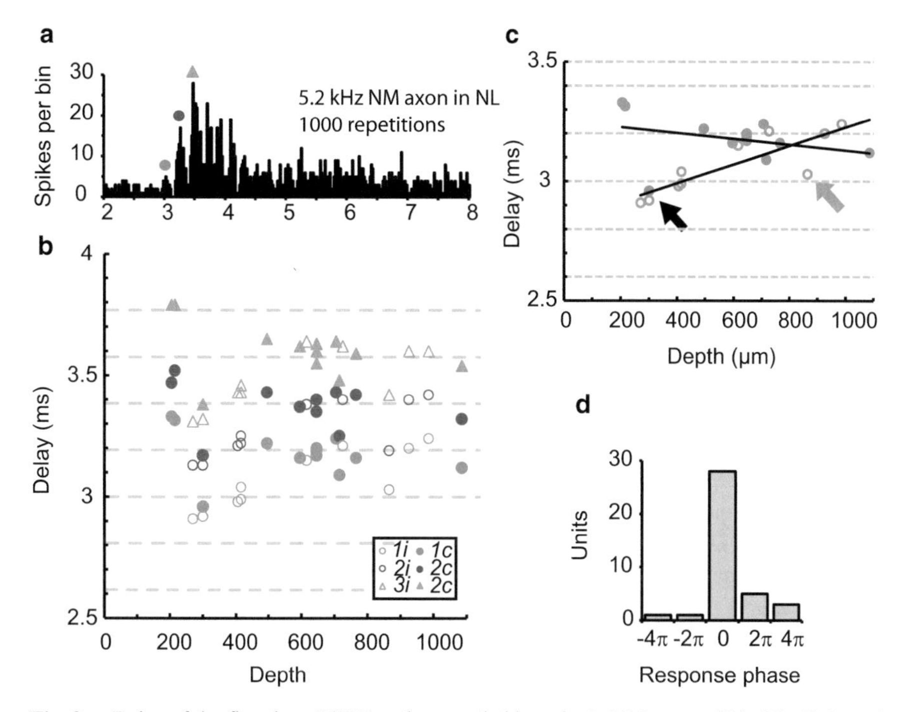

Fig. 2 **a** Delay of the first three PSTH peaks recorded in a single NM axon within NL. Coloured symbols mark peaks in the PSTH. *Blue circle* = first peak, or signal-front delay, *red circle* = second peak and *green triangle* = third peak. **b** Click latencies in 22 NM afferents recorded in a single penetration through the 5.2 kHz region of left NL. Both ipsilateral (*open symbols*) and contralateral (*closed symbols*) responses were encountered, symbols as in a. The latencies of the first 3 PSTH peaks for these afferents varied by multiples of 1/BF, shown as *horizontal dashed lines* 192 μs apart. The first three PSTH peaks for each recording were shown with respect to recording depth as a vertical array of *blue circles* (first peak), *red circles* (second peak, +2π) and *green triangles* (third peak, +4π). **c** Latencies of the ipsilateral NM axons increased with depth (*open symbols*, y = 0.4 s/m x + 2.84 ms, r^2 = 0.7), while recordings from contralateral NM axons in NL decreased in latency with depth (*closed symbols*, y = −0.12 s/m x + 3.25, r^2 = 0.10). **d** Most data from 3 penetrations through NL (*n* = 38 units, 2 owls) showed systematic shifts in latency with recording depth, consistent with models of delay (see 2b, lumped in 0 μs bin). Other units were 2π or 4π or later than neighbouring units (c, *arrows*)

3.03, while the adjacent ipsilateral recordings at 725 and 925 μm had latencies of 3.21 and 3.20 ms (Fig. 2c, grey arrow). Most measurements had similar phase *and* latency (Fig. 2d, bar marked 0), but some occurred 2π earlier, and others 2π or 4π later than the rest of the population (Fig. 2d).

In general, changes in delay with depth were consistent with the NM axons acting as delay lines (Carr and Konishi 1990), and with modelled conduction velocities in McColgan et al. (2014). Ipsilateral recordings showed an increase in latency with recording depth in NL (open symbols in Fig. 2), while contralateral recordings (closed symbols) showed a decrease in latency with depth, consistent with the delay line geometry. The regression lines in Fig. 2c predicted velocities of +2.5 and –8.3 m/s, consistent with modelling results (McColgan et al. 2014) and measurements from previous studies (Carr and Konishi 1990). Nevertheless, we point out that while these depth-dependent changes in delay would be expected from multiple recordings from the same NM axon, they were not required for ITD sensitivity. Sensitivity to ITDs should be instead conferred by the differences in phase.

Both extracellular neurophonic recordings and intracellular NM axonal recordings in NL revealed maps of ITD, created by precisely regulated conduction delays from intrinsically noisy elements. The microsecond precision needed to construct these maps of ITD appears to be achieved through precisely regulated phase delays, rather than by regulation of absolute latency.

4 Discussion

Delays from each side must be matched for coincidence detection to occur. How? Coincidence requires microsecond precision, which may be achieved by a two-step process, where latencies from each ear are matched during development, and then precisely matched modulo 2π. This was first proposed by Gerstner et al. (1996), who postulated that the necessary degree of coherence in the signal arrival times could be attained during development by an unsupervised Hebbian learning rule that selects connections with matching delays from a broad distribution of axons with random delays. Our data support this matching hypothesis, with coarse delays laid down through normal development, and precise, i.e. ± 20 μs modulo 2π, delays regulated by some as yet unknown activity dependent processes (Seidl et al. 2010).

Anatomical data are consistent with coarsely matched latencies. For each frequency, latencies in barn owl NM and in the midline cross fibre tract are similar, consistent with a common path from the ear into the brain (see Fig. 7 in Köppl 1997). Our previous measures of latency did not have the precision of the current intracellular recordings, with ipsilateral and contralateral delays at the dorsal and ventral borders of NL of 2.82 ± 0.24 and 2.87 ± 0.21 ms respectively (Carr and Konishi 1990). Note the standard deviations of about 200 μs in these early recordings of latency. Our current recordings support the hypothesis that within NL, latencies vary by multiples of 2π, and may be precisely regulated at a fine time scale, in order to create a cycle by cycle representation of the stimulus at the point(s) of coincidence detection (Funabiki et al. 2011; Ashida et al. 2013). Modulation by multiples of 2π is also consistent with cross-correlation and spike timing dependent plasticity models (Gerstner et al. 1996; Kempter et al. 1998; Pena and Konishi 2000; Fischer et al. 2008). Variability in response latency also characterizes mammalian auditory nerve and cochlear nucleus recordings (Sanes and Constantine-Paton 1985; Carney and Yin 1988; Young et al. 1988).

Acknowledgments This research was sponsored by NIH DC00436 to CEC, by NIH P30 DC04664 to the University of Maryland Center for the Comparative and Evolutionary Biology of Hearing, by the German Research Foundation (DFG, Wa-606/12, Ke-788/1-3, 4) and the Bundesministerium für Bildung und Forschung (BMBF: 01GQ0972 and 01GQ1001A, Bernstein Collaboration Temporal Precision, 01GQ07101 to HW and 01GQ07102 to RK,), and by the cluster of Excellence, "Hearing4all" at the University of Oldenburg (GA).

References

Anderson DJ, Rose JE, Hind JE, Brugge J (1970) Temporal position of discharges in single auditory nerve fibers within the cycle of a sine-wave stimulus: frequency and intensity effects. J Acoust Soc Am 49(2):1131–1139

Ashida G, Funabiki K, Carr CE (2013) Biophysical basis of the sound analog membrane potential that underlies coincidence detection in the barn owl. Front Comput Neurosci 7:102. doi:10.3389/fncom.2013.00102

Carney L, McDuffy M, Shekhter I (1999) Frequency glides in the impulse responses of auditory-nerve fibers. J Acoust Soc Am 105:2384

Carney LH, Yin TCT (1988) Temporal coding of resonances by low-frequency auditory nerve fibers: single-fiber responses and a population model. J Neurophysiol 60:1653–1677

Carr CE, Konishi M (1990) A circuit for detection of interaural time differences in the brain stem of the barn owl. J Neurosci 10(10):3227–3246

Carr CE, Shah S, Ashida G, McColgan T, Wagner H, Kuokkanen PT et al (2013) Maps of ITD in the nucleus laminaris of the barn owl. Adv Exp Med Biol 787:215–222. doi:10.1007/978-1-4614-1590-9_24

Christianson GB, Peña JL (2006) Noise reduction of coincidence detector output by the inferior colliculus of the barn owl. J Neurosci 26(22):5948–5954

Fischer BJ, Christianson GB, Peña JL (2008) Cross-correlation in the auditory coincidence detectors of owls. J Neurosci 28(32):8107–8115

Funabiki K, Ashida G, Konishi M (2011) Computation of interaural time difference in the owl's coincidence detector neurons. J Neurosci 31(43):15245–15256

Gerstner W, Kempter R, van Hemmen JL, Wagner H (1996) A neuronal learning rule for sub-millisecond temporal coding. Nature 383(6595):76–78. doi:10.1038/383076a0

Joris PX, Carney LH, Smith PH, Yin TCT (1994) Enhancement of neural synchronization in the anteroventral cochlear nucleus. I. Responses to tones at the characteristic frequency. J Neurophysiol 71(3):1022–1036

Kempter R, Gerstner W, van Hemmen JL, Wagner H (1998) Extracting oscillations: Neuronal coincidence detection with noisy periodic spike input. Neural Comput 10(8):1987–2017

Köppl C (1997) Frequency tuning and spontaneous activity in the auditory nerve and cochlear nucleus magnocellularis of the barn owl Tyto alba. J Neurophysiol 77:364–377

Köppl C, Carr CE (2008) Maps of interaural time difference in the chicken's brainstem nucleus laminaris. Biol Cybern 98(6):541–559. doi:10.1007/s00422-008-0220-6

Kuokkanen PT, Ashida G, Carr CE, Wagner H, Kempter R (2013) Linear summation in the barn owl's brainstem underlies responses to interaural time differences. J Neurophysiol 110(1):117–130. doi:10.1152/jn.00410.2012

Kuokkanen PT, Wagner H, Ashida G, Carr CE, Kempter R (2010) On the origin of the extracellular field potential in the nucleus laminaris of the barn owl (Tyto alba). J Neurophysiol 104(4):2274–2290. doi:10.1152/jn.00395.2010

Mc Laughlin M, Verschooten E, Joris PX (2010) Oscillatory dipoles as a source of phase shifts in field potentials in the mammalian auditory brainstem. J Neurosci 30(40):13472–13487. doi:10.1523/JNEUROSCI.0294-10.2010

McColgan T, Shah S, Köppl C, Carr CE, Wagner H (2014) A functional circuit model of interaural time difference processing. J Neurophysiol 112(11):2850–2864. doi:10.1152/jn.00484.2014

Pena JL, Konishi M (2000) Cellular mechanisms for resolving phase ambiguity in the owl's inferior colliculus. Proc Natl Acad Sci U S A 97(22):11787–11792. doi:10.1073/pnas.97.22.11787

Recio A, Rich N, Narayan S, Ruggero M (1998). Basilar-membrane responses to clicks at the base of the chinchilla cochlea. J Acoust Soc Am 103(4):1972–1989.

Ruggero M, Robles L, Rich N, Costalupes J (1986) Basilar membrane motion and spike initiation in the cochlear nerve. In: Moore BCJ, Patterson RD (eds) Auditory frequency selectivity (119 ed., Vol. 189). Auditory Frequency Selectivity, London

Sachs M, Young E, Lewis R (1974) Discharge patterns of single fibers in the pigeon auditory nerve. Brain Res 70(3):431–447

Sanes DH, Constantine-Paton M (1985) The sharpening of frequency tuning curves requires patterned activity during development in the mouse, Mus musculus. J Neurosci 5(5):1152–1166

Seidl AH, Rubel EW, Harris DM (2010) Mechanisms for adjusting interaural time differences to achieve binaural coincidence detection. J Neurosci 30(1):70–80. doi:10.1523/JNEUROSCI.3464-09.2010

Sullivan WE, Konishi M (1986) Neural map of interaural phase difference in the owl's brainstem. Proc Natl Acad Sci U S A 83:8400–8404

Wagner H, Brill S, Kempter R, Carr CE (2005) Microsecond precision of phase delay in the auditory system of the barn owl. J Neurophysiol 94(2):1655–1658

Wagner H, Brill S, Kempter R, Carr CE (2009) Auditory responses in the barn owl's nucleus laminaris to clicks: impulse response and signal analysis of neurophonic potential. J Neurophysiol 102(2):1227–1240. doi:10.1152/jn.00092.2009

Wagner H, Takahashi T, Konishi M (1987) Representation of interaural time difference in the central nucleus of the barn owl's inferior colliculus. J Neurosci 7(10):3105–3116

Young E, Robert J, Shofner W (1988) Regularity and latency of units in ventral cochlear nucleus: implications for unit classification and generation of response properties. J Neurophysiol 60(1):1–29

Objective Measures of Neural Processing of Interaural Time Differences

David McAlpine, Nicholas Haywood, Jaime Undurraga and Torsten Marquardt

Abstract We assessed neural sensitivity to interaural time differences (ITDs) conveyed in the temporal fine structure (TFS) of low-frequency sounds and ITDs conveyed in the temporal envelope of amplitude-modulated (AM'ed) high-frequency sounds. Using electroencephalography (EEG), we recorded brain activity to sounds in which the interaural phase difference (IPD) of the TFS (or the modulated temporal envelope) was repeatedly switched between leading in one ear or the other. When the amplitude of the tones is modulated equally in the two ears at 41 Hz, the interaural phase modulation (IPM) evokes an IPM following-response (IPM-FR) in the EEG signal. For low-frequency signals, IPM-FRs were reliably obtained, and largest for an IPM rate of 6.8 Hz and when IPD switches (around 0°) were in the range 45–90°. IPDs conveyed in envelope of high-frequency tones also generated IPM-FRs; response maxima occurred for IPDs switched between 0° and 180° IPD. This is consistent with the interpretation that distinct binaural mechanisms generate the IPM-FR at low and high frequencies, and with the reported physiological responses of medial superior olive (MSO) and lateral superior olive (LSO) neurons in other mammals. Low-frequency binaural neurons in the MSO are considered maximally activated by IPDs in the range 45–90°, consistent with their reception of excitatory inputs from both ears. High-frequency neurons in the LSO receive excitatory and inhibitory input from the two ears receptively—as such maximum activity occurs when the sounds at the two ears are presented out of phase.

D. McAlpine (✉) · N. Haywood · J. Undurraga · T. Marquardt
UCL Ear Institute, 332 Gray's Inn Road, London WC1X 8EE, UK
e-mail: d.mcalpine@ucl.ac.uk

N. Haywood
e-mail: n.haywood@ucl.ac.uk

J. Undurraga
e-mail: j.undurraga@ucl.ac.uk

T. Marquardt
e-mail: t.marquardt@ucl.ac.uk

P. van Dijk et al. (eds.), *Physiology, Psychoacoustics and Cognition in Normal and Impaired Hearing,* Advances in Experimental Medicine and Biology 894,
DOI 10.1007/978-3-319-25474-6_21

Keywords Interaural phase modulation · Objective measures · Frequency-following response · Electroencephalography

1 Introduction

1.1 Advantages of Binaural Listening

Binaural hearing confers considerable advantages in everyday listening environments. By comparing the timing and intensity of a sound at each ear, listeners can locate its source on the horizontal plane, and exploit these differences to hear out signals in background noise—an important component of 'cocktail party listening'. Sensitivity to interaural time differences (ITDs) in particular has received much attention, due in part to the exquisite temporal performance observed; for sound frequencies lower than about 1.3 kHz, ITDs of just a few tens of microseconds are discriminable, observed both at the neural and behavioural levels. Sensitivity to ITDs decreases with ageing (Herman et al. 1977; Abel et al. 2000; Babkoff et al. 2002), and is typically impaired in hearing loss, albeit variability across individuals and aetiologies (see Moore et al. 1991, for review), with reduced sensitivity most apparent when trying to localise sources in background noise (Lorenzi et al. 1999). ITD processing may also remain impaired following surgery to correct conductive hearing loss (Wilmington et al. 1994), and sensitivity to ITDs is typically poor in users of hearing technologies such as bilateral cochlear implants (CIs). In terms of restoring binaural function in therapeutic interventions, there is considerable benefit to be gained from developing objective measures to assess binaural processing.

1.2 Objective Measures of Binaural Hearing

Direct measures of ITD sensitivity, attributed to multiple generator sites in the thalamus and auditory cortex (Picton et al. 1974; Hari et al. 1980; Näätänen and Picton 1987; Liégeois-Chauvel et al. 1994), have been demonstrated in auditory-evoked 'P1-N1-P2' responses, with abrupt changes in either the ITD (e.g. McEvoy 1991), or the interaural correlation of a noise (e.g. Chait et al. 2005), eliciting responses with typical latencies of 50, 100, and 175 ms, respectively. Although these studies demonstrate the capacity to assess binaural processing of interaural temporal information, they provide for only a relatively crude measure of ITD sensitivity, comparing, for example, EEG responses to diotic (identical) sounds at the two ears with responses to sounds that are statistically independent at the two ears. A more refined measure of ITD sensitivity would provide the possibility of assessing neural mechanisms of binaural processing in the human brain, aiding bilateral fitting of CIs in order to maximise binaural benefit.

2 Methods

Eleven NH listeners took part in the low-frequency EEG experiment (6 male; mean age = 24.4 years, range = 18–34 years). Four listeners took part in the high-frequency experiment (3 male; mean age = 30.0 years, range = 27–34 years). All subjects demonstrated hearing levels of 20 dB hearing level or better at octave frequencies between 250 and 8000 Hz, and reported a range of musical experience. Subjects were recruited from poster advertisements. All experiments were approved by the UCL Ethics Committee. Subjects provided informed consent, and were paid an honorarium for their time. For EEG recordings assessing sensitivity to ITDs in the low-frequency TFS, 520-Hz carrier tones were modulated with 100 % sinusoidal AM (SAM), at a rate of 41 Hz. Stimuli were presented continuously for 4 min and 48 s (70 epochs of 4.096 s) at 80 dB SPL. Carrier and modulation frequencies were set so that an integer number of cycles fitted into an epoch window. The carrier was presented with an IPD (in the range ± 11.25°–± 135° IPD around 0°), whilst the modulation envelope remained diotic at all times. In order to generate IPM, the overall magnitude of the carrier IPD was held constant throughout the stimulus, but the ear at which the signal was leading in phase was periodically modulated between right and left. Each IPM was applied instantaneously at a minimum in the modulation cycle in order to minimize the (monaural) salience of the instantaneous phase transition (see Fig. 1). Stimuli were AM'ed at a rate of 41 Hz. The IPD cycled

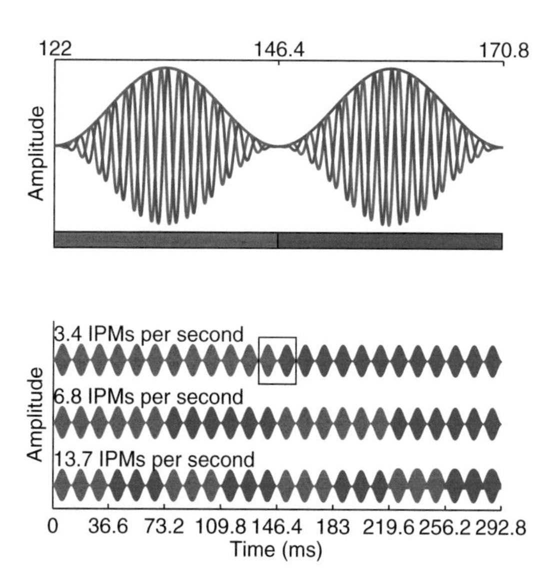

Fig. 1 AM stimuli presented to each ear. *Top panel*, *red* and *blue* correspond to sounds presented to *right* and *left* ears, respectively. Filled *horizontal bars* indicate the ear with leading phase. The IPD of this example corresponds to ±90° (switching from −45 to 45°, and vice versa, in the two ears). IPD transitions are introduced when the stimulus amplitude is zero. *Bottom panel*, three IPM rates employed. *Red* regions illustrate an IPD of +90° IPD, whereas *blue* regions illustrate an IPD of −90° IPD. IPM rate is controlled by the number of AM cycles where the IPD is held constant

periodically at three different rates: 1.7, 3.4 and 6.8 Hz, referred as IPD cycle rates. However, since responses were expected to be elicited by each IPD transition, we refer to the stimulus largely in terms of IPM rate, which indicates the total number of IPD transitions per second, irrespective of direction. Thus, the three IPM rates tested were: 3.4, 6.8 and 13.7 Hz. These IPM rates corresponded to an IPD transition every 12, 6 and 3 AM cycles, respectively (Fig. 1).

For EEG recordings to ITDs conveyed in the stimulus envelope, 3000-Hz tones were modulated with a transposed envelope (Bernstein and Trahiotis 2002) at 128 Hz, and a second-order AM (41 Hz) applied diotically. IPDs of the 128-Hz envelopes were switched between ±90° IPD (around 0°) or from systematically between 0° and 180° (i.e. transposed phase of −90° in one ear and +90° in the other).

Stimuli were created in Matlab, and presented by an RME Fireface UC sound card (24 bits, 48 kHz sampling rate) connected to Etymotic Research ER-2 insert earphones. Sound level was verified with a 2-cc B&K artificial ear. EEG responses were recorded differentially from surface electrodes; the reference electrode was placed on the vertex (Cz), and the ground electrode on the right clavicle. Two recording electrodes were placed on the left and right mastoid (TP9 and TP10). Electrode impedances were kept below 5 kΩ. Responses were amplified with a 20x gain (RA16LI Tucker-Davis Technologies), and digitalized at a rate of 24.414 kHz, and a resolution of 16 bits/sample (Medusa RA16PA Tucker-Davis Technologies). The cutoff frequencies of the internal bandpass filter were 2.2 and 7.5 kHz, respectively (6 dB per octave). Recordings were next stored on a RX5 Pentusa before being passed to hard disk via custom software. Recordings were processed off-line using Matlab.

During the experiment, subjects sat in a comfortable chair in an acoustically isolated sound booth, and watched a subtitled film of their choice. Subjects were encouraged to sit as still as possible, and were offered a short break every 15–20 min. Epochs of each measurement were transformed to the frequency domain (fast Fourier transform (FFT) of 100,000 points; 0.24 Hz of resolution) and FFTs from all epochs were averaged. Six frequency bins were tested for significance; corresponding to the IPD cycle rate and the next four harmonics, as well as the frequency bin corresponding to the AM rate.

3 Results

3.1 *Sensitivity to IPDs Conveyed in the Temporal Fine Structure of Low-Frequency Sounds*

EEG recordings were obtained for three IPM rates (3.4, 6.8 and 13.7 Hz) and seven IPMs (±11.25°, ±22.5°, ±45°, ±67.5°, ±90°, ±112.5° and ±135°), generating a total of 30 conditions (21 dichotic, 9 diotic), each lasting ≈5 min, giving a total recording time of 2.5 h. Figure 2 plots the spectral magnitude of a typical recording for a dichotic condition with IPM of ±67.8° and IPM rate of 6.8 Hz,, in which a

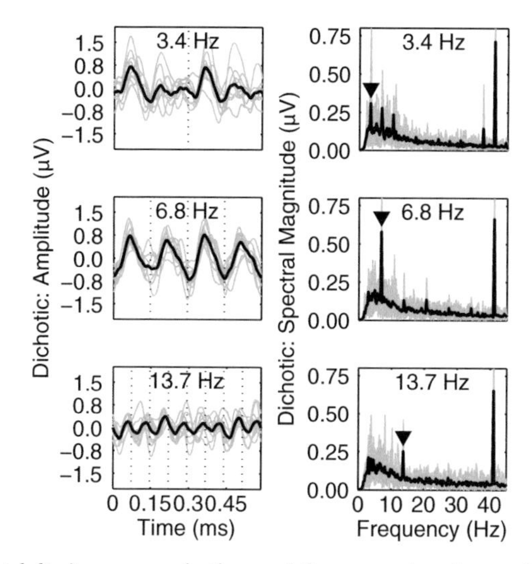

Fig. 2 Average (*thick line*) responses in time and frequency domains to dichotic stimuli, with individual responses shown in *thin gray lines*. Time domain responses are shown in *left panels* and frequency responses in *right panels*. Responses to *low*, *middle* and *high IPM* rates (3.4, 6.8 and 13.7 Hz) in *top*, *middle*, and *bottom* rows. All responses correspond to the ±67.8° IPD condition. *Dashed vertical lines* in time-domain responses indicate the time of the IPD transition. Black markers in the frequency domain indicate IPM rate

significant response was observed for the frequency bin corresponding to the IPM rate (black arrow), and corresponding diotic conditions are shown in Fig. 3—for which no response to the IPM rate was observed (see Ross 2008). Note that in both conditions the ASSR to the AM rate (41 Hz) was clearly observed.

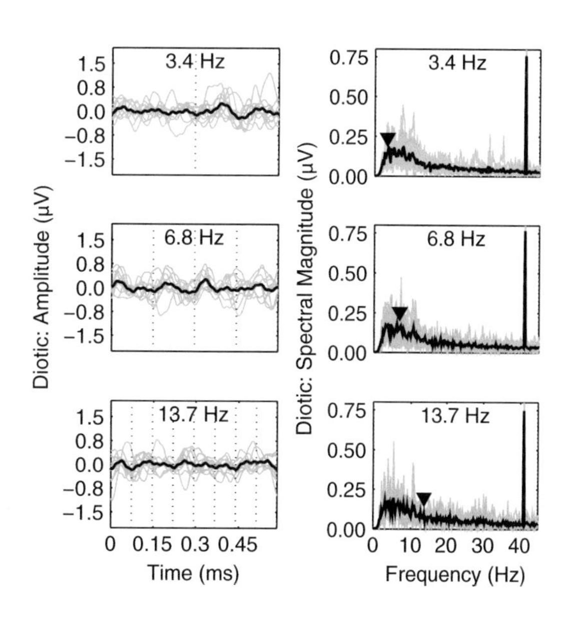

Fig. 3 Average (*thick line*) responses in time and frequency domains to diotic stimuli. Data are plotted in a manner consistent with Fig. 2

Figure 2 shows responses analysed in the time domain (left panels) and the frequency domain (right) for a single IPM condition (±67.8°). Individual (gray lines) and group mean (black) for the slowest IPM rate (3.4 Hz; Fig. 3, *top*), typically displayed a P1-N1-P2 response to each IPD transition, with component latencies of approximately 50, 100 and 175 ms, respectively. The next two harmonics (6.8 and 10.2 Hz) were also observed clearly in the response to the 3.4-Hz IPM. Evoked responses to the intermediate IPM rate (6.8 Hz; Fig. 2) became steady state at the same rate as the IPM, and so we term this response the IPM-FR. In the time domain (left-middle), the peak amplitude of the IPM-FR is broadly the same as that for the slowest IPM rate (3.2 Hz), but responses were more steady state than those at the low rate, being more consistent in phase and amplitude across epochs (evidenced by a more favourable SNR). This is confirmed by comparing the variability of the individual responses in the time domain as well as the frequency domain (Fig. 2), where the spectral magnitude was almost twice larger than that the largest magnitude obtained at the lowest IPM rate. Moreover, the average spectral magnitude of the next harmonics was almost twice smaller than those obtained at the lowest IPM rate. Finally, the IPM-FR is observed at the most rapid IPM rate (13.7 Hz, bottom panels of Fig. 3). As for the intermediate IPM rate, responses show a steady-state pattern in the time domain, albeit with a reduced amplitude. Frequency domain analysis revealed the following response occurred primarily at the frequency bin corresponding to the IPM rate. Harmonics were no longer observed in the grand averaged data.

For analysis of data in the frequency domain, Hotelling's T2 tests were applied to the frequency bins corresponding to the IPM rate and the next four harmonics. Responses were classified as significant if either the left (Tp9), right (Tp10), or both electrodes observed a significant response for a given frequency bin. The frequency bin corresponding to the IPM rate (second harmonic) elicited the highest number of significant responses. Indeed, responses obtained at IPM rates of 6.8 and 13.7 Hz were observed for all ten subjects for the ±45° and ±135° IPD conditions, respectively. This was not the case for the lowest IPM rate where significant responses were observed for only seven subjects. In terms of spectral magnitude, responses obtained at the intermediate IPM rate showed the largest magnitude at the second harmonic, i.e. the harmonic corresponding to the IPM rate, consistent with the hypothesis that a following-response (FR) is evoked by IPM.

Finally, we assessed the frequency bin corresponding to the IPM rate. A three-way non-parametric repeated measures ANOVA with factors of electrode position (left or right mastoid), IPM rate, and IPD revealed that factors IPM rate ($p<0.001$), IPD ($p<0.001$) as well as the interaction between IPM rate and IPD ($p=0.01$) were significant. Responses obtained at 6.8 Hz IPM rate were maximal for IPDs spanning ±45–±90°, whereas responses obtained at 13.4 Hz IPM rate increased monotonically with increasing IPD.

3.2 Sensitivity to IPDs Conveyed in the Temporal Envelope of High-Frequency Sounds

In contrast to low-frequency tones, the magnitude of the IPM-FR for IPDs conveyed in the modulated envelopes of high-frequency (3-kHz) tones was relatively low for IPMs switching between ±90° IPD. Therefore, we also assessed EEG responses to IPMs switching between 0° (in phase) and 180° (anti-phasic) IPD conditions. These latter switches evoked considerably larger EEG responses (Fig. 4, *left*). The magnitude of the IPM-FR decreased with increasing carrier frequency, and was sensitive to an offset in the carrier frequency between the ears (Fig. 4, *right*).

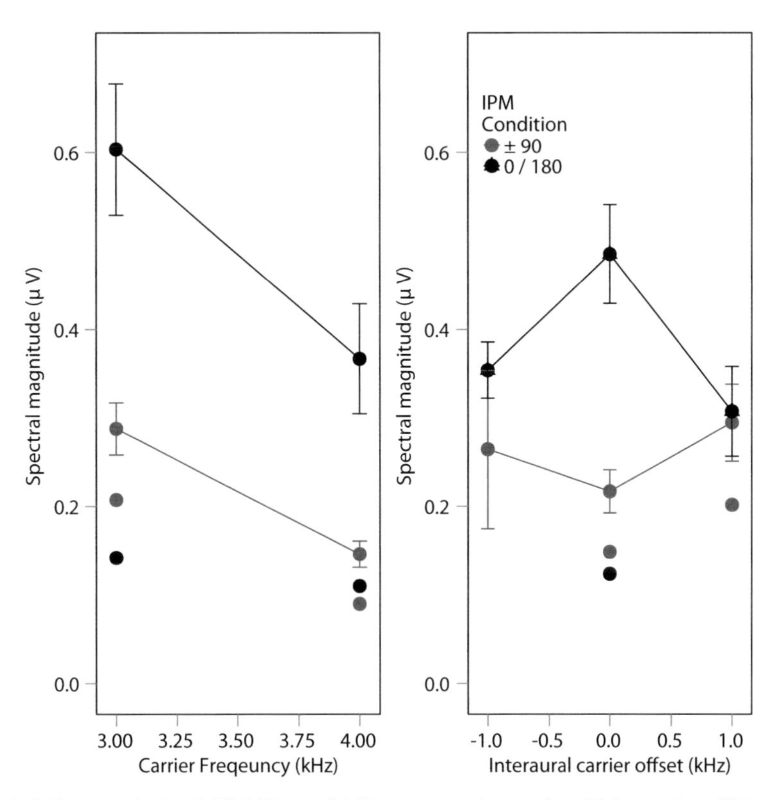

Fig. 4 *Left*—magnitude of IPM-FRs to 3-kHz transposed tones in which envelope IPDs were switched between ±90°, or between 0° and 180° (transposed phase of 90° at one and −90° at the opposite ear). *Right*—magnitude of IPM-FR as a function of carrier frequency mismatch. In both plots, unconnected symbols indicate response magnitude in equivalent diotic control conditions

4 Discussion

We have demonstrated that an IPM-FR can be reliably obtained and used as an objective measure of binaural processing. Depending on the IPM rate, periodic IPMs can evoke a robust steady-state following response. This is consistent with previous reports that an abrupt change in carrier IPD evokes a P1-N1-P2 transient response (Ross et al. 2007a, b; Ross 2008), and with evidence that abrupt periodic changes in interaural correlation evoked steady-state responses (Dajani and Picton 2006; Massoud et al. 2011). The data also demonstrate that the magnitude of the IPM-FR varies with both the IPM rate and the magnitude of the carrier IPD. As diotic carrier phase modulation did not elicit detectable neural responses, the observed responses to IPM must reflect binaural processing rather than the monaural encoding of abrupt phase shifts (Ross et al. 2007a, b; Ross 2008). The time-domain response waveforms are consistent with evoked responses occurring in cortical and sub-cortical sources along the auditory pathway, and that they originate from superposition of transient middle-latency responses (Galambos et al. 1981).

The difference in the preferred IPDs for EEG signals conveying IPDs in the low-frequency TFS and the modulated envelopes of high-frequency tones is consistent with the underlying physiological mechanisms of neurons in the brainstem pathways sub-serving these two cues. Low-frequency binaural neurons in the MSO and midbrain (inferior colliculus) are maximally activated by IPDs in the range 45–90° (see Grothe et al. 2010 for review). This is consistent with their delay sensitivity being generated by largely excitatory inputs from the two ears. Conversely, those in the LSO appear to be excited by one ear, and inhibited by the other (Grothe et al. 2010). Consequently, maximum activity is generated in this (high-frequency) pathway when the sounds at the two ears are out of phase with each other (here, when the signal is switched from 0° to 180°). The IPM-FR, therefore, appears able to distinguish between these two pathways based on the preferred IPDs employed to evoke the response. This provides a possible means of distinguishing which of the binaural pathways is activated in bilateral CI users, as well as a means of objective fitting of devices. Work is currently underway to assess the IPM-FR in bilateral CI users. The EEG recording of steady-state responses in CI users is made challenging by the electrical artefacts created by stimulus pulses, but artefact rejection techniques have been successfully applied for the recording of AM-evoked steady-state responses (e.g., Hofmann and Wouters 2010). Such techniques should also be applicable for the recording of the IPM-FR.

Acknowledgments This work is supported by the European Commission under the Health Theme of the 7th Framework Programme for Research and Technological Development.

References

Abel SM, Giguère C, Consoli A, Papsin BC (2000) The effect of aging on horizontal plane sound localization. J Acoust Soc Am 108:743–52

Babkoff H, Muchnik C, Ben-David N, Furst M, Even-Zohar S, Hildesheimer M (2002) Mapping lateralization of click trains in younger and older populations. Hear Res 165:117–27

Bernstein LR, Trahiotis C (2002) Enhancing sensitivity to interaural delays at high frequencies by using transposed stimuli. J Acoust Soc Am 122:1026–1036

Chait M, Poeppel D, de Cheveigné A, Simon JZ (2005) Human auditory cortical processing of changes in interaural correlation. J Neurosci 25:8518–27

Dajani HR, Picton TW (2006) Human auditory steady-state responses to changes in interaural correlation. Hear Res 219:85–100

Galambos R, Makeig S, Talmachoff PJ (1981) A 40-Hz auditory potential recorded from the human scalp. Proc Natl Acad Sci U S A 78:2643–2647

Grothe B, Pecka M, McAlpine D (2010) Mechanisms of sound localization in mammals. Physiol Rev 90:983–1012

Hari R, Aittoniemi K, Järvinen ML, Katila T, Varpula T (1980) Auditory evoked transient and sustained magnetic fields of the human brain. Localization of neural generators. Exp Brain Res 40:237–240

Herman GE, Warren LR, Wagener JW (1977) Auditory lateralization: age differences in sensitivity to dichotic time and amplitude cues. J Gerontol 32(2):187–191. doi: 10.1093/geronj/32.2.187

Hofmann M, Wouters J (2010) Electrically evoked auditory steady state responses in cochlear implant users. J Assoc Res Otolaryngol 11:267–282

Liégeois-Chauvel C, Musolino A, Badier JM, Marquis P, Chauvel P (1994) Evoked potentials recorded from the auditory cortex in man: evaluation and topography of the middle latency components. Electroencephalogr Clin Neurophysiol 92:204–214

Lorenzi C, Gatehouse S, Lever C (1999) Sound localization in noise in hearing-impaired listeners. J Acoust Soc Am 105:3454–3463

Massoud S, Aiken SJ, Newman AJ, Phillips DP, Bance M (2011) Sensitivity of the human binaural cortical steady state response to interaural level differences. Ear Hear 32:114–120

McEvoy LK, Picton TW, Champagne SC (1991) Effects of stimulus parameters on human evoked potentials to shifts in the lateralization of a noise. Audiology 30:286–302

Moore DR, Hutchings ME, Meyer SE (1991) Binaural masking level differences in children with a history of otitis media. Audiology 30:91–101

Näätänen R, Picton T (1987) The N1 wave of the human electric and magnetic response to sound: a review and an analysis of the component structure. Psychophysiology 24:375–425

Picton TW, Hillyard SA, Krausz HI, Galambos R (1974) Human auditory evoked potentials. I. Evaluation of components. Electroencephalogr Clin Neurophysiol 36:179–190

Ross B (2008) A novel type of auditory responses: temporal dynamics of 40-Hz steady-state responses induced by changes in sound localization. J Neurophysiol 100:1265–1277

Ross B, Fujioka T, Tremblay KL, Picton TW (2007a) Aging in binaural hearing begins in mid-life: evidence from cortical auditory-evoked responses to changes in interaural phase. J Neurosci 27:11172–11178

Ross B, Tremblay KL, Picton TW (2007b) Physiological detection of interaural phase differences. J Acoust Soc Am 121:1017

Wilmington D, Gray L, Jahrsdoerfer R (1994) Binaural processing after corrected congenital unilateral conductive hearing loss. Hear Res 74(1-2):99–114

Minimum Audible Angles Measured with Simulated Normally-Sized and Oversized Pinnas for Normal-Hearing and Hearing-Impaired Test Subjects

Filip M. Rønne, Søren Laugesen, Niels S. Jensen and Julie H. Pedersen

Abstract The human pinna introduces spatial acoustic cues in terms of direction-dependent spectral patterns that shape the incoming sound. These cues are specifically useful for localization in the vertical dimension. Pinna cues exist at frequencies above approximately 5 kHz, a frequency range where people with hearing loss typically have their highest hearing thresholds. Since increased thresholds often are accompanied by reduced frequency resolution, there are good reasons to believe that many people with hearing loss are unable to discriminate these subtle spectral pinna-cue details, even if the relevant frequency region is amplified by hearing aids.

One potential solution to this problem is to provide hearing-aid users with artificially enhanced pinna cues—as if they were listening through oversized pinnas. In the present study, it was tested whether test subjects were better at discriminating spectral patterns similar to enlarged-pinna cues. The enlarged-pinna patterns were created by transposing (T) generic normal-sized pinna cues (N) one octave down, or by using the approach (W) suggested by Naylor and Weinrich (System and method for generating auditory spatial cues, United States Patent, 2011). The experiment was cast as a determination of simulated minimum audible angle (MAA) in the median saggital plane. 13 test subjects with sloping hearing loss and 11 normal-hearing test subjects participated. The normal-hearing test subjects showed similar discrimination performance with the T, W, and N-type simulated pinna cues, as expected. However, the results for the hearing-impaired test subjects showed only marginally lower MAAs with the W and T-cues compared to the N-cues, while the

F. M. Rønne (✉) · S. Laugesen · N. S. Jensen · J. H. Pedersen
Eriksholm Research Centre, Rørtangvej 20, 3070 Snekkersten, Denmark
e-mail: fimr@eriksholm.com

S. Laugesen
e-mail: slau@eriksholm.com

N. S. Jensen
e-mail: nsje@eriksholm.com

J. H. Pedersen
e-mail: jpd@eriksholm.com

P. van Dijk et al. (eds.), *Physiology, Psychoacoustics and Cognition in Normal and Impaired Hearing,* Advances in Experimental Medicine and Biology 894,
DOI 10.1007/978-3-319-25474-6_22

overall discrimination thresholds were much higher for the hearing-impaired compared to the normal-hearing test subjects.

Keywords Pinna cues · MAA · Localization · Vertical localization · Discrimination · Saggital plane

1 Introduction

Spatial hearing relies on spatial acoustic cues. The binaural cues (interaural time and level differences, ITD and ILD) are reasonably well preserved with hearing aids, while monaural spectral cues can be substantially changed. The most striking example is that the spectral filtering imposed by the pinna is completely bypassed with behind-the-ear (BTE) hearing aids, because the microphones sit on top of the pinna. Pinna cues are useful for localisation out of the horizontal plane, and for resolving front-back confusions.

Previous attempts at providing hearing-aid users with natural pinna cues have only been partly successful (Jensen et al. 2013). A likely explanation is that the fine high-frequency spectral details (present from 5 kHz and upwards) of natural pinna cues are at least partly imperceptible by people with hearing impairment, due to reduced frequency selectivity.

Naylor and Weinrich (2011) described a hearing-aid configuration incorporating a method to create directional spectral cues with similar characteristics to pinna cues, but in a lower frequency region around 2–6 kHz, whereby they would be easier to perceive for hearing-impaired test subjects. The hypothesis was that hearing-aid users could learn to utilise these cues and thereby retrieve lost spatial hearing abilities. Hypothesised end-user benefits of such a method could be improved localisation out of the horizontal plane and in the front-back dimension, improved spatial unmasking, and improved spatial awareness in general.

A subsequent question is whether test subjects can learn to exploit a modified set of pinna cues. Hofman et al. (1998) and Van Wanrooij and Van Opstal (2005) reported about experiments where test subjects had their pinnae modified by concha moulds. Results show that while localisation performance was disrupted immediately after mould insertion, subjects regained their localisation ability over the course of weeks. Typically after 6 weeks of learning they had reached an asymptote close to their previous performance.

1.1 Main Research Question

In brief, the outcome measure of this study's experiment was the Minimum Audible Angle (MAA) in the vertical plane. This is the minimum angular change away from a given reference angle, which can be discriminated in a simulated set-up with sound stimuli presented via headphones. The monaural spatial information is provided by different sets of pinna cues, either simulated normally-sized pinna cues or enhanced pinna cues either transposed one octave down or cues similar to what Naylor and Weinrich (2011) suggested. Thus, the MAA will be determined with normally-sized (N), transposed (T) or Weinrich (W) pinna cues.

The main research question was: Are the enhanced pinna cues (T or W) easier to discriminate than natural pinna cues (N) for hearing-impaired test subjects? Thus, is the MAA obtained with natural pinna cues (MAA_N) significantly larger than the MAA obtained with the two types of enhanced pinna cues (MAA_T, MAA_W)?

2 Method and Material

2.1 Pinna Cues

Three different cue-sets were used in the experiment, all artificially generated. The normally-sized pinna cues are designed to mimic key trends of recorded pinna cues found in literature (e.g., Shaw et al. 1997). Figure 1 illustrates the response of the normally-sized (N) pinna cue system to different incidence angles in the median saggital plane. Worth noticing is the low-frequency notch that moves upwards as

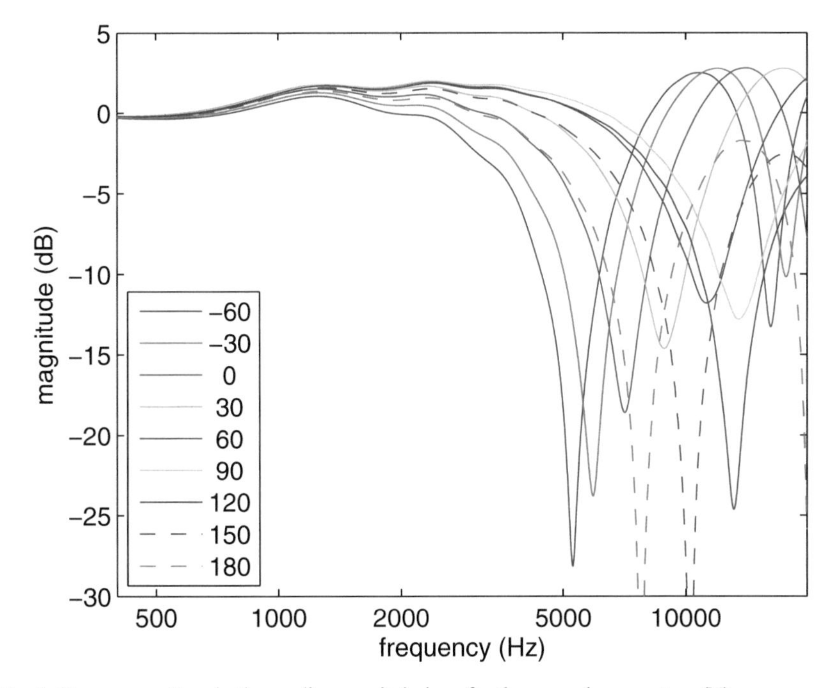

Fig. 1 Response pattern in the median saggital plane for the normal-cue system (N)

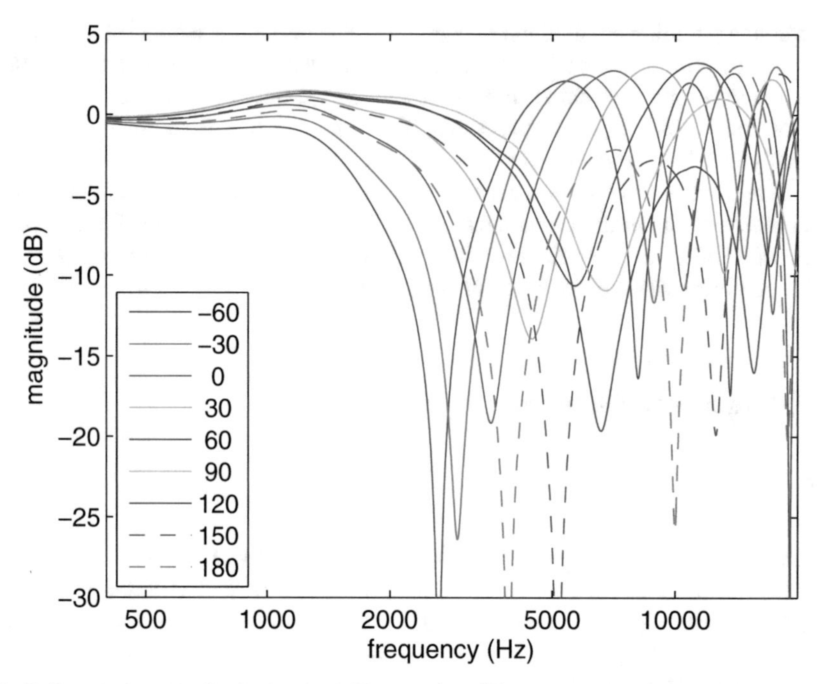

Fig. 2 Response pattern for the transposed-cues system (T)

the incidence angle moves away from 0° (0° = straight ahead, 90° = straight up). Figure 2 shows the transposed system (T), which was designed to shift the entire N-cue-structure one octave down. Figure 3 shows the (Naylor and Weinrich 2011) cue system (W).

2.2 Test Subjects

13 HI test subjects (33–79 years old, mean 64 years, SD 15 years) with no history of cognitive problems participated in the study. It was required that the uncomfortable level (UCL) was at least 15 dB higher than the hearing threshold (HTL), to ensure enough headroom for the hearing-loss compensation used to make stimuli audible.

11 test subjects with expected normal hearing were recruited for the NH group (21–52 years, mean 39 years, SD 11 years).

All test-subject audiograms were re-measured if older than 1 year, or if there was a suspicion that changes had occurred. The test subjects' individual hearing threshold levels for the test ear are plotted in Figs. 4 and 5.

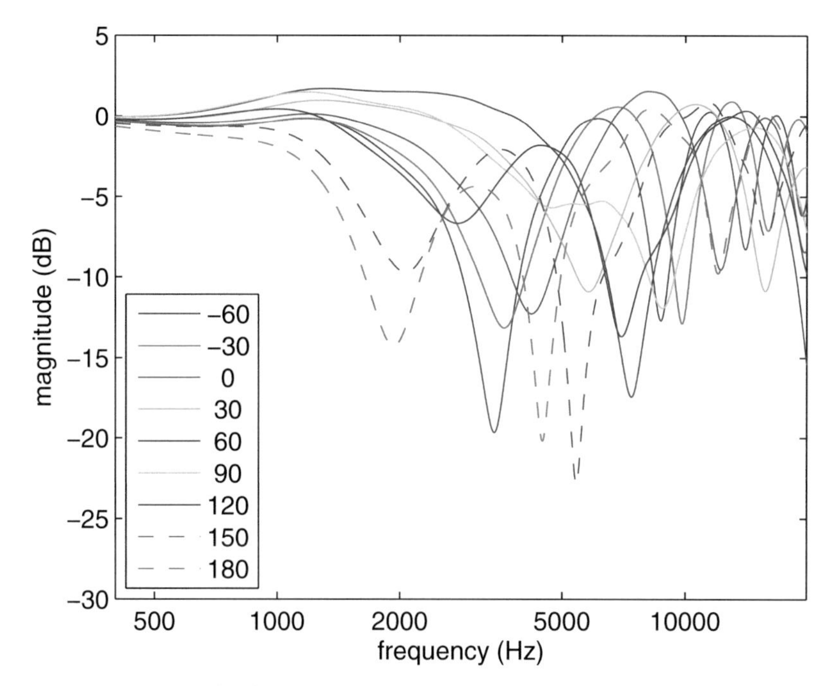

Fig. 3 Response pattern for the (Naylor and Weinrich 2011) cue system (W)

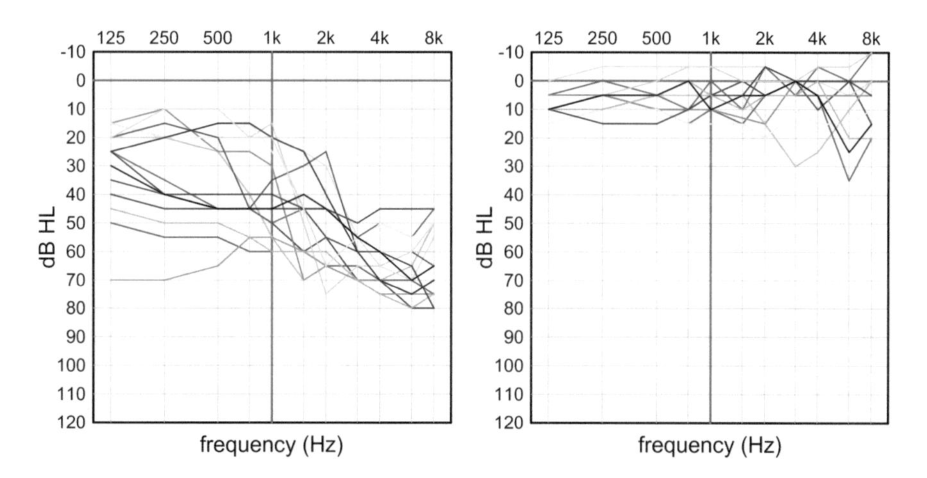

Fig. 4 Test-ear hearing thresholds of hearing impaired test subjects

Fig. 5 Test-ear hearing thresholds of normal hearing test subjects

2.3 Psychoacoustic Experiment

An adaptive three-interval, three-alternative forced-choice (3-AFC) discrimination procedure, coupled with a one-up-two-down staircase rule tracking the 70.7 %-correct point on the underlying psychometric function, was used.

In each trial, three consecutive stimuli were presented. Two represented the reference incidence angle (0° or 120°) and one represented the variable angle (the 'target') determined by the adaptive procedure. The test subject's task was to determine the diverging stimulus by pressing one of three buttons on a touch screen, on which the target stimulus was randomly assigned to one of the buttons in each trial.

Following each response, the test subject was given visual feedback (about the correct answer) via the screen. In the first trial within a run, the initial incidence angle was easy to discriminate. Pilot testing suggested that appropriate values of the adaptive angle were 60° for the 0° reference and 180° for the 120° reference. The step-size was decreased along the run, with initial step size set to 8°. After each down-up reversal, the step size was halved, until a final step size of 2° was reached. After the final step size was reached, another eight reversals were completed.

The final MAA threshold estimate was calculated based on the median of the last three lower reversal points during a run. On average, this method provides a threshold value, which is lower than that produced by e.g. Levitt (1971) or Neher et al. (2011). However, data inspection indicated that both of these alternate methods could lead to misleading results if a test subject had been performing unreliably during parts of the run. Further, since this study investigates differences between conditions (cues) rather than the absolute MAA values, the consequently lower MAA was accepted.

2.4 Set-Up

The test took place in a single-walled, sound-attenuated test booth. The test software was run on a personal computer, which was connected to a 14" touch screen and a RME Hammerfall DSP Multiface II 24-bit soundcard running at 96 kHz sampling frequency. A Lake People phone-amp G103 was used to drive a pair of Sennheiser HDA 200 headphones.

2.5 Stimuli

All test stimuli were based on white noise samples generated by the test software and steeply band pass filtered to 0.4–16 kHz. The same frozen noise sample was used within each 3AFC triplet, while a new white noise sample was generated for each new triplet. In each triplet, the simulated pinna cues were then added on the fly in Matlab.

Compensation for hearing loss was applied. In short, a level corresponding to 60 dB SPL(C) free-field was selected as the presentation level to be used with NH test subjects. For the HI test subjects, it was then ensured that in each one-third octave band the RMS level of the stimulus was at least 15 dB above hearing threshold before pinna-cue shaping. The required gain was added by designing an IIR filter in the test software to shape the stimulus. Note that the + 15-dB gain strategy was terminated at 9 kHz. In higher frequency bands, the same gain as found for 9 kHz was applied, because of the lack of audiogram information. A similar approach was used towards the lowest frequencies.

2.6 Test Protocol

Both the NH and the HI test subjects completed nine test conditions: The MAA test in all combinations of two reference angles (0°/120°) and three cues (N/T/W), and further a front-back discrimination test with all three cues (these results are not reported in this paper).

The order of conditions was counterbalanced for the NH subjects, who completed all nine measurements within one visit. The testing of HI test subjects included two visits. The first visit included audiometry, if the existing audiogram was more than 1 year old, and the three MAA tests with 120° as reference angle. The second visit included the MAA tests with 0° as reference angle, the front-back tests, and a retest measurement (MAA at 0° with T cues). All visits included the same training conditions, T0 and N120. Within each visit, the order of test conditions was counterbalanced.

The blocking of the 120° measurements introduced a bias (visit effect) in the comparison of the 0° and 120° measurements, but this was preferred over a bias in the comparison between different types of cues, which was the essential comparison in this study.

3 Results

3.1 Data Inspection and Removal

Prior to analysis, the collected data were inspected. The subjective visual inspection of the adaptive MAA measurements pointed out 6 test runs, from which no reliable result could be extracted, as a performance plateau was never reached. These data points were removed.

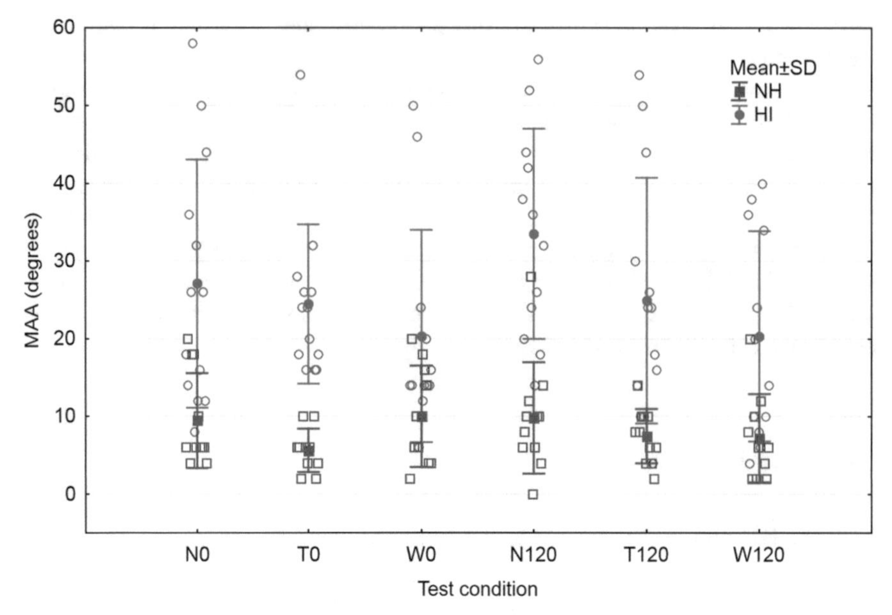

Fig. 6 Raw Minimum-Audible-Angle (MAA) data and mean ± SD for the normal hearing (*NH*) and hearing impaired (*HI*) groups for each of the six test conditions

3.2 Raw Data and Descriptive Statistics

The MAA data for the NH and the HI groups are plotted in Fig. 6. The plot show raw and mean (± SD) data for each of the six MAA test conditions.

The main observations which can be made from Fig. 6 are:

- Substantially better (lower MAA) performance in the NH group.
- Large individual variation, especially in the HI group.
- Very small deviations between test conditions in the NH group.
- Some deviations between test conditions in the HI group, with better performance with W-cues than with N-cues (with T-cues in between), most pronounced for the 120° conditions.

A mixed-model ANOVA was used to analyze the NH MAA data. Besides a significant test-subject effect ($p=0.002$) (the random variable "test subject" was only included as a main effect), the various effects in the model came out as being non-significant, as already indicated by the rather similar mean values shown in Fig. 6. Thus, the NH-group performance did not depend on either cue type or reference angle.

Another mixed-model ANOVA was used to analyze the HI MAA data. As could be expected based on the large individual variance displayed in Fig. 6, a highly significant test-subject effect ($p<<0.001$) was observed. Furthermore, a significant effect of cue type was observed ($p=0.002$), while the effect of reference angle was non-significant. No interactions were significant.

Since the difference between normal and enhanced cues was the essence of the research question, the cue-type effect was further explored by conducting two planned comparisons, N vs. T and N vs. W (across the two reference angles). These analyses showed a significant difference between N and W cues (mean MAA difference of around 10°, $p=0.0005$) and a non-significant difference between N and T cues (mean MAA difference of around 5°, $p=0.06$).

3.3 Training Effects and Test-Retest Variability (Reliability)

Possible training effects, i.e. improved performance over time, were investigated by introducing 'test run number' as a variable. Neither the NH nor the HI subject analysis showed any significant within-session training effects.

The HI tests were conducted during two separate visits, thus enabling a possible between-visit training effect, but with two different sets of test conditions included at the two visits (MAA-120° conditions at first visit, MAA-0° and FB conditions at second visit), it was not possible to assess the between-visit effect. However, the fact that we found no reference-angle effects suggests that the between-visit effect, if present, was limited.

4 Discussion

4.1 Research Question

The hypothesis of the main research question was that the MAA is larger for normal cues than for enhanced cues in a group of HI test subjects. The results showed a significant MAA difference between N and W cues, but not between N and T cues. Thus, the hypothesis was confirmed for the W cues, but not confirmed for the T cues (even though the overall trend was the same in these data).

For the NH group, no differences in performance between cues were observed. This was as expected.

The NH data may be used as a reference in the discussion of the HI data. When the two data sets are compared (see Fig. 6), it is noteworthy that even though the use of enhanced cues improves the HI performance in the MAA measurements (from around 30° to around 20°), the mean performance with enhanced cues is still substantially poorer than NH mean performance, which is around 10° with all three types of cues. It may be questioned whether improving the MAA from 30° to 20° will result in any perceivable real-life benefit, but answering this question will obviously require further testing including training test subjects to exploit modified pinna cues.

4.2 Stimuli Level Differences—A Potential Discrimination Cue?

This study investigated whether the frequency-shaping differences caused by changes in pinna-cue filtering across different incidence angles were discriminable by test subjects. However, the subjects did not only have the frequency shaping to listen for, but consequently also a level cue. To quantify whether the level cue was large enough for the test subjects to take advantage of it, the rms-levels were calculated for the three different cue sets. Unfortunately, the specific hearing loss of a specific test subject, and the frequency range one includes in the analysis, very much affects the rms difference between different incidence angles, making it difficult to make conclusions. However, at large angular differences it was clearly observed that the level cue becomes prominent, whereas the level cue decreased with decreasing incidence angle. At average MAAs (<35° for HI) level-cue magnitudes (< approximately 1 dB) were found to be smaller than the level just-noticeable-differences for HI test subjects reported to be 1–2 dB (Moore 2003). This indicates that most subjects likely used the frequency-shaping cue rather than the level cue.

The test design could have addressed the potential level-cue issue by applying level roving. However, this would have made the test more difficult to perform, because subjects would have been forced to ignore these added level differences when performing the test. Thus, level roving was rejected to avoid the much more extensive training required, cf. Moore et al. (1989).

5 Conclusions

This study compared the ability to discriminate simulated normally-sized pinna cues with the ability to discriminate two different sets of simulated enhanced pinna cues, in a group of NH test subjects and in a group of HI test subjects, respectively. The results showed that the NH group discriminated equally well with all three sets of pinna cues, showing a mean MAA of around 10° across cue types and reference angle. In the HI group, use of W-cues improved the mean MAA from around 30° to around 20°. This improvement was significant and twice as large as the non-significant mean improvement (from around 25° to 20°) offered by the T-cues.

While the HI group showed an improved ability to discriminate enhanced pinna cues as compared to normal pinna cues, the improvements were not as large as expected, and performance was still substantially poorer than that observed in the NH group.

References

Hofman PM, Van Riswick JG, Van Opstal AJ (1998) Relearning sound localization with new ears. Nat Neurosci 1(5):417–421

Jensen NS, Neher T, Laugesen S, Johannesson RB, Kragelund L (2013) Laboratory and field study of the potential benefits of pinna cue-preserving hearing aids. Trends Amplif 17(3):171–188

Levitt H (1971) Transformed up-down methods in psychoacoustics. J Acoust Soc Am 49:467–477

Moore BC (2003). An introduction to the psychology of hearing, vol 5. Academic, San Diego

Moore BC, Oldfield SR, Dooley GJ (1989) Detection and discrimination of spectral peaks and notches at 1 and 8 kHz. J Acoust Soc Am 85:820–836

Naylor G, Weinrich SG (2011, May 3) System and method for generating auditory spatial cues. United States Patent

Neher T, Laugesen S, Jensen NS, Kragelund L (2011) Can basic auditory and cognitive measures predict hearing-impaired listeners' localization and spatial speech recognition abilities? J Acoust Soc Am 130(3):1542–1558

Shaw E, Anderson T, Gilkey R (1997) Binaural and spatial hearing in real and virtual environments. In: Gilkey RH, Anderson TR (eds) Acoustical features of human ear. Lawrence Erlbaum Associates, Mahwah, pp 25–47

Van Wanrooij MM, Van Opstal AJ (2005) Relearning sound localization with a new ear. J Neurosci 25(22):5413–5424

Moving Objects in the Barn Owl's Auditory World

Ulrike Langemann, Bianca Krumm, Katharina Liebner, Rainer Beutelmann and Georg M. Klump

Abstract Barn owls are keen hunters of moving prey. They have evolved an auditory system with impressive anatomical and physiological specializations for localizing their prey. Here we present behavioural data on the owl's sensitivity for discriminating acoustic motion direction in azimuth that, for the first time, allow a direct comparison of neuronal and perceptual sensitivity for acoustic motion in the same model species. We trained two birds to report a change in motion direction within a series of repeating wideband noise stimuli. For any trial the starting point, motion direction, velocity (53–2400°/s), duration (30–225 ms) and angular range (12–72°) of the noise sweeps were randomized. Each test stimulus had a motion direction being opposite to that of the reference stimuli. Stimuli were presented in the frontal or the lateral auditory space. The angular extent of the motion had a large effect on the owl's discrimination sensitivity allowing a better discrimination for a larger angular range of the motion. In contrast, stimulus velocity or stimulus duration had a smaller, although significant effect. Overall there was no difference in the owls' behavioural performance between "inward" noise sweeps (moving from lateral to frontal) compared to "outward" noise sweeps (moving from frontal to lateral). The owls did, however, respond more often to stimuli with changing motion direction in the frontal compared to the lateral space. The results of the behavioural experiments are discussed in relation to the neuronal representation of motion cues in the barn owl auditory midbrain.

G. M. Klump (✉) · U. Langemann · B. Krumm · K. Liebner · R. Beutelmann
Cluster of Excellence "Hearing4all", Animal Physiology and Behaviour Group, Department for Neuroscience, University of Oldenburg, Oldenburg, Germany
e-mail: ulrike.langemann@uni-oldenburg.de

B. Krumm
e-mail: bianca.krumm@uni-oldenburg.de

K. Liebner
e-mail: katharina.liebner@uni-oldenburg.de

R. Beutelmann
e-mail: rainer.beutelmann@uni-oldenburg.de

G. M. Klump (✉)
e-mail: georg.klump@uni-oldenburg.de

P. van Dijk et al. (eds.), *Physiology, Psychoacoustics and Cognition in Normal and Impaired Hearing,* Advances in Experimental Medicine and Biology 894,
DOI 10.1007/978-3-319-25474-6_23

Keywords Auditory motion discrimination · Motion-direction sensitivity · Sound localization · Bird

1 Introduction

Our auditory world is by no means stationary. Natural sound sources often change position and also listeners move in their environment. Predators like the barn owl (*Tyto alba*) will routinely hunt moving prey, and they perform in-flight corrections (Hausmann et al. 2008). Thus, with respect to sound source localization, real world situations confront the auditory system with varying binaural cues. Possible motion-related dynamic binaural cues are the change in interaural time differences (ITDs), in interaural level differences (ILDs) or in spectral peaks and notches. Generally, ITD cues identify sound source azimuth, ILD cues azimuth and/or elevation and peaks and notches mostly elevation (e.g., Grothe et al. 2010). It has been suggested that auditory motion processing relies on specialized circuits composed of directionally tuned and motion sensitive neurons in the midbrain (reviewed by Wagner et al. 1997). Time constants for processing of non-stationary dynamic cues must be short to allow tracking of moving sound sources (Wagner 1991). However, time constants for binaural processing have found to depend on measuring procedures (e.g., Shackleton and Palmer 2010).

Auditory motion perception has been investigated simulating motion by switching between neighbouring free-field loudspeakers (e.g., Wagner and Takahashi 1992) or by varying interaural cues in headphone presentation (reviewed by Middlebrooks and Green 1991). Neurophysiological correlates to human auditory motion perception have been investigated in animal models with both free-field and headphone stimulation (e.g., McAlpine et al. 2000; Wagner et al. 1997). What is missing, however, is the direct comparison of neuronal and perceptual sensitivity for acoustic motion in the same model species.

Here we report auditory motion perception in the barn owl, a species for which data on the neuronal representation of motion stimuli in the inferior colliculus (IC) are available (e.g., Wagner and Takahashi 1992; Kautz and Wagner 1998; Wagner and von Campenhausen 2002; Wang and Peña 2013). Specifically, Wagner and Takahashi (1992) showed that the angular velocity and stimulus duration affected the neurons' response. Here, we investigate how stimulus velocity, the size of the angle of sound incidence (i.e., the angular range), and stimulus duration affect the owls' perceptual performance. Furthermore, we analyse whether the perception of changes in motion direction differs for stimuli presented in the frontal or in the lateral space. Finally, we will discuss the behavioural data with reference to the responses of motion direction sensitive neurons in the barn owl auditory midbrain.

2 Methods

We trained two barn owls to report a change in auditory motion within a series of repeating band-pass noise stimuli (500–6000 Hz) presented in the horizontal plane. Motion was simulated by sequentially activating up to eight loudspeakers (Vifa XT25TG30-04), fading in and out the sound between adjacent loudspeakers (spacing 12°; Fig. 1). The experiments were performed in a sound-attenuating echo-reduced chamber (IAC 1203-A, walls with sound absorbing foam). Signals were generated by a 24-bit sound card (Hammerfall DSP Multiface II, RME, Germany) and the loudspeakers were driven by a multichannel amplifier (RMB-1048, Rotel, Japan). The barn owls had learned to sit on a waiting perch during the repeated reference stimuli and to fly to a report perch when a test stimulus was presented (Go/NoGo paradigm). Within a trial the reference stimuli (repetition period 1300 ms) had the same direction, while the test was a single stimulus presented with opposite motion direction. Each stimulus started at a random position between −42° and +42° in azimuth. This prevented the owls from using location *per se* rather than motion direction to solve the task. The owls were also trained to generally orient their head in the waiting position towards 0° in azimuth. Correct responses were rewarded. The waiting intervals before a test stimulus were randomized. For any trial the direction of the noise sweeps, the velocity (53–2400°/s) and the effective duration (30–225 ms, with 10 ms Hanning ramps) and angular range of the motion (12–72°) of the reference stimuli were randomly selected from a distribution of combinations of parameters. Each combination of parameters was presented 20

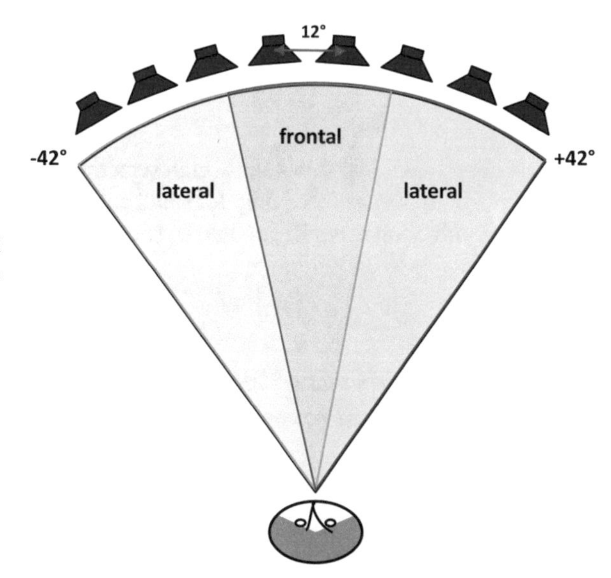

Fig. 1 Stimuli were classified as moving either "inward" (from lateral to frontal) or "outward" (from frontal to lateral). "Frontal" stimuli were moving at least 50 % of the time in the range between + and − 15° from the midline. "Lateral" stimuli were moving more than 50 % of the time in one of the two hemispheres (including up to 15° of the other hemisphere)

times. Experimental sessions with 44 trials each (36 test trials and 8 catch trials having no change in motion direction) were included in the analysis if the owls responded at least ten times (10 "Hits").

We performed two experimental series. Series 1 had effective stimulus durations of 75, 150 and 225 ms and provided stimulus velocities of between 53 and 960°/s. Series 2 had effective stimulus durations of 30, 60 and 90 ms and provided stimulus velocities of between 133 and 2400°/s. Note that stimulus duration and stimulus velocity are interdependent variables. The relation between sensitivity and the parameters stimulus duration, angular range, and stimulus velocity was analysed by linear regression. We will use the regressions' coefficient of determination (R^2) and associated ANOVA results to describe the contribution of each of the three parameters to the perception of a change in motion direction.

We made the following assumptions for data analysis. (1) Noise sweeps moving from the owl's lateral space toward the frontal space were classified as moving "inward", noise sweeps moving from the frontal space toward the lateral space were classified as moving "outward". (2) "Lateral" stimuli were noise sweeps that stayed in one hemisphere more the 50 % of the time, but were allowed to cross the midline up to 15° to the contralateral side, "frontal" stimuli were noise sweeps moving more the 50 % of the time in the range between plus and minus 15° from the midline (Fig. 1). The reasoning for this approach is that the owls' space maps in the left and right IC overlap by about this amount (Knudsen and Konishi 1978a, b). For the lateral/frontal classification, we limited our analysis to noise sweeps of an extent of 36° and 48° to avoid floor and ceiling effects. We then compared the owl's performance for inward and outward stimuli and for stimuli between the frontal and lateral space using X^2 tests.

3 Results

In a behavioural task, barn owls show high sensitivity for auditory motion discrimination. Sensitivity was reduced only for small angular ranges. Furthermore auditory motion discrimination was still good at the shortest stimulus duration (30 ms) and at high velocities (up to 2400°/s).

3.1 Effects of Stimulus Velocity, Size of the Angular Range, and Stimulus Duration on Auditory Motion Discrimination

Our data show that the owls' sensitivity for auditory motion discrimination mainly increased with increasing angular range. The amount of variance accounted for (R^2) by "angular range" was between 0.70 and 0.80 for each of the conditions shown in Figs. 2 and 3. Stimulus duration and stimulus velocity had only a small, though

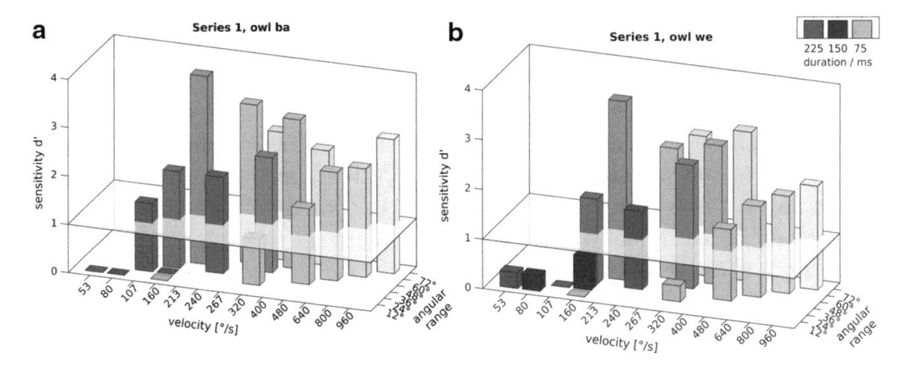

Fig. 2 Series 1 with stimulus velocities up to 960°/s, results from two owls (A, B): Sensitivity increases with larger angular range and longer stimulus duration. Sensitivity index (d') is plotted as a function of stimulus velocity, angular range and duration

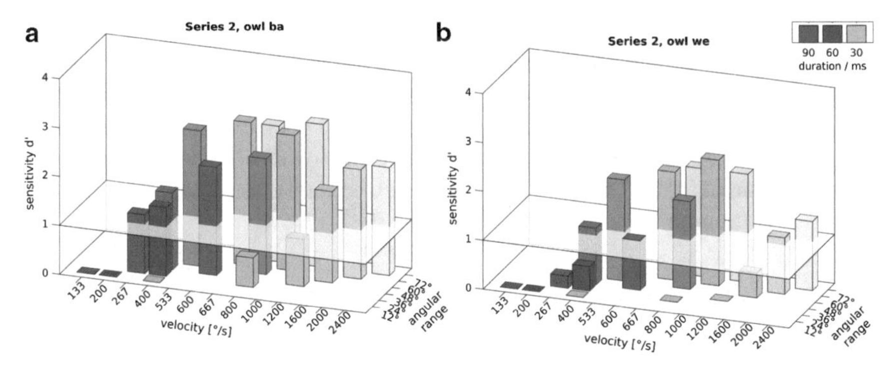

Fig. 3 Series 2 with stimulus velocities up to 2400°/s, results from two owls (A, B): Sensitivity is hardly affected when stimulus velocity is further increased. Sensitivity index (d') is plotted as a function of stimulus velocity, angular range and duration

significant effect on sensitivity. When comparing sensitivity at similar angular ranges, sensitivity moderately increased for longer stimuli and decreased for faster moving stimuli (R^2 values for "duration" were between 0.03 and 0.13, R^2 values for "velocity" were between 0.07 and 0.21, all $P \leq 0.001$). The same pattern was represented by the ANOVA results: the factor "angular range" had F-values of at least 854, F-values for "duration" were between 12 and 55, F-values for "velocity" were between 25 and 95 (all $P \leq 0.001$). The owls' sensitivity was still rather high, if stimulus duration was further decreased in series 2 (Fig. 3). When comparing sensitivity for noise sweeps of similar velocities at different durations (Fig. 2, 3), we had expected to find them to be similar across conditions. In contrast to our expectation, stimulus velocity was a rather bad predictor of the owls' behavioural performance. This can be explained, however, by the interaction of duration and angular range that determines stimulus velocity.

3.2 *Effects of Stimulus Position on Auditory Motion Discrimination*

Responses to inward and outward sweeps were equally frequent: the owls had relative hit rates of between about 35 and 43 % for each of the two directions (X^2 tests n.s., Table 1). The owls responded more often to stimuli with a change in direction moving in the frontal range compared to those moving in the lateral range (Table 2): Owl "ba" had a relative Hit rate of about 67 % in the frontal space and of only 38 % in the lateral space ($X^2=15.7$, $P<0.01$) while owl "we" had relative hit rates of about 52 and 39 % for the frontal and lateral range, respectively ($X^2=3.1$, $P=0.077$).

4 Discussion

Using simulated acoustic motion stimuli, we have demonstrated how the sensitivity for discriminating auditory motion direction in the horizontal plane is related to stimulus velocity, stimulus duration, and the size of the angular range covered by the moving sound. Furthermore, we describe how motion discrimination is affected by the location in the owl's auditory space. Here we relate these data to results from studies of motion-direction sensitive neurons in the barn owl.

Table 1 The absolute "Hit" and "Miss" rates combining Series 1 and 2. Test stimuli were moving either inward or outward

Owl ba	*Inward*	*Outward*	
Hit	29	28	$X^2=0.009$
Miss	52	49	$p=0.926$ $N=158$
Owl we	*Inward*	*Outward*	
Hit	28	38	$X^2=0.862$
Miss	53	51	$p=0.353$ $N=170$

Table 2 The absolute "Hit" and "Miss" rates combining Series 1 and 2. Noise sweeps were presented from either lateral or frontal

Owl ba	*Lateral*	*Frontal*	
Hit	51	52	$X^2=15.743$
Miss	85	26	$p<0.001$ $N=214$
Owl we	*Lateral*	*Frontal*	
Hit	66	38	$X^2=3.131$
Miss	104	35	$p=0.077$ $N=243$

Motion direction sensitive neurons have been observed in the barn owl IC and optic tectum (Wagner et al. 1997, Wagner and von Campenhausen 2002). Neurophysiological studies on directional sensitivity in auditory motion processing by the owl have applied a wide range of angular velocities. Witten et al. (2006) presented simulated motion based on ITD change at a speed of 32°/s observing shifts of auditory spatial receptive fields in relation to motion direction that could be explained by adaptation. Similarly, Wang and Peña (2013) showed that adaptation affected direction selectivity for simulated motion of 470°/s in free-field stimulation. Wagner and Takahashi (1992) observed a broad tuning of motion direction selective units with respect to stimulus velocity when presenting simulated motion with an angular velocity ranging between 125 and 1200°/s. They suggest that inhibition accounts for the directional sensitivity. Our behavioural data obtained with noise sweeps in the free field and with an angular velocity between 53 and 2400°/s are in accord with the observed broad angular velocity tuning. Stimulus velocity had a moderate effect on motion direction sensitivity provided that the stimulus duration was not too short.

Duration of the noise sweeps per se had only a moderate effect on the behavioural sensitivity. Sensitivity was lower for short than for long stimulus durations. We could observe the lowest sensitivity for stimuli of 30 ms duration. This result is in accord with data describing the barn owl's binaural temporal window for ITD processing ranging from 4 to 42 ms (Wagner 1991). Inhibitory neuronal processing has been observed to be effective already between 20 and 40 ms after stimulus onset (Wagner 1990). Inhibition could thus contribute to the dynamic sharpening of the ITD representation in the barn owl's IC and to the duration of the binaural temporal window.

The angular range covered by the noise sweeps in our experiments exerted the strongest effect on the owl's sensitivity. To solve the task, the angular range had to be at least 24°. The angular range necessary for motion direction discrimination was much larger than the mean error for locating points in azimuth as indicated by the head orienting response towards the sound source elicited by broadband noise bursts (Knudsen et al. 1979) or by the owl's minimum audible angle (Bala et al. 2007). Using a pupillary dilation response, Bala et al. (2007) measured a minimum audible angle of 3° in azimuth. In the horizontal plain, Knudsen et al. (1979) observed a localization error being as low as 2° for broadband sounds in the frontal direction (open-loop conditions). The azimuth localization error increased to 6° if the target location was 50° lateral. The small localization error in comparison to the minimum angular range needed for motion direction discrimination suggests that at least partially different neuronal circuits are involved in the location of stationary and of moving sound sources. Wagner and Takahashi (1992) suggested a motion detector circuit in the barn owl auditory midbrain that included inhibition and a delayed interaction between neurons representing different locations in space (Wagner et al. 1997; Kautz and Wagner 1998). This circuit is interwoven with the computational auditory space map in the barn owl midbrain and relies on the space map providing input (Wagner et al. 1997). Thus, the barn owl brain uses separate but intermingled systems for representing auditory motion and spatial location. Furthermore, the large difference between the localization accuracy and the minimum angular range needed for motion direction discrimination renders it unlikely that the owls solve the motion direction discrimination task by a "snapshot" mechanism. The snapshot hypothesis suggests that the processing of motion simply relies on comparing starting and endpoint of the moving stimulus (e.g., Perrot and Marlborough 1989). The snapshot hypothesis was also dismissed in human psychophysical studies determining the minimum audible movement angle (reviewed by Middlebrooks and Green 1991). The difference between the barn owl's sensitivity for comparing angular locations of static sounds (Bala et al. 2007) and the large angular range needed for direction discrimination (present study) suggests that the snapshot hypothesis does not apply to barn owls.

Neurophysiological recordings in the barn owl midbrain have revealed an interaction between the starting point of the motion and the preferred direction of motion direction sensitive units (Wagner and von Campenhausen 2002; Wang and Peña 2013). However, the results from the two studies were contradictory. Wagner and von Campenhausen (2002) observed that neurons responded preferentially to outward motion whereas Wang and Peña (2013) observed a preference to inward motion. The behavioural data obtained in the present study are not suited to resolve the discrepancy. Our owls were equally sensitive in detecting inward and outward moving noise sweeps.

Previous barn owl studies indicated that the frontal space is represented by a larger population of neurons than the lateral space (e.g., as reflected by the distribution of best ITDs in the IC; Wagner et al. 2007). Furthermore, the space maps in each IC represent the contralateral space plus an angular range of about 15° from the midline in the ipsilateral space (Knudsen and Konishi 1978a, 1978b). This results in an overlap in the representation of the frontal space (−15° to 15°) by IC neurons from both sides of the brain. Finally, neurons in the IC tuned to locations in the frontal space have more confined spatial receptive fields (Knudsen and Konishi 1978b). The neurophysiological results have prompted Knudsen and Konishi (1978b) to predict that localization accuracy should be larger in the frontal space than in the lateral space. As expected, Knudsen et al. (1979) observed a smaller localization error in the frontal field than in the lateral field when measuring the barn owls' head orienting response towards a stationary sound source. Our results are consistent with these findings. The barn owls' motion direction discrimination was more sensitive for stimuli being presented in the frontal space compared to stimuli presented in the lateral space.

Acknowledgments This study was funded by a grant from the Deutsche Forschungsgemeinschaft (TRR 31).

References

Bala AD, Spitzer MW, Takahashi TT (2007) Auditory spatial acuity approximates the resolving power of space-specific neurons. PLoS ONE 2(8):e675
Grothe B, Pecka M, McAlpine D (2010) Mechanisms of sound localization in mammals. Physiol Rev 90:983–1012
Hausmann L, Plachta DTT, Singheise M, Brill S, Wagner H (2008) In-flight corrections in free-flying barn owls (*Tyto alba*) during sound localization tasks. J Exp Biol 211:2976–2988
Kautz D, Wagner H (1998) GABAergic inhibition influences auditory motion-direction sensitivity in barn owls. J Neurophysiol 80:172–185
Knudsen EI, Konishi M (1978a) A neural map of auditory space in the owl. Science 200:795–797
Knudsen EI, Konishi M (1978b) Space and frequency are represented separately in auditory midbrain of the owl. J Neurophysiol 41(4):870–884
Knudsen EI, Blasdel GG, Konishi M (1979) Sound localization by the barn owl (*Tyto alba*) measured with the search coil technique. J Comp Physiol A 133:1–11
McAlpine D, Jiang D, Shackleton TM, Palmer AR (2000) Responses of neurons in the inferior colliculus to dynamic interaural phase cues: evidence for a mechanism of binaural adaptation. J Neurophysiol 83:1356–1365
Middlebrooks JC, Green DM (1991) Sound localization by human listeners. Ann Rev Psychol 42:135–159
Perrott DR, Marlborough K (1989) Minimum audible movement angle: marking the end points of the path traveled by a moving sound source. J Acoust Soc Am 85(4):1773–1775
Shackleton TM, Palmer AR (2010) The time course of binaural masking in the Inferior colliculus of Guinea pig does not account for binaural sluggishness. J Neurophysiol 104:189–199
Wagner H (1990) Receptive fields of neurons in the owl's auditory brainstem change dynamically. Eur J Neurosci 2:949–959
Wagner H (1991) A temporal window for lateralization of interaural time difference by barn owls. J Comp Physiol A 169:281–289
Wagner H, Takahashi T (1992) Influence of temporal cues on acoustic motion-direction sensitivity of auditory neurons in the owl. J Neurophysiol 68(6):2063–2076
Wagner H, von Campenhausen M (2002) Distribution of auditory motion-direction sensitive neurons in the barn owl's midbrain. J Comp Physiol A 188:705–713
Wagner H, Kautz D, Poganiatz I (1997) Principles of acoustic motion detection in animals and man. Trends Neurosci 20(12):583–588
Wagner H, Asadollahi A, Bremen P, Endler F, Vonderschen K, von Campenhausen M (2007) Distribution of interaural time difference in the barn owl's inferior colliculus in the low- and high-frequency ranges. J Neurosci 27(15):4191–4200
Wang Y, Peña JL (2013) Direction selectivity mediated by adaption in the owl's inferior colliculus. J Neurosci 33(49):19167–19175
Witten IB, Bergan JF, Knudsen EI (2006) Dynamic shifts in the owl's auditory space map predict moving sound location. Nat Neurosci 9(11):1439–1445

Change Detection in Auditory Textures

Yves Boubenec, Jennifer Lawlor, Shihab Shamma and Bernhard Englitz

Abstract Many natural sounds have spectrotemporal signatures only on a statistical level, e.g. wind, fire or rain. While their local structure is highly variable, the spectrotemporal statistics of these auditory textures can be used for recognition. This suggests the existence of a neural representation of these statistics. To explore their encoding, we investigated the detectability of changes in the spectral statistics in relation to the properties of the change.

To achieve precise parameter control, we designed a minimal sound texture—a modified cloud of tones—which retains the central property of auditory textures: solely statistical predictability. Listeners had to rapidly detect a change in the frequency marginal probability of the tone cloud occurring at a random time.

The size of change as well as the time available to sample the original statistics were found to correlate positively with performance and negatively with reaction time, suggesting the accumulation of noisy evidence. In summary we quantified dynamic aspects of change detection in statistically defined contexts, and found evidence of integration of statistical information.

Keywords Auditory textures · Change detection · Sound statistics

Y. Boubenec (✉) · J. Lawlor · S. Shamma · B. Englitz
Laboratoire des Systèmes Perceptifs, CNRS UMR 8248, 29 rue d'Ulm, 75005 Paris, France
e-mail: boubenec@ens.fr

Y. Boubenec · J. Lawlor · S. Shamma · B. Englitz
Département d'études cognitives, Ecole normale supérieure PSL Research University, 29 rue d'Ulm, 75005 Paris, France

S. Shamma
Neural Systems Laboratory, University of Maryland in College Park, MD, USA

B. Englitz
Department of Neurophysiology, Donders Centre for Neuroscience, Radboud Universiteit Nijmegen, Nijmegen, The Netherlands

P. van Dijk et al. (eds.), *Physiology, Psychoacoustics and Cognition in Normal and Impaired Hearing,* Advances in Experimental Medicine and Biology 894,
DOI 10.1007/978-3-319-25474-6_24

1 Introduction

For many natural sounds, such as wind, fire, rain, water or bubbling, integration of statistical information under continuous conditions is critical to assess the predictability of changes. While their spectrogram is dominated by small, recognizable elements (e.g. gust, crackles, drops), their occurrence is not predictable *exactly*, but they occur randomly in time and frequency, yet constrained by certain probabilities. As shown previously, these probabilities/statistics define the identity of such sounds, so called auditory textures (McDermott et al. 2013). McDermott and others could demonstrate this point directly, by recreating auditory textures from samples, using only the spectrotemporal statistics of those sounds (e.g. McDermott and Simoncelli 2011).

Detecting relevant changes in auditory textures is hence complicated due to the lack of deterministic spectrotemporal predictability, and instead relies on the prediction of stimulus statistics (and their dynamics). The aim of this study is to determine how changes in stimulus statistics are detected under continuous presentation of stimulus. Concretely, we assessed how first-order non-uniform sound statistics are integrated and how listeners disentangle a change in these from the intrinsic stochastic variations of the stimulus.

We designed a broadband stimulus, composed of tone-pips, whose occurrence is only constrained by a frequency marginal probability. While the central 'textural' property of solely statistical predictability is maintained, it is devoid of most other systematic properties, such as frequency modulation, across channel or temporal correlations. A change is introduced by modifying the marginal distribution at a random time during the stimulus presentation, which subjects are asked to report.

This change detection task allows us to explicitly address the integration of statistical information under continuous conditions, which models more closely the real-world challenge of detecting changes in complex ongoing auditory scenes. In this context, we studied the dependence of detection on the time and size of the change. We found evidence for integration of statistical information to predict a change in stimulus statistics, both based on hit rate as well as on reaction times. The main predictors appear to be the time available for integration of evidence.

2 Methods

2.1 Participants

Fifteen subjects (mean age 24.8, 8 females, higher education: undergraduate and above) participated in the main study in return for monetary compensation. All subjects reported normal hearing, and no history of psychiatric disorders.

2.2 Experimental Setup

Subjects were seated in front of a screen with access to a response box in sound-attenuated booth (Industrial Acoustics Company GmbH). Acoustic stimulus presentation and behavioural control were performed using custom software package written in MATLAB (BAPHY, NSL, University of Maryland). The acoustic stimulus was sampled at 100 kHz, and converted to an analog signal using an IO board (National Instruments NI PCIe-6353) before being sent to diotic presentation using headphones (Sennheiser HD380, calibrated flat, i.e. ± 5 dB within 100 Hz–20,000 Hz). Reaction times were measured via a custom-built response box and collected by the same IO card with a sampling rate of 1 kHz.

2.3 Stimulus Design

We used a simplified sound texture model, which retained the property of being predictable only from the statistical properties of its complex spectrotemporal structure (Fig. 1a). The texture was a tone cloud consisting of a sequence of 30-ms, temporally overlapping pure tones whose frequency covered a range of 2.2 octaves (400–1840 Hz) divided in 8 frequency bins. The frequency resolution of the tone distribution was 12 semitones per octave, starting at 400 Hz. This allowed 26 tone frequencies in total with 3–4 frequency values in each frequency bin.

The minimal temporal unit of the stimulus was a 30-ms chord in which the number of tones in a particular frequency range was drawn according to a marginal distribution for that frequency range. On average, for each chord duration the mean number of tones per octave was 2. The marginal distribution of occurrence probability was obtained by modifying a uniform distribution in each frequency bin. The probability in each of these is called $P_{uniform} = \frac{1}{8} = 0.125$. To generate different initial frequency marginals, we perturbed the marginals randomly by adding/subtracting fixed value Δ corresponding to 50% of the probability (which would be observed for a uniform distribution ($\Delta = P_{uniform}/2 = 0.0625$)). The resulting marginal distribution was thus pseudo-random with 3 bins at $P_{uniform}+\Delta$, 3 bins at $P_{uniform}-\Delta$ and 2 bins left intact at $P_{uniform}$. This implies that the average initial probability in 2 bins can take 5 different values, namely Δ, $3\Delta/2$, 2Δ, $5\Delta/2$, and 3Δ.

The change consisted in an increment of the marginal distribution in the selected frequency bins at a random point in time (referred to as *change time*) during stimulus presentation. We chose to use an increment of the marginal distribution in two adjacent frequency bins on the basis that an appearance is more salient than its opposite in a complex acoustic stimulus (Constantino et al. 2012). After the change, the second stimulus continued for up to 2 s or until the subject made an earlier decision, whichever happened first.

The increment size, referred to as *change size*, was drawn from a set of discrete values [50, 80, 110, 140] %, relative to the single bin probability in a uniform

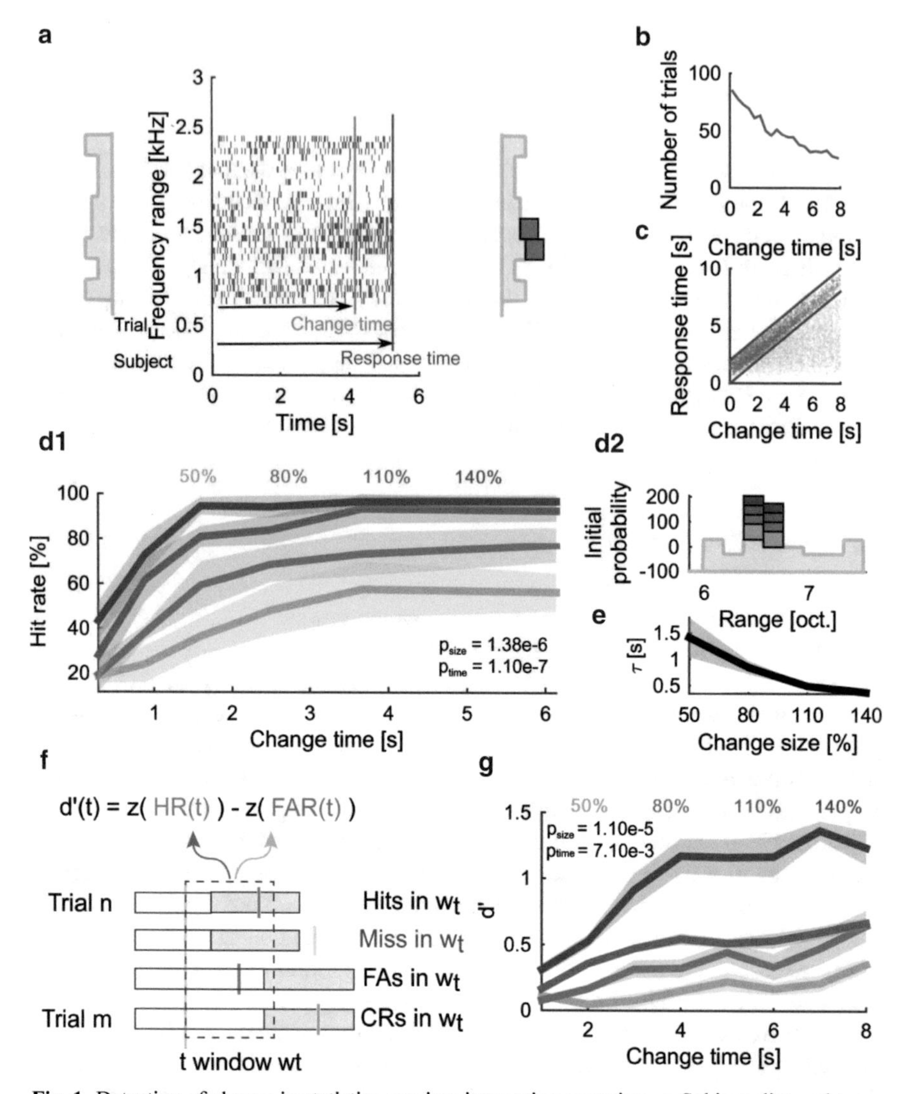

Fig. 1 Detection of change in statistics requires integration over time. **a** Subjects listened to an acoustic textural stimulus, whose sole predictability was governed by its marginal frequency distribution (*grey curve, left*). Tones in individual frequency bins were all drawn independent consistent with the marginal (*middle*). Listeners were instructed to report changes by a button press. The frequency marginal was modified after a randomly chosen point in time ('change time', *red* bars on the *right*). The probability in two frequency bins separated by a random distance was modified at a time. All other bins were slightly reduced in probability to maintain the overall loudness constant. **b** Distribution of change times across all subjects (n=10). Change times were drawn from an exponential distribution. **c** Response times occurred before (false alarms) and after the change time (hits). Subjects usually responded only after an initial time of listening, allowing them to acquire the acoustic statistics. **d** Hit rate of change detection depended significantly on change time (x-axis) and change size (colour). **e** Four different change sizes were used (colour-coded). **f** The dynamics of the hit rate curve also varied with change size, indicated by the fitted parameter τ of a cumulative Erlang distribution (see *Methods*). **g** To control for false alarms, d' was also computed as a function of time into the trial (see *Methods* for details), where within each trial the statistics change where the gray area starts. **h** Similar to hit rate, d' increased significantly over time into the trial, and with change size. Errorbars indicate confidence bounds two SEM based on a bootstrap across subjects

distribution (for 8 bins, the single bin uniform probability is 1/8th, and thus a 50 % change size would be 1/16th). In order to maintain the overall number of tones per chord before and after the change, the occurrence probability in the six remaining frequency bins was decreased accordingly.

The time at which the change occurred (*change time*) was drawn randomly from an exponential distribution (mean: 3.2 s) limited to the interval [0, 8] s. This choice of distribution prevents subjects from developing a timing strategy, as the instantaneous probability of a change in the next time step is constant.

2.4 Procedure

The experiment was separated into three phases: instruction, training, and main experiment. After reading the instructions, subjects went through 10 min of training (60 trials), where they were required to obtain at least 40 % performance. The training comprised only stimuli of the two greatest change sizes (110 %, 140 %). Three subjects did not attain the criterion level of performance and were not tested further.

The main experiment was composed of two sessions of about 70 min each, comprising a total of 930 trials, corresponding to 60 repetitions of each condition. The two sessions were never more than 2 days apart. After reading the instructions, subjects were aware that the change could arise at any moment on each trial and that their task was to detect it within the 2 s window.

Visual feedback was always displayed on the screen in front of them; either a red square was displayed when the button was pressed before the change (false alarm), or when the button was not pressed within the 2 s time window after the change (miss). A green square was displayed when the button was pressed after the change but within the 2 s window.

In addition, sound level was roved from trial to trial; it was chosen randomly between 60 and 80 dB SPL (sound pressure level). This procedure is classically applied to prevent subjects from adopting any absolute level strategy. The inter-trial interval was ~ 1 s with a small, random jitter (<0.1 s) depending on computer load.

2.5 Data Analysis

We quantified the ability of the subjects to detect the change in stimulus statistics using two measures, hit rate and d-prime (d'). We also found reaction times to depend on the difficulty.

These measures were computed as a function of change size and change time. Since change times were distributed continuously but with an exponential distribution, the set of change times was binned with approximately exponentially increasing bin size (in order to achieve comparable numbers of trials in each bin).

To control for inattentive subjects, we set a 35 % threshold for the total false alarm rate. Two subjects were discarded according to this criterion leaving a total of 10 subjects for the data analysis, with false alarm rates around 25 %.

2.5.1 Hit Rate and Reaction Times

We computed a subject's hit rate as the fraction between successful detection (hits) out of the total trials for which the change occurred before the subject's response (hits + misses). False alarms were excluded from the hit rate computation, since they occurred before the subject was exposed to the change (see d' below for an inclusion of false alarms). We obtained reaction times by subtracting the change time from the response time in each trial.

2.5.2 d' Analysis

We computed d' values to assess the ability to detect changes, while taking their false alarm rate into account. Due to the present task structure, d' was computed as a function of time from stimulus onset (see Fig. 1e for an illustration), approximated as d'(t) = Z(HR(t)) − Z(FAR(t)), where Z(p) is the inverse of the cumulative Gaussian distribution. HR(t) is the hit rate as a function of time t since stimulus onset. Hit rate was computed as the fraction of correct change detections, in relation to the number of trials with changes occurring at t. Similarly, the false alarm rate FAR(t) was computed as the number of false alarms that occurred over all 2 s windows (starting at t), in which no change in statistics occurred. The window of 2 s was chosen to be compatible with the hit rates in the 2 s decision window. d' was computed separately for different times and change sizes, yielding only a limited number of trials per condition. To avoid degenerate cases (i.e. d' would be infinite for perfect scores), the analysis was not performed separately by subject, but over the pooled data. Confidence bounds (95 %) were then estimated by grouping data from all subjects. The analysis was verified on surrogate data from a random responder (binomial with very low p at each point in time), providing d' very close to 0 on a comparable number of trials.

2.5.3 Hit rate Dynamics

In order to compare the hit rate dynamics for different change sizes, we fitted (least-square non-linear minimization) a cumulative Erlang distribution to the data according to:

$$P(\Delta_c, t_c) = P_0(\Delta_c) + P_{max}(\Delta_c) * \gamma(k, t_c / \tau(\Delta_c)) / (k-1)!$$

where P_0 is the minimal hit-rate, P_{max} is the maximal hit rate, t_c is change time, Δ_c the change size, γ the incomplete gamma function, τ the function rate, and k controls the function shape. k was kept constant across subjects and change sizes, assuming the shape of the hit rate curves is invariant, which appeared to be the case in our sample.

2.6 Statistical Analysis

In the statistical analysis only non-parametric tests were used. One-way analysis of variance were computed with Kruskal-Wallis' test, the two-way were computed using Friedman's test. Unless is indicated otherwise, error bars correspond to twice the standard error of the mean (SEM). All statistical analyses were performed using Matlab (The Mathworks, Natick).

3 Results

Human listeners ($n=15$, 10 performed to criterion level) were presented with a continuous acoustic texture and required to detect a change in stimulus statistics, which could occur at a random time, after which they had 2 s to respond. Changes occurred in the probability of occurrence of tones in randomly chosen spectral regions (Fig. 1a).

Several parameters of the change could determine its salience, such as its size, timing. These parameters were tested in a single experiment with ~1000 trials over two sessions per subject.

3.1 Detection of Changes in Statistics is Consistent with Integration

The ability of subjects to detect a change in stimulus statistics (Fig. 1a) improved with the time they were allowed to listen to the initial statistics of the texture (Fig. 1d, progression along ordinate, performance measured as hit rate, $p<10^{-6}$, Friedman test). Hit rate increased monotonically to an asymptotic level for all change sizes (4 levels, [50, 80, 110, 140] %). Asymptotic hit rate depended on the change size, with bigger changes in marginal probability leading to greater asymptotic hit rate (Fig. 1d, different colours). Although asymptotic hit rate was above chance (26.6 %) for all change sizes, the increase with change size was large and significant (from 50 to 95 %, $p<10^{-7}$, Friedman test).

Change size also influenced the shape of the dependence on change time, such that greater change sizes lead to improved hit rate already for shorter change times

(Fig. 1d). This translates to a combined steepening and leftward shift of the hit rate curves with change size. Significance of this effect was assessed by fitting sigmoidal hit rate curves of individual subjects with a parametric function (see *Methods*) in order to extract the change size-dependent time constant (Fig. 1f). Hit rate time constants τ significantly decreased with respect to change sizes ($p<10^{-6}$, Friedman test).

The results above stay true if false alarms are taken into account (Fig. 1h). For this purpose, we computed a time-dependent d' measure (see Fig. 1g for an illustration, and *Methods*). Classical d' analysis does not apply in the present dynamic decision paradigm, since the eventual outcome of a trial is not accessible at early times (i.e. when computing d'(0.5 s), a trial with a hit at 5 s should not be counted as a hit, but as a 'correct reject up to 0.5 s; but when computing d'(4 s), the same trial will be counted as a hit). In accordance with the perceived difficulty of the detection task, d' values are significantly positive (Fig. 1h, based on the 95 % confidence bounds) and increase with change time (Friedman test, $p<0.01$), but stay relatively low especially for 110 % and below, indicates how challenging the task was, leading to a substantial number of false alarms.

The above results are consistent with predictions from statistical integration of information. First, unsurprisingly, change detection should improve with greater change sizes. Second, change detection should improve, but saturate, as the estimation of the statistics of the first stimulus converges, i.e. if longer observation time is available. Third, change detection of bigger changes should require less precision in the estimation, translating to higher hit rate already for shorter observation times.

3.2 *Reaction Times are Consistent with Integration*

The dependence of performance on change time suggests a dynamical integration mechanism to play a role in the decision process. While the dependence on change time indicates an ongoing estimation of the initial statistics, it does not provide much insight into the estimation of the post-change statistics. We hypothesized that reaction times here could be informative. Reaction times had previously been demonstrated to depend on task difficulty (Kiani et al. 2014), and difficulty in the present paradigm correlates inversely with the availability of a greater difference in change size, i.e. statistics (Fig. 2a).

Reaction time distributions changed both in time and shape as a function of change size (Fig. 2b). Median reaction time correlated negatively with change sizes ($p<0.001$; Fig. 2c), in a manner that corresponded to the inverse of the increase in hit rate with larger change sizes. In addition, the onset time of the distributions decreased with change size, together suggesting that certainty was reached earlier for bigger step-sizes.

As a function of change time, the reaction time distribution changed in a qualitatively similar manner, however, less pronounced (Fig. 2e). Median reaction times correlated negatively with change times, mirroring dependence of hit rate on change

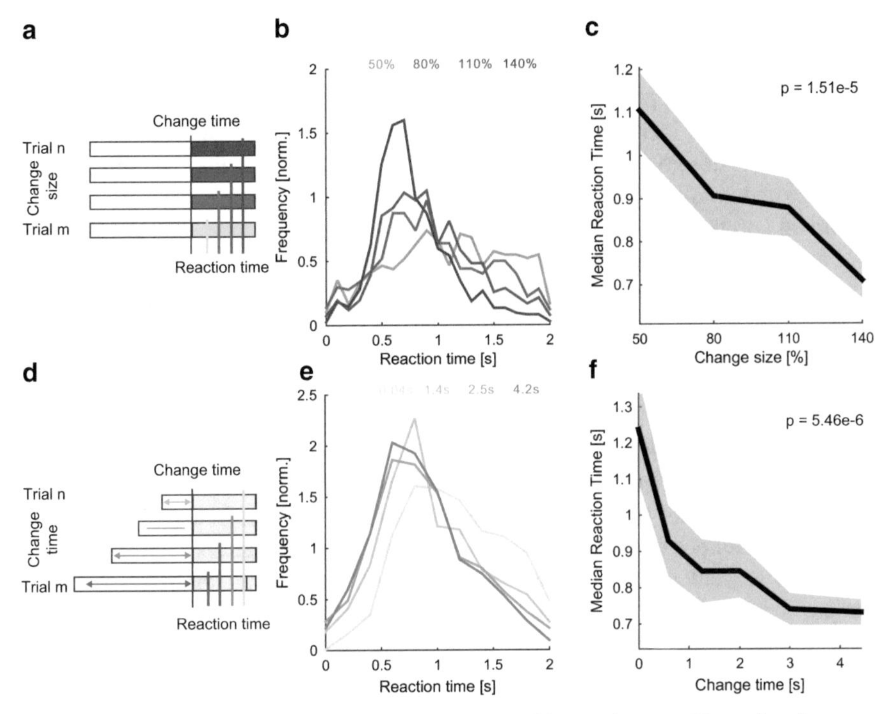

Fig. 2 Reaction times reflect task difficulty and temporal integration. **a–c** If reaction times are separated by change size (**a**), one observes a progression in the shape (**b**) as well as median response time (**c**). The shape of the reaction time distribution sharpens with increasing step-size and the median response time decreases significantly (p=1.51e-05, Kruskal-Wallis). **d–f** If reaction times are separated by the duration of the change time before the change in statistics occurred (**d**), the distribution of reaction times again sharpens with longer change times (**e**) and the median reaction time (**f**) decreases significantly (p=5.46e-06, Kruskal-Wallis)

times ($p<0.001$; Fig. 2f). This dependence can already be seen in the raw data (Fig. 1c), where hit trials (black) for longer change times exhibit shorter reaction times. Again the onset time decreased correspondingly with longer change time, suggesting more accurate estimation of the initial statistics.

4 Discussion

In the present study we used a simplified acoustic texture to study detecting changes in the statistics of complex acoustic stimuli. The results suggest that humans are able to integrate stimulus statistics, as expressed in their performance and reaction times depending on the parameters of the task.

4.1 Dynamic Representation of Spectral Statistics

The stimulus of the present paradigm can be considered a stochastic version of the appearance or increase in loudness of a sound source within a natural auditory scene. Its design is a compromise between complexity of the spectrotemporal structure and simplicity in specification since the parameter-space is reduced to a minimum and parameters are straightforward to interpret.

The present paradigm builds on research from the spectral and the temporal domain, e.g. profile analysis (Green and Berg 1991) or some paradigms in mismatch negativity. For example in profile analysis subjects assess the spectral shape of stimuli. In this context it has also been demonstrated that subjects can compare spectra, however, the results were not interpretable in a statistical sense, due to several reasons. First, the time scale of presentation is much longer in the present paradigm than typically in profile analysis, where the stimulus is presented for multiple seconds rather than subsecond presentations. Second, in the present context the stimulus is composed of randomly timed occurrences of tones in certain frequency bins (which defines the spectral profile over time), rather than static profiles from constantly presented tones or filtered noise. Third, the comparison between different profiles is performed dynamically, rather than in a two stimulus comparison task. As described above this allowed us to study the dynamic integration of stochastic information, and enables the collection of reaction times as another indicator of the underlying processes.

4.2 Future Directions

The present study lays some groundwork for future studies, by investigating some of the basic, previously unaddressed questions related to the integration of statistics in complex auditory stimuli, such as auditory textures. Further studies are required to address variations of the present paradigm, such as reductions in probability and continuous stimulus presentation. Attention was not controlled for in the present setting, although subjects were instructed to attend to the stimulus. Some subjects reported improved detection of change, when *not* attending to the stimulus, suggesting that a dual-task paradigm would be interesting where attention can be assessed and controlled.

Acknowledgments Funding was provided through the Advanced ERC ADAM 295603, the ANR-10-LABX-0087-IEC and the ANR-10-IDEX-0001-02-PSL*.

References

Constantino FC, Pinggera L, Paranamana S, Kashino M, Chait M (2012) Detection of appearing and disappearing objects in complex acoustic scenes. PLoS ONE. doi:10.1371/journal.pone.0046167

Green DM, Berg BG (1991) Spectral weights and the profile bowl. Q J Exp Psychol Sect A 43(3):449–458

Kiani R, Corthell L, Shadlen MN (2014) Choice certainty is informed by both evidence and decision time. Neuron 84(6):1329–1342

McDermott JH, Simoncelli EP (2011) Sound texture perception via statistics of the auditory periphery: evidence from sound synthesis. Neuron 71(5):926–940

McDermott JH, Schemitsch M, Simoncelli EP (2013) Summary statistics in auditory perception. Nat Neurosci 16(4):493–498

The Relative Contributions of Temporal Envelope and Fine Structure to Mandarin Lexical Tone Perception in Auditory Neuropathy Spectrum Disorder

Shuo Wang, Ruijuan Dong, Dongxin Liu, Luo Zhang and Li Xu

Abstract Previous studies have demonstrated that temporal envelope (E) is sufficient for speech perception, while fine structure (FS) is important for pitch perception for normal-hearing (NH) listeners. Listeners with sensorineural hearing loss (SNHL) have an impaired ability to use FS in lexical tone perception due to the reduced frequency resolution. Listeners with auditory neuropathy spectrum disorder (ANSD) may have deficits in temporal resolution. Little is known about how such deficits may impact their ability to use E and FS to perceive lexical tone, and whether it is the deficit in temporal resolution or frequency resolution that may lead to more detrimental effects on FS processing in pitch perception. Three experiments were conducted in the present study. Experiment I used the "auditory chimera" technique to investigate how SNHL and ANSD listeners would achieve lexical tone recognition using either the E or the FS cues. Experiment II tested their frequency resolution as measured with their psychophysical tuning curves (PTCs). Experiment III tested their temporal resolution as measured with the temporal gap detection (TGD) threshold. The results showed that the SNHL listeners had reduced frequency selectivity, but intact temporal resolution ability, while the ANSD

*Parts of this paper may be published elsewhere.

S. Wang (✉) · R. Dong · D. Liu · L. Zhang
Otolaryngology—Head & Neck Surgery, Beijing Tongren Hospital,
Beijing Institute of Otolaryngology, Capital Medical University, Beijing, China
e-mail: shannonwsh@aliyun.com

R. Dong
e-mail: drj_789@163.com

D. Liu
e-mail: liudx891209@163.com

L. Zhang
e-mail: dr.luozhang@139.com

L. Xu
Communication Sciences and Disorders, Ohio University, Athens, OH, USA
e-mail: xul@ohio.edu

P. van Dijk et al. (eds.), *Physiology, Psychoacoustics and Cognition in Normal and Impaired Hearing,* Advances in Experimental Medicine and Biology 894,
DOI 10.1007/978-3-319-25474-6_25

listeners had degraded temporal resolution ability, but intact frequency selectivity. In comparison with the SNHL listeners, the ANSD listeners had severely degraded ability to process the FS cues and thus their ability to perceive lexical tone mainly depended on the ability to use the E cues. These results suggested that, in comparison with the detrimental impact of the reduced frequency selectivity, the impaired temporal resolution may lead to more degraded FS processing in pitch perception.

Keywords Auditory neuropathy spectrum disorder (ANSD) · Sensorineural hearing loss (SNHL) · Auditory chimera · Lexical tone · Pitch · Temporal cues · Fine structure · Temporal envelope · Temporal resolution · Frequency resolution

1 Introduction

Acoustic signals can be decomposed into temporal envelope (E) and fine structure (FS). Previous studies have demonstrated that E is sufficient for speech perception (Shannon et al. 1995; Apoux et al. 2013), while FS is important for pitch perception (Smith et al. 2002) and lexical tone perception (Xu and Pfingst 2003). Mandarin Chinese is a tone language with four phonologically distinctive tones that are characterized by syllable-level fundamental frequency (F0) contour patterns. These pitch contours are described as high-level (tone 1), rising (tone 2), falling-rising (tone 3), and falling (tone 4). It has been demonstrated that FS is also important for lexical tone perception in listeners with sensorineural hearing loss (SNHL), but as their hearing loss becomes more severe, their ability to use FS becomes degraded so that their lexical tone perception relies increasingly on E rather than FS cues (Wang et al. 2011). The reduced frequency selectivity may underlie the impaired ability to process FS information for SNHL listeners. Auditory neuropathy spectrum disorder (ANSD) is an auditory disorder characterized by dys-synchrony of the auditory nerve firing but normal cochlear amplification function. Several studies have demonstrated that ANSD listeners have a dramatically impaired ability for processing temporal information (Zeng et al. 1999; Narne 2013). It is important to assess the ability of lexical tone perception for ANSD listeners and to investigate how they use FS and E cues to perceive lexical tone, as these results may shed light on the effects of frequency resolution and temporal resolution on FS processing in pitch perception.

2 Lexical Tone Perception

Experiment I was aimed to investigate the ability of lexical tone perception for both SNHL and ANSD listeners, and to assess the relative contributions of FS and E to their lexical tone perception using the "auditory chimera" technique (Smith et al. 2002; Xu and Pfingst 2003). As described in Wang et al. (2011), the chimeric tokens were generated in a condition with 16 channels. These 16 FIR band-pass filters with nearly rectangular response were equally spaced on a cochlear frequency map, ranging from 80 to 8820 Hz. In order to avoid the blurriness between E and FS in the chimeric stimuli, we adopted a lowpass filter (cut-off at 64 Hz) for extraction of the envelopes from the filters. By analysing the tone responses to the chimeric tone tokens, we could understand the competing roles of FS and E in tone perception in various groups of listeners.

2.1 Methods

Three groups of adult subjects, including 15 NH subjects (8 females and 7 males), 16 patients with SNHL (10 females and 6 males), and 27 patients with ANSD (9 females and 18 males), participated in the present study. The subjects with SNHL had hearing loss ranging from moderate to severe degree, and the ANSD subjects had hearing loss ranging from mild to severe. The original tone materials consisted of 10 sets of Chinese monosyllables with four tone patterns for each, resulting in a total of 40 commonly used Chinese words. These original words were recorded using an adult male and an adult female native Mandarin speaker. The tone tokens in which the durations of the four tones in each monosyllable were within a 5-ms precision were chosen as the original tone tokens. These tone tokens were processed through a 16-channel chimerizer in which the FS and E of any two different tone tokens of the same syllable were swapped. For example, two tokens of the same syllable but with different tone patterns were passed through 16 band-pass filters to split each sound into 16 channels. The output of each filter was then divided into its E and FS using a Hilbert transform. Then, the E of the output in each filter band was exchanged with the E in that band for the other token to produce the single-band chimeric wave. The single-band chimeric waves were summed across all channels to generate two chimeric stimuli. Finally, a total of 320 tokens were used in the tone perception test, including 80 original unprocessed tone tokens and 240 chimeric tone tokens. A four-alternative, forced-choice procedure was used for the tone perception test.

2.2 Results

Figure 1 (left panel) plots the accuracy of tone perception to the original tone tokens for the NH, SNHL, and ANSD subjects. The median tone perception performance was 97.2, 86.5, and 62.8 % correct, respectively for the three groups. The performance slightly reduced for the subjects with SNHL, while the ANSD subjects had dramatic decreases in their tone perception performance with a very large individual variability. The right panel of Fig. 1 plots the mean percentages and standard deviations of tone perception responses to the chimeric tone tokens. The

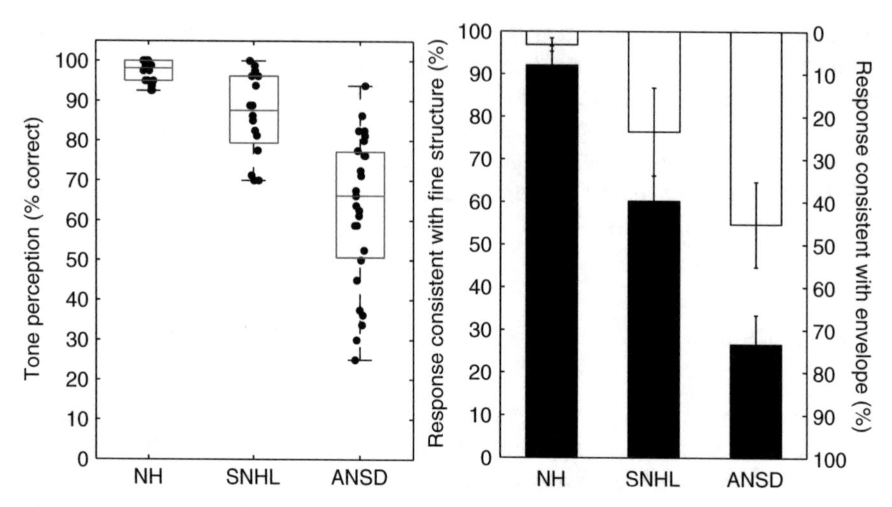

Fig. 1 Tone perception performance. *Left*: Boxplots show the tone perception performance with the original, unprocessed tone tokens for the subjects with NH, SNHL, and ANSD. The box represents the 25th to 75th percentiles and the horizontal line in the box represents the median. The whiskers represent the range of the data. *Right*: Mean and standard deviation of tone perception with the chimeric tone tokens for the subjects with NH, SNHL, and ANSD. *Black* bars represent responses consistent with FS (right ordinate) and open bars represent responses consistent with E (left ordinate)

subjects with NH relied on FS nearly entirely to perceive lexical tones regardless of E. In comparison with the NH subjects, the subjects with SNHL depended more on E cues to perceive tones, as they had impaired ability to use FS. In contrast to the finding in the NH and SNHL subjects, only 26.5 % of the tone responses were consistent with the FS of the chimeric tone tokens for the ANSD subjects, indicating that the ANSD subjects had more severely impaired ability to use FS in tone perception. On the other hand, the ANSD subjects relied more heavily on E cues in perceiving lexical tones, indicating that the ability of the ANSD subjects to use E cues for lexical tone perception still remained at a reasonable level.

3 The Frequency Resolution and Temporal Resolution

Studies have shown that listeners with SNHL may have reduced frequency resolution (Kluk and Moore 2006), but probably normal temporal resolution (Glasberg et al. 1987). In contrast, listeners with ANSD may have a dramatically impaired ability for processing temporal information (Narne 2013), while their frequency selectivity may be close to normal (Vinay and Moore 2007). To clarify whether it is the deficit in temporal resolution or frequency resolution that may lead to more detrimental effects on FS processing in lexical tone perception, we also measured the frequency resolution using the mean $Q_{10\,dB}$ values of the psychophysical tuning curves (PTCs), and the temporal resolution using the temporal gap detection (TGD) threshold for the SNHL and ANSD subjects.

3.1 Methods

Three groups of subjects who were recruited in Experiment I also participated in Experiment II in which we measured the psychophysical tuning curves (PTCs) of the auditory system using a fast method called SWPTC, developed by Sek and Moore (2011). Each subject was required to detect a sinusoidal signal, which was pulsed on and off repeatedly in the presence of a continuous noise masker. The PTCs at 500 and 1000 Hz in each ear were tested separately. The sinusoidal signals were presented at 15 dB sensation level (SL). The masker was a narrowband noise, slowly swept in frequency. The frequency at the tip of the PTC was estimated using a four-point moving average method (4-PMA), and the values of $Q_{10\ dB}$ (i.e., signal frequency divided by the PTC bandwidth at the point 10 dB above the minimum level) was used to assess the sharpness of the PTC. The greater the $Q_{10\ dB}$ values, the sharper the PTC.

Experiment III tested the temporal gap detection (TGD) threshold for all subjects using a TGD program developed by Zeng et al. (1999). The test stimuli were generated using a broadband (from 20 to 14,000 Hz) white noise. The noise had a duration of 500 ms with 2.5-ms cosine-squared ramps. A silent gap was produced in the center of the target noise. The other two reference noises were uninterrupted. The TGD threshold of left and right ear was measured separately for each subject using a three-alternative, forced-choice procedure. The intensity level of the white noise was set at the most comfortable loudness level for all subjects.

3.2 Results

The values of $Q_{10\ dB}$ were determined in all 15 NH subjects, 7 of the 16 SNHL subjects, and 24 of the 27 subjects with ANSD. Note that for 9 of the 16 subjects with SNHL, the PTCs were too broad so that the $Q_{10\ dB}$ values could not be determined. Since no significant differences were found between the $Q_{10\ dB}$ values at 500 and 1000 Hz, the $Q_{10\ dB}$ values averaged across the two test frequencies are plotted in Fig. 2. Statistically significant differences of the $Q_{10\ dB}$ values were found between the subjects with NH and those with SNHL and between the subjects with ANSD and those with SNHL, while no statistically significant differences were present between the subjects with NH and those with ANSD. This indicates that a majority of the subjects with ANSD showed close-to-normal frequency resolution, whereas the SNHL subjects had poor frequency resolution in the present study.

Figure 3 shows the averaged TGD thresholds of both ears for the three groups of subjects. Welch's test in a one-way ANOVA found significant differences of

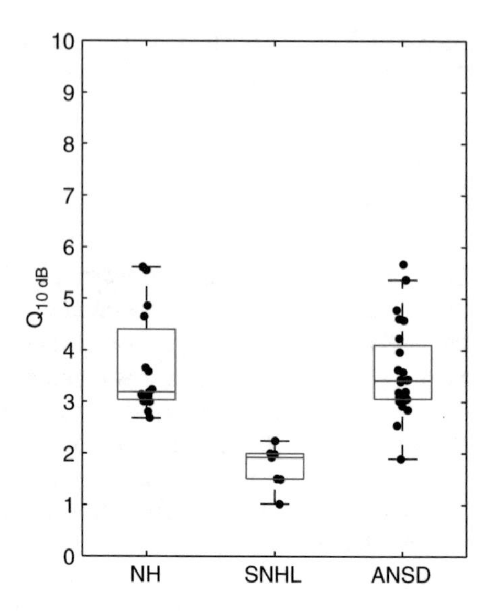

Fig. 2 The $Q_{10\ dB}$ values of psychophysical tuning curves for subjects with NH, with SNHL, and with ANSD. The box represents the 25–75th percentiles and the horizontal line in the box represents the median. The whiskers represent the range of the data. Data points outside of the whiskers are outliers. Note that for 9 of the 16 subjects with SNHL, the PTCs were too broad so that the Q_{10dB} values could not be determined

TGD thresholds among the three groups of subjects [$F(2,56)=9.9$, $p<0.001$]. Post hoc Tamhane' T2 correction test analysis indicated that the mean TGD thresholds for both ears was significantly higher in the listeners with ANSD (11.9 ms) than in the listeners with SNHL (4.0 ms; $p<0.001$) and in the NH (3.9 ms; $p<0.001$) listeners; while the subjects with SNHL had the TGD thresholds close to normal. It is notable that the variability in TGD thresholds for the ANSD subjects was very

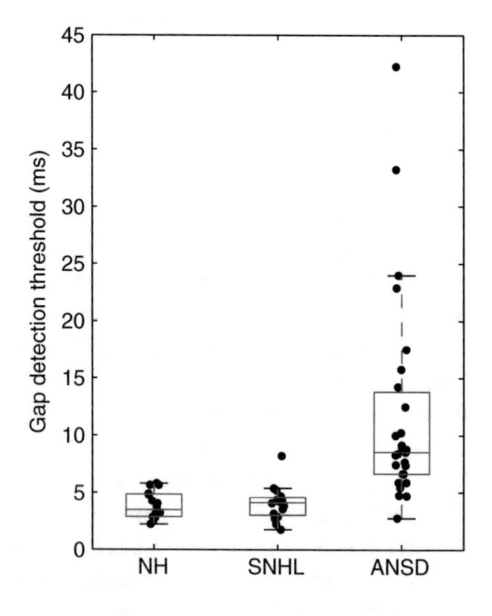

Fig. 3 Temporal gap detection thresholds for subjects with NH, SNHL, and ANSD. The box represents the 25th to 75th percentiles and the horizontal line in the box represents the median. The whiskers represent the range of the data. Data points outside of the whiskers are outliers

large, ranging from normal limits to 10 times the norm. This result indicates that the SNHL subjects had normal temporal resolution as long as the audibility was compensated, but a majority of the subjects with ANSD had poor temporal resolution in the present study.

4 Discussion and Conclusions

The fine structure (FS) cues in the speech signals play a dominant role in lexical tone perception (Xu and Pfingst, 2003; Wang et al. 2011). The temporal envelope (E) cues contribute to tone perception when the FS cues are not available (Xu et al. 2002; Kong and Zeng 2006; Xu and Zhou 2011). Experiment I of the present study demonstrated that the listeners with ANSD had great deficits in using the FS cues for lexical tone perception, which resulted in great difficulties in perceiving lexical tones for the ANSD listeners. Even though the ability of the listeners with ANSD to use E cues for lexical tone perception remained at a reasonable level, this level of performance was much lower in comparison with that of the NH listeners. The NH listeners could utilize the FS cues and achieve perfect tone perception performance. The tone perception performance in the listeners with ANSD was also lower than that in the listeners with SNHL who were still able to use the FS cues for tone perception to some extent. An earlier report has shown that the ability to use the FS cues for tone perception is negatively correlated with the degree of hearing loss in listeners with SNHL (Wang et al. 2011).

Experiments II and III evaluated the frequency resolution and temporal resolution, respectively for the subjects with SNHL and ANSD. Consistent with previous studies, the listeners with SNHL in the present study had poor frequency resolution whereas a majority of the listeners with ANSD had normal frequency resolution. On the other hand, a majority of the listeners with ANSD had poor temporal resolution whereas the listeners with SNHL had normal temporal resolution. This may imply that poor temporal resolution rather than the frequency resolution exerts the major detrimental effects on FS cue processing for pitch perception.

Acknowledgements We thank all the subjects for participating in this study and our colleagues at the Clinical Audiology Center in Beijing Tongren Hospital for helping recruit the hearing-impaired subjects. This work was funded in part by grants from the National Natural Science Foundation of China (81200754), the 2012 Beijing Nova Program (Z121107002512033), the Capital Health Research and Development of Special from the Beijing Municipal Health Bureau (2011-1017-04).

References

Apoux F, Yoho SE, Youngdahl CL, Healy EW (2013) Role and relative contribution of temporal envelope and fine structure cues in sentence recognition by normal-hearing listeners. J Acoust Soc Am 134:2205–2212

Glasberg BR, Moore BCJ, Bacon SP (1987) Gap detection and masking in hearing-impaired and normal-hearing subjects. J Acoust Soc Am 81:1546–1556

Kluk K, Moore BCJ (2006) Detecting dead regions using psychophysical tuning curves: a comparison of simultaneous and forward masking. Int J Audiol 45:463–476

Kong YY, Zeng FG (2006) Temporal and spectral cues in Mandarin tone recognition. J Acoust Soc Am 120:2830–2840

Narne VK (2013) Temporal processing and speech perception in noise by listeners with auditory neuropathy. Plos ONE 8:1–11

Sek A, Moore BCJ (2011) Implementation of a fast method for measuring psychophysical tuning curves. Int J Audiol 50:237–242

Shannon RV, Zeng FG, Wygonski J (1995) Speech recognition with primarily temporal cues. Science 270:303–304

Smith ZM, Delgutte B, Oxenham AJ (2002) Chimeric sounds reveal dichotomies in auditory perception. Nature 416:87–90

Vinay, Moore BCJ (2007) Ten(HL)-test results and psychophysical tuning curves for subjects with auditory neuropathy. Int J Audiol 46:39–46

Wang S, Mannell R, Xu L (2011) Relative contributions of temporal envelope and fine structure cues to lexical tone recognition in hearing-impaired listeners. J Assoc Res Otolaryngol 12:783–794

Xu L, Pfingst BE (2003) Relative importance of temporal envelope and fine structure in lexical-tone perception. J Acoust Soc Am 114:3024–3027

Xu L, Zhou N (2011) Tonal languages and cochlear implants. In: Zeng F-G, Popper AN, Fay RR. (eds) Auditory prostheses: new horizons. Springer Science + Business Media, LLC, New York, pp 341–364

Xu L, Tsai Y, Pfingst BE (2002) Features of stimulation affecting tonal-speech perception: implications for cochlear prostheses. J Acoust Soc Am 112:247–258

Zeng FG, Oba S, Grade S, Sininger Y, Starr A (1999) Temporal and speech processing deficits in auditory neuropathy. Neuroreport 10:3429–3435

Interaction of Object Binding Cues in Binaural Masking Pattern Experiments

Jesko L.Verhey, Björn Lübken and Steven van de Par

Abstract Object binding cues such as binaural and across-frequency modulation cues are likely to be used by the auditory system to separate sounds from different sources in complex auditory scenes. The present study investigates the interaction of these cues in a binaural masking pattern paradigm where a sinusoidal target is masked by a narrowband noise. It was hypothesised that beating between signal and masker may contribute to signal detection when signal and masker do not spectrally overlap but that this cue could not be used in combination with interaural cues. To test this hypothesis an additional sinusoidal interferer was added to the noise masker with a lower frequency than the noise whereas the target had a higher frequency than the noise. Thresholds increase when the interferer is added. This effect is largest when the spectral interferer-masker and masker-target distances are equal. The result supports the hypothesis that modulation cues contribute to signal detection in the classical masking paradigm and that these are analysed with modulation bandpass filters. A monaural model including an across-frequency modulation process is presented that account for this effect. Interestingly, the interferer also affects dichotic thresholds indicating that modulation cues also play a role in binaural processing.

Keywords Binaural processing · Binaural masking level difference · Beating · Modulation processing · Across-frequency process · Modulation detection interference · Off-frequency masking · Object binding cues · Modulation filterbank · Auditory modelling

J. L.Verhey (✉) · B. Lübken
Department of Experimental Audiology, Otto von Guericke University Magdeburg, Leipziger Straße 44, 39120 Magdeburg, Germany
e-mail: jesko.verhey@med.ovgu.de

S. van de Par
Acoustics group, Cluster of Excellence “Hearing4All”, Carl von Ossietzky Universität Oldenburg, 26111 Oldenburg, Germany
e-mail: steven.van.de.par@uni-oldenburg.de

P. van Dijk et al. (eds.), *Physiology, Psychoacoustics and Cognition in Normal and Impaired Hearing,* Advances in Experimental Medicine and Biology 894,
DOI 10.1007/978-3-319-25474-6_26

1 Introduction

In complex acoustical environments, the auditory system makes use of various cues to group sound components from one particular sound source. These cues, sometimes referred to as object binding cues, are often associated with characteristics of the sound such as coherent level fluctuations in different frequency regions (as, e.g., observed in speech), or the location of the sources in space. Differences in location result, among others, in interaural differences. Psychoacousticians developed experiments to study how the human auditory system makes use of the information. For example, the ability to use interaural disparities is shown in experiments on binaural masking level differences (BMLD). The BMLD is a well investigated phenomenon that signifies the reduction in masked thresholds when a tonal signal and masker have disparate interaural cues. For a diotic narrowband masker, a tonal target can be detected at a 25 dB lower level when the interaural target phase is changed from 0 to π (e.g. van de Par and Kohlrausch 1999). This large BMLD can only be observed when masker and target are spectrally overlapping. Several studies have shown that the BMLD reduces to rather small values for off-frequency targets (e.g., van de Par et al. 2012).

This lack of a BMLD could be interpreted as a lack of effective binaural processing in off-frequency conditions. Alternatively, it could result from monaural processing which is much better in off-frequency as compared to on-frequency conditions. The narrowband noise masker has an envelope spectrum that predominantly has components in the range from 0 Hz to the masker bandwidth. Thus, the presence of an on-frequency tonal target signal does not substantially change this envelope spectrum. When, however, the tonal signal is presented off-frequency, new, higher-frequency spectral modulation components are introduced due to the beating between tone and masker that are only masked to a small degree by the inherent masker fluctuations. This modulation cue provides for an additional lowering of off-frequency monaural detection thresholds in excess of the peripheral filtering effect. The modulation frequency selectivity of the auditory system shown in previous studies (e.g., Dau et al. 1997) indicates that the inherent modulations can be dissociated from those due to beating. The contribution of modulation cues in off-frequency masking has been demonstrated by Nitschmann and Verhey (2012) using a modulation filterbank model.

In a previous study (van de Par et al. 2012), this monaural explanation was tested by attempting to reduce the effectiveness of the modulation cues by creating a Modulation Detection Interference (MDI) effect. To this end a tonal interferer was provided at the opposite spectral side of the masker compared to where the target was, but with the same spectral distance between target and masker as between masker and interferer. This extra interferer created beating frequencies that were equal to those that should be created by the presence of the target tone. Although these beating frequencies are then present in different auditory channels, the across-frequency nature of MDI (Yost et al. 1989) implies that they should still interfere with the modulation cues used for detecting the target tone. The level of the interfering tone was chosen to be 6 dB lower than the masker level to ensure that no energetic masking should occur. Figure 1 shows the data of van de Par et al. (2012) as masking patterns with (filled symbols) and without (open symbols) interferer for the diotic (circles) and the dichotic (triangles) conditions. As hypothesized the presence of the interferer increased the diotic thresholds substantially and had less effect on the dichotic thresholds.

In the present contribution, we investigated two aspects of the interference effect described above. First it is known that MDI exhibits some degree of frequency selectivity (Yost et al. 1989). By altering the frequency offset between the interferer tone and the narrowband masker, the interference on the fixed sinusoidal signal should change, giving the largest interference effect when the frequency differences between the target and interferer are the same relative to the masker centre frequency. Secondly, we will investigate to what extent the interference effect can be modelled.

2 Methods

2.1 Psychoacoustic Experiment

Three masking conditions were considered. The masker of the first masking condition was a 25-Hz wide bandpass-filtered Gaussian noise with an overall level of 65 dB SPL generated using a brick-wall filter in the frequency domain. The arithmetic centre frequency f_c was 700 Hz. Apart from this noise-only condition, two other conditions were used, where an additional pure tone (interferer) was presented

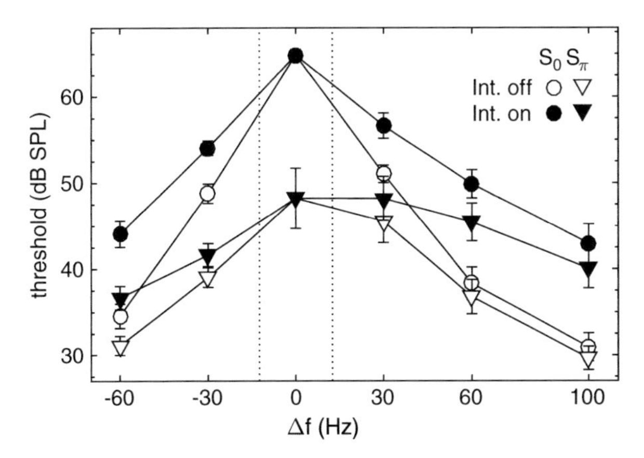

Fig. 1 Masked thresholds as a function of the spectral distance between target signal and masker *centre* frequency. Open symbols indicate data for the classical masking pattern experiment with a *narrowband* diotic noise masker and a sinusoidal signal. Filled symbols indicate data of a modified masking pattern experiment with an additional tonal interferer. *Circles* denote thresholds for the diotic condition, *triangles* those of the dichotic condition. The data were taken from our previous study on the effect of an interferer on masking patterns (van de Par et al. 2012)

at 670 Hz (i.e., 30 Hz below f_c) or 600 Hz (i.e., 100 Hz below f_c). All masker and interferer components were presented diotically. The target was a pure tone that was either presented diotically (S_0) or interaurally out of phase (S_π). The target frequency was either 730 Hz (30 Hz above f_c) or 800 Hz (100 Hz above f_c). The duration of masker, interferer and target was 500 ms including 50-ms raised cosine-ramps at onset and offset.

Masked thresholds were measured using an adaptive three-interval three-alternative forced-choice procedure. In each of the three 500-ms stimulus intervals of a trial the masker was presented. Stimulus intervals were separated by 500 ms of silence. In one randomly chosen interval of a trial, the target signal was added. The subject had to indicate which interval contained the target. Feedback about correctness of the answer was provided. A two-down one-up adaptive staircase procedure was used. The initial step size of 6 dB was reduced to 3 dB after the first upper reversal and to 1 dB after the second upper reversal. With this minimum step size, the run continued for another eight reversals of which the mean was used as threshold estimate. Four threshold estimates were obtained and averaged leading to the final threshold estimate for this signal and masking condition. The conditions were run in random order.

The stimuli were generated digitally with a sampling rate of 44100 Hz, converted to analogue signals using a 32 bit D/A converter (RME Fireface UC) and presented via headphones (Sennheiser HD650) within a sound insulated booth. Eight experienced subjects (Five male and three female) aged from 21 to 30 years participated. All subjects had pure tone thresholds of 10 dB HL or lower for the standard audiogram frequencies.

2.2 Simulations

For the simulations a model was developed based on the model used by Nitschmann and Verhey (2012) to predict their masking patterns. The monaural front end was (i) a first-order band-pass filter with cut-off frequencies of 0.5 and 5.3 kHz to simulate the filtering of the outer and middle ear, (ii) a bank with fourth-order gammatone filters (centre frequencies 313–1055 Hz) with a bandwidth of one equivalent rectangular bandwidth (ERB) and a spectral distance of adjacent filters of one ERB, (iii) white noise with a level of 7 dB SPL added to the output of each filter to simulate thresholds in quiet, (iv) half-wave rectification and a first order low-pass filter with a cut-off frequency of 1 kHz and (v) five consecutive adaptation loops to model adaptation and compression effects in the auditory system. In the monaural pathway the output of the adaptation loops was analysed by a modulation filterbank, where the highest centre frequency of the modulation filters was restricted to a quarter of the centre frequency of the auditory filter. To simulate across-channel (AC) processes, two different versions of the monaural part of the model were used.

The first version is based on the approach of Piechowiak et al. (2007). In this approach a weighted sum of modulation filters at the output of off-frequency auditory filters was subtracted from the corresponding modulation filters at the output of the on-frequency filter. In contrast to Piechowiak et al. (2007), a multi-auditory-channel version of this approach was used. Since the weighting is essentially based on the energy in the auditory filters it will be referred to as the energy-based AC model in the following.

The second version of across-channel processing is based on the assumption that modulation in off-frequency auditory filters channels hampers the processing of the modulation filters tuned to the corresponding modulation frequency in the on-frequency auditory channel. This detrimental effect is realised by assuming that modulation frequencies adjacent to the modulation created by the masker-interferer distance could not be processed. This was implemented in an extreme form by setting the output of the modulation filters tuned to frequencies adjacent to the interferer-masker distance induced modulation frequency to zero. This approach is referred to as the modulation-based AC model.

The binaural pathway of the model is the same in both models. It is realised as an equalisation-cancellation process followed by a low-pass filtering with a first-order low-pass filter with a cut-off frequency of 8 Hz. The final stage is an optimal detector, which calculates the cross correlation between a temporal representation of the signal at a supra-threshold level (template) with the respective actual activity pattern (Dau et al. 1997). For the simulations the same adaptive procedure was used as for the psychoacoustic experiment with the model as an artificial observer. Simulated thresholds are calculated as the mean of the estimates from 24 runs.

3 Results

Figure 2 shows average thresholds and standard errors of the eight subjects participating in the psychoacoustic experiment. Each panel shows thresholds of one signal as indicated in the top of the panel. Thresholds are shown for the three interferer conditions: "no interferer", "interferer 30 Hz below f_c" and "interferer 100 Hz below f_c". For the diotic 730-Hz target (top left panel), thresholds increase by 4.5 dB when an interferer was added at a distance of 30 Hz. For the spectral distance interferer-masker of 100 Hz, the threshold was 1.5 dB lower than for the spectral distance interferer-masker of 30 Hz. For the diotic 800 Hz target (top right panel), adding an interferer had a higher impact on threshold than for the lower signal frequency, as shown in Fig. 1. Threshold was highest when the spectral distance masker-target and interferer-masker threshold was the same. It was 4.5 dB lower when the interferer was positioned 30 Hz instead of 100 Hz below the masker. A similar trend was observed for the dichotic 800-Hz target (bottom right panel) although the difference between the highest threshold for an interferer at 100 Hz below the masker and the lowest threshold without interferer was only 8 dB whereas it was 11.5 dB for the diotic 800-Hz target. For the dichotic 730-Hz target (bottom left panel) thresholds were about the same for the two conditions with interferer. The threshold without interferer was 1.5 dB lower than thresholds with interferer.

Figure 3 shows model predictions. Since the binaural pathway is the same in both models, predictions of the two models do not differ for the dichotic target sig-

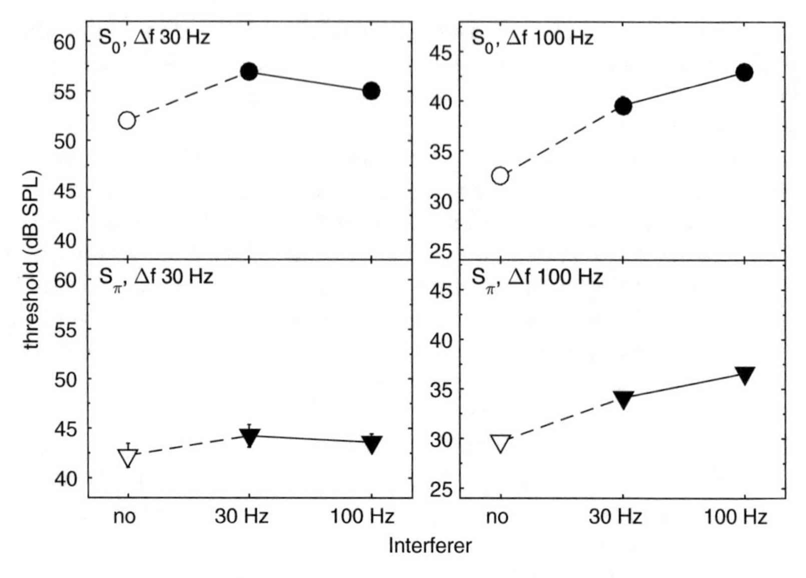

Fig. 2 Measured thresholds for three different interferer conditions: without interferer (no) and with interferers at 30 or 100 Hz *below* the masker. Each panel shows the average thresholds and standard errors of the eight subjects for one target signal. The signal frequency was either 730 Hz (masker-target Δf=30 Hz, *left* panels) or 800 Hz (Δf=100 Hz, *right* panels). The target was either presented diotically (*top* panels) or dichotically (*bottom* panels)

nals (bottom panels). Both models predict no effect of the interferer on threshold, in contrast to the data. For the diotic 730-Hz target (top left panel), the energy-based AC model predicted no change in threshold when the interferer was added. In contrast, the modulation-based AC model predicted an increase of about 5 dB when the interferer was added 30 Hz below the masker compared to the condition without interferer. For an interferer 100 Hz below the masker, threshold was about 2 dB lower than for the interferer 30 Hz below the masker. Thus, only the modulation-based AC model predicts the effect of the interferer on thresholds for the diotic 730-Hz target. For diotic 800-Hz target, both models predicted highest thresholds for the interferer 100 Hz below the masker and lowest for the condition without interferer. The energy-based AC model predicted an effect of 4.5 dB, i.e., considerably smaller than the 11.5 dB observed in the data. The modulation-based model predicted an effect of 14 dB. For the interferer 30 Hz below the masker, the predicted threshold was similar to the thresholds without interferer, whereas the difference between these two thresholds was 7 dB in the measured data.

4 Discussion

The behavioural results of the present study support the hypothesis that modulation cues affect detection of a diotic target in masking pattern experiments and that an additional interfering tone with the same distance below (or above) the masker as

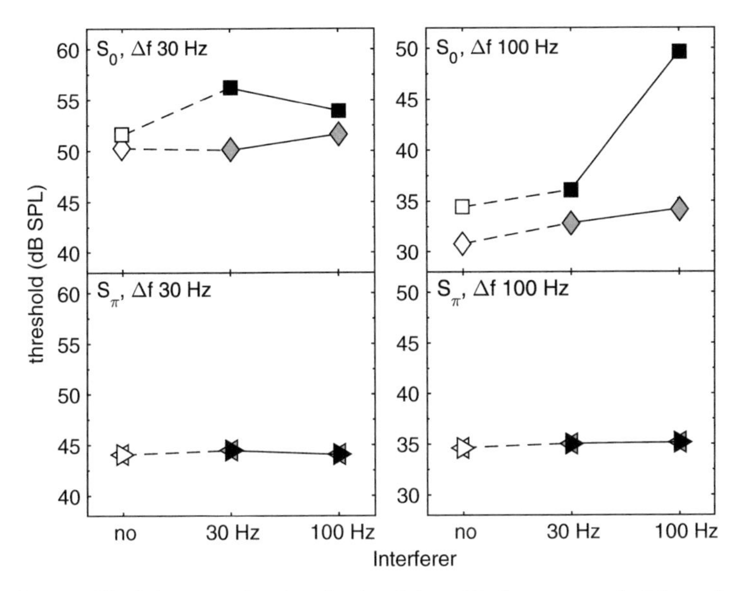

Fig. 3 Same as Fig. 2 but now showing simulated data. *Black squares* and *rightward*-pointing *triangles* indicate predictions of the modulation-based AC model. *Grey diamonds* and *leftward*-pointing *triangles* show the predictions of the energy-based AC model

the spectral distance of the signal spectrally above (or below) the masker makes the beating cue between signal and masker less usable for detection. The present study shows that the same spectral distance masker-target and interferer-masker increases thresholds more than an interferer at a different spectral distance. This result points towards a modulation spectral selectivity as previously observed in MDI experiments.

The present study tested an extended (energy-based AC) model of the modulation filterbank model originally designed to predict CMR by Piechowiak et al. (2007). Although this model predicted the trend for the 800-Hz target the effect was too small. For the 730-Hz target, no effect was predicted, in contrast to the data. Thus this energy-based AC model was unable to predict the effect of the interferer on the diotic thresholds.

The second model assumed that modulation cues with frequencies adjacent to the beating frequency between interferer and masker could not be used for signal detection. This was realised in the modulation-based AC model by setting the output of the corresponding modulation filters to zero. This approach predicted general trends of the diotic data of the present study. Among others it predicted a larger effect of the interferer for the 800-Hz signal than for the 730-Hz target, in agreement with the data. However, there was a tendency of an overestimation of the effect for the same spectral distance masker-signal and interferer-masker and an underestimation of the effect of the interferer with a spectral distance not equal to that of signal and masker, at least for the 800-Hz target. This indicates that the present approach was too simplistic, i.e., modulation cues may only be reduced and not abolished by the presence of the interferer. In addition, not only modulation frequencies close to the beating frequency may be affected by the presence of the interferer but also other modulation frequencies. This is also consistent with MDI results where an MDI is also observed for modulation frequencies of the interferer that differ from the target modulation frequency (Yost et al. 1989). A better match between data and simulations could be achieved by assuming weights to the output of the modulation filters that gradually increase as the distance between beating frequency (masker—interferer) increases.

As already shown in van de Par (2012), dichotic thresholds are also affected by the presence of an interferer. The present data for the dichotic 800-Hz target indicate a similar effect of interferer-masker distance as for the diotic target, although less pronounced. Thus, also dichotic thresholds may be influenced by a frequency-specific modulation processing. This was already hypothesized in Nitschmann and Verhey (2012). Binaural modulation frequency selectivity seems to be less accurate than monaural modulation frequency selectivity (Thompson and Dau 2008). Thus, if binaural modulation frequency selectivity affects the dichotic off-frequency masked thresholds the mismatch of beating frequency between masker-target and interferer-masker should have less effect on threshold than for the diotic condition. The present data for the 800-Hz target support this hypothesis.

Acknowledgments This work was supported by the Deutsche Forschungsgemeinschaft (DFG, sfb trr 31).

References

Dau T, Kollmeier B, Kohlrausch A (1997) Modeling auditory processing of amplitude modulation. I. Detection and masking with narrow-band carriers. J Acoust Soc Am 102(5):2892–2905

Nitschmann M, Verhey JL (2012) Modulation cues influence binaural masking-level difference in masking-pattern experiments. J Acoust Soc Am Exp Lett 131(3):EL223–EL228

Piechowiak T, Ewert SD, Dau T (2007) Modeling comodulation masking release using an equalization-cancellation mechanism. J Acoust Soc Am 121(4):2111–2126

Thompson ER, Dau T (2008) Binaural processing of modulated interaural level differences. J Acoust Soc Am 123(2):1017–1029

van de Par S, Kohlrausch A (1999) Dependence of binaural masking level differences on center frequency, masker bandwidth and interaural parameters. J Acoust Soc Am 106(4):1940–1947

van de Par S, Luebken B, Verhey JL, Kohlrausch A (2012) Off-frequency BMLD: the role of monaural processing. In: Moore BCJ, Patterson RD, Winter IM, Carlyon RP, Gockel HE (eds) Basic aspects of hearing: physiology and perception. Springer, New York, pp 293–301

Yost WA, Sheft S, Opie J (1989) Modulation interference in detection and discrimination of amplitude modulation. J Acoust Soc Am 86:2138–2147

Speech Intelligibility for Target and Masker with Different Spectra

Thibaud Leclère, David Théry, Mathieu Lavandier and John F. Culling

Abstract The speech intelligibility index (SII) calculation is based on the assumption that the effective range of signal-to-noise ratio (SNR) regarding speech intelligibility is [− 15 dB; +15 dB]. In a specific frequency band, speech intelligibility would remain constant by varying the SNRs above + 15 dB or below − 15 dB. These assumptions were tested in four experiments measuring speech reception thresholds (SRTs) with a speech target and speech-spectrum noise, while attenuating target or noise above or below 1400 Hz, with different levels of attenuation in order to test different SNRs in the two bands. SRT varied linearly with attenuation at low-attenuation levels and an asymptote was reached for high-attenuation levels. However, this asymptote was reached (intelligibility was not influenced by further attenuation) for different attenuation levels across experiments. The − 15-dB SII limit was confirmed for high-pass filtered targets, whereas for low-pass filtered targets, intelligibility was further impaired by decreasing the SNR below − 15 dB (until − 37 dB) in the high-frequency band. For high-pass and low-pass filtered noises, speech intelligibility kept improving when increasing the SNR in the rejected band beyond + 15 dB (up to 43 dB). Before reaching the asymptote, a 10-dB increase of SNR obtained by filtering the noise resulted in a larger decrease of SRT than a corresponding 10-dB decrease of SNR obtained by filtering the target (the slopes SRT/attenuation were different depending on which source was filtered). These results question the use of the SNR range and the importance function adopted by the SII when considering sharply filtered signals.

Keywords Speech intelligibility index · Speech in noise · Speech intelligibility measurement

T. Leclère (✉) · D. Théry · M. Lavandier
Laboratoire Génie Civil et Bâtiment, ENTPE, Université de Lyon, Rue Maurice Audin, 69518 Vaulx-en-Velin, France
e-mail: thibaud.leclere@entpe.fr

D. Théry
e-mail: david.thery@entpe.fr

M. Lavandier
e-mail: mathieu.lavandier@entpe.fr

J. F. Culling
School of psychology, Cardiff University, Tower Building, Park Place, Cardiff CF10 AT, UK
e-mail: cullingj@cardiff.ac.uk

P. van Dijk et al. (eds.), *Physiology, Psychoacoustics and Cognition in Normal and Impaired Hearing,* Advances in Experimental Medicine and Biology 894,
DOI 10.1007/978-3-319-25474-6_27

1 Introduction

Speech intelligibility in noise is strongly influenced by the relative level of the target compared to that of the noise, referred to as signal-to-noise ratio (SNR). High SNRs lead to better speech recognition than low SNRs. But, although SNRs can take infinite values, it is not the case for speech intelligibility. Some SNR floors and ceilings must exist, such that intelligibility would not be influenced by varying the SNR above or below these limits. Several speech intelligibility models are based on SNRs computations in frequency bands (Rhebergen and Versfeld 2005; Beutelmann et al. 2010; Collin and Lavandier 2013). In the presence of amplitude-modulated or filtered noise, these models could compute infinite SNRs, and then conduct to infinite intelligibility. In order to keep the predictions realistic, a limitation of SNRs must be introduced. Collin and Lavandier (2013) proposed a 10-dB ceiling limit in their model. Other models for intelligibility in modulated noise (Rhebergen and Versfeld 2005; Beutelmann et al. 2010) are based on the speech intelligibility index (SII) calculation (ANSI S3.5 1997) which adopts the SNR values of −15 dB and+15 dB as floor and ceiling limits respectively. This range has its origins in the work of Beranek (1947) who reported that the total dynamic range of speech is about 30 dB (in any frequency band) by interpreting the short-term speech spectrum measurements reported by Dunn and White (1940). The aim of this study was to determine floor and ceiling values based on four speech intelligibility experiments and to compare them to those proposed by the SII.

Speech Reception Thresholds (SRTs, SNR yielding 50% intelligibility) were measured in the presence of a speech target and a speech-shaped noise. Target or noise was attenuated above or below 1400 Hz at different levels of attenuation in order to vary the SNR in the low or high frequency regions. The SRT is expected to increase along with the attenuation level in the case of a filtered target. Conversely, when the noise is filtered, the SRT should decrease while the attenuation level is increased. In both cases, SRTs are expected to remain unchanged and form an asymptote beyond a certain attenuation level: variations of SNR should not influence speech intelligibility any longer. The floor and ceiling values (attenuation from which SRTs are no longer influenced) obtained in each experiment will be compared to those adopted by the SII standard [−15 dB;+15 dB].

General methods of the experiments are presented first, detailing the conditions and stimuli tested in this study, then followed by the results of each experiment. These results are finally discussed in the last section.

2 Methods

2.1 Stimuli

2.1.1 Target Sentences

The speech material used for the target sentences was designed by Raake and Katz (2006) and consisted of 24 lists of 12 anechoic recordings of the same male voice digitized and down-sampled here at 44.1 kHz with 16-bit quantization. These recordings were semantically unpredictable sentences in French and contained four key words (nouns, adjectives, and verbs). For instance, one sentence was "la LOI BRILLE par la CHANCE CREUSE" (the LAW SHINES by the HOLLOW CHANCE).

2.1.2 Maskers

Maskers were 3.8-s excerpts (to make sure that all maskers were longer than the longest target sentence) of a long stationary speech-shaped noise obtained by concatenating several lists of sentences, taking the Fourier transform of the resulting signal, randomizing its phase, and finally taking its inverse Fourier transform.

2.1.3 Filters

Digital finite impulse response filters of 512 coefficients were designed using the host-windowing technique (Abed and Cain 1984). High-pass (HP) and low-pass (LP) filters were used on the target or the masker at different attenuations (0 to 65 dB) depending on the experiment. The cut-off frequency was set to 1400 Hz for both HP and LP filters to achieve equal contribution from the pass and stop bands according to the SII band importance function (ANSI 1997; Dubno et al. 2005).

2.2 *Procedure*

Four experiments were conducted to test each filter type (HP or LP) on each source (target or masker). Except for experiment 1 (HP target), each experiment was composed of two sub-experiments of eight conditions because no asymptote was reached with the first set of eight conditions. Experiment 1 tested only eight conditions. In each sub-experiment, each SRT was measured using a list of twelve target sentences and an adaptive method (Brand and Kollmeier 2002). The twelve sentences were presented one after another against a different noise excerpt corresponding to the same condition. Listeners were instructed to type the words they heard on a computer keyboard after each presentation. The correct transcript was then displayed on a monitor with the key words highlighted in capital letters. Listeners identified and self-reported their score (number of correct key words they perceived). For the first sentence of the list, listeners had the possibility to replay the stimuli, producing an increase in the broadband SNR of 3 dB, which was initially very low (−25 dB). Listeners were asked to attempt a transcript as soon as they believed that they could hear half of the words in the sentence. No replay was possible for the following sentences, for which the broadband SNR was varied across sentences by modifying the target level while the masker level was kept constant at 74 dB_A SPL. For a given sentence, the broadband SNR was increased if the score obtained at the previous sentence was greater than 2, it was decreased if the score was less than 2 and it remained unchanged if the score was 2. The sound level of the kth ($2<k<12$) sentence of the list (L_k, expressed in dB SPL) was determined by Eq. 1 (Brand and Kollmeier 2002):

$$L_k = L_{k-1} - 10 \times 1.41^{-i} \times ((SCORE_{k-1}/4) - 0.5)$$

where $SCORE_{k-1}$ is the number of correct key words between 1 and 4 for the sentence k-1 and i is the number of times ($SCORE_{k-1}$/4)-0.5 changed sign since the beginning of the sentence list. The SRT was taken as the mean SNR in the pass band across the last eight sentences. In each sub-experiment, the SRT was measured for eight conditions presented in a pseudorandom order, which was rotated for successive listeners to counterbalance the effects of condition order and sentence lists, which were presented in a fixed sequence. Each target sentence was thus presented only once to every listener in the same order and, across a group of eight listeners, a complete rotation of conditions was achieved. In each experiment, listeners began the session with two practice runs, to get used to the task, followed by eight runs with break after four runs.

2.3 Equipment

Signals were presented to listeners over Sennheiser HD 650 headphones in a double walled soundproof booth after having been digitally mixed, D/A converted, and amplified using a Lynx TWO sound card. A graphical interface was displayed on a computer screen outside the booth window. A keyboard and a computer mouse were inside the booth to interact with the interface and gather the transcripts.

3 Listeners

Listeners self-reported normal hearing and French as their native language and were paid for their participation. Eight listeners took part in each sub-experiment. Within each experiment, no listener participated at both sub-experiments since the target sentences used in each sub-experiment were the same.

4 Results

4.1 HP Target

Figure 1 presents the SRTs measured in the presence of a HP-filtered target as a function of the filter attenuation in the low-frequency region. SRTs first increased linearly from 0 to 15-dB attenuation and then remained constant for further attenu-

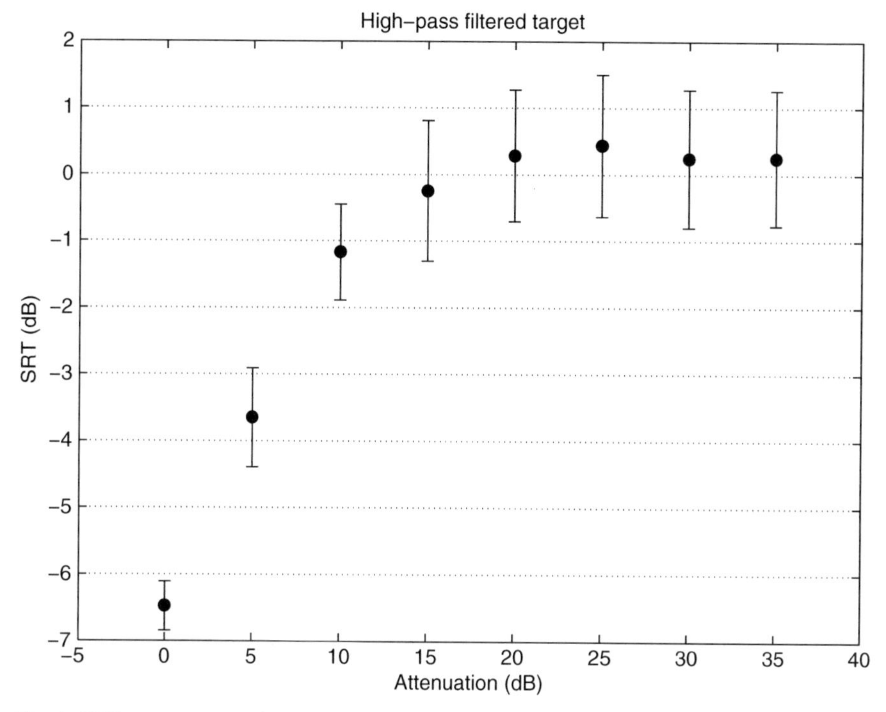

Fig. 1 SRT measurements for a high-pass filtered target as a function of the filter attenuation

ations. Intelligibility was not disrupted any longer after filtering out the target by 15-dB attenuation. A one-factor repeated measures analysis of variance (ANOVA) was performed on the experimental data, showing a main effect of the filter attenuation [$F(7,7)=10.58$; $p<0.001$]. Tukey pairwise comparisons were performed on the data: none of the SRTs from 10-dB attenuation to 35-dB were significantly different from each other. By fitting a broken stick function on the experimental data, the floor value was determined at 13-dB attenuation and the slope of the linear increase of SRT was 0.53 dB SRT/dB attenuation. The same fitting process has been used in each experiment to determine the slope and the floor (or ceiling) value.

4.2 LP Target

SRT measurements in the presence of a LP-filtered target are plotted in Fig. 2 as a function of the filter attenuation. As in the HP case, SRTs increased linearly with a slope of 0.48 dB SRT/dB attenuation, but unlike the HP case, SRTs kept increasing until a floor value of 37-dB attenuation. Intelligibility remained at a SRT of about 10 dB for further attenuations. A one-factor repeated measures ANOVA was performed on each sub-experiment independently. In both sub-experiments, a main

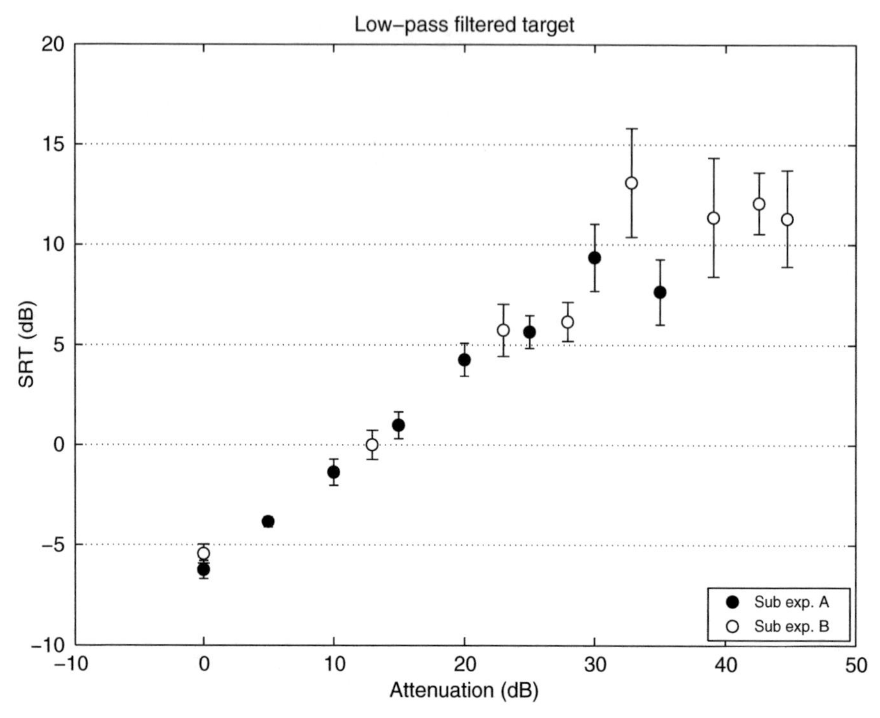

Fig. 2 SRT measurements for a low-pass filtered target as a function of the filter attenuation. Filled circles correspond to the first sub-experiment while open circles correspond to the second

effect of attenuation was observed [$F(7,7)>16.2$; $p<0.01$]. Post-hoc Tukey pairwise comparisons were performed on the data of each sub-experiment. In sub-experiment A (filled circles), the four SRTs at the highest level of attenuation (20, 25, 30 and 35 dB) are not significantly different from each other. In sub-experiment B (open circles), the six SRTs at the highest attenuation level (23, 28, 33, 39, 42 and 45 dB) are not significantly different from each other.

4.3 HP Masker

SRTs measured with a HP filtered masker are presented as a function of the filter attenuation in Fig. 3. SRTs decreased linearly with a slope of −0.65 dB SRT/dB attenuation indicating an improvement of speech intelligibility by filtering out the low frequencies in the masker signal. At 43-dB attenuation (ceiling value), SRTs stopped decreasing and presented an asymptote at about −35 dB. A one-factor repeated measures ANOVA indicated a significant main effect of the filter attenuation on speech intelligibility [$F(7,7)>41.8$; $p<0.01$ in each sub-experiment]. Tukey pairwise comparisons were performed on the dataset of sub-experiments A and B.

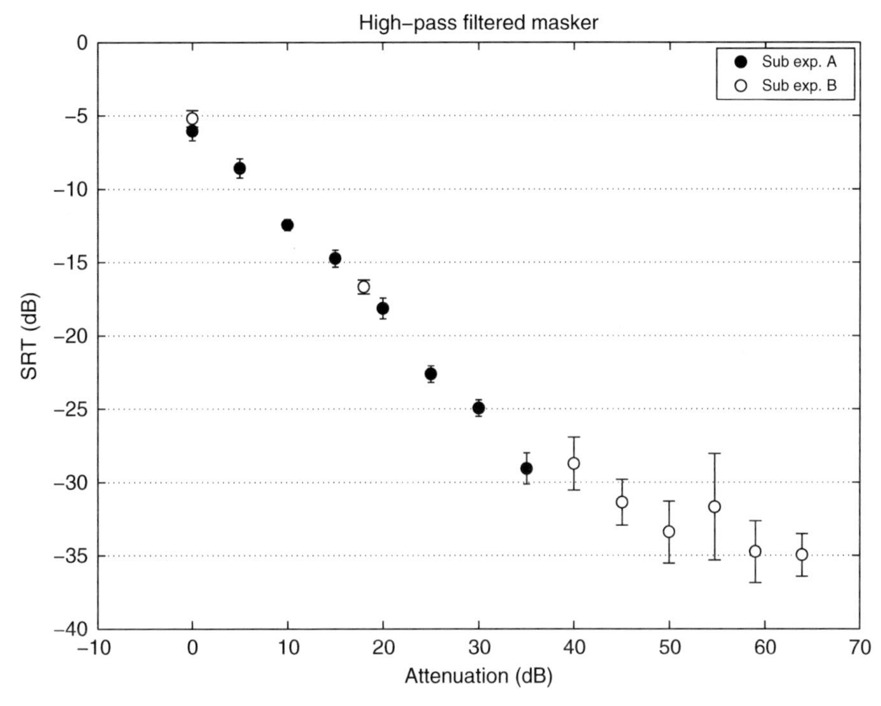

Fig. 3 SRT measurements for high-pass filtered masker as a function of the filter attenuation. Filled circles correspond to the first sub-experiment while open circles correspond to the second

In sub-experiment A (filled circles), only the pairs of SRTs at the attenuations 30/25, 10/15 and 0/5 were not significantly different. All the other pairs of SRTs were significantly different from each other. In sub-experiment B (open circles), SRTs for 0 and 20-dB attenuation were significantly different from each other and all SRTs obtained for 40-dB attenuation at 40 dB and above were not significantly different from each other.

4.4 LP Masker

Figure 4 presents the SRTs measurements obtained in the presence of a LP filtered masker as a function of the filter attenuation. As in the HP case, SRTs linearly decreased with attenuation until the ceiling value of 36-dB attenuation. For further attenuations, SRTs were constant at about −35 dB. The slope of the linear decrease of SRTs was −0.76 dB SRT/dB attenuation. A one-factor repeated measures ANOVA was performed on the data from each sub-experiment independently. A main effect of the filter attenuation was found [$F(7,7)>44.5$; $p<0.01$], which was further investigated by performing Tukey pairwise comparisons on the dataset from each sub-experiment. In sub-experiment A (filled circles), all pairs of SRTs were differ-

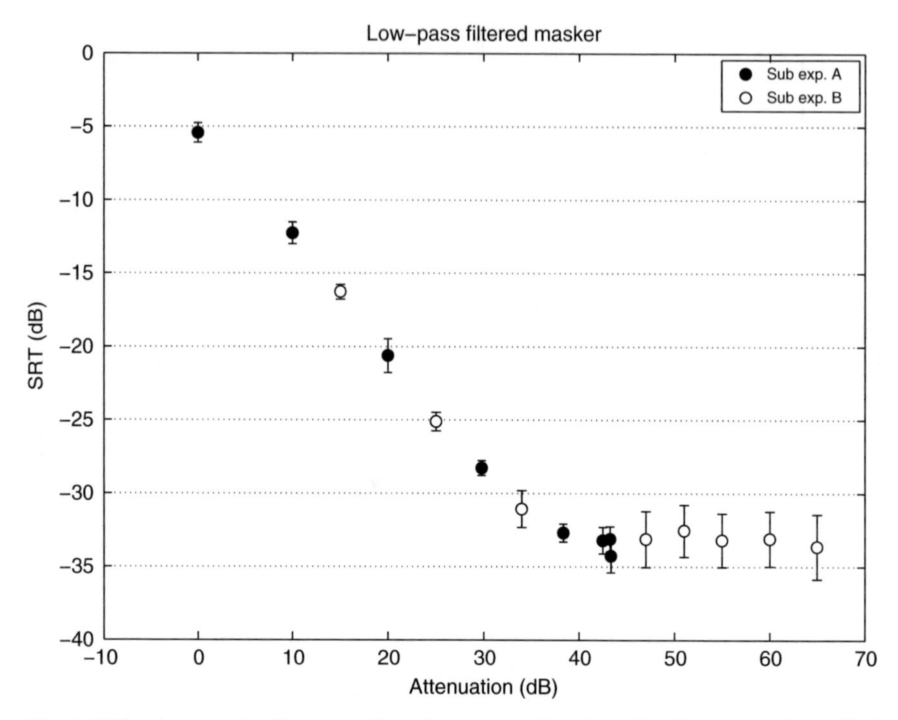

Fig. 4 SRT measurements of low-pass filtered masker as a function of the filter attenuation. Filled circles correspond to the first sub-experiment while open circles correspond to the second

ent from each other except for those corresponding to an attenuation at 38 dB and above. In sub-experiment B (open circles), none of the SRTs between 34-dB and 65-dB attenuation were significantly different. SRTs obtained for lower attenuations were all significantly different from each other.

5 Discussion

In each experiment, speech intelligibility was strongly influenced by the SNR in a specific band. However, SRTs were not affected any longer beyond a certain attenuation level (floor or ceiling values) and remained constant. The floor/ceiling values observed in these experiments are not in agreement with the SII assumptions, except when the target was HP-filtered. In this specific case, the results suggested a floor value at 13-dB attenuation. The results from the other experiments suggested a larger range of SNRs contributing to speech intelligibility [−37 dB; 43 dB]. When the target is filtered, the floor value seemed to be frequency-dependent in contrast to the SII approach which proposes the same SNR range in each frequency band. It is worth noting that the SII standard was not designed for sharply filtered speech or sharply filtered noise such as the stimuli used in the present study. The SII is then not expected to predict the data presented here but other values for the ceiling and floor SNRs than those proposed by the standard are needed to describe speech intelligibility of sharply filtered signals.

Changing the filter type (HP or LP) had a small influence on the slope of the linear increase (or decrease) of SRTs before reaching the asymptote, confirming that the chosen low and high frequency regions contributed equally to speech intelligibility. But, the results showed that filtering the masker seemed to have a larger impact on speech intelligibility than filtering the target. SRTs were not symmetrical as the SNR was varied up or down with the same amount. The benefit was greater when the filtered part of the noise decreased rather than when the filtered part of the target increased. This result questions the uniformly distributed importance over the [−15;+15] interval adopted by the SII and suggests that a greater importance should be attributed to positive SNRs compared to negative SNRs.

Studebaker and Sherbecoe (2002) derived intensity importance functions from intelligibility scores in five frequency bands. The functions they obtained differed from the one used in the SII calculation. Their functions are frequency-dependent and nonlinear along SNRs, which is in agreement with the findings of the present study. However, the derivation of their results yielded importance functions defined between SNRs of −15 dB and +45 dB. The results observed here suggested a floor SNR of −13 dB only when the target was HP-filtered. For the other conditions, lower SNRs need to be considered. The importance function of Studebaker and Sherbecoe (2002) also attributes negative contribution to very high SNRs (>30 dB on average across frequency bands) regarding speech intelligibility. This negative contribution might be due to high sound levels used in their study which could have led to poorer word recognition scores (Dubno et al. 2005). Their derived importance functions take into account both the effect of SNR and absolute speech level. It would be preferable to separate the influence of these two parameters. Their approach could however be inspiring in order to propose simple frequency-dependent SNR-importance functions, allowing modelling the SRTs measured in the present study.

6 Conclusion

Speech reception thresholds were measured in four speech intelligibility experiments. Target or noise was either high-pass filtered or low-pass filtered with different SNRs in the rejected band by varying the attenuation level of the filter. As expected, it was observed in each experiment that SRTs remained constant beyond a certain attenuation level (i.e. a certain SNR in the rejected band). In general, the SNR value from which SRT was not influenced any longer differed from previous values reported in the literature, especially in the SII standard. These results provide ceiling and floor values of SNR for wide frequency bands based on experimental measurements. They do not question the validity of the SII (which was not designed for sharply filtered sources) but they rather point out the need of nonlinear SNR-importance functions in speech intelligibility models based on SNR weightings to predict SRTs, especially if they aim to account for filtered sources. Further work needs to be done to determine ceiling and floor values in narrower frequency bands. To model these results, SNR-importance functions also need to be built, for example following the approach of Studebaker and Sherbecoe (2002) who proposed a two-dimensional importance function.

Acknowledgments The authors would like to thank all listeners who took part in the experiments. This work was performed within the Labex CeLyA (ANR-10-LABX-0060/ANR-11-IDEX-0007).

References

Abed A-EHM, Cain GD (1984) The host windowing technique for FIR digital filter design. IEEE Trans Acoust Speech Signal Process (ASSP) 32:683–694

Beranek LL (1947) The design of speech communication systems. Proc Inst Radio Eng 35:880–890

Beutelmann et al (2010) Revision, extension, and evaluation of a binaural speech intelligibility model. J Acoust Soc Am 127:2479–2497

Brand T, Kollmeier B (2002) Efficient adaptive procedures for thresholds and concurrent slope estimates for psychophysics and speech intelligibility tests. J Acoust Soc Am 111(6):2801–2810

Collin B, Lavandier M (2013) Binaural speech intelligibility in rooms with variations in spatial location of sources and modulation depth of noise interferers. J Acoust Soc Am 134:1146–1159

Dubno JR, Horwitz AR, Ahlstrom JB (2005) Recognition of filtered words in noise at higher-than-normal levels: decreases in scores with and without increases in masking. J Acous Soc Am 118(2):923–933

Dunn HK, White SD (1940) Statistical measurements on conversational speech. J Acoust Soc Am 11:278–288

Raake A, Katz BFG (2006). SUS-based method for speech reception threshold measurement in French. Proc Lang Resour Eval Conf 2028–2033

Studebaker GA, Sherbecoe RL (2002) Intensity-importance functions for bandlimited monosyllabic words. J Acoust Soc Am 111:1422–1436

Versfeld R (2005) A Speech Intelligibility Index-based approach to predict the speech reception threshold for sentences in fluctuating noise for normal-hearing listeners. J Acoust Soc Am 117:2181–2192

ANSI S3.5 (1997). Methods for calculation of the speech intelligibility index. American National Standards Institute, New York

Dynamics of Cochlear Nonlinearity

Nigel P. Cooper and Marcel van der Heijden

Abstract Dynamic aspects of cochlear mechanical compression were studied by recording basilar membrane (BM) vibrations evoked by tone pairs ("beat stimuli") in the 11–19 kHz region of the gerbil cochlea. The frequencies of the stimulus components were varied to produce a range of "beat rates" at or near the characteristic frequency (CF) of the BM site under study, and the amplitudes of the components were balanced to produce near perfect periodic cancellations, visible as sharp notches in the envelope of the BM response. We found a compressive relation between instantaneous stimulus intensity and BM response magnitude that was strongest at low beat rates (e.g., 10–100 Hz). At higher beat rates, the amount of compression reduced progressively (i.e. the responses became linearized), and the rising and falling flanks of the response envelope showed increasing amounts of hysteresis; the rising flank becoming steeper than the falling flank. This hysteresis indicates that cochlear mechanical compression is not instantaneous, and is suggestive of a gain control mechanism having finite attack and release times. In gain control terms, the linearization that occurs at higher beat rates occurs because the instantaneous gain becomes smoothened, or low-pass filtered, with respect to the magnitude fluctuations in the stimulus. In terms of peripheral processing, the linearization corresponds to an enhanced coding, or decompression, of rapid amplitude modulations. These findings are relevant both to those who wish to understand the underlying mechanisms and those who need a realistic model of nonlinear processing by the auditory periphery.

Keywords Basilar membrane · Hair cell · Compression · Gain control

1 Introduction

Basilar-membrane (BM) vibrations in healthy cochleae grow compressively with the intensity of stationary sounds such as tones and broad band noise (de Boer and Nuttall 1997; Rhode 1971; Robles and Ruggero 2001). Similarly compressive

M. van der Heijden (✉) · N. P. Cooper
Erasmus MC, PO Box 2040, 3000 CA Rotterdam, Nederland
URL: http://beta.neuro.nl/research/vanderheijden

P. van Dijk et al. (eds.), *Physiology, Psychoacoustics and Cognition in Normal and Impaired Hearing,* Advances in Experimental Medicine and Biology 894,
DOI 10.1007/978-3-319-25474-6_28

nonlinearities also affect mechanical responses to pulsatile stimuli such as clicks (e.g. Robles et al. 1976), and in two-tone experiments (Robles et al. 1997), where they reputedly act on a cycle-by-cycle basis to generate distinctive patterns of intermodulation distortion products. The extent to which a compressive input-output relation holds when the intensity of a stimulus fluctuates in time has not been studied systematically, however, and the possibility that the cochlea operates more like an automatic gain control (AGC) system than an instantaneously nonlinear system remains feasible (van der Heijden 2005). In the present study, we investigate the dynamics of the cochlea's compressive mechanism(s) by observing BM responses to balanced, beating pairs of low-level tones near the recording site's characteristic frequency (CF).

2 Methods

Sound-evoked BM vibrations were recorded from the basal turn of the cochlea in terminally anesthetised gerbils, using techniques similar to those reported previously (Versteegh and van der Heijden 2012). Acoustic stimuli included: (i) inharmonic multi-tone complexes (zwuis stimuli), which were used to characterise the tuning properties of each site on the BM (including the site's CF); and (ii) "beating" pairs of inharmonic near-CF tones, with component levels adjusted to produce near-perfect periodic-cancellations in the response envelopes at individual sites on the BM (cf. Fig. 1).

3 Results

3.1 Time-Domain Observations

BM responses to well balanced, "beating" two-tone stimuli had heavily compressed envelopes, as illustrated in Fig. 1. The amount of compression was quantified by comparing the shapes of the observed response envelopes with those that would be predicted in a completely linear system (illustrated by the dashed blue and red lines in Figs. 1b and c, respectively). When scaled to have similar peak magnitudes, the observed envelopes exceeded the linear predictions across most of the beat cycle (cf. Figs. 1b, c). Expressing the ratios of the observed and predicted envelopes as "normalised gains" (Fig. 1d), the envelope's maximum gain always occurred near (but not exactly at) the instant of maximal cancellation between the two beating tones. This is as expected in a compressive system, where (by definition) "gain" decreases with increasing intensity, or increases with decreasing intensity.

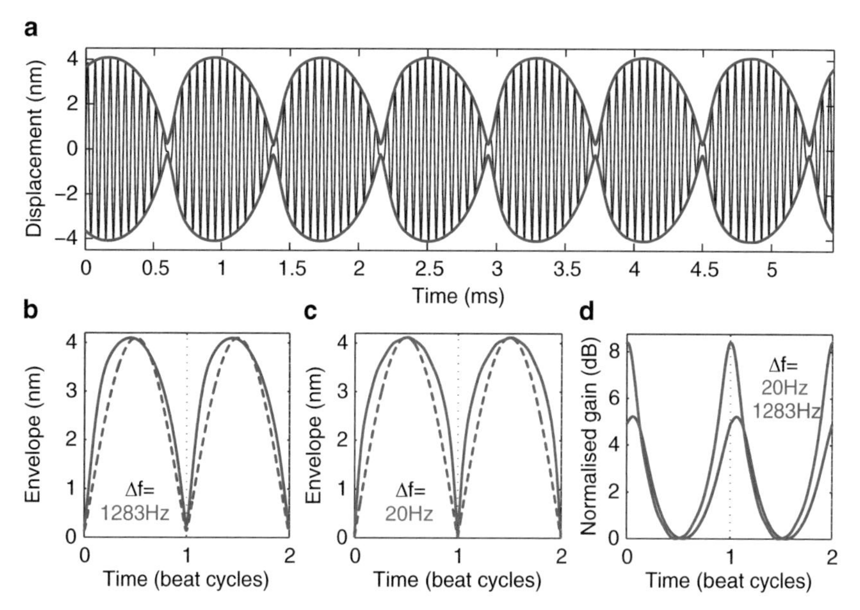

Fig. 1 **a** BM responses to near-CF beat stimuli. $f_1=17{,}373$ Hz, 30dB SPL, $f_2=18{,}656$ Hz, 33dB SPL, $\Delta f=f_2-f_1=1283$ Hz. **b** Observed (*solid blue line*, from A) and linearly predicted (*dashed blue line*) response envelopes for $\Delta f=1283$ Hz. **c** Observed (*solid red line*) and predicted (*dashed red line*) response envelopes for $\Delta f=20$ Hz. **d** Normalised gain (=observed/linearly-predicted envelope) for $\Delta f=20$ Hz (*red*) and 1283 Hz (*blue*). Experiment RG15641, CF=18 kHz

3.2 Temporal Asymmetry

At low beat rates (e.g., for $\Delta f=f_2-f_1\sim 10$–160 Hz), observed response envelopes appeared symmetric in the time domain (cf. Fig. 1c). At higher rates, however, the rising flanks of the response envelopes became steeper than the falling flanks (cf. Figs. 1a, b). This temporal asymmetry reflects hysteresis, and is reminiscent of the output of a gain control system with separate attack and decay time-constants.

3.3 "Gain Control" Dynamics

Three further response characteristics (in addition to the hysteresis) became more pronounced at higher beat rates: (1) the point at which the observed envelopes exhibited their maximal gains (relative to the linearly predicted responses) occurred later and later in the beat cycle, (2) the overall amount of envelope gain decreased, and (3) the peakedness of this gain within the beat-cycle decreased (compare the red and blue curves in Fig. 1d). These characteristics were confirmed and quantified by spectral analysis of the envelopes' temporal gain curves, as illustrated in Fig. 2.

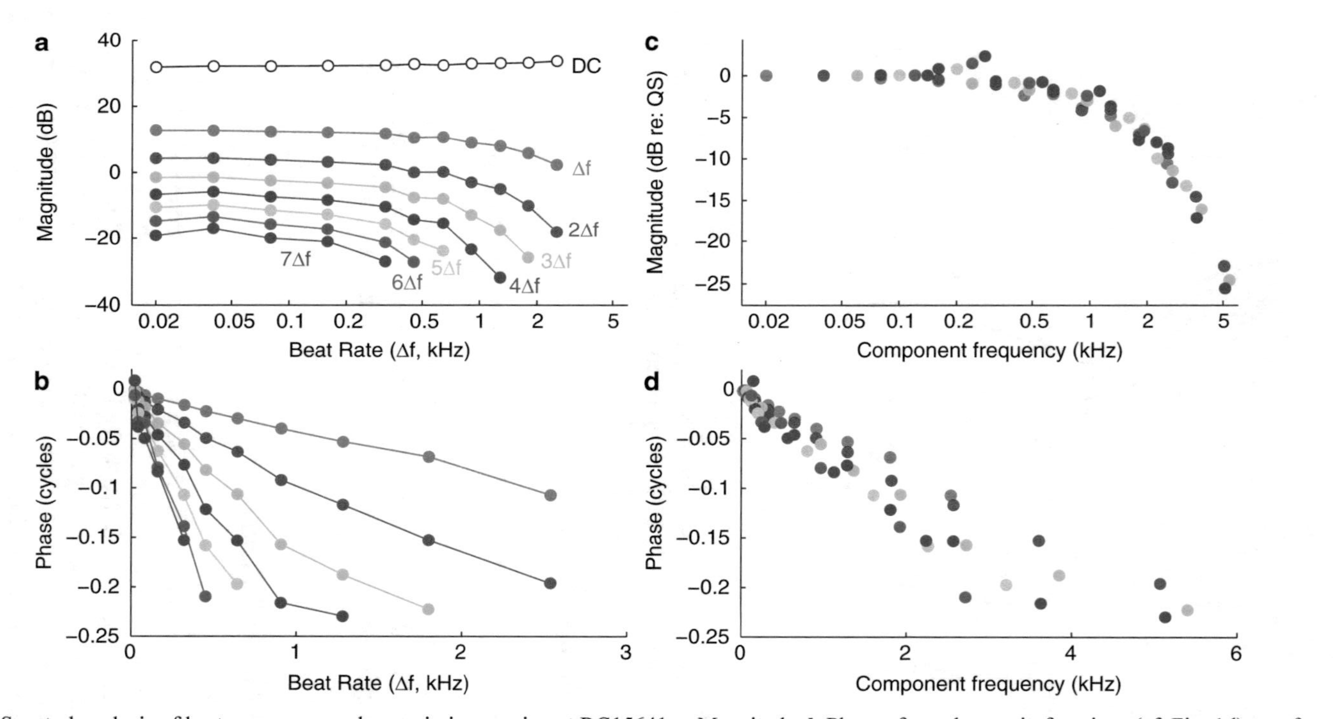

Fig. 2 Spectral analysis of beat response envelope gain in experiment RG15641. **a** Magnitude. **b** Phase of envelope gain functions (cf. Fig. 1d) as a function of beat rate ($\Delta f = f_2 - f_1$). *Coloured curves* represent the contributions of the different harmonics of Δf (labelled in A) to the envelope gain functions at each beat rate. **c** Magnitude. **d** Phase of the envelope transfer function derived from A and B. *Coloured symbols* from A and B are plotted at the actual frequencies of the spectral components (i.e. at f=n.Δf) for the whole family of beat rates tested (20-2560 Hz), and magnitudes are normalized re: the quasi-static (QS, Δf=20 Hz) harmonics

The left panels of Fig. 2 show spectral decompositions of the envelope gain functions (cf. Fig. 1d) for a wide range of stimulus beat rates. Response energy at each harmonic of the beat rate decreases as the beat rate increases (Fig. 2a), suggesting that envelope gain is subjected to a low-pass filtering (or smoothing) process before it is observed on the BM: this is the spectral counterpart to the decreased gain at higher beat rates (point 2) referred to above. Response energy decreases more rapidly with increasing frequency for the higher harmonic numbers in Fig. 2 (i.e., the coloured curves in Fig. 2a are more closely spaced at low beat rates, and more widely spaced at higher rates): this is the spectral counterpart of the decreased peakedness of the temporal gain curves (point 3) referred to above.

The low-pass filtering apparent in the spectral analysis of the beat responses can be characterised by deriving a compound transfer function for the response envelope gain, as illustrated in Figs. 2c and d. This shows the amount of energy in the envelope gain function at each harmonic of the beat rate, after normalization by the BM's quasi-static (i.e., $\Delta f \rightarrow 0$) input-output function. In the case of Fig. 2c, we have used the 20 Hz beat rate data to normalize the data at higher frequencies, and plotted each of the harmonics at its own absolute frequency (i.e. at $f = n.\Delta f$, where n is the harmonic number). This reveals a low-pass "gain control" transfer function with a −3 dB cut-off at ~1 kHz.

The energy at each harmonic of the beat rate in the envelope gain function also occurs later in the beat cycle as the beat rate increases, as shown in Figs. 2b and d (negative phases in Fig. 2 indicate phase-lags). The phases in Fig. 2b scale almost proportionally with harmonic number, suggesting that (on average) the observed gain curves lag the effective stimuli (the linearly predicted envelopes) by a constant amount of time. This delay is approximately 50 μs for most of our data (i.e., equating to 0.2 cycles of phase-lag in a 4 kHz range, as shown in Fig. 2d). These observations are the spectral counterparts of the delays observed in the instants of maximum envelope gain mentioned in Sect. 3.1 (cf. Fig. 1d, where both the red and blue curves peak 50 μs after the null in the observed response envelope).

3.4 Linearization of Function

The observed decreases in envelope gain and in gain peakedness with increasing beat rate act synergistically to reduce the amount of compression exhibited in the BM response envelopes at high beat rates. This "linearization" is illustrated directly in Fig. 3 by plotting the response envelopes as a function of the instantaneous intensity of the beating stimulus. The decreases in gain are observed as clockwise rotations in the linear coordinates of Fig. 3a, and as downward shifts on the logarithmic plots of Fig. 3b. The linearization that accompanies the decreases in gain is seen as an increase in the threshold above which compression is seen (e.g. the blue 20 Hz data of Fig. 3b start to deviate from linearity above ~10 dB SPL and ~0.3 nm, whereas the steepest, rising flank of the black 1280 Hz data remains nearly linear up to ~25 dB SPL and 2 nm). One consequence of this linearization is that BM's

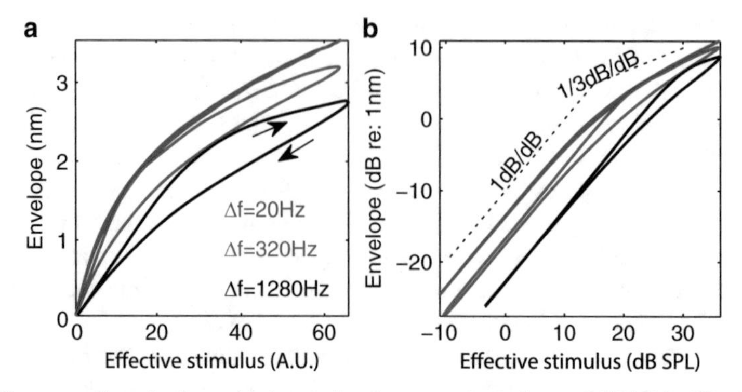

Fig. 3 Response linearization with increasing beat rate (experiment RG14612, CF = 15 kHz). Instantaneous envelope input-output functions for three beat stimuli on **a** linear, and **b** logarithmic scales. *Arrows* in A distinguish rising and falling phases of 1280 Hz data. *Dashed lines* in B show linear (1 dB/dB) and compressive (0.333 dB/dB) growth rates

response dynamics can actually enhance the encoding of dynamic stimuli, within certain limits. However, this "enhancement" comes at the price of reduced overall sensitivity.

4 Discussion

The mechanical nonlinearity of the mammalian cochlea is clearly fast, but not instantaneous. The results of the current investigation show that dynamic changes in stimulus intensity can only be followed accurately up to a rate of ~ 1 kHz in the basal turn of the gerbil cochlea. This limitation appears to be produced by a low-pass gain control filter of some kind: this could be a mechanical filter, imparted by the load on a force generating element (such as a motile outer hair cell, or OHC), or an electrical filter (such as the OHC's basolateral membrane, cf. Housley and Ashmore 1992). The filter could even involve a combination of multiple, coupled mechanisms: preliminary analysis suggests that the gain at one site on the BM may be controlled remotely, presumably by elements (e.g., OHCs) that are distributed more basally along the BM.

Our observation of hysteresis in the BM's response to a dynamic stimulus is not entirely new—similarly asymmetric envelope responses have been reported in studies using amplitude modulated tonal stimuli (e.g., Rhode and Recio 2001). Our observation of a distinct delay to the appearance of compression (or gain) in the response envelopes is new, however. We currently believe that it is impossible to produce such a delay without resorting to a non-instantaneous form of gain control, such as an AGC. Other (non-AGC) types of system which produce level-dependent group delays can easily mimic the envelope asymmetry (hysteresis) effects that we observe, but (to the best of our knowledge) they cannot produce an actual delay in the "effect" of the nonlinearity. If this delay really does signify the presence of an AGC in the cochlea, it may well prove significant for the coding of a wide range of stimuli, and have consequences for numerous psychophysical phenomena.

Acknowledgments Supported the Netherlands Organization for Scientific Research, ALW 823.02.018.

References

De Boer E, Nuttall AL (1997) The mechanical waveform of the basilar membrane. I. Frequency modulations ("glides") in impulse responses and cross-correlation functions. J Acoust Soc Am 101(6):3583–3592. http://doi.org/10.1121/1.418319

Housley GD, Ashmore JF (1992) Ionic currents of outer hair cells isolated from the guinea-pig cochlea. J Physiol 448:73–98

Rhode WS (1971). Observations of the vibration of the basilar membrane in squirrel monkeys using the Mössbauer technique. J Acoust Soc Am, 49(4), Suppl 2:1218–1231

Rhode WS, Recio A (2001) Multicomponent stimulus interactions observed in basilar-membrane vibration in the basal region of the chinchilla cochlea. J Acoust Soc Am 110(6):3140–3154

Robles L, Ruggero MA (2001) Mechanics of the mammalian cochlea. Physiol Rev 81(3):1305–1352

Robles L, Rhode WS, Geisler CD (1976) Transient response of the basilar membrane measured in squirrel monkeys using the Mössbauer effect. J Acoust Soc Am 59(4):926–939

Robles L, Ruggero MA, Rich NC (1997) Two-tone distortion on the basilar membrane of the chinchilla cochlea. J Neurophysiol 77(5):2385–2399

Van der Heijden M (2005). Cochlear gain control. J Acoust Soc Am, 117(3 Pt 1), 1223–1233

Versteegh CPC, van der Heijden M (2012) Basilar membrane responses to tones and tone complexes: nonlinear effects of stimulus intensity. J Assoc Res Otolaryngol 13(6):785–798. http://doi.org/10.1007/s10162-012-0345-0

Responses of the Human Inner Ear to Low-Frequency Sound

Markus Drexl, Eike Krause, Robert Gürkov and Lutz Wiegrebe

Abstract The perceptual insensitivity to low frequency (LF) sound in humans has led to an underestimation of the physiological impact of LF exposure on the inner ear. It is known, however, that intense, LF sound causes cyclic changes of indicators of inner ear function after LF stimulus offset, for which the term "Bounce" phenomenon has been coined.

Here, we show that the mechanical amplification of hair cells (OHCs) is significantly affected after the presentation of LF sound. First, we show the Bounce phenomenon in slow level changes of quadratic, but not cubic, distortion product otoacoustic emissions (DPOAEs). Second, Bouncing in response to LF sound is seen in slow, oscillating frequency and correlated level changes of spontaneous otoacoustic emissions (SOAEs). Surprisingly, LF sound can induce new SOAEs which can persist for tens of seconds. Further, we show that the Bounce persists under free-field conditions, i.e. without an in-ear probe occluding the auditory meatus. Finally, we show that the Bounce is affected by contralateral acoustic stimulation synchronised to the ipsilateral LF sound. These findings clearly demonstrate that the origin of the Bounce lies in the modulation of cochlear amplifier gain. We conclude that activity changes of OHCs are the source of the Bounce, most likely caused by a temporary disturbance of OHC calcium homeostasis. In the light of these findings, the effects of long-duration, anthropogenic LF sound on the human inner ear require further research.

M. Drexl (✉) · E. Krause · R. Gürkov
German Center for Vertigo and Balance Disorders (IFB), Department of Otorhinolaryngology, Head and Neck Surgery, Grosshadern Medical Centre, University of Munich, 81377 Munich, Germany
e-mail: markus.drexl@med.uni-muenchen.de

E. Krause
e-mail: eike.krause@med.uni-muenchen.de

R. Gürkov
e-mail: robert.guerkov@med.uni-muenchen.de

L. Wiegrebe
Division of Neurobiology, Dept. Biology II, University of Munich, 82152 Martinsried, Germany
e-mail: lutzw@lmu.de

P. van Dijk et al. (eds.), *Physiology, Psychoacoustics and Cognition in Normal and Impaired Hearing,* Advances in Experimental Medicine and Biology 894,
DOI 10.1007/978-3-319-25474-6_29

Keywords Cochlea · Calcium homeostasis · Cochlear amplifier · Outer hair cells · Inner hair cells · Bounce phenomenon · Otoacoustic emissions

1 Introduction

For decades, low-frequency sound, i.e. sound with frequencies lower than 250 Hz (Berglund et al. 1996), has been considered to largely bypass the inner ear even at intense levels, simply because human hearing thresholds for frequencies below 250 Hz are relatively high. Recent evidence from animal models shows that physiological cochlear responses to LF sound are even larger than those evoked by equal-level, higher frequencies in the more sensitive range of hearing (Salt et al. 2013). No data for human subjects are available, but, considering the higher sensitivity of humans for LF sounds, similar results can be expected (Salt et al. 2013).

Hirsh and Ward (1952) observed temporary deteriorations of human absolute thresholds about 2 min after presenting subjects with an intense, non-traumatic LF tone. Later on, the term 'Bounce' was used to describe bimodal changes in absolute thresholds starting with a sensitisation period followed by an about equal-duration temporary desensitisation (Hughes 1954).

Perceptual thresholds essentially reflect the sensitivity of inner hair cells (IHCs), which are functionally coupled to inner ear fluids (Nowotny and Gummer 2006; Guinan 2012). IHCs are therefore sensitive to basilar-membrane velocity, which decreases with decreasing frequency. OHCs, in contrast, are mechanically linked to both the basilar membrane and the tectorial membrane. OHCs are therefore sensitive to basilar membrane displacement (Dallos et al. 1982; Dallos 1986), which does not decrease with decreasing stimulus frequency. Thus, OHCs are more sensitive to LF sound than IHCs and it is this difference in LF sensitivity, which contributes to the LF limit of sound perception (Salt and Hullar 2010; Salt et al. 2013). In humans, non-invasive recordings of DPOAEs allow indirect access to OHC function while SOAEs represent, for ears that exhibit them, a more direct and very sensitive marker of OHC function. SOAEs are narrowband acoustic signals which are spontaneously emitted by the inner ear in the absence of acoustic stimulation. Human SOAEs persist over years and are relatively stable in both frequency and level (Burns 2009).

Here, we use both DPOAE and SOAE measurements to assess LF-induced changes of cochlear physiology and active sound amplification. Specifically, we monitored the sound level and frequency of DPOAEs and SOAEs before and after the exposure to a 90 s LF sinusoid with 30 Hz and a level of 80 dBA (120 dB SPL). Both the sound level and the exposure duration were controlled to be within the exposure limits for normal working conditions as regulated by the European Commission Noise at Work Directive 2003/10/EC.

2 Methods

Data were collected from young adult normal hearing subjects. The ethics committee of the University Hospital of the Ludwig-Maximilians University Munich, Germany, in agreement with the Code of Ethics of the World Medical Association (Declaration of Helsinki) for experiments involving humans, approved the procedures, and all subjects gave their written informed consent. An Etymotic Research 10C DPOAE probe system was used for recording of OAEs. The LF tone (30 Hz sine wave, 120 dB SPL, 90 s, including 0.1 s raised-cosine ramps) was supplied by a separate loudspeaker (Aurasound NSW1-205-8A). This loudspeaker was connected via a 50 cm polyethylene tube (inner diameter 1 mm) and the foam ear tip of the ER-10C DPOAE probe so that it faced the tympanic membrane. The loudspeaker was driven by a Rotel RB-960BX power amplifier. Stimulation to the contralateral ear was provided by an Etymotic Research 4PT earphone, which was sealed into the contralateral ear canal with foam ear tips. The earphone was driven by the headphone amplifier of the audio interface (RME audio Fireface UC, fs = 44.1 kHz) which was programmed (using MatLab and the HörTech SoundMexPro audio toolbox) for synchronous stimulation and recording of all required inputs and outputs.

3 Results

3.1 Effect of LF Sound Exposure on DPOAEs

The effect of the 80 dBA LF exposure on quadratic (QDP) and cubic (CDP) OAEs is shown in Fig. 1a and b, respectively. In 14 out of 20 tested subjects, the LF exposure induced a subsequent increase of the QDP level lasting for about 60 to 90 s (see Fig. 1a). QDP levels increased with a median of 3.4 dB. In most cases, this QDP increase was followed by a similar QDP decrease (median: −2.4 dB), at about 120–150 s post-exposure. This decrease slowly recovered to pre-exposure QDP levels. The median duration of the overall oscillatory change of the QDP level was 214 s.

In many cases it was also possible to extract CDP levels from the same recording (albeit f2/f1 ratios were optimized to achieve maximum QDP levels). Typically, we observed no significant changes of CDP level after LF sound exposure (Fig. 1b).

3.2 Effect of LF Sound Exposure on SOAEs

We recorded 80 SOAEs from 27 ears of 16 young, normal-hearing subjects. The median SOAE sound levels were 0.6 dB SPL (first and third quartiles, −4.5 dB SPL; 4.0 dB SPL) with a signal-to-noise ratio of 16.6 dB (11.6 dB, 23.5 dB).

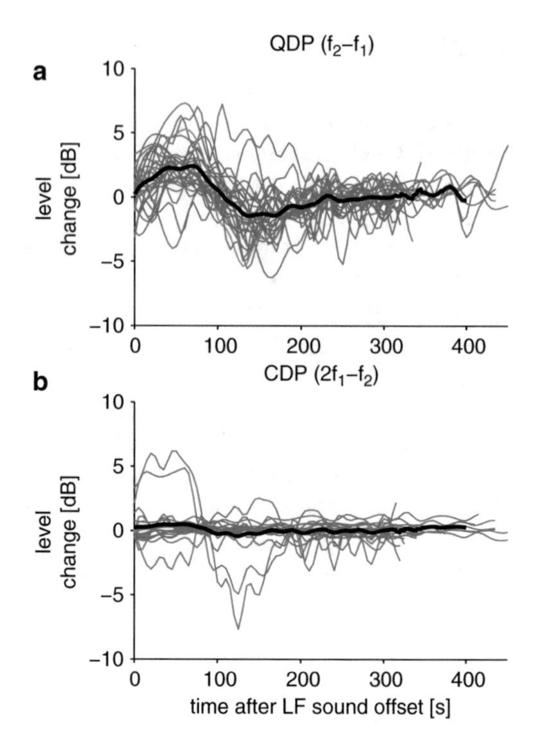

Fig. 1 Effect of LF sound exposure on quadratic distortion products (QDP, A, $N=28$ from 14 subjects) and cubic distortion products (CDP, B, $N=22$ from 11 subjects). QDPs oscillate significantly after LF offset (**a**) whereas CDPs are rarely affected, **b** *bold black lines* represent median DPOAE measures, *fine grey lines* are individual DPOAE measures. (Modified after Drexl et al. (2014), with permission)

After LF sound stimulation, 56 of these 80 SOAEs increased in both sound level and frequency. This increase was followed by a decrease of both level and frequency relative to pre-exposure (see Fig. 2a). In 10 of the 80 pre-exposure SOAEs, we observed an inverted pattern with an initial level and frequency decrease, followed by a level and frequency increase.

SOAE level- and frequency oscillations were fitted with an (inverted-phase) underdamped sinusoidal oscillation. The period of the fitted sinusoid was 257 s (202 s, 294 s) for the level time course and 252 s (215 s, 367 s) for the frequency time course. The time constant of the damped sinusoid for level changes was 120 s (76 s, 157 s) and for frequency changes 94 s (58 s, 141 s). SOAE frequency changes amounted to 5 Cent (4 Cent, 9 Cent) with peak values of 25 Cent. Relative to the SOAE frequency in the control condition, the frequency showed initial maximum increases of 4 Cent (3 Cent, 7 Cent), followed by maximum decreases of 1 Cent (0 Cent, 2 Cent).

17 of 21 tested subjects revealed an overall of 56 new SOAEs, which had not been measurable before LF stimulation (see Fig. 2b). These new SOAEs were characterized by an initial level and frequency increase, qualitatively similar to the pre-existing SOAEs. Comparable to the enhancing half cycle of Bouncing SOAEs, their level and frequency oscillated before they disappeared into the noise floor. The duration of the level and frequency changes was 67.5 s (47.5 s, 90 s). New SOAEs started to arise within 12.5 s (5 s, 25 s) after LF sound offset and reached a level maximum at 50 s (35 s, 62.5 s) after LF offset. The maximum SOAE level was

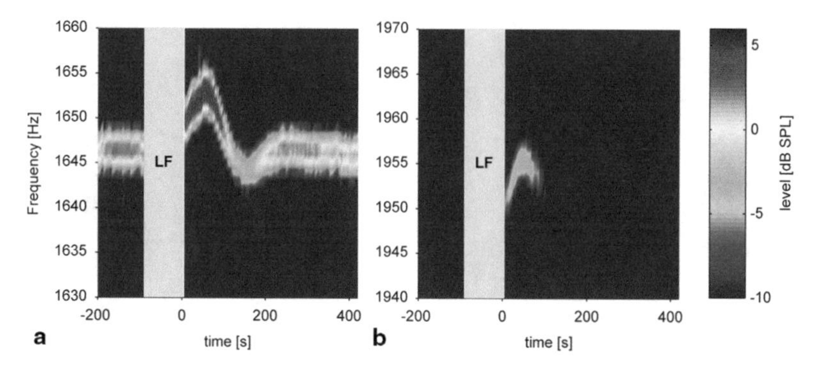

Fig. 2 **a** Exemplary level and frequency changes of an SOAE recorded pre- and post LF sound exposure. The *grey bar* indicates the presentation of the LF stimulus (30 Hz, 80 dBA, 90 s). **b** Same as in a, but for a new SOAE which only appears for a short period after LF exposure. (Modified after Kugler et al. (2014), licensed under CC BY 4.0 (https://creativecommons.org/licenses/by/4.0/))

−0.3 dB SPL (−4.1 dB SPL, 4.9 dB SPL) with a signal to noise ratio of 13.8 dB (11.9 dB, 17.6 dB). The difference between the new SOAE frequency maximum and minimum was 4 Cent (1 Cent, 6 Cent). The time course of level and frequency changes was almost identical and maximum level and frequency changes coincided.

3.3 *SOAE Bouncing in the free Sound Field*

Although the observed pattern of synchronized SOAE frequency- and amplitude changes is incompatible with the SOAE bouncing being elicited by changes in middle ear impedance, it is conceivable that bouncing may be only seen in the closed sound field where the auditory meatus is blocked by the OAE probe. Here, we recorded SOAEs in the open meatus using an ER10C probe microphone fitted to the meatus via an about 8 cm silicon tube (2.8 mm outer diameter) which did not block the meatus. The tip of the tube was positioned about 10 mm in front of the tympanum. LF exposure was provided by two powerful custom-made subwoofers. Subjects lay on a deck chair in a long basement corridor at a point where standing waves in the corridor maximised the sound level at 30 Hz. LF exposure was 118 dB SPL for 90 s. Photos of LF stimulation apparatus and probe-tube placement are shown in Fig. 3a, b.

Both ears of 45 young, normal-hearing subjects were screened for SOAEs. 33 subjects showed at least one SOAE in one ear. Overall we could record in the open meatus about 52 % of those SOAEs detectable in a sound-attenuated room and a closed-ear recording technique. The remaining 48 % were not significantly above the much higher noise floor of the free-field, open-meatus measurement.

Exemplary measurements of both permanent and transient SOAEs are shown in Fig. 3c and d in the same format as Fig. 2. Indeed many of those 48 % SOAEs

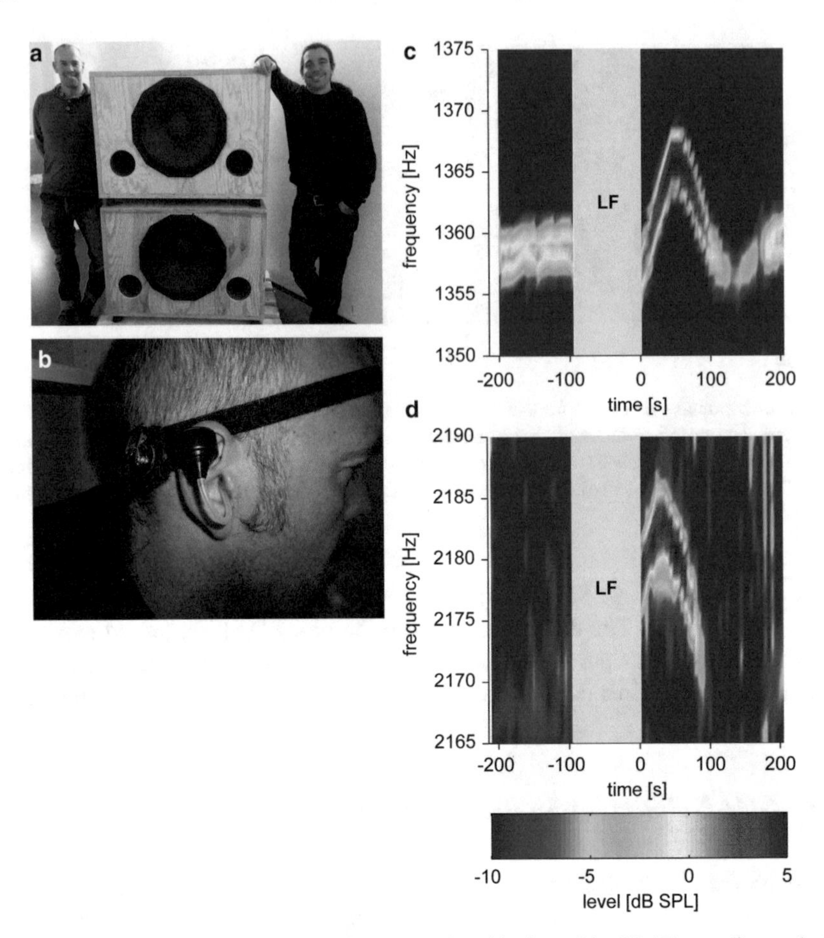

Fig. 3 LF free-field stimulation subwoofers (**a**), and positioning of the SOAE recording probe in the open ear canal (**b**), Exemplary recordings in the open meatus of a pre-existing SOAE (**c**), and a transient SOAE (**d**) after free-field exposure to a 118 dB SPL, 30 Hz, 90 s LF sound (*grey bar*)

that had been initially identified in the closed-meatus measurements, but could no longer be detected in the open meatus, appeared directly after the LF exposure for a short time period, before falling again below the noise floor (Fig. 3d). These data clearly show that Bouncing of SOAEs can indeed be elicited by free-field exposure to LF sound sources of natural or anthropogenic origin.

3.4 *Effect of Contralateral Acoustic Stimulation (CAS)*

Patuzzi (2011) suggested that large receptor-potentials elicited by low-frequency stimulation produce a net Ca^{2+} influx. The Bounce presumably reflects an under-damped, homeostatic readjustment of increased Ca^{2+} concentrations and related gain changes after low-frequency sound offset. Here, we tested this hypothesis by activating the medial olivocochlear (MOC) efferent system during presentation of the Bounce-evoking LF sound. The MOC system is known to modulate OHC Ca^{2+} concentrations (Sridhar et al. 1997) and receptor potentials (Fex 1967) and therefore it should modulate the characteristics of the Bounce. Here, CAS was provided simultaneously to the (ipsilateral to the observed SOAE) LF exposure. The CAS consisted of a 90 s, bandpass-filtered Gaussian noise (100 Hz—8 kHz) presented at 65 or 70 dB SPL.

CAS is well known to suppress ipsilaterally recorded SOAEs during presentation and SOAEs quickly recover after CAS offset within less than 1 s (Zhao and Dhar 2010, 2011). Due to the duration of our analysis segments (5 s), the SOAEs already fully recovered from the CAS exposure within the first analysis segment. Consequently, we found no SOAEs fulfilling our criteria for the Bounce or indeed any other significant changes in the CAS control recordings. When the CAS was presented simultaneously with the ipsilateral LF tone, however, Bouncing of permanent SOAEs after LF offset changed significantly. Exemplary time courses of a preexisting and a transient SOAE are shown in Fig. 4a and b, respectively. While in the reference recording (red) the preexisting SOAE showed a significant biphasic Bounce, presentation of a 65 dB SPL (blue) or 70 dB SPL (green) CAS together with the ipsilateral LF tone clearly affected the magnitude of the Bounce.

Overall, temporary level reductions of preexisting SOAEs were less pronounced with CAS than without (Wilcoxon signed rank test, $p = 0.085$ and 0.007 for CAS

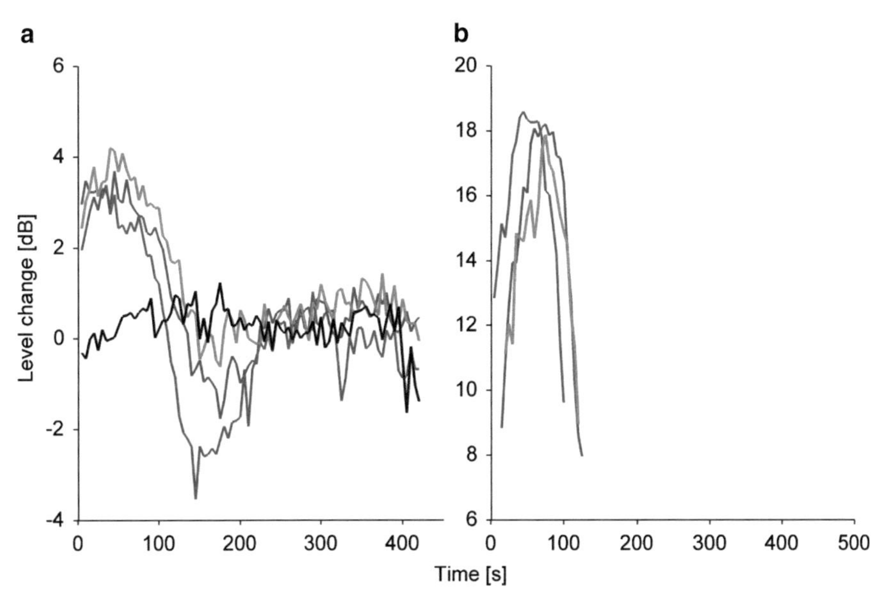

Fig. 4 A preexisting SOAE (**a**), and a new, transient SOAE (**b**) after LF exposure (*red*) and after LF exposure witch simultaneous CAS at 65 dB SPL (*blue*) or 70 dB SPL (*green*), and after CAS exposure alone (*black*)

levels of 65 and 70 dB, respectively), while the SOAE level increases remained fairly unchanged. This resulted in a less symmetrical shape of the SOAE Bounce compared to the reference recording. Consequently, the time constants of the fitted function with CAS shortened significantly.

4 Discussion

The current biophysical experiments reveal a significant effect of LF sound exposure on the inner ear, and specifically on the function of OHCs. The observed effects are reflected in significant changes of quadratic distortion products, but importantly, in changes of SOAEs, which (for the ears that exhibit them) allows for a least-invasive evaluation of inner ear function.

As the low-frequency stimuli used in this study are intense, mechanisms associated with acoustic overexposure and recovery thereof could in principle be responsible for the Bounce phenomenon. Temporary or permanent damage to OHCs represent an unlikely cause, as recovery from acoustic overexposure with sounds in the sensitive range of hearing is typically monotonic, does not oscillate and no hypersensitivity can be seen. It has been shown, however, that intense sound stimulation of the cochlea in the isolated temporal bone preparation increases intracellular Ca^{2+} level of hair cells and supporting cells (Fridberger et al. 1998; Jacob et al. 2013). Acoustic injury consists of a plethora of structural and metabolic changes to the cochlea, with structural damage possibly masking more subtle (and possibly oscillating) metabolic changes of cochlear sensitivity caused by the rise of intracellular Ca^{2+} levels.

The observed effects support the hypothesis that OHC Ca^{2+} homeostasis is the source of the Bounce: Patuzzi (2011) suggested that the Bounce is a direct result of OHC activation by LF sound, rather than a secondary effect caused by modulation of stria vascularis activity, as the LF sound-induced endocochlear potential modulation during the Bounce (Kirk and Patuzzi 1997) could suggest. Patuzzi (2011) hypothesized that large, LF-induced receptor potentials in OHCs are the underlying cause of cochlear sensitivity oscillations associated with the Bounce phenomenon. These LF OHC receptor potentials are not attenuated by the low-pass characteristic of the OHC membrane. Patuzzi (2011) postulated that large, LF receptor potentials activate voltage-sensitive, L-type Ca^{2+} channels in the OHC membrane. This results in an increase of intracellular Ca^{2+} in OHCs. Ca^{2+}-induced Ca^{2+} release and -uptake mechanisms, with different time courses, can then cause slow, underdamped oscillations of OHC cytosolic Ca^{2+} concentrations, modulating the gain of the cochlear amplifier (Patuzzi 2011).

The increased damping (corresponding to decreased decay time constants) we observed in the CAS experiments indicates that processes re-adjusting the Ca^{2+} overshoot may accelerate due to the activation of the medial olivo-cochlear bundle by the CAS. Even while the LF sound is on, the LF-induced Ca^{2+} concentration changes presumably also undergo oscillations (Patuzzi 2011). In contrast, the slow efferent effect (Sridhar et al. 1997) can cause a constant Ca^{2+} release from OHC internal stores, the lateral cisternae, while CAS is on. We hypothesize that this constant influx of Ca^{2+} may help to accelerate Ca^{2+}-dependent Ca^{2+}-uptake and thus, a quicker recovery of the system.

In summary, the current experiments reveal a pronounced effect of LF exposure on the active mechanisms in the inner ear, as they are mediated by OHCs. Considering that the current LF exposure was limited to 90 s, it is unclear how a longer-duration LF exposure may affect the system.

Acknowledgments The authors wish to acknowledge the contributions of Margarete Überfuhr, David Laubender and Kathrin Kugler, who carried out the experimental procedures this chapter is based on.

This work was funded by a grant (EO 0901) from the German Ministry of Science and Education to the German Center for Vertigo and Balance Disorders (IFB), project TR-F9.

References

Berglund B, Hassmen P, Job RF (1996) Sources and effects of low-frequency noise. J Acoust Soc Am 99(5):2985–3002

Burns EM (2009) Long-term stability of spontaneous otoacoustic emissions. J Acoust Soc Am 125(5):3166–3176. doi:10.1121/1.3097768

Dallos P (1986) Neurobiology of cochlear inner and outer hair cells: intracellular recordings. Hear Res 22:185–198

Dallos P, Santos-Sacchi J, Flock A (1982) Intracellular recordings from cochlear outer hair cells. Science 218(4572):582–584

Drexl M, Uberfuhr M, Weddell TD, Lukashkin AN, Wiegrebe L, Krause E, Gurkov R (2014) Multiple indices of the 'bounce' phenomenon obtained from the same human ears. J Assoc Res Otolaryngol 15(1):57–72. doi:10.1007/s10162-013-0424-x

Fex J (1967) Efferent inhibition in the cochlea related to hair-cell dc activity: study of postsynaptic activity of the crossed olivocochlear fibres in the cat. J Acoust Soc Am 41(3):666–675

Fridberger A, Flock A, Ulfendahl M, Flock B (1998) Acoustic overstimulation increases outer hair cell Ca^{2+} concentrations and causes dynamic contractions of the hearing organ. Proc Natl Acad Sci U S A 95(12):7127–7132

Guinan JJ (2012) How are inner hair cells stimulated? Evidence for multiple mechanical drives. Hear Res 292(1–2):35–50

Hirsh I, Ward W (1952) Recovery of the auditory threshold after strong acoustic stimulation. J Acoust Soc Am 24:131

Hughes JR (1954) Auditory Sensitization. J Acoust Soc Am 26(6):1064–1070. doi:10.1121/1.1907450

Jacob S, Johansson C, Fridberger A (2013) Noise-induced alterations in cochlear mechanics, electromotility, and cochlear amplification. Pflügers Archiv—European. J Physiol 465(6):907–917. doi:10.1007/s00424-012-1198-4

Kirk DL, Patuzzi RB (1997) Transient changes in cochlear potentials and DPOAEs after low-frequency tones: the 'two-minute bounce' revisited. Hear Res 112(1–2):49–68

Kugler K, Wiegrebe L, Grothe B, Kössl M, Gürkov R, Krause E, Drexl M (2014). Low-frequency sound affects active micromechanics in the human inner ear. R Soc Open Sci 1(2):140–166

Nowotny M, Gummer AW (2006) Nanomechanics of the subtectorial space caused by electromechanics of cochlear outer hair cells. Proc Natl Acad Sci U S A 103(7):2120–2125. doi:10.1073/pnas.0511125103

Patuzzi R (2011) Ion flow in cochlear hair cells and the regulation of hearing sensitivity. Hear Res 280(1–2):3–20. doi:10.1016/j.heares.2011.04.006

Salt AN, Hullar TE (2010) Responses of the ear to low frequency sounds, infrasound and wind turbines. Hear Res 268(1–2):12–21. doi:S0378-5955(10)00312-6 [pii] 10.1016/j.heares.2010.06.007

Salt AN, Lichtenhan JT, Gill RM, Hartsock JJ (2013) Large endolymphatic potentials from low-frequency and infrasonic tones in the guinea pig. J Acoust Soc Am 133(3):1561–1571. doi:10.1121/1.4789005

Sridhar TS, Brown MC, Sewell WF (1997) Unique postsynaptic signaling at the hair cell efferent synapse permits calcium to evoke changes on two time scales. J Neurosci 17(1):428–437

Zhao W, Dhar S (2010) The effect of contralateral acoustic stimulation on spontaneous otoacoustic emissions. J Assoc Res Otolaryngol 11(1):53–67. doi:10.1007/s10162-009-0189-4

Zhao W, Dhar S (2011) Fast and slow effects of medial olivocochlear efferent activity in humans. PloS ONE 6(4):e18725. doi:10.1371/journal.pone.0018725

Suppression Measured from Chinchilla Auditory-Nerve-Fiber Responses Following Noise-Induced Hearing Loss: Adaptive-Tracking and Systems-Identification Approaches

Mark Sayles, Michael K. Walls and Michael G. Heinz

Abstract The compressive nonlinearity of cochlear signal transduction, reflecting outer-hair-cell function, manifests as suppressive spectral interactions; e.g., two-tone suppression. Moreover, for broadband sounds, there are multiple interactions between frequency components. These frequency-dependent nonlinearities are important for neural coding of complex sounds, such as speech. Acoustic-trauma-induced outer-hair-cell damage is associated with loss of nonlinearity, which auditory prostheses attempt to restore with, e.g., "multi-channel dynamic compression" algorithms.

Neurophysiological data on suppression in hearing-impaired (HI) mammals are limited. We present data on firing-rate suppression measured in auditory-nerve-fiber responses in a chinchilla model of noise-induced hearing loss, and in normal-hearing (NH) controls at equal sensation level. Hearing-impaired (HI) animals had elevated single-fiber excitatory thresholds (by ~20–40 dB), broadened frequency tuning, and reduced-magnitude distortion-product otoacoustic emissions; consistent with mixed inner- and outer-hair-cell pathology. We characterized suppression using two approaches: adaptive tracking of two-tone-suppression threshold (62 NH, and 35 HI fibers), and Wiener-kernel analyses of responses to broadband noise (91 NH, and 148 HI fibers). Suppression-threshold tuning curves showed sensitive low-

M. G. Heinz (✉) · M. Sayles · M. K. Walls
Department of Speech, Language, & Hearing Sciences, Purdue University, Lyles-Porter Hall, 715 Clinic Drive, West Lafayette, IN 47907-2122, USA
e-mail: mheinz@purdue.edu

M. G. Heinz
Weldon School of Biomedical Engineering, Purdue University, 206 S. Martin Jischke Drive, West Lafayette, IN 47907-2032, USA

M. Sayles
Laboratory of Auditory Neurophysiology, Katholieke Universiteit Leuven, Campus Gasthuisberg—O&N II, Herestraat 49–bus 1021, 3000 Leuven, Belgium
e-mail: sayles.m@gmail.com

M. K. Walls
e-mail: mkwalls@purdue.edu

P. van Dijk et al. (eds.), *Physiology, Psychoacoustics and Cognition in Normal and Impaired Hearing,* Advances in Experimental Medicine and Biology 894,
DOI 10.1007/978-3-319-25474-6_30

side suppression for NH and HI animals. High-side suppression thresholds were elevated in HI animals, to the same extent as excitatory thresholds. We factored second-order Wiener-kernels into excitatory and suppressive sub-kernels to quantify the relative strength of suppression. We found a small decrease in suppression in HI fibers, which correlated with broadened tuning. These data will help guide novel amplification strategies, particularly for complex listening situations (e.g., speech in noise), in which current hearing aids struggle to restore intelligibility.

Keywords Auditory nerve · Suppression · Frequency-dependent nonlinearity · Chinchilla · Spike-triggered neural characterization · Wiener kernel · Singular-value decomposition · Threshold tracking · Frequency tuning · Hearing impairment · Noise exposure · Acoustic trauma · Cochlear hearing loss · Auditory-brain-stem response · Otoacoustic emissions

1 Introduction

Frequency-dependent cochlear signal-transduction nonlinearities manifest as suppressive interactions between acoustic-stimulus components (Sachs and Kiang 1968; de Boer and Nuttall 2002; Versteegh and van der Heijden 2012, 2013). The underlying mechanism is thought to be saturation of outer-hair-cell receptor currents (Geisler et al. 1990; Cooper 1996). Despite the importance of suppression for neural coding of complex sounds (e.g., Sachs and Young 1980), relatively little is known about suppression in listeners with cochlear hearing loss (Schmiedt et al. 1990; Miller et al. 1997; Hicks and Bacon 1999). Here we present preliminary data on firing-rate suppression, measured with tones, and with broadband noise, from ANFs in chinchillas (*Chinchilla laniger*), following noise-induced hearing loss. Quantitative descriptions of suppression in the hearing-impaired auditory periphery will help guide novel amplification strategies, particularly for complex listening situations (e.g., speech in noise), in which current hearing aids struggle to improve speech intelligibility.

2 Methods

2.1 Animal Model

Animal procedures were under anesthesia, approved by Purdue University's IACUC, and followed NIH-issued guidance. Evoked-potential and DPOAE measures, and noise exposures, were with ketamine (40 mg/kg i.p.) and xylazine (4 mg/kg s.c.). Single-unit neurophysiology was done with sodium pentobarbital (boluses: 5–10 mg/hr. i.v.).

2.1.1 Noise Exposure & Hearing-Loss Characterization

Two groups of chinchillas are included: normal-hearing controls (NH), and hearing-impaired (HI) animals with a stable, permanent, sensorineural hearing loss. Prior to noise exposure, we recorded auditory brainstem responses (ABRs) and distortion-product otoacoustic emissions (DPOAEs) from the HI group, verifying "normal" baseline status.

ABRs were recorded in response to pure-tone bursts, at octave-spaced frequencies between 0.5 and 16 kHz. Threshold was determined according to statistical criteria (Henry et al. 2011). DPOAE stimulus primaries (f_1 and f_2) were presented with equal amplitude (75-dB SPL), and with a constant f_2/f_1 ratio of 1.2. f_2 varied between 0.5 and 12 kHz, in 2-semitone steps.

Animals were exposed to either a 500-Hz-centered octave-band noise at 116-dB SPL for 2 h, or a 2-kHz-centered 50-Hz-wide noise at 114-115-dB SPL for 4 h. They were allowed to recover for 3–4 weeks before an acute single-unit experiment. At surgery, ABR and DPOAE measures were repeated prior to any additional intervention (for NH animals, these were their only ABR and DPOAE measurements).

2.1.2 Single-Unit Neurophysiology

The auditory nerve was approached via a posterior-fossa craniotomy. ANFs were isolated using glass pipettes with 15–25 MΩ impedance. Spike times were recorded with 10-μs resolution. Stimulus presentation and data acquisition were controlled by MATLAB programs interfaced with hardware modules (TDT and National Instruments). For HI animals, ANF characteristic frequency (CF) was determined by the high-side-slope method of Liberman (1984).

2.2 Stimuli

2.2.1 Adaptive-Tracking: Suppression-Threshold Tuning

This two-tone suppression (2TS) adaptive-tracking technique is based on Delgutte (1990). Stimuli were sequences of 60-ms duration supra-threshold tones at CF, with or without a second tone at the suppressor frequency (FS; Fig. 1). For each FS frequency-level combination, 10 repetitions of the two-interval sequence were presented. The decision to increase or decrease sound level was made based on the mean of the 2nd through 10th of these two-interval comparisons. The algorithm sought the lowest sound level of FS which *reduced* the response to the CF tone by 20 spikes s^{-1}.

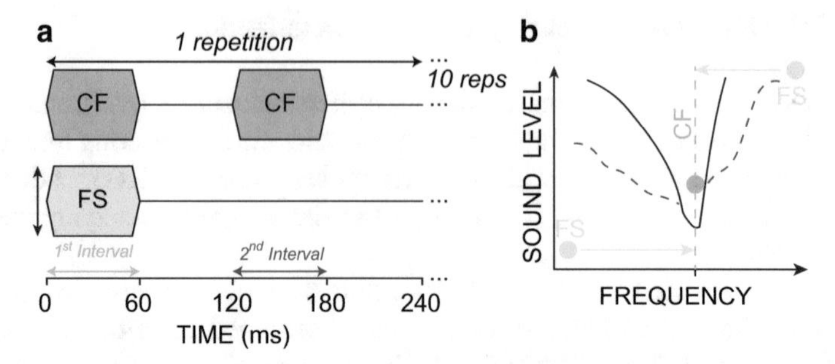

Fig. 1 Schematized stimulus paradigm: adaptive tracking. (B) *Black line*, excitatory tuning curve; *Red* and *blue dashed lines*, high- and low-side suppression-threshold curves, respectively

2.2.2 Systems-Identification Approach

Using spike-triggered characterization, we probed suppressive influences on firing rate in response to broadband-noise stimulation (e.g., Lewis and van Dijk 2004; Schwartz et al. 2006). Our implementation is based on singular-value decomposition of the second-order Wiener kernel (h_2) in response to 16.5-kHz bandwidth, 15-dB SL noise (Lewis et al. 2002a, 2002b; Recio-Spinoso et al. 2005). We collected ~10–20 K spike times per fiber.

2.3 Analyses

2.3.1 Second-Order Wiener Kernels

For each fiber, h_2 was computed from the second-order cross-correlation between the noise-stimulus waveform $x(t)$ and N spike times. The spike-triggered cross correlation was sampled at 50 kHz, and with maximum time lag τ of 10.2 ms ($m=512$ points) for CFs > 3 kHz, or 20.4 ms ($m=1024$ points) for CFs < 3 kHz. $h_2(\tau_1, \tau_2)$ is calculated as

$$\frac{N_0}{A^2}\left[R_2(\tau_1, \tau_2) - \phi_{xx}(\tau_2 - \tau_1)\right]$$

where τ_1 and τ_2 are time lags, A the instantaneous noise power, N_0 the mean firing rate in spikes s^{-1}, and $R_2(\tau_1, \tau_2)$ the second-order reverse-correlation function calculated as

$$\frac{1}{N}\sum_{i=1}^{N} x(t_i - \tau_1)x(t_i - \tau_2)$$

with t_i the ith spike time and $\phi_{xx}(\tau_2 - \tau_1)$ the stimulus autocorrelation matrix. So computed, h_2 is an m-by-m matrix with units of spikes·s^{-1}·Pa^{-2} (Recio-Spinoso et al. 2005). We used singular-value decomposition to parse h_2 into excitatory ($h_{2\varepsilon}$) and suppressive ($h_{2\sigma}$) sub-kernels (Lewis et al. 2002a, 2002b; Lewis and van Dijk 2004; Rust et al. 2005; Sneary and Lewis 2007).

2.3.2 Excitatory and Suppressive Sub-Kernels

Using the MATLAB function *svd*, h_2s were decomposed as

$$h_2 = \mathrm{USV}$$

where U, S and V are m-by-m matrices. The columns of U and rows of V are the *left* and *right singular vectors*, respectively. S is a diagonal matrix, the nonzero values of which are the weights of the corresponding-rank vectors. The decomposition can be rephrased as

$$h_2 = \sum_{j=1}^{m} k_j u_j v_j$$

where u_j and v_j are column vector elements of U and V, respectively, and k_j is the signed weight calculated as

$$k_j = sgn\left[u_j(j)v_j(j)\right]s_j$$

where *sgn* is the signum function, $u_j(j)$ is the jth element of the jth left singular vector, $v_j(j)$ is the jth element of the jth right singular vector, and s_j is the jth element of the nonzero diagonal of S.

Positively and negatively weighted vectors are interpreted as excitatory and suppressive influences, respectively (Rust et al. 2005; Schwartz et al. 2006; Sneary and Lewis 2007). However, mechanical suppression, adaptation, and refractory effects may all contribute to what we term "suppressive" influences.

To determine statistical significance of each weighted vector, we re-computed h_2 from 20 different spike:train randomizations, conserving the first-order inter-spike-interval distribution. Based on this bootstrap distribution, weights were expressed as z-scores, and any vector with $|z| > 3$ and rank≤ 20 was considered significant. We calculated a normalized excitatory-suppressive ratio as

$$R_{\varepsilon,\sigma} = \frac{\sum_{j=1}^{N_\varepsilon}\left|Z_{j\varepsilon}\right| - \sum_{j=1}^{N_\sigma}\left|Z_{j\sigma}\right|}{\sum_{j=1}^{N_\varepsilon}\left|Z_{j\varepsilon}\right| + \sum_{j=1}^{N_\sigma}\left|Z_{j\sigma}\right|}$$

for N_ε significant excitatory vectors and N_σ significant suppressive vectors. This normalized ratio varies from 1 (only excitation, no suppression), through 0 (equal excitation and suppression), to −1 (only suppression and no excitation: in practice $R_{\varepsilon,\sigma} < 0$ does not occur).

2.3.3 Spectro-Temporal Receptive Fields

Based on Lewis and van Dijk (2004), and Sneary and Lewis (2007), we estimated the spectro-temporal receptive field (STRF) from h_2. These STRFs indicate the timing of spectral components of the broadband-noise stimulus driving either increases or decreases in spike rate. Moreover, the STRF calculated from the whole h_2 kernel is the sum of excitatory and suppressive influences. Therefore, we also determined STRFs separately from $h_{2\varepsilon}$ and $h_{2\sigma}$ to assess the tuning of excitation and suppression (Sneary and Lewis 2007).

3 Results

3.1 Hearing-Loss Characterization

Noise exposure elevated ABR threshold, and reduced DPOAE magnitude, across the audiogram; indicating a mixed inner- and outer-hair-cell pathology (Fig. 2).

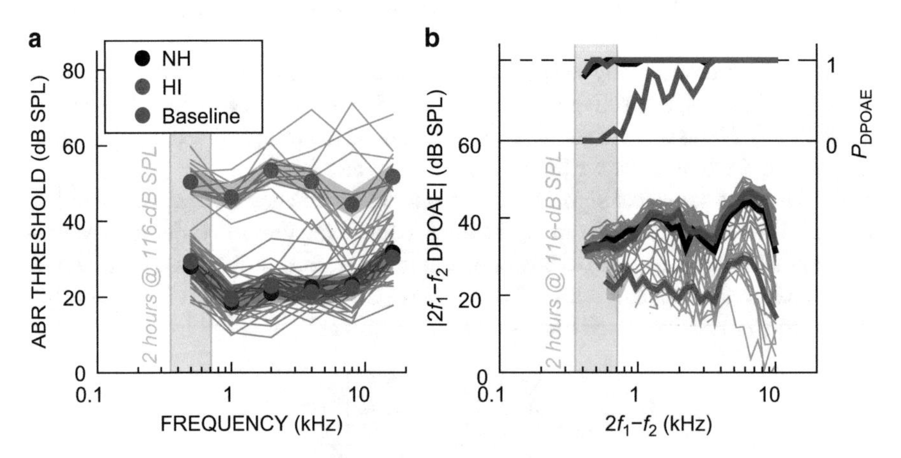

Fig. 2 Audiometric characterization. **a** ABR thresholds. *Thin lines*, individual-animal data; symbols, within-group least-squares-mean values; shading, S.E.M; *green* area, noise-exposure band. **b** *Lower plot*, DPOAE magnitude; *Upper plot*, probability of observing a DPOAE above the noise floor; thick lines, within-group least-squares means; shading, S.E.M

3.2 Suppression Threshold

Two-tone-suppression tuning curves were obtained from 62 NH fibers, and 35 HI fibers (Fig. 3a and b). HI excitatory tuning curves show threshold elevation and broadened tuning (Fig. 3c). For NH fibers, high-side 2TS was always observed (Fig. 3a). However, in 6 of 35 HI fibers, we could not detect significant high-side 2TS (Fig. 3b). These fibers have very broadened excitatory tuning (Fig. 3c). In HI fibers with detectable high-side 2TS, the dB difference between excitatory threshold and suppressive threshold was not greater than observed in NH fibers (Fig. 3d). For many fibers high-side-2TS threshold was within 20 dB of on-CF excitatory threshold. Low-side 2TS-threshold estimates are surprisingly low (Fig. 3a and b). All fibers had low-side suppressive regions, regardless of hearing status, typically in the region of 0- to 20-dB SPL.

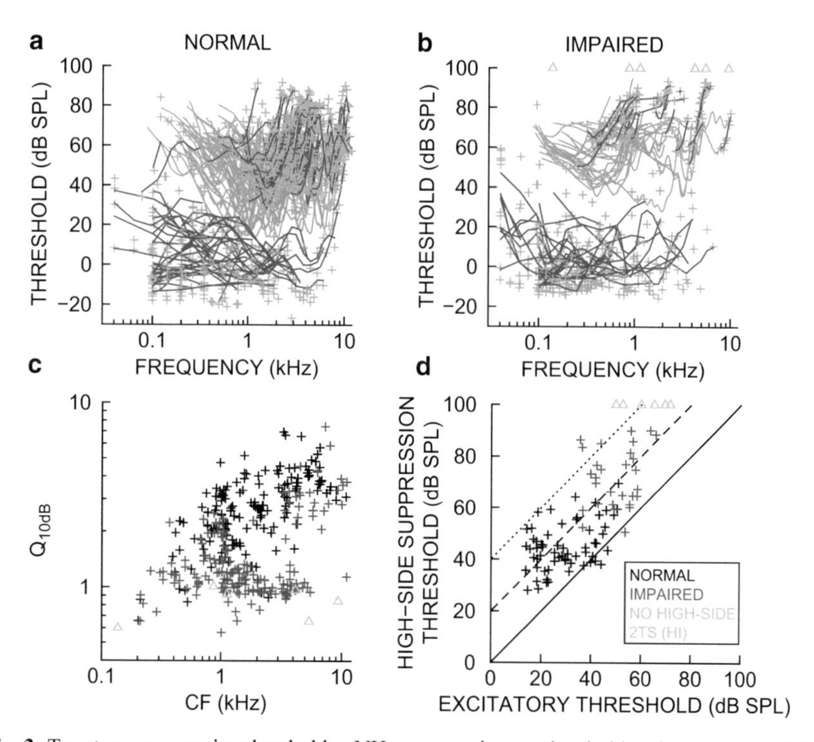

Fig. 3 Two-tone-suppression threshold. **a** NH: *gray*, excitatory-threshold tuning curves; *red lines* (fits) and *crosses* (data), high-side suppression-threshold tuning curves; *blue lines* and *crosses*, low-side suppression-threshold. **b**, HI data. **c** Q_{10dB}. **d** Solid line, equality; *dashed* and *dotted* lines, 20- and 40-dB shifts in suppressive threshold, respectively. **b–d**, *Green triangles* at 100-dB SPL, CFs of HI fibers with no high-sided suppression

3.3 Wiener-Kernel Estimates of Suppression

Figure 4 shows second-order Wiener-kernel and STRF analyses from the responses of a single medium-spontaneous-rate (<18 spikes s^{-1}) ANF, with CF 4.2 kHz, recorded in a NH chinchilla. The h_2 kernel is characterized by a series of parallel diagonal lines representing the non-linear interactions driving the ANF's response to noise (Fig. 4a; Recio-Spinoso et al. 2005). The STRF derived from h_2 shows sup-

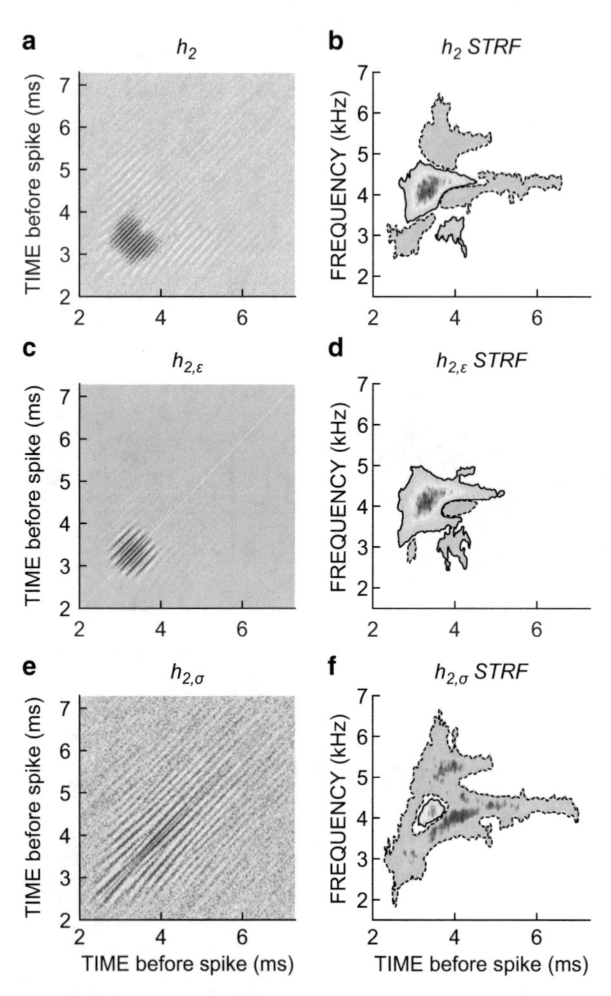

Fig. 4 Second-order Wiener-kernel, excitatory and suppressive sub-kernels, and their spectro-temporal receptive-field equivalents calculated from the responses of a single medium-spontaneous-rate ANF in a NH chinchilla: CF=4.2 kHz, θ=20-dB SPL, Q_{10}=3.6, SR=6.0 spikes s^{-1}. 21,161 driven spike times are included in the analysis. *Right-hand column*, warm colors and solid lines indicate areas of excitation, cool colors and dashed lines areas of suppression, with z-score >3 w.r.t. the bootstrap distribution

pressive regions flanking the main excitatory region along the frequency axis, with a high-to-low frequency glide (early-to-late) consistent with travelling-wave delay (Fig. 4b). Decomposing h_2 into its excitatory and suppressive sub-kernels, we observe a broad area of suppressive tuning which overlaps with the excitatory tuning in time and frequency (Fig. 4d and f). There is also an "on-CF" component to the suppression, occurring before the excitation.

There is substantial variability in the relative contribution of suppression to ANF responses (Fig. 5a), similar to that observed with tone stimuli in the "fractional-response" metric (e.g., Miller et al. 1997). There is a group of fibers (both NH and HI) across the CF axis, which show no significant suppression ($R_{\varepsilon,\sigma} = 1$; Fig. 5a). These are mainly, but not exclusively, of the high-spontaneous-rate class (≥ 18 spikes s^{-1}). HI fibers in the region of greatest damage (~2–5 kHz) tend to have reduced suppression. Plotting $R_{\varepsilon,\sigma}$ *vs*. CF-normalized 10-dB bandwidth, we find a significant correlation between broadened tuning and loss of suppression (Fig. 5b). Considering only CFs >2 kHz in this linear regression increased the variance explained from 9.1 to 18.4%.

4 Discussion

Using tones and broadband noise, we found significant changes in the pattern of suppression in the responses of ANFs following noise-induced hearing loss, likely reflecting outer-hair-cell disruption. Previous studies have examined the relationship between ANF 2TS and chronic low-level noise exposure coupled with ageing in the Mongolian gerbil (Schmiedt et al. 1990; Schmiedt and Schultz 1992), and following an acute intense noise exposure in the cat (Miller et al. 1997). Ageing in a noisy environment was associated with a loss of 2TS. Schmiedt and colleagues often found complete absence of high-side 2TS, with sparing of low-side 2TS, even in cochlear regions with up to 60% outer-hair-cell loss; suggesting potentially different mechanisms for low- and high-side suppression. Our 2TS data are in broad agreement with these earlier findings: HI chinchillas had elevated high-side 2TS thresholds, but retained sensitivity to low-side suppressors. Miller et al. (1997) related a reduction in 2TS in ANFs to changes in the representation of voiced vowel sounds. Weakened compressive nonlinearities contributed to a reduction in "synchrony capture" by stimulus harmonics near vowel-formant peaks. These effects likely contribute to deficits in across-CF spatio-temporal coding of temporal-fine-structure information in speech for HI listeners (e.g., Heinz et al. 2010).

Although the 2TS approach provides important insights on cochlear nonlinearities, it is important to consider the effects of hearing impairment on cochlear nonlinearity in relation to broadband sounds. Our Wiener-kernel approach demonstrated reduced suppression in ANF responses from HI animals, which was correlated with a loss of frequency selectivity. These HI animals also exhibited reduced magnitude DPOAEs: additional evidence of reduced compressive nonlinearity. However, the analyses presented here (Fig. 5) only quantify the overall relative suppressive influ-

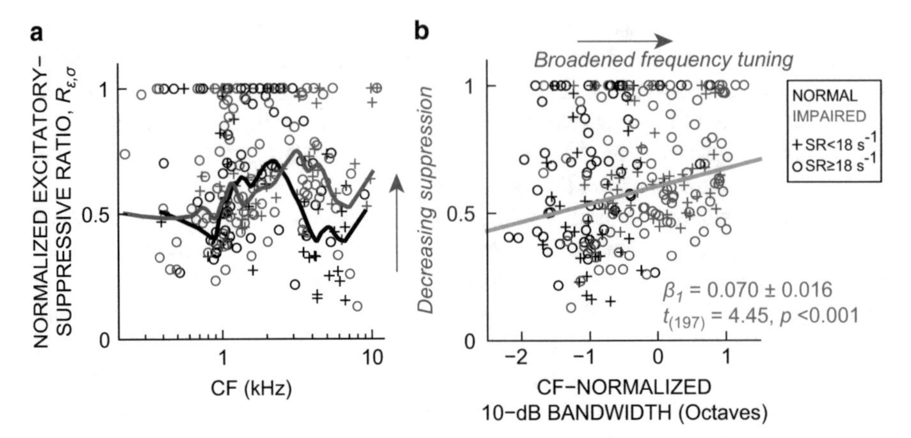

Fig. 5 Normalized excitatory-suppressive ratio *vs*. CF (**a**), and *vs*. CF-normalized 10-dB bandwidth (**b**). **a**, Solid lines, *lowess* fits to values of $R_{\varepsilon,\sigma} < 1$. **b**, Horizontal axis expresses 10-dB bandwidth in octaves relative to the 95th percentile of NH-chinchilla ANF data (Kale and Heinz 2010). Gray line and text, least-squares linear fit to values of $R_{\varepsilon,\sigma} < 1$

ence on ANF responses. Using the STRFs derived from h_2s, we aim to characterize the timing and frequency tuning of suppression in the HI auditory periphery in response to broadband sounds. Moreover, these same techniques can be exploited to address potential changes in the balance between excitation and inhibition ("central gain change") in brainstem nuclei following damage to the periphery. Such approaches have previously proved informative across sensory modalities (e.g., Rust et al. 2005), and will likely yield results with tangible translational value.

Acknowledgements Funded by NIH R01-DC009838 (M.G.H.) and Action on Hearing Loss (UK-US Fulbright Commission scholarship; M.S.). We thank Sushrut Kale, Kenneth S. Henry, and Jon Boley for re-use of ANF noise-response data, and Bertrand Fontaine and Kenneth S. Henry for helpful discussions on spike-triggered neural characterization.

References

Cooper NP (1996) Two-tone suppression in cochlear mechanics. J Acoust Soc Am 99(5):3087–3098
de Boer E, Nuttall AL (2002) The mechanical waveform of the basilar membrane. IV. Tone and noise stimuli. J Acoust Soc Am 111(2):979–989
Delgutte B (1990) Two-tone rate suppression in auditory-nerve fibers: dependence on suppressor frequency and level. Hear Res 49(1–3):225–246
Geisler CD, Yates GK, Patuzzi RB, Johnstone BM (1990) Saturation of outer hair cell receptor currents causes two-tone suppression. Hear Res 44(2–3):241–256
Heinz MG, Swaminathan J, Boley J, Kale S (2010) Across-fiber coding of temporal fine-structure: effects of noise-induced hearing loss on auditory nerve responses. In: Lopez-Poveda EA, Palmer AR, Meddis R (eds) The neurophysiological bases of auditory perception. Springer, New York, pp 621–630
Henry KS, Kale S, Scheidt RE, Heinz MG (2011) Auditory brainstem responses predict auditory nerve fiber thresholds and frequency selectivity in hearing impaired chinchillas. Hear Res 280(1–2):236–244
Hicks ML, Bacon SP (1999) Effects of aspirin on psychophysical measures of frequency selectivity, two-tone suppression, and growth of masking. J Acoust Soc Am 106(3):1436–1451
Kale S, Heinz MG (2010) Envelope coding in auditory nerve fibers following noise-induced hearing loss. J Assoc Res Otolaryngol 11(4):657–673. doi:10.1007/s10162-010-0223-6
Lewis ER, van Dijk P (2004) New variation on the derivation of spectro-temporal receptive fields for primary auditory afferent axons. Hear Res 189(1–2):120–136
Lewis ER, Henry KR, Yamada WM (2002a) Tuning and timing in the gerbil ear: wiener-kernel analysis. Hear Res 174(1–2):206–221
Lewis ER, Henry KR, Yamada WM (2002b) Tuning and timing of excitation and inhibition in primary auditory nerve fibers. Hear Res 171(1–2):13–31
Liberman MC (1984) Single-neuron labeling and chronic cochlear pathology. I. Threshold shift and characteristic-frequency shift. Hear Res 16(1):33–41
Miller RL, Schilling JR, Franck KR, Young ED (1997) Effects of acoustic trauma on the representation of the vowel "eh" in cat auditory nerve fibers. J Acoust Soc Am 101(6):3602–3616
Recio-Spinoso A, Temchin AN, van Dijk P, Fan YH, Ruggero MA (2005) Wiener-kernel analysis of responses to noise of chinchilla auditory-nerve fibers. J Neurophysiol 93(6):3615–3634
Rust NC, Schwartz O, Movshon JA, Simoncelli EP (2005) Spatiotemporal elements of macaque V1 receptive fields. Neuron 46(6):945–956
Sachs MB, Kiang NY (1968) Two-tone inhibition in auditory-nerve fibers. J Acoust Soc Am 43(5):1120–1128
Sachs MB, Young ED (1980) Effects of nonlinearities on speech encoding in the auditory nerve. J Acoust Soc Am 68(3):858–875
Schmiedt RA, Schultz BA (1992) Physiologic and histopathologic changes in quiet- and noise-aged gerbil cochleas. In: Dancer AL, Henderson D, Salvi RJ, Hamernik RP (eds) Noise-induced hearing loss. Mosby, St. Louis, pp 246–256
Schmiedt RA, Mills JH, Adams JC (1990) Tuning and suppression in auditory nerve fibers of aged gerbils raised in quiet or noise. Hear Res 45(3):221–236
Schwartz O, Pillow JW, Rust NC, Simoncelli EP (2006) Spike-triggered neural characterization. J Vis 6(4):484–507
Sneary MG, Lewis ER (2007) Tuning properties of turtle auditory nerve fibers: evidence for suppression and adaptation. Hear Res 228(1–2):22–30
Versteegh CP, van der Heijden M (2012) Basilar membrane responses to tones and tone complexes: nonlinear effects of stimulus intensity. J Assoc Res Otolaryngol 13(6):785–798. doi:10.1007/s10162-012-0345-0
Versteegh CP, van der Heijden M (2013) The spatial buildup of compression and suppression in the mammalian cochlea. J Assoc Res Otolaryngol 14(4):523–545. doi:10.1007/s10162-013-0393-0

Does Signal Degradation Affect Top–Down Processing of Speech?

Anita Wagner, Carina Pals, Charlotte M. de Blecourt, Anastasios Sarampalis and Deniz Başkent

Abstract Speech perception is formed based on both the acoustic signal and listeners' knowledge of the world and semantic context. Access to semantic information can facilitate interpretation of degraded speech, such as speech in background noise or the speech signal transmitted via cochlear implants (CIs). This paper focuses on the latter, and investigates the time course of understanding words, and how sentential context reduces listeners' dependency on the acoustic signal for natural and degraded speech via an acoustic CI simulation.

In an eye-tracking experiment we combined recordings of listeners' gaze fixations with pupillometry, to capture effects of semantic information on both the time course and effort of speech processing. Normal-hearing listeners were presented with sentences with or without a semantically constraining verb (e.g., crawl) preceding the target (baby), and their ocular responses were recorded to four pictures, including the target, a phonological (bay) competitor and a semantic (worm) and an unrelated distractor.

A. Wagner (✉) · C. Pals · D. Başkent
Department of Otorhinolaryngology/Head and Neck Surgery, University of Groningen, University Medical Center Groningen, Groningen, The Netherlands
e-mail: a.wagner@umcg.nl

C. Pals
e-mail: c.pals@alumnus.rug.nl

D. Başkent
e-mail: d.baskent@umcg.nl

A. Wagner · C.M. de Blecourt · A. Sarampalis · D. Başkent
Graduate School of Medical Sciences (Research School of Behavioural and Cognitive Neurosciences), University of Groningen, Groningen, The Netherlands

C.M. de Blecourt
e-mail: cdeblecourt@hotmail.com

A. Sarampalis
Department of Psychology, University of Groningen, Groningen, The Netherlands
e-mail: a.sarampalis@rug.nl

P. van Dijk et al. (eds.), *Physiology, Psychoacoustics and Cognition in Normal and Impaired Hearing,* Advances in Experimental Medicine and Biology 894,
DOI 10.1007/978-3-319-25474-6_31

The results show that in natural speech, listeners' gazes reflect their uptake of acoustic information, and integration of preceding semantic context. Degradation of the signal leads to a later disambiguation of phonologically similar words, and to a delay in integration of semantic information. Complementary to this, the pupil dilation data show that early semantic integration reduces the effort in disambiguating phonologically similar words. Processing degraded speech comes with increased effort due to the impoverished nature of the signal. Delayed integration of semantic information further constrains listeners' ability to compensate for inaudible signals.

Keywords Speech perception · Degraded speech · Cochlear implants

1 Introduction

Processing of speech, especially in one's native language, is supported by world knowledge, the contextual frame of the conversation, and the semantic content. As a consequence, listeners can understand speech even under adverse conditions, where it is partially masked or degraded. Access to these signal-independent sources of information can, however, be compromised if the entire speech signal is degraded, rather than parts of it. This is the case for profoundly hearing impaired listeners who rely on the signal transmitted via a cochlear implant (CI) for verbal communication. Though CIs allow listeners to perceive speech, this remains an effortful task for them.

In optimal conditions, effortless processing of speech depends on the integration of analyses along a hierarchy of processing stages, as they are described in models of speech perception. These models differ in the way they view the spread of information across various analysis stages (e.g. TRACE: McClelland and Elman 1986; Shortlist: Norris 1994; Shorlist B: Norris and McQueen 2008), but they do agree on the presence of lexical competition. Lexical competition is the process through which listeners consider all the mental representations that overlap with the heard signal as candidates for the word intended by the speaker. Before making a lexical decision listeners thus subconsciously consider multiple words, including homonyms (e.g., 'pair' and 'pear') and lexical embeddings (e.g., *paint* in *paint*ing). In optimal conditions, lexical competition is resolved (i.e. phonologically similar words are disambiguated) very early in the course of speech perception because listeners can rely on a plethora of acoustic cues that mark the difference between phonologically overlapping words (e.g., Salverda et al. 2003), and further also benefit from semantic information in sentences (Dahan and Tanenhaus 2004).

These models are based on data on natural speech perception in optimal conditions, so the question of how analysis of speech is affected by constant degradation of the signal remains unanswered. The present study investigates the time course of lexical competition and semantic integration when processing degraded speech. Furthermore this study will also query whether semantic integration can reduce the mental effort involved in lexical competition in natural and degraded speech. This question has not been studied before since understanding speech in optimal conditions is commonly perceived as effortless. To address these questions we will adapt the approach of Dahan and Tanenhaus (2004), and perform an eye tracking experiment in which listeners are presented with natural and degraded speech. We will further combine the recordings of gaze fixations with pupillometry to obtain a measure of processing effort.

Eye-tracking has been used to study the time course of lexical competition (e.g., Allopenna et al. 1998), since listeners' gazes to pictures on the screen reflect their lexical considerations during lexical access as they gradually match the heard signal to an object on the screen. To study the effort involved in processing speech we will record also listeners' change in pupil size. Pupil dilation is a measure that has been used to study effort involved in solving various cognitive tasks (e.g., Hoeks and Levelt 1993). An increase in pupil dilation has also been shown for listeners presented with degraded speech relative to highly intelligible speech (e.g., Zekveld et al. 2014). Pupil dilation reflects next to adaptations to changes in luminance or lightness, occurring within the timescale of 200–500 ms, also a slower evolving response to mental effort, in the timescale of about 900 ms (Hoeks and Levelt 1993).

2 Methods

2.1 Participants

Twenty-eight native speakers of Dutch, aged between 20 and 30 years (mean = 26), participated in this experiment. None of the participants reported any known hearing or learning difficulties. Their hearing thresholds were normal, i.e. below 20 dB HL on audiometric frequencies from 500 to 8000 kHz. All the participants signed a written consent form for this study as approved by the Medical Ethical Committee of the University Medical Center Groningen. The volunteers received either course credit or a small honorarium for their participation.

2.2 Stimuli

The set of stimuli consisted of the materials used by Dahan and Tanenhaus (2004), and an additional set constructed analogously, resulting in a total of 44 critical items. The critical items were quadruplets of nouns, which were presented together as pictures on the screen. To study the time course of lexical competition we created pairs of critical Dutch words with phonological overlap at the onset, e.g., the target 'pijp' [pipe] was combined with the phonological competitor 'pijl' [arrow]. To study whether disambiguating semantic context reduces lexical competition between acoustically similar words, the two phonologically similar items were presented within sentences, in which a verb that was coherent with only one of these two nouns (e.g. 'rookte' [smoked]) either preceded or followed the noun. The critical pair was presented as pictures together with two Dutch nouns, of which one was semantically viable to follow the verb (e.g., 'kachel' [heater]), the semantic distractor, and the other a phonologically and semantically unrelated distractor ('mossel' [mussel]).

Next to the critical items we constructed 60 sets of filler items. The verbs used in all of these filler sentences were coherent with two nouns, the target and the semantic distractor. The filler items were also presented in quadruplets, and the two remaining distractor nouns were not semantically coherent subjects for the verb. To create a balance between the critical and the filler items, in 20 of the filler items the distractor nouns were phonologically overlapping at the onset. The remaining 40 sets of distractors were phonologically unrelated.

All sentences began with a prepositional phrase, such as "Never before..." or "This morning..." The sentences were recorded from a male native speaker of Dutch. Black and white drawings were created as display pictures, specifically for the purpose of this study.

Two listening conditions were used in the experiment; natural speech (NS) and degraded speech (DS). The degraded stimuli were created using a noise-band-vocoder to simulate CI processing. The stimuli were first bandlimited to 80–6000 Hz, and were subsequently bandpass-filtered into 6 channels. Sixth order Butterworth filters were used, with a spacing equal to the distances in the cochlea as determined using the Greenwood function. The slow-varying amplitude envelopes were extracted from each channel via lowpass filtering, and these envelopes were then used to modulate carrier wideband noise, the resulting 6 channels were finally bandpass filtered once more using the same 6 bandpass filters. The processed stimuli were the summed signals from the output of all channels. This manipulation lead to stimuli with unnatural spectrotemporally degraded form, hence stimuli that simulate the signal conveyed via CIs.

2.3 Procedure

Before data collection, participants were familiarized with the pictures and the nouns that refer to the pictures. They were then seated in a comfortable chair facing the monitor, and an Eyelink 500 eye-tracker was mounted and calibrated. This head mounted eye-tracker contains two small cameras, which can be aligned with the participants' pupil to track the pupil's movements and size continuously during the experiment. Pupil size was recorded together with gaze fixations using a sampling rate of 250 Hz.

The stimuli were presented via a speaker in sound attenuated room. The lighting in this room was kept constant throughout the experiment to avoid effects of ambient light intensity on the pupil diameter. The participants' task was to listen to the stimuli and to click on the picture corresponding to the target noun in the sentence.

Each participant was presented with stimuli blocked into an NS and DS condition. Before the DS condition, the participants were familiarized with the degradation used in this study by listening to 30 degraded sentences and selecting the correct one from a set of sentences presented on the screen.

Each experimental item was presented only once in either the context or neutral sentence, and in either NS or DS. Between the two blocks (NS and DS) there was a break. Four practice trials preceded each block (using filler items), and a block consisted of 48 experimental items; 22 critical items and 26 filler items. The order of the presentation between blocks and items was quasi-random.

2.4 Analysis

Trials in which participants clicked on the wrong picture were excluded from the analysis. Trials with eye blinks longer than 300 ms were also excluded. Shorter blinks were corrected for by means of linear interpolation.

2.4.1 Gaze Fixations

To address the question of how semantic context affects lexical competition between phonologically similar words the statistical analyses focus on listeners' gaze fixations towards the phonological competitor and the semantic distractor. The probabilities of gaze fixations towards this competitor and this distractor were statically analyzed by means of growth curves (Mirman 2014). R (R Core team 2013) with lme4 package (Bates et al. 2014) was used to model the time curves of fixations as 4th order polynomials within the time window of 200–2000 ms after word onset. Two logistic-regression multi-level models were used, with fixations to either the phonological competitor or the semantic distractor, coded as a binomial response. The time course curves were described in four terms: intercept, the overall slope of the curve, the width of the rise and fall around the inflection, and the curvature in the tails. The probability of fixations along the time course was modeled as a function of Context (neutral versus context), Presentation (NS versus DS) and the possible three-way interactions between these two factors and all four terms describing the curves. As random effect, we included individual variation among participants and items on all four terms describing the time curve. Model comparison was used to estimate the contribution of individual predictors to the fit of the model. For this, individual fixed effects were sequentially added, and the change in the model fit was evaluated by means of likelihood ratio test.

2.4.2 Pupil Dilation

To investigate the effort involved in the process of lexical competition with and without semantic context, the pupil dilation data per participant were baseline-corrected to the 200 ms preceding the presentation of the experimental item. The baseline-corrected data were normalized to correct for individual differences in pupil size, according to the equation:

$$\%Event\,Related\,Pupil\,Dilation = (observation - baseline)\,/\,baseline * 100.$$

For the statistical analysis, the pupil size data, as captured by the event-related pupil dilation (ERPD), were analyzed analogously to the fixation data, as time curves of pupil dilation. The time-course functions were analyzed as 3rd -order polynomials, since, during fitting, the fourth order turned out to be redundant to the description of these curve functions. The terms describing the curves are: intercept, the slope of the curve, and a coefficient for the curvature around the inflection point. These time curves were analyzed by means of multi-level nonlinear regression model. The statistical models contained in addition to the terms describing the curves per participant also random effects on these three terms per participant, and for the phonological competitor model also random effects per item.

3 Results

3.1 Gaze Fixations

Figure 1 displays the time curves of fixations to all four pictures displayed within the NS blocks for C (a), and N (b), and for the DS blocks for C (c) and N (d). These figures show proportions of fixations to the four pictures displayed, averaged across participants, and the 95 % confidence intervals for the fixations to the target and competitor.

Of particular interest for this study are the three-way interactions between Context (C versus N) and Presentation (NS versus DS) and the terms describing the course of the curves. For the fixations to the phonological competitor, as significant emerged the three way interactions with the first term (the intercept) of the curve ($\chi 2(18)=28476, p<0.001$, the interaction with the quadratic term (the slope), ($\chi 2(18)=28184$, $p<0.001$), the interaction between the cubic term (rise and fall around the central inflection), ($\chi 2(18)=27632, p<0.001$), and the quartic term (curvature in the tails), ($\chi 2(18)=27651$, $p<0.05$). The interaction with the intercept shows that the context sentences reduced the area under the fixation curves to the competitor for NS (red lines in Fig. 1a versus b), and that this reduction was smaller for DS (red lines in Fig. 1c versus d). The interaction with the slope shows that the growth of fixations to the competitor is shallower for DS in the neutral context than it is for NS in neutral context. The interaction with the cubic term reflects that the location of the peak of fixations towards the competitor in DS is delayed for about 300 ms relative to the location of the peak for NS, and that the course of this curve is more symmetric than for NS, and this mainly for the items presented in neutral

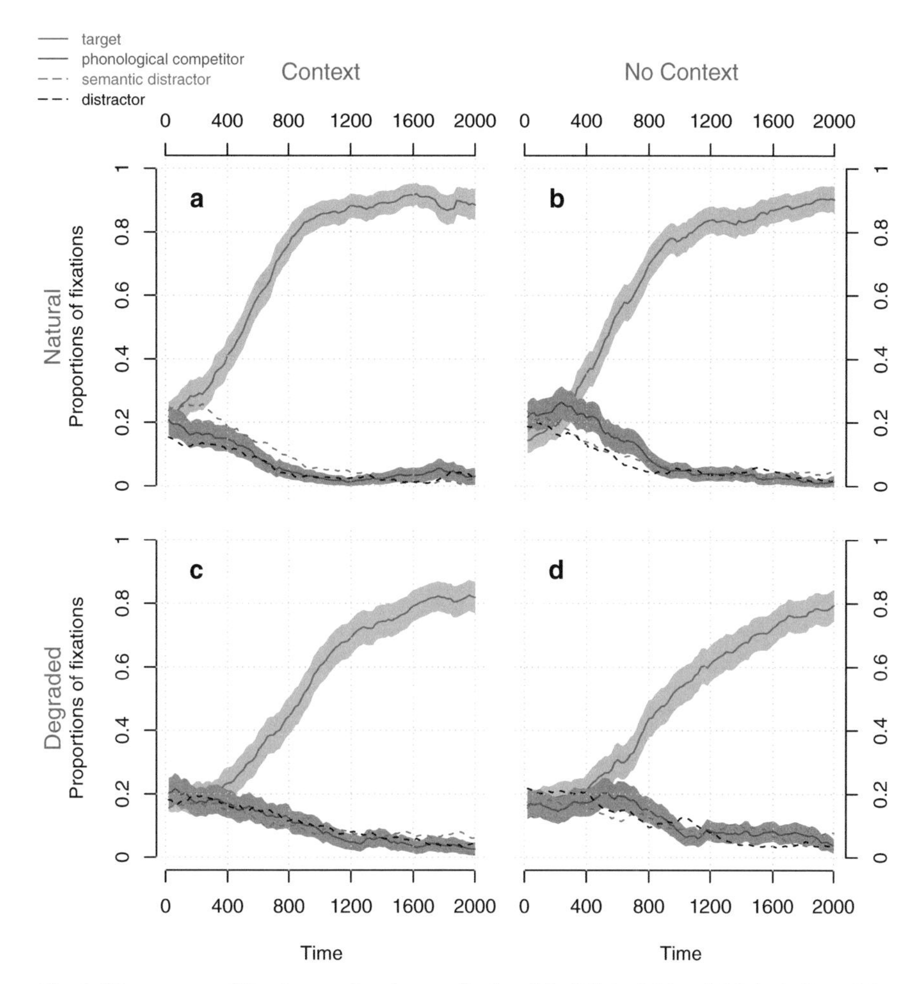

Fig. 1 Time curves of fixations to the pictures displayed for NS (**a** & **b**) and DS (**c** & **d**), and for items presented in context sentences (**a** & **c**), and neutral sentences (**b** & **d**)

context. The interaction with the quartic term reflects a slower decline of fixations towards the competitor in DS versus NS, and shallower for items in context than in neutral sentences.

For the fixations to the semantic distractor, as significant emerged the interactions between Context and Presentation and the intercept of the curve ($\chi2(3)=2268.6$ $p<0.001$), the interaction with the quadratic term ($\chi2(3)=337.25$, $p<0.001$), the interaction between the cubic term, ($\chi2(3)=69.41$, $p<0.001$), and the quartic term ($\chi2(3)=19.09$, $p<0.05$). These interactions reflect what can also be seen in a comparison of between NS and DS in Fig. 1. Namely that in NS, listeners fixate the semantic competitor more often in the context sentences than in neutral context. This effect is absent for DS.

3.2 Pupil Dilation

Figure 2 displays the time course of pupil dilation for NS and DS and for the two contexts C and N.

The curves for NS and DS in the neutral condition show a constant increase in pupil size over time as a function of lexical competition. The curves for the context condition show a decline in pupil size growth starting at around 800 ms after the onset of the target word. The statistical analysis revealed significant three way interactions with Context (N versus C) and Presentation (NS versus DS) on all terms describing the curves: Intercept ($\chi 2$ (3)=301.90, $p<0.001$), slope ($\chi 2$ (3)=145.3, $p<0.001$), and the cubic term, the curvature around the peak ($\chi 2$ (3)=272.52, $p<0.001$). This implies that pupil dilation was sensitive in capturing the reduced effect of lexical competition in the context sentences versus neutral context, but this effect was delayed and smaller in DS than in NS.

A look at this figure suggests that the effort involved in lexical competition for DS was overall smaller for DS than for NS. This overall smaller increase in pupil dilation can be explained by the fact that these curves are normalized to a baseline

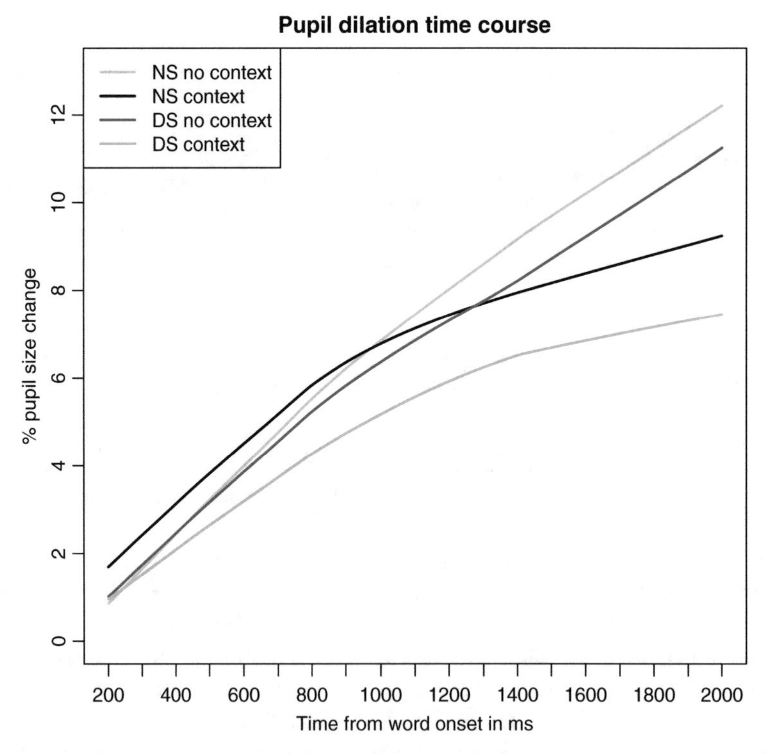

Fig. 2 Time curves of pupil dilation, averaged across participants for NS (*black* and *grey*) and DS (*red* and *green*), and for items presented in context sentences (*black* and *green*), and neutral sentences (*red* and *grey*)

of 200 ms preceding the presentation of each critical item per trial, participant and condition. Listening to degraded speech is by itself more effortful than listening to natural speech (e.g., Winn et al. 2015), and therefore there is a difference in the baseline between DS and NS. These differences in the baseline can be explained by the difference in processing degraded versus natural speech, and are independent of the effects of semantic integration on lexical competition.

4 Discussion

This present study examined the effect of semantic integration on the time course of lexical competition, and on the effort involved in solving lexical competition in natural and degraded speech. Our results show that processing natural speech comes with a timely integration of semantic information, which in turn reduces lexical competition. Listeners are then able to pre-select a displayed target based on its semantic coherence with the context, and this allows listeners to reduce the effort involved in lexical competition. When processing degraded speech the integration of semantic information is delayed, as is also lexical competition. This implies that semantic integration is not able to reduce lexical competition, which by itself is longer and occurs later. These results were also mirrored by the pupil dilation data, in which a release from lexical competition was visible but delayed. Mapping of degraded speech to mental representations is more effortful due the mismatch between the actual signal and its mental representation, and lexical context is not able to release listeners from this effort on time. In natural situations, in which words are being heard in succession, and the speech signal evolves quickly over time, such a difference in processing speed of degraded speech will accumulate effort, and draw more strongly on resources in working memory.

Acknowledgments We would like to thank Dr. Paolo Toffanin for technical support, and Prof. Frans Cornelissen (University Medical Center Groningen) for providing the eye-tracker for this study. This work was supported by a Marie Curie Intra-European Fellowship (FP7-PEOPLE-2012-IEF 332402). Support for the second author came from a VIDI Grant from the Netherlands Organization for Scientific Research (NWO), the Netherlands Organization for Health Research and Development (ZonMw) Grant No. 016.093.397. The study is part of the research program of our department: Healthy Aging and Communication.

References

Allopenna PD, Magnuson JS, Tanenhaus MK (1998) Tracking the time course of spoken word recognition using eye movements: evidence for continuous mapping models. J Memory Lang 38:419–439

Bates D, Maechler M, Bolker B, Walker S (2014). lme4: Linear mixed-effects models using Eigen and S4. R package version 1.1-7. http://CRAN.R-project.org/package=lme4

Dahan D, Tanenhaus MK (2004) Continuous mapping from sound to meaning in spoken-language comprehension: immediate effects of verb-based thematic constraints. J Exp Psychol Learn Mem Cogn 30:498–513

Hoeks B, Levelt W (1993) Pupillary dilation as a measure of attention: a quantitative system analysis. Behav Res Methods 25(1):16–26

McClelland JL, Elman JL (1986). The TRACE model of speech perception. Cognitive Psychology 18:1–86

Mirman D (2014) Growth curve analysis and visualization using R. Chapman and Hall/CRC, Florida

Norris D (1994) Shortlist: a connectionist model of continuous speech recognition. Cognition 52:189–234

Norris D, McQueen JM (2008) Shortlist B: a Bayesian model of continuous speech recognition. Psychol Rev 115(2):357–395

R Core Team (2013) R: a language and environment for statistical computing. R Foundation for Statistical Computing, Vienna. http://www.R-project.org/

Salverda AP, Dahan D, McQueen JM (2003) The role of prosodic boundaries in the resolution of lexical embedding in speech comprehension. Cognition 90:51–89

Winn MB, Edwards JR, Litovsky RY (2015). The impact of auditory spectral resolution on listening effort revealed by pupil dilation. Ear and Hear (ahead of print)

Zekveld AA, Heslenfeld DJ, Johnsrude IS, Versfeld N, Kramer SE (2014) The eye as a window to the listening brain: neural correlates of pupil size as a measure of cognitive listening load. Neuroimage 101:76–86

The Effect of Peripheral Compression on Syllable Perception Measured with a Hearing Impairment Simulator

Toshie Matsui, Toshio Irino, Misaki Nagae, Hideki Kawahara and Roy D. Patterson

Abstract Hearing impaired (HI) people often have difficulty understanding speech in multi-speaker or noisy environments. With HI listeners, however, it is often difficult to specify which stage, or stages, of auditory processing are responsible for the deficit. There might also be cognitive problems associated with age. In this paper, a HI simulator, based on the dynamic, compressive gammachirp (dcGC) filterbank, was used to measure the effect of a loss of compression on syllable recognition. The HI simulator can counteract the cochlear compression in normal hearing (NH) listeners and, thereby, isolate the deficit associated with a loss of compression in speech perception. Listeners were required to identify the second syllable in a three-syllable "nonsense word", and between trials, the relative level of the second syllable was varied, or the level of the entire sequence was varied. The difference between the Speech Reception Threshold (SRT) in these two conditions reveals the effect of compression on speech perception. The HI simulator adjusted a NH listener's compression to that of the "average 80-year old" with either normal compression or complete loss of compression. A reference condition was included where the HI simulator applied a simple 30-dB reduction in stimulus level. The

T. Matsui (✉) · T. Irino · M. Nagae · H. Kawahara
Faculty of Systems Engineering, Wakayama University, Sakaedani 930, Wakayama-city 640-8510, Japan
e-mail: tmatsui@sys.wakayama-u.ac.jp

T. Irino
e-mail: irino@sys.wakayama-u.ac.jp

M. Nagae
e-mail: s155029@center.wakayama-u.ac.jp

H. Kawahara
e-mail: kawahara@sys.wakayama-u.ac.jp

R. D. Patterson
Department of Physiology, Development and Neuroscience, Centre for the Neural Basis of Hearing, University of Cambridge, Downing Street, Cambridge CB2 3EG, UK
e-mail: rdp1@cam.ac.uk

P. van Dijk et al. (eds.), *Physiology, Psychoacoustics and Cognition in Normal and Impaired Hearing,* Advances in Experimental Medicine and Biology 894,
DOI 10.1007/978-3-319-25474-6_32

results show that the loss of compression has its largest effect on recognition when the second syllable is attenuated relative to the first and third syllables. This is probably because the internal level of the second syllable is attenuated proportionately more when there is a loss of compression.

Keywords Hearing impairment simulator · Cochlear compression · Dynamic compressive gammachirp filterbank · Audiogram

1 Introduction

Age related hearing loss (presbycusis) makes it difficult to understand speech in noisy environments and multi-speaker environments (Moore 1995; Humes and Dubno 2010). There are several factors involved in presbycusis: loss of frequency selectivity, recruitment, and loss of temporal fine structure (Moore 2007). The isolation and measurement of these different aspects of hearing impairment (HI) in elderly listeners is not straightforward, partly because they often have multiple auditory problems, and partly because they may also have more central cognitive problems including depression. Together the problems make it difficult to obtain sufficient behavioral data to isolate different aspects of HI (Jerger et al. 1989; Lopez et al. 2003). We have developed a computational model of HI and an HI simulator that allow us to perform behavioural experiments on cochlear compression—experiments in which young normal hearing (NH) listeners act as "patients" with isolated, sensory-neural hearing losses of varying degree.

The HI simulator is essentially a high-fidelity synthesizer based on the dynamic, compressive gammachirp filter bank (dcGC-FB) (Irino and Patterson 2006). It analyses naturally recorded speech sounds with a model of the auditory periphery that includes fast-acting compression and then resynthesizes the sounds in a form that counteracts the compression in a NH listener. They hear speech with little distortion or background noise (Irino et al. 2013) but without their normal compression. This makes it possible to investigate recruitment phenomenon that are closely related to loss of cochlear compression (Bacon et al. 2004) using NH listeners who have no other auditory problems. Specifically, we describe an experiment designed to examine how the loss of cochlear compression affects the recognition of relatively soft syllables occurring in the presence of louder, flanking syllables of varying levels. The aim is to reveal the role of cochlear compression in multi-speaker environments or those with disruptive background noise.

2 Method

Listeners were presented three-syllable "nonsense words" and required to identify the second syllable. Between trails, the level of the second syllable within the word was varied ("dip" condition) or the level of the entire word was varied ("constant" condition). The difference between second-syllable recognition in the two conditions reveals the effect of compression. The HI simulator is used to adjust syllable level to simulate the hearing of 80 year old listeners with “normal” presbycusis (80 year 100 %) or a complete loss of compression (80 year 0 %).

2.1 Participants

Ten listeners participated in the experiment (four males; average age: 23.6 years; SD: 4.8 years). None of the listeners reported any history of hearing impairment. The experiment was approved by the ethics committee of Wakayama University and all listeners provided informed consent. Participants were paid for their participation, except for the first and third authors who also participated in the experiment.

2.2 Stimuli and Procedure

The stimuli were generated in two stages: First, three-syllable nonsense words were composed from the recordings of a male speaker, identified as “MIS” (FW03) in a speech sound database (Amano et al. 2006), and adjusted in level to produce the dip and constant conditions of the experiment. Second, the HI simulator was used to modify the stimuli to simulate the hearing of two 80 year old listeners, one with average hearing for an 80 year old and one with a complete loss of compression. There was also a normal-hearing control condition in which the level was reduced a fixed 30 dB.

In the first stage, nonsense words were composed by choosing the second syllable at random from the 50 Japanese syllables presented without parentheses in Table 1. This is the list of syllables (57-S) recommended by the Japanese Audiological Society (2003) for use in clinical studies to facilitate comparison of behavioural data across studies. The first and third syllables in the words were then selected at random from the full set of 62 syllables in Table 1. The set was enlarged to reduce the chance of listeners recognizing the restriction on the set used for the target

Table 1 The 62 Japanese syllables used in the experiment using the international phonetic alphabet. The syllables in parentheses are also Japanese syllables but not included in list 57-S. All 62 were used in the draw for the first and third syllables. The draw for the second syllable was restricted to the 50 syllables not enclosed in parentheses

a	ka	sa	ta	na	ha	ma	ja	ɾa	wa	ga	(dza)	da	ba
i	kʲi	ʃi	tʃi	ɲi	çi	mi		ɾʲi		(gʲi)	ʤi		(bʲi)
ɯ	kɯ	sɯ	tsɯ	(nɯ)	ɸu	mɯ	jɯ	ɾɯ		(gɯ)	dzɯ		(bɯ)
e	ke	se	te	ne	he	me		(ɾe)		(ge)	(dze)	de	(be)
o	ko	so	to	no	ho	mo	jo	ɾo		go	(dzo)	do	(bo)

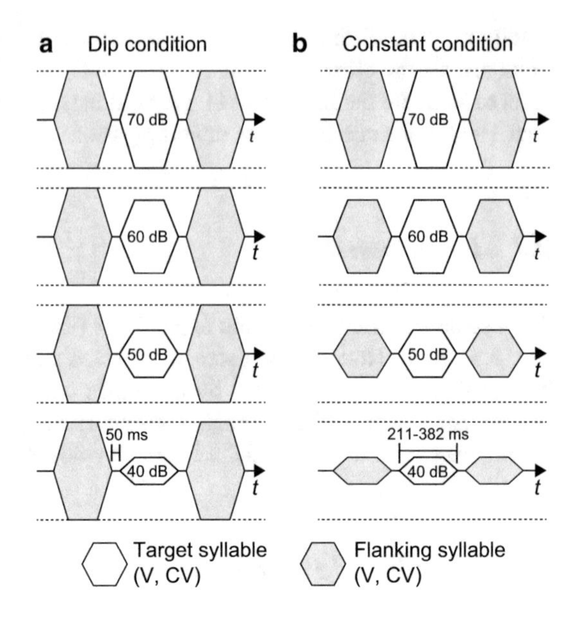

Fig. 1 Schematic of the experimental paradigm. The levels of the nonsense words were adjusted to produce stimuli for **a** the Dip condition and **b** the Constant condition. The 70-dB stimuli were the same in the two conditions

syllable. Syllable duration ranged from 211 to 382 ms. There was a 50-ms silence between adjacent syllables within a word.

The nonsense words were then adjusted to produce the two level conditions illustrated in Fig. 1: In the "dip" condition (left column), the level of the second syllable was 70, 60, 50 or 40 dB, while the first and third syllables were fixed at 70 dB. In the "constant" condition (right column), all three syllables were set to 70, 60, 50 or 40 dB.

In the second stage, the stimuli were processed to simulate:

1. the average hearing level of an 80 year old person (audiogram ISO7029, ISO/TC43 2000) simulated with 100 % compression (80 year 100 %),
2. the average hearing level of an 80-year-old simulated with 0 % compression (80 year 0 %),
3. normal adult hearing with a 30-dB reduction across the entire frequency range (NH −30 dB).

To realize the 80 year 100 % condition, the output level of each channel was decreased until the audiogram of the dcGC-FB (normal hearing) matched the average audiogram of an 80 year old. To realize the 80 year 0 % condition, an audiogram was derived with the dcGC-FB having no compression whatsoever, and then the output level of each channel was decreased until the audiogram matched the audiogram of an average 80 year old. In the compression-cancellation process, loss of hearing level was limited so that the audiogram did not fall below the hearing level of an average 80 year old. The compression was applied using the signal processing system described in Nagae et al. (2014).

Each participant performed all conditions—a total of 1050 trials of nonsense words: 50 words × 7 dip/constant conditions × 3 hearing impairment simulations. The participants were required to identify the second syllable using a GUI on a computer screen. The stimuli were manipulated, and the results collected using MATLAB. The sounds were presented diotically via a DA converter (Fostex, HP-A8) over headphones (Sennheiser, HD-580). The experiment was carried out in a sound-attenuated room with a background level of about 26 dB L_{AEq}.

3 Results

For each participant, the four percent-correct values for each level condition (dip or constant) were fitted with a cumulative Gaussian psychometric function to provide a Speech Reception Threshold (SRT) value for that condition, separately in each of the three simulated hearing-impairment conditions. The cumulative Gaussian was fitted by the bootstrap method (Wichmann and Hill 2011a; Wichmann and Hill 2011b) and SRT was taken to be the sound pressure level associated with 50 % correct recognition of the second syllable. The SRT in the constant condition with the NH − 30 dB hearing impairment was taken as standard performance on the task. The difference between this standard SRT and the SRTs of the other conditions are the experimental results (ΔSRTs) for each participant. Figure 2a shows the mean and standard deviation over participants of ΔSRT for each condition.

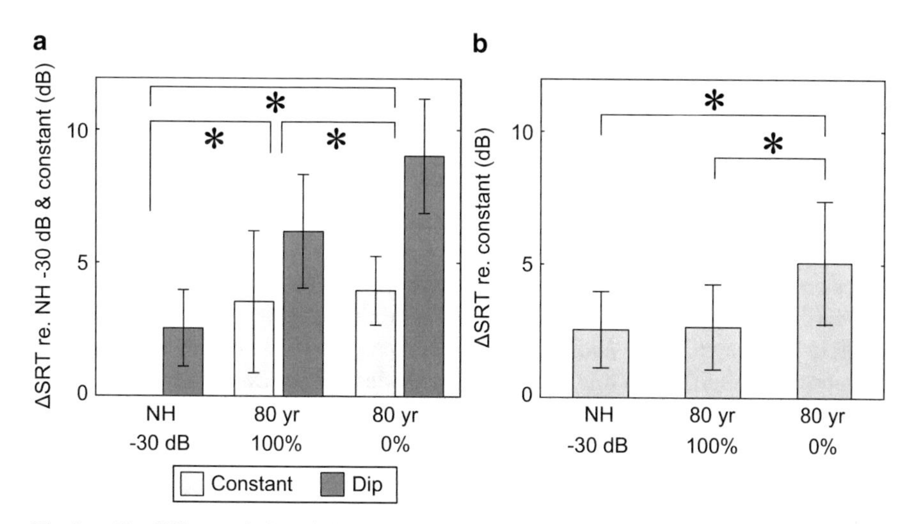

Fig. 2 **a** The difference in SRT (ΔSRT) between the constant condition of the NH − 30 dB simulation and the SRTs of the other conditions. **b** The SRT difference between the dip and constant conditions for each hearing impairment condition. Asterisks (*) show significant differences ($\alpha = 0.05$) in a multiple comparison, Tukey-Kramer HSD test. Error bars show the standard deviation

A two-way ANOVA was performed for the dip/constant conditions and the hearing impairment conditions combined, and it showed that the manipulations had a significant effect on recognition performance (ΔSRT) ($F(5, 54)=23.82$, $p<0.0001$). Significant, separate main effects were also observed for the dip/constant condition and the hearing impairment condition ($F(1, 54)=41.99$, $p<0.0001$; $F(2, 54)=36.11$, $p<0.0001$, respectively). The interaction between the two factors was not significant ($\alpha=0.05$). Multiple comparisons with the Tukey-Kramer HSD test were performed. There were significant differences between all of the hearing impairment conditions ($\alpha=0.05$).

ΔSRT provides a measure of the difficulty of the task, and the statistical analysis shows that the different hearing impairment simulations vary in their difficulty. In addition, ΔSRT for the dip conditions is larger than ΔSRT for constant conditions for all of the hearing impairment simulations. This difference in difficulty could be the result of forward masking since the gap between offset of the first syllable and the onset of the target syllable is less than 100 ms, and the third syllable limits any post processing of the second syllable. The dip condition would be expected to produce more masking than the constant condition, especially when the target syllable is sandwiched between syllables whose levels are 20 or 30 dB greater. The constant condition only reveals differences between the different hearing impairment simulations. This does, however, include differences in audibility.

Figure 2b shows the difference between SRT in the dip and constant conditions; that is, ΔSRT relative to the constant condition. It provides a measure of the amount of any forward masking. A one-way ANOVA was performed to confirm the effect of hearing impairment simulation ($F(2, 29)=6.14$, $p=0.0063$). A multiple comparisons test (Tukey-Kramer HSD) showed that ΔSRT in the 80 year 0 % condition was significantly greater than in the 80 year 100 % condition, or the NH −30 dB condition ($\alpha=0.05$).

4 Discussion

It is assumed that forward masking occurs during auditory neural processing, and that it operates on peripheral auditory output (i.e. neural firing in the auditory nerve). Since the compression component of the peripheral processing is normal in both the 80 year 100 % condition and the NH −30 dB condition, the level difference between syllable 2 and syllables 1 and 3 should be comparable in the peripheral output of any specific condition. In this case, ΔSRT provides a measure of the more central masking that occurs during neural processing. When the compression is cancelled, the level difference in the peripheral output for syllable 2 is greater in the dip conditions. If the central masking is constant, independent of the auditory representation, the reduction of the output level for syllable 2 should be reflected in the amount of masking, i.e. ΔSRT. This is like the description of the growth of masking by Moore (2012), where the growth of masking in a compressive system is observed to be slower than in a non-compressive system. Moore was describing experiments where both the signal and the masker were isolated sinusoids. The results of the current study show that the effect generalizes to speech perception where both the target and the interference are broadband, time-vary sounds.

There are, of course, other factors that might have affected recognition performance: The fact that the same speaker provided all of the experimental stimuli might have reduced the distinctiveness of the target syllable somewhat, especially since syllable 2 was presented at a lower stimulus level in the dip conditions. The fact that listeners had to type in their responses might also have decreased recognition performance a little compared to everyday listening. It is also the case that our randomly generated nonsense words might accidentally produce a syllable sequence that is similar to a real word with the result that the participant answers with reference to their mental dictionary and makes an error. But, all of these potential difficulties apply equally to all of the HI conditions, so it is unlikely that they would affect the results.

This suggests that the results truly reflect the role of the loss of compression when listening to speech in everyday environments for people with presbycusis or other forms of hearing impairment. It shows that this new, model-based approach to the simulation of hearing impairment can be used to isolate the compression component of auditory signal processing. It may also assist with isolation and analysis of other aspects of hearing impairment.

5 Summary

This paper reports an experiment performed to reveal the effects of a loss of compression on the recognition of syllables presented with fluctuating level. The procedure involves using a filterbank with dynamic compression to simulate hearing impairment in normal hearing listeners. The results indicate that, when the compression of normal hearing has been cancelled, it is difficult to hear syllables at low sound levels when they are sandwiched between syllables with higher levels. The hearing losses observed in patients will commonly involve multiple auditory problems and they will interact in complicated ways that make it difficult to distinguished the effects of any one component of auditory processing. The HI simulator shows how a single factor can be isolated using an auditory model, provided the model is sufficiently detailed.

Acknowledgments This work was supported in part by JSPS Grants-in-Aid for Scientific Research (Nos 24300073, 24343070, 25280063 and 25462652).

References

Amano S, Sakamoto S, Kondo T, Suzuki Y (2006) NTT-Tohoku University familiarity controlled word lists (FW03). Speech resources consortium. National Institute of Informatics, Japan

Bacon SP, Fay R, Popper AN (eds) (2004) Compression: from cochlea to cochlear implants. Springer, New York

Humes LE, Dubno JR (2010) Factors affecting speech understanding in older adults. In: Gordon-Salant S, Frisina RD, Popper AN, Fay RR (eds) The aging auditory system. Springer, New York, pp 211–257

Irino T, Patterson RD (2006) A dynamic compressive gammachirp auditory filterbank. IEEE Trans Audio Speech Lang Process 14(6):2222–2232

Irino T, Fukawatase T, Sakaguchi M, Nisimura R, Kawahara H, Patterson RD (2013) Accurate estimation of compression in simultaneous masking enables the simulation of hearing impairment for normal hearing listeners. In: Moore BCJ, Patterson RD, Winter IM, Carlyon RP, Gockel HE (eds) Basic aspects of hearing, physiology and perception. Springer, New York, pp 73–80

ISO/TC43 (2000) Acoustics—statistical distribution of hearing thresholds as a function of age. International Organization for Standardization, Geneva

Japan Audiological Society (2003) The 57-S syllable list. Audiol Jpn 46:622–637

Jerger J, Jerger S, Oliver T, Pirozzolo F (1989) Speech understanding in the elderly. Ear Hear 10(2):79–89

Lopez OL, Jagust WJ, DeKosky ST, Becker JT, Fitzpatrick A, Dulberg C, Breitner J, Lyketsos C, Jones B, Kawas C, Carlson M, Kuller LH (2003) Prevalence and classification of mild cognitive impairment in the cardiovascular health study cognition study: part 1. Arch Neurol 60(10):1385–1389

Moore BCJ (1995) Perceptual consequences of cochlear damage. Oxford University Press, New York

Moore BCJ (2007) Cochlear hearing loss: physiological, psychological and technical issues. Wiley, Chichester

Moore BCJ (2012) An introduction to the psychology of hearing, 6th edn. Emerald Group Publishing Limited, Bingley

Nagae M, Irino T, Nishimura R, Kawahara H, Patterson RD (2014). Hearing impairment simulator based on compressive gammachirp filter. Proceedings of APSIPA ASC

Wichmann FA, Hill NJ (2011a) The psychometric function: I. Fitting, sampling, and goodness of fit. Percept Psychophys 63(8):1293–1313

Wichmann FA, Hill NJ (2011b) The psychometric function: II. Bootstrap-based confidence intervals and sampling. Percept Psychophys 63(8):1314–1329

Towards Objective Measures of Functional Hearing Abilities

Hamish Innes-Brown, Renee Tsongas, Jeremy Marozeau and Colette McKay

Abstract *Aims* People with impaired hearing often have difficulties in hearing sounds in a noisy background. This problem is partially a result of the auditory systems reduced capacity to process temporal information in the sound signal. In this study we examined the relationships between perceptual sensitivity to temporal fine structure (TFS) cues, brainstem encoding of complex harmonic and amplitude modulated sounds, and the ability to understand speech in noise. Understanding these links will allow the development of an objective measure that could be used to detect changes in functional hearing before the onset of permanent threshold shifts.

Methods We measured TFS sensitivity and speech in noise performance (Quick-SIN) behaviourally in 34 normally hearing adults with ages ranging from 18 to 63 years. We recorded brainstem responses to complex harmonic sounds and a 4000 Hz carrier signal modulated at 110 Hz. We performed cross correlations between the stimulus waveforms and scalp-recorded brainstem responses to generate a simple measure of stimulus encoding accuracy, and correlated these measures with age, TFS sensitivity and speech-in-noise performance.

Results Speech-in-noise performance was positively correlated with TFS sensitivity, and negatively correlated with age. TFS sensitivity was also positively correlated with stimulus encoding accuracy for the complex harmonic stimulus, while

H. Innes-Brown (✉) · R. Tsongas · J. Marozeau · C. McKay
Bionics Institute, Melbourne, Australia
e-mail: hinnes-brown@bionicsinstitute.org

R. Tsongas
e-mail: r.tsongas@student.unimelb.edu.au

J. Marozeau
Department of Electrical Engineering, Technical University of Denmark, Kongens Lyngby, Denmark
e-mail: jemaroz@elektro.dtu.dk

C. McKay
e-mail: cmckay@bionicsinstitute.org

H. Innes-Brown
Research Group Experimental Oto-laryngology, KU Leuven, Leuven, Belgium

P. van Dijk et al. (eds.), *Physiology, Psychoacoustics and Cognition in Normal and Impaired Hearing,* Advances in Experimental Medicine and Biology 894,
DOI 10.1007/978-3-319-25474-6_33

increasing age was associated with lower stimulus encoding accuracy for the modulated tone stimulus.

Conclusions The results show that even in a group of people with normal hearing, increasing age was associated with reduced speech understanding, reduced TFS sensitivity, and reduced stimulus encoding accuracy (for the modulated tone stimulus). People with good TFS sensitivity also generally had less faithful brainstem encoding of a complex harmonic tone.

Keywords Speech in noise · Objective measures · Temporal fine structure · Aging · Electrophysiology · Envelope-following response · Frequency-following response

1 Introduction

The benefits of amplification are greatest when hearing interventions are made as early as possible. There is therefore great interest in the clinical audiology community in the development of objective techniques to measure various hearing abilities that do not require behavioural responses from the patient and are able to determine fitting parameters for hearing aids and cochlear implants. While the use of cortical responses is in development (Billings et al. 2007, 2011; Carter et al. 2010; Billings et al. 2012; Chang et al. 2012), transient brainstem responses and steady state responses (Luts et al. 2004, 2006; Alaerts et al. 2010) are already used in clinical practice to objectively assess hearing thresholds in young children. However, while these measures provide an estimate of *audibility* with the prescribed hearing aid gain, they do not provide any indication of the expected *hearing ability* of the patient.

In order to develop an objective measure of hearing ability, it is necessary to establish the links between speech perception, psychophysical measures of perceptual sensitivity to the acoustic cues that underlie effective speech perception, and the proposed objective measures. This paper describes our initial investigations in this direction.

The overall aim of this study was to examine the relationships between perceptual sensitivity to temporal fine-structure cues, brainstem encoding of complex harmonic and amplitude-modulated sounds, and the ability to understand speech in noise. Understanding these links will allow the development of an objective measure that could be used to detect changes in functional hearing before the onset of permanent threshold shifts.

2 Methods

2.1 *Participants*

Thirty-four participants (14 men and 20 women) aged between 18–63 years took part in the experiment. All participants had normal hearing bilaterally, defined by four frequency average hearing loss thresholds (500 Hz, 1 kHz, 2 kHz and 4 kHz) of less than 25 dB HL. Pure tone hearing thresholds for the 34 participants are shown in Fig. 1. Thresholds were measured using an Otometrics MADSEN Itera II audiometer with TDH-39 Stereo Headphones. The study was approved by Royal Victorian Eye & Ear Hospital Human Research Ethics Committee. Consent was written and informed.

2.2 *Temporal Fine Structure Sensitivity*

The TFS1 test (Moore and Sek 2009, 2012) was used to measure participants' temporal fine structure (TFS) sensitivity. The test was carried out using the TFS1 test software (http://hearing.psychol.cam.ac.uk/) and was based on the standard TFS1 test protocol. The task was performed on the participant's better ear as determined by the audiogram.

One practice run was given prior to the test. If participants could perform the task and attain a threshold they were given three real runs. For some participants the staircase procedure saturated (reached the 'easy' limit). Instead of receiving a threshold, these participants received a percent-correct score obtained from 40 trials of the 2AFC task at the easiest level of the staircase. Both the threshold and the

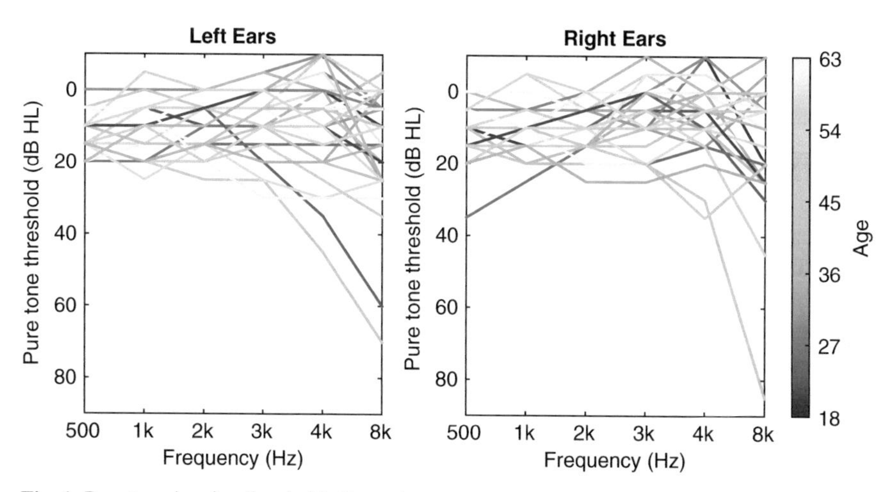

Fig. 1 Pure tone hearing thresholds for each ear

percent-correct scores were converted to a *d'* sensitivity measure using the method outlined by Hopkins and Moore (2007).

2.3 Speech in Noise Tests (QuickSIN)

Speech-in-noise performance was assessed behaviourally using the QuickSIN test (Killion et al. 2004). Six sets of six sentences with five key words per sentence were presented to both ears in four-talker babble noise. The sentences were presented with decreasing signal-to-noise ratios (SNRs) from +25 dB SNR to 0 dB SNR in 5 dB steps (+25, +20, +15, +10, +5 and 0 dB). The average SNR loss was calculated across the six lists. This score indicates the increase in SNR required for the participant to understand the sentence compared with a normal hearing person. Any score less than 4 dB is considered normal, and a lower SNR loss score reflects better speech-in-noise performance.

2.4 Electrophysiology

2.4.1 Stimuli

Envelope-following responses (EFRs) from the brainstem were elicited in response to two stimulus types—a complex harmonic tone, and a 4 KHz sinusoidal carrier tone modulated at 110 Hz. Both sounds were 100 ms in duration with 5 ms linear onset and offset ramps. The complex harmonic tone had an F0 of 180 Hz and 20 harmonics of equal amplitude and random phase. The modulated tone had a modulation depth of 50%. Each stimulus was presented with alternating polarities. The first 20 ms of each stimulus are shown in Fig. 2.

Stimuli were controlled via custom software in MAX/MSP (Cycling '74), played through an RME Fireface 400 audio interface and Etymotic ER3-A insert-phones.

2.4.2 EEG Recordings and Pre-processing

EEG data were recorded from the scalp via Ag-AgCl electrodes, using a BioSemi ActiveTwo EEG System. Electrode offsets were ±40 mV. The EEG data were collected in continuous mode at a sampling rate of 16.384 kHz. The −3 dB point of the antialiasing lowpass filter was 3276 Hz.

The EEG recordings were segmented into epochs of −50 to 150 ms, separately for each stimulus type. The epochs were artefact rejected using the pop_autorej function from the EEGLAB toolbox (Delorme and Makeig 2004) using MATLAB software. EFRs were computed by adding responses to the positive and negative stimulus polarities (EFR=(Pos+Neg)/2). All subsequent correlational analyses were conducted on the averaged EFR waveforms.

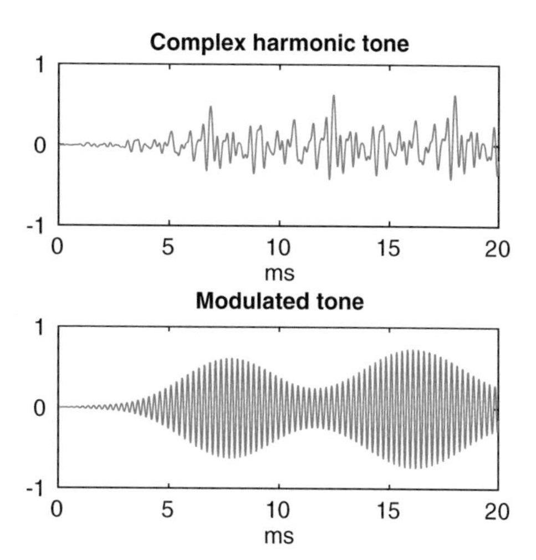

Fig. 2 The first 20 ms of the complex harmonic tone (*top*) and modulated tone (*bottom*)

The signal-to-noise ratio (SNR) of the EFR responses was calculated as 20log10(RMSpost/RMSpre), where RMSpost and RMSpre were the root mean square of the amplitude in the 'response' period (defined as 25–75 ms post-stimulus) and the root mean square of the pre-stimulus period (defined as 50 ms pre-stimus until stimulus onset at 0 ms) respectively. Three participants were removed from the EEG analysis as they had SNRs < 1.5 dB due to movement and other muscle artefact.

Stimulus-to-response cross-correlations (*r*-values) were generated using the Brainstem Toolbox 2013 (Skoe and Kraus 2010). The maximum cross-correlation values were chosen irrespective of the lag. Cross-correlations were performed against the Hilbert envelope of the stimulus, as recorded through the transducer and an artificial ear (GRAS Type 43-AG). All data transformation and statistical tests (correlational analyses) were conducted using Matlab and Minitab®. Spearman's rank correlations were performed where data were skewed.

3 Results

3.1 Hearing Thresholds and Age

Hearing sensitivity generally declined slightly with age. The correlation between age and the pure-tone average hearing loss (at 500 Hz, 1 kHz, 2 kHz and 4 kHz) for the best ear was significant, $r=0.43$, $p=0.001$.

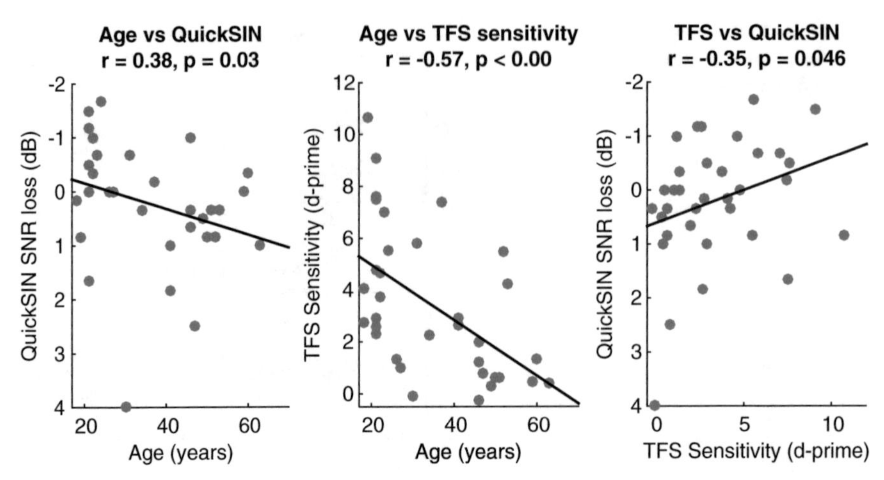

Fig. 3 *Left panel*: Speech in noise performance (QuickSIN SNR loss) as a function of age (*left*). Note that the direction of the y-axis has been reversed so that better performance is up. *Middle panel*: Sensitivity to TFS as a function of age. *Right panel*: Speech in noise performance (QuickSIN) as a function of TFS sensitivity. TFS sensitivity is expressed using a *d'* measure. Note that the direction of the y-axis has been reversed so that better performance is up. The *grey line* in all three panels indicates a least-squares linear regression

3.2 Correlations: QuickSIN, TFS Sensitivity, Age and Speech in Noise

Figure 3 shows the relationships between age, speech in noise perception, and TFS sensitivity. Both speech in noise scores ($r=-0.38$, $p<0.03$) and TFS sensitivity ($r=-0.57$, $p<0.001$) were significantly negatively correlated with age. Older participants generally had worse speech scores, and worse TFS sensitivity, with a stronger relationship in the case of TFS sensitivity. Speech in noise performance was moderately and significantly related to TFS sensitivity ($r=-0.34$, $p=0.046$). Participants who had good TFS sensitivity generally had good speech in noise scores (good scores are negative).

3.3 Electrophysiology

Grand average responses for the complex harmonic and modulated tones are shown in Fig. 4. The responses show clear phase-locking to periodicity in the stimuli.

A measure of stimulus encoding strength was generated by calculating the cross-correlation between the stimulus envelope (as measured through the transducer and artificial ear) and the brainstem response. Figure 5 shows this process for one listener.

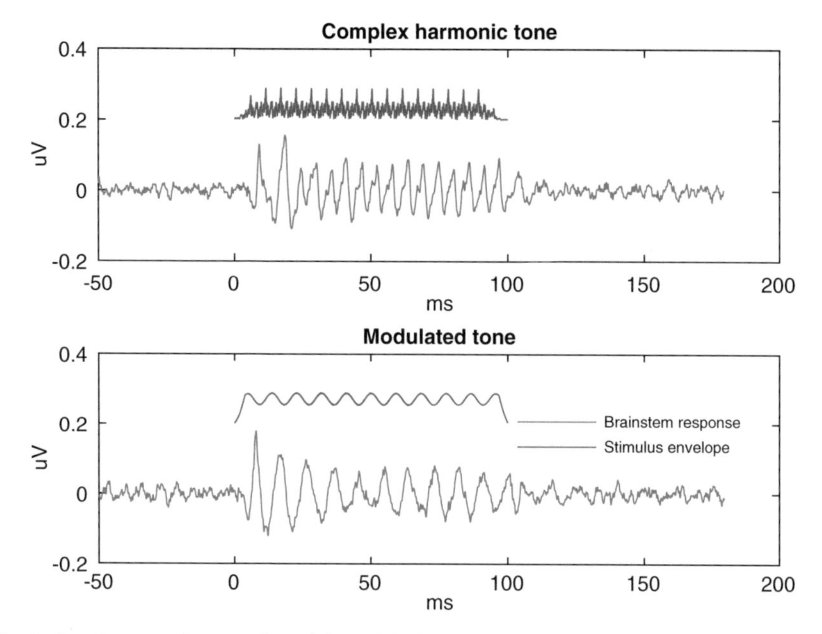

Fig. 4 Grand average (across all participants) brainstem responses for the complex harmonic tone (*top panel*) and modulated tone (*bottom panel*). The Hilbert envelope of the stimulus waveform is shown in *red*

3.4 Correlations: Stimulus Encoding Accuracy with Age, TFS Sensitivity and QuickSIN

The maximum cross-correlation value obtained from each participant and stimulus type was correlated with age, the TFS sensitivity and speech in noise scores. The top row of Fig. 6 shows that increasing age was associated with decreasing stimulus encoding accuracy, but only for the modulated tone stimulus ($r = -0.53$, $p < 0.001$). The middle row of Fig. 6 also shows a striking relationship between increased TFS sensitivity and reduced stimulus encoding for the complex harmonic tone ($r = -0.56$, $p = 0.004$). Interestingly, there was no such relationship for the modulated tone. There were also no significant relationships between stimulus encoding accuracy and the speech in noise scores (bottom row).

3.5 Regression Analysis

In order to determine which psychophysical and/or EEG measure best predicted the speech scores, best subsets regression was performed. The QuickSIN scores were entered as the response variable. Age, four-frequency pure-tone hearing thresholds (in the better ear), TFS sensitivity, and stimulus encoding accuracy for both the

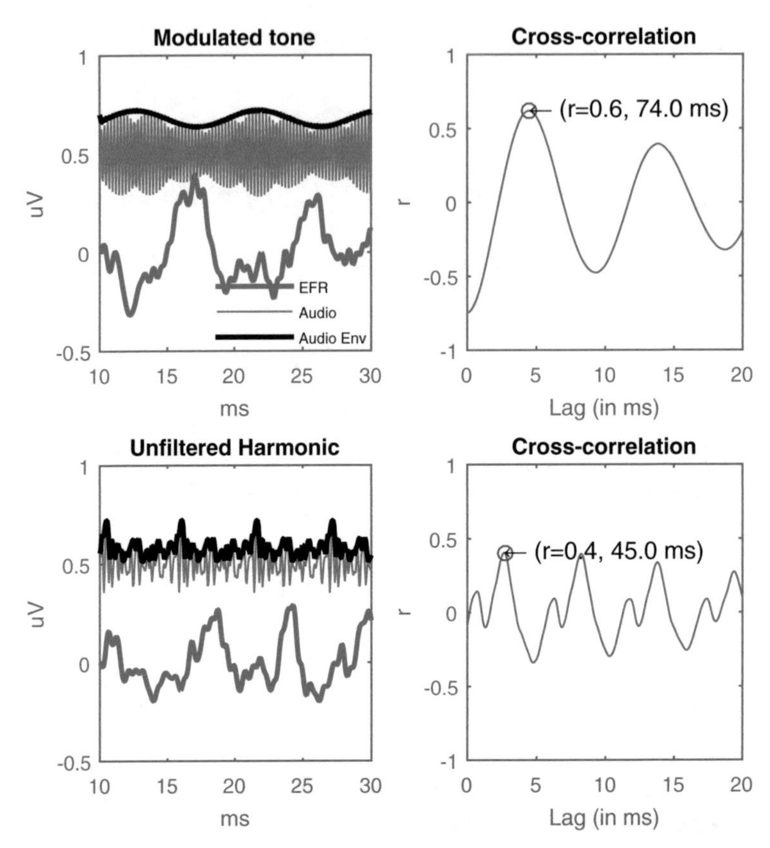

Fig. 5 An illustration for a single listener of the calculation of the cross-correlation values for the modulated tone (*top panels*) and complex harmonic tone (*bottom panels*). The *left panels* show the brainstem responses (in *blue*) and the stimulus waveform (*red*) and its Hilbert envelope (*black*). Note the stimulus waveform here is not the electrical signal: rather it has been recorded through the transducer and an artificial ear. The *right panels* show the cross-correlation values between the stimulus envelope and the brainstem response as a function of the lag between them. The maximum of the cross-correlation function was determined for further analysis

complex harmonic and modulated tones were entered as predictors. The regression indicated that QuickSIN scores were best predicted by a combination of all variables except the hearing thresholds ($R^2 = 64.4$, $F_{(4,26)} = 6.32$, $p = 0.004$). A follow-up standard linear regression using the four predictors identified by the best-subsets procedure echoed these results, although TFS sensitivity was not a significant factor. The main model was significant ($R^2 = 41.8$, $F_{(4,26)} = 4.49$, $p = 0.007$), with age accounting for most of the variance in the model (18.1 %), followed by the stimulus encoding accuracy variables contributing 13.7 and 10.0 % for the modulated and complex harmonic tones respectively. When the pure-tone average hearing thresholds were added to the model, they contributed only 2 % variance.

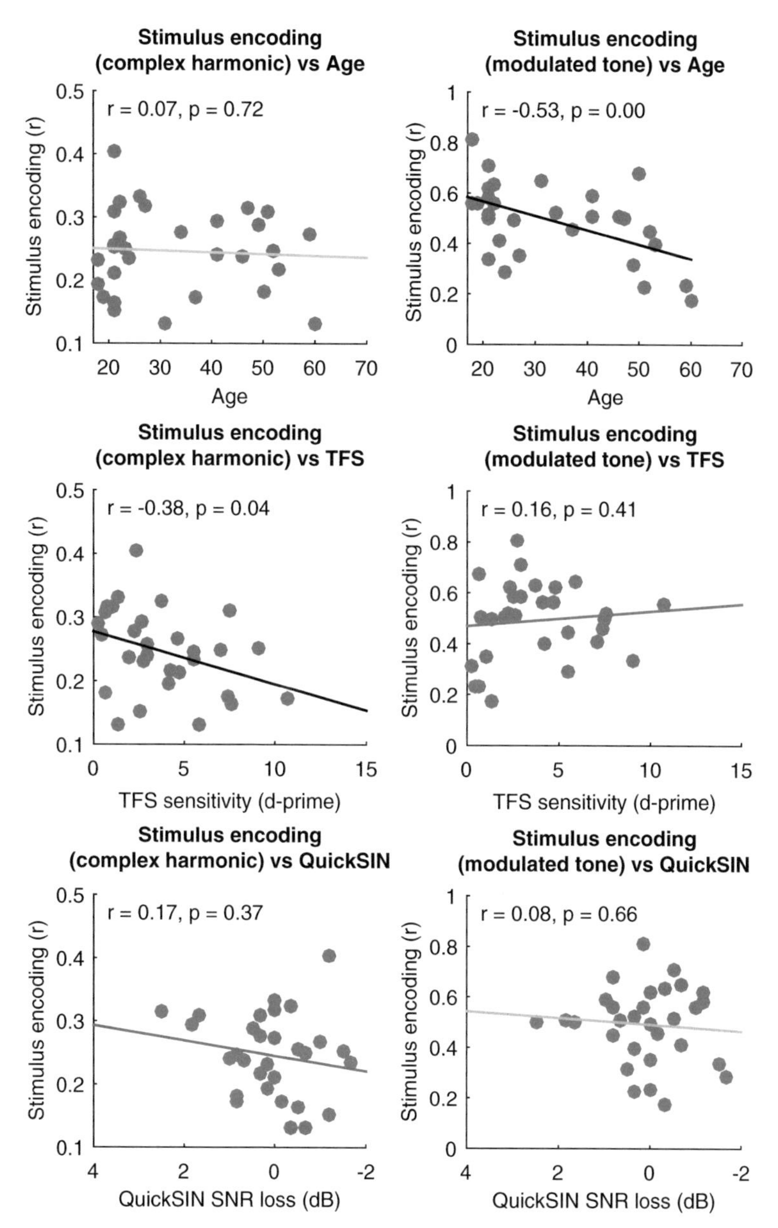

Fig. 6 Stimulus-to-response cross-correlation values between the complex harmonic (*left*) and modulated (*right*) stimuli and brainstem response as a function of age (*top row*), TFS sensitivity (*middle row*) and speech in noise scores (*bottom row*)

4 Discussion

In summary, we found that in our group of normally-hearing adults (with a variety of ages), speech-in-noise performance was negatively correlated with TFS sensitivity and age. TFS sensitivity was also positively correlated with stimulus encoding accuracy for the complex harmonic stimulus, while increasing age was associated with lower stimulus encoding accuracy for the modulated tone stimulus. Surprisingly, we found that better speech in noise understanding was associated with worse stimulus encoding accuracy. Despite this unexpected direction of correlation the measures did contribute modestly to a regression model predicting the speech in noise scores. A regression analysis found that age and the combination of the two stimulus encoding accuracy measures had roughly equal contributions to the model.

Further work in this area should consider other psychophysical predictors that are known to be associated with speech understanding, such as measures of temporal modulation sensitivity, and EEG measures that more closely match the stimuli used in the psychophysics.

Acknowledgments Supported provided by the National Health and Medical Research Council of Australia (Peter Doherty Early Career Research Fellowship #1069999 to HIB). The Bionics Institute acknowledges the support it receives from the Victorian Government through its Operational Infrastructure Support Program. HIB is currently supported by a FWO Pegasus Marie Curie Fellowship at KU Leuven, Belgium.

References

Alaerts J, Luts H, Van Dun B, Desloovere C, Wouters J (2010) Latencies of auditory steady-state responses recorded in early infancy. Audiol Neuro-Otol 15(2):116–127

Billings CJ, Papesh MA, Penman TM, Baltzell LS, Gallun FJ (2012) Clinical use of aided cortical auditory evoked potentials as a measure of physiological detection or physiological discrimination. Int J Otolaryngol 14

Billings CJ, Tremblay KL, Souza PE, Binns MA (2007) Effects of hearing aid amplification and stimulus intensity on cortical auditory evoked potentials. Audiol Neuro-Otol 12(4):234–246. doi:10.1159/000101331

Billings CJ, Tremblay KL, Miller CW (2011) Aided cortical auditory evoked potentials in response to changes in hearing aid gain. Int J Audiol 50(7):459–467. doi:10.3109/14992027.2011.568011

Carter L, Golding M, Dillon H, Seymour J (2010) The detection of infant cortical auditory evoked potentials (CAEPs) using statistical and visual detection techniques. J Am Acad Audiol 21(5):347–356. doi:10.3766/jaaa.21.5.6

Chang H-W, Dillon H, Carter L, Van Dun B, Young S-T (2012) The relationship between cortical auditory evoked potential (CAEP) detection and estimated audibility in infants with sensorineural hearing loss. Int J Audiol 51:663–670

Delorme A, Makeig S (2004) EEGLAB: an open source toolbox for analysis of single-trial EEG dynamics including independent component analysis. J Neurosci Methods 134(1):9–21. [10.1016/j.jneumeth.2003.10.009]

Hopkins K, Moore BC (2007) Moderate cochlear hearing loss leads to a reduced ability to use temporal fine structure information. J Acoust Soc Am 122(2):1055–1068. doi:10.1121/1.2749457

Killion MC, Niquette PA, Gudmundsen GI, Revit LJ, Banerjee S (2004). Development of a quick speech-in-noise test for measuring signal-to-noise ratio loss in normal-hearing and hearing-impaired listeners. J Acoust Soc Am 116(4 Pt 1), 2395–2405

Luts H, Desloovere C, Kumar A, Vandermeersch E, Wouters J (2004) Objective assessment of frequency-specific hearing thresholds in babies. Int J Pediatr Otorhinolaryngol 68(7):915–926

Luts H, Desloovere C, Wouters J (2006) Clinical application of dichotic multiple-stimulus auditory steady-state responses in high-risk newborns and young children. Audiol Neuro-Otol 11(1):24–37

Moore BCJ, Sek A (2009) Development of a fast method for determining sensitivity to temporal fine structure. Int J Audiol 48:161–171

Moore BCJ, Sek A (2012) Implementation of two tests for measuring sensitivity to temporal fine structure. Int J Audiol 51:58–63

Skoe E, Kraus N (2010) Auditory brain stem response to complex sounds: a tutorial. Ear Hear 31(3):302–324

Connectivity in Language Areas of the Brain in Cochlear Implant Users as Revealed by fNIRS

Colette M. McKay, Adnan Shah, Abd-Krim Seghouane, Xin Zhou, William Cross and Ruth Litovsky

Abstract Many studies, using a variety of imaging techniques, have shown that deafness induces functional plasticity in the brain of adults with late-onset deafness, and in children changes the way the auditory brain develops. Cross modal plasticity refers to evidence that stimuli of one modality (e.g. vision) activate neural regions devoted to a different modality (e.g. hearing) that are not normally activated by those stimuli. Other studies have shown that multimodal brain networks (such as those involved in language comprehension, and the default mode network) are altered by deafness, as evidenced by changes in patterns of activation or connectivity within the networks. In this paper, we summarise what is already known about brain plasticity due to deafness and propose that functional near-infra-red spectroscopy (fNIRS) is an imaging method that has potential to provide prognostic and diagnostic information for cochlear implant users. Currently, patient history factors account for only 10% of the variation in post-implantation speech understanding, and very few post-implantation behavioural measures of hearing ability correlate with speech understanding. As a non-invasive, inexpensive and user-friendly imaging method, fNIRS provides an opportunity to study both pre- and post-implantation brain function. Here, we explain the principle of fNIRS measurements and illustrate its use in studying brain network connectivity and function with example data.

C. M. McKay (✉) · A. Shah · X. Zhou · W. Cross
The Bionics Institute of Australia, Melbourne, Australia
e-mail: cmckay@bionicsinstitute.org

C. M. McKay · X. Zhou
Department of Medical Bionics, The University of Melbourne, Melbourne, Australia

A. Shah · A.-K. Seghouane
Department of Electrical and Electronic Engineering, The University of Melbourne, Melbourne, Australia

W. Cross
Department of Medicine, The University of Melbourne, Melbourne, Australia

R. Litovsky
Waisman Center, The University of Wisconsin-Madison, Madison, USA

P. van Dijk et al. (eds.), *Physiology, Psychoacoustics and Cognition in Normal and Impaired Hearing,* Advances in Experimental Medicine and Biology 894,
DOI 10.1007/978-3-319-25474-6_34

Keywords fNIRS · Cochlear implants · Deafness · Brain plasticity · Connectivity in brain networks

1 Introduction

1.1 Deafness, Language and Brain Plasticity: Evidence from Imaging Studies

Speech understanding involves complex multimodal networks involving vision, hearing and sensory motor areas as well as memory and frontal lobe functions mostly in the left hemisphere (LH), and encompasses a range of elements such as phonology, semantics, and syntactics. The right hemisphere (RH) has fewer specialised functions for language processing, and its role is mostly in the evaluation of the communication context (Vigneau et al. 2011). Imaging studies have shown that adults who have undergone periods of profound post-lingual deafness demonstrate changes in brain activity and function in language-associated brain areas that are not observed in normally-hearing individuals, and that further functional plasticity occurs as a result of cochlear implantation.

Lee et al. (2003) used positron emission tomography (PET) to compare resting-state activity in 9 profoundly deaf individuals and 9 age-matched normal-hearing controls. They found that glucose metabolism in some auditory areas was lower than in normally-hearing people, but significantly increased with duration of deafness, and concluded that plasticity occurs in the sensory-deprived mature brain. Later, they showed that children with good speech understanding 3 years after implantation had enhanced metabolic activity in the left prefrontal cortex and decreased metabolic activity in right Heschl's gyrus and in the posterior superior temporal sulcus before implantation compared to those with poor speech understanding (Lee et al. 2007). They argued that increased activity in the resting state in auditory areas may reflect cross modal plasticity that is detrimental to later success with the cochlear implant (CI). Recently, Dewey and Hartley (2015) used functional near infrared spectroscopy (fNIRS) to demonstrate that the auditory cortex of deaf individuals is abnormally activated by simple visual stimuli.

Not all studies have suggested detrimental effects of brain changes due to deafness on the ability to adapt to listening with a CI. Two studies by Giraud et al used PET to study the activity induced by speech stimuli in CI users. The first showed that, compared to normal-hearing listeners, they had altered functional specificity of the superior temporal cortex, and exhibited contribution of visual regions to sound recognition (Giraud et al. 2001a). Secondly, the contribution of the visual cortex to speech recognition *increased* over time after implantation (Giraud et al. 2001b), suggesting that the CI users were actively using enhanced audio-visual integration to facilitate their learning of the novel speech sounds received through the CI. In contrast, Rouger et al. (2012) suggested a negative impact of cross modal plasticity: they found that the right temporal voice area (TVA) was abnormally activated in CI users by a *visual* speech-reading task and that this activity declined over time after implantation while the activity in Broca's area (normally activated by speech reading) increased over time after implantation. Coez et al. (2008) also used PET to study activation by *voice* stimuli of the TVA in CI users with poor and good speech understanding. The voice stimuli induced bilateral activation of the TVA along the superior temporal sulcus in both normal hearing listeners and CI users with good speech understanding, but not in CI users with poor understanding. This result is consistent with the proposal of Rouger et al that the TVA is 'taken over' by visual speech reading tasks in CI users who do not understand speech well. Strelnikov et al. (2013) measured PET resting state activity and activations elicited by auditory and audio-visual speech in CI users soon after implantation and found that good speech understanding after 6 months of implant use was predicted by a higher activation level of the right occipital cortex and a lower activation in the right middle superior temporal gyrus. They suggested that the pre-implantation functional changes due to reliance on lip-reading were *advantageous* to development of good speech understanding through a CI via enhanced audio-visual integration. In summary, functional changes that occur during deafness can both enhance and degrade the ability to adapt to CI listening, perhaps depending on the communication strategy each person used while deaf.

Lazard et al. have published a series of studies using functional magnetic resonance imaging (fMRI), in which they related pre-implant data with post-implant speech understanding. They found that, when doing a rhyming task with written words, CI users with later good speech understanding showed an activation pattern that was consistent with them using the normal phonological pathway to do the task, whereas those with poor outcomes used a pathway normally associated with lexical-semantic understanding (Lazard et al. 2010). They further compared activity evoked by speech and non-speech imageries in the right and left posterior superior temporal gyrus/supramarginal gyrus (PSTG/SMG) (Lazard et al. 2013). These areas are normally specialised for phonological processing in the left hemisphere and environmental sound processing in the right hemisphere. Their results suggested the abnormal recruitment of the right PSTG/SMG region for phonological processing.

In summary, studies have shown functional changes due to periods of deafness, some of which are detrimental and some advantageous to post-implant speech understanding. It is evident that the functional changes involve not just the auditory cortex, but the distributed multimodal language networks and that reliance on lip-reading while deaf may be a major factor that drives functional changes.

1.2 Functional Near Infra-red Spectroscopy (fNIRS)

Near infra-red (NIR) light (wavelengths 650–1000 nm) is relatively transparent to human tissues. The absorption spectra of oxygenated and de-oxygenated haemoglobin (HbO and HbR respectively) for NIR light differ in that HbO maximally absorbs light of longer wavelength (900–1000 nm) whereas HbR maximally absorbs light of shorter wavelength. These differential absorption spectra make it possible to separate out changes in the concentration of HbO and HbR. In response to neural activity in the brain, an increase in oxygenated blood is directed to the region, resulting in a drop in de-oxygenated blood. Thus HbO and HbR concentrations change in opposite directions in response to neural activity (Shah and Seghouane 2014). Changes in HbO and HbR resulting from a neural response to a stimulus or due to resting state activity can be analysed to produce activation patterns and also to derive connectivity measures between regions of interest.

In the fNIRS imaging system optodes are placed at various locations on the scalp. Each fNIRS channel consists of a paired source and detector. The source emits a light beam, directed perpendicular to the scalp surface, and the detector detects the light emerging from the brain. The detected light beam has scattered in a banana-shaped pathway through the brain reaching a depth of approximately half the distance between the source and detector. In a montage of multiple optodes, each source can be associated with a number of surrounding detectors to form a multi-channel measurement system. In continuous wave systems, two frequencies of light are emitted by the diodes, and the signals at each source are frequency-modulated at different rates to facilitate separation of the light from different sources at the same detector position.

For studying language areas in the brain in CI users, fNIRS offers some advantages over other imaging methods: compared to PET it is non-invasive; in contrast to fMRI it can be used easily with implanted devices, is silent, and is more robust to head movements; compared to EEG/MEG it has much greater spatial resolution, and in CI users is free from electrical or magnetic artifacts from the device or the stimuli. The portability, low-cost, and patient-friendly nature of fNIRS (similar to EEG) makes it a plausible method to contribute to routine clinical management of patients, including infants and children.

However, there are also limitations of fNIRS. The spatial resolution is limited by the density of the optodes used, and is not generally as good as that found with fMRI or PET. The depth of imaging is limited, so that it is only suitable for imaging areas of the cortex near the surface [although other designs of fNIRS systems that use lasers instead of diodes and use pulsatile stimuli can derive 3-dimensional images of the brain, at least in infants (Cooper et al. 2014)]. However, fNIRS has been successfully used in a range of studies to determine effects on language processing in hearing populations of adults and children (Quaresima et al. 2012), therefore it shows promise for assessing language processing in deaf populations and those with cochlear implants.

In this paper we describe methods and present preliminary data for two investigations using fNIRS to compare CI users and normally-hearing listeners. In the first experiment, we measure resting state connectivity, and in the second experiment we compare the activation of cortical language pathways by visual and auditory speech stimuli.

2 Methods

2.1 *fNIRS Equipment and Data Acquisition*

Data were acquired using a multichannel 32 optode (16 sources and 16 detectors) NIRScout system. Each source LED emitted light of two wavelengths (760 and 850 nm). The sources and detectors were mounted in a pre-selected montage using an EASYCAP with grommets to hold the fNIRS optodes, which allowed registration of channel positions on a standard brain template using the international 10–20 system. To ensure optimal signal detection in each channel, the cap was fitted first, and the hair under each grommet moved aside before placing the optodes in the grommets. For CI users, the cap was fitted over the transmission coil with the speech processor hanging below the cap. The data were exported into MatLab or nirsLAB for analysis.

2.2 *Resting-State Connectivity in CI Users Using fNIRS*

In this study we compare the resting state connectivity of a group of normal hearing listeners compared to a group of experienced adult CI users. Brain function in the resting state, or its default-mode activity (Raichle and Snyder 2007) is thought to reflect the ability of the brain to predict changes to the environment and track any deviation from the predicted. Resting state connectivity (as measured by the correlation of resting state activity between different cortical regions) is influenced by functional organisation of the brain, and hence is expected to reflect plastic changes such as those due to deafness. In this study, we hypothesised that CI users would exhibit resting-state connectivity that was different from that of normally-hearing listeners.

We used a 4×4 montage of 8 sources and detectors (24 channels) in each hemisphere (Fig. 1d), which covered the auditory and somatosensory regions of the brain. Sources and detectors were separated by 3 cm. Data in each channel were pre-processed to remove 'glitches' due to head movements and 'good' channels were identified by a significant cross-correlation ($r > 0.75$) between the data for the two wavelengths. Any 'poor' channels were discarded. The data were then converted to HbO and HbR based on a modified Beer-Lambardt Law (Cope et al. 1988). Since resting state connectivity is based upon correlations between activity in relevant channels, it is important to carefully remove any aspects of the data (such as fluctuations from heartbeat, breathing, or from movement artifacts that are present in all channels) that would produce a correlation but is not related to neural activity. These processing steps to remove unwanted signals from the data included: minimizing regional drift using a discrete cosine transform; removing global drift using PCA 'denoising' techniques, low-pass filtering (0.08 Hz) to remove heart and breathing fluctuations. Finally the connectivity between channels was calculated using Pearson's correlation, r. To account for missing channels, the data were reduced to 9 regions of interest in each hemisphere by averaging the data from groups of 4 channels.

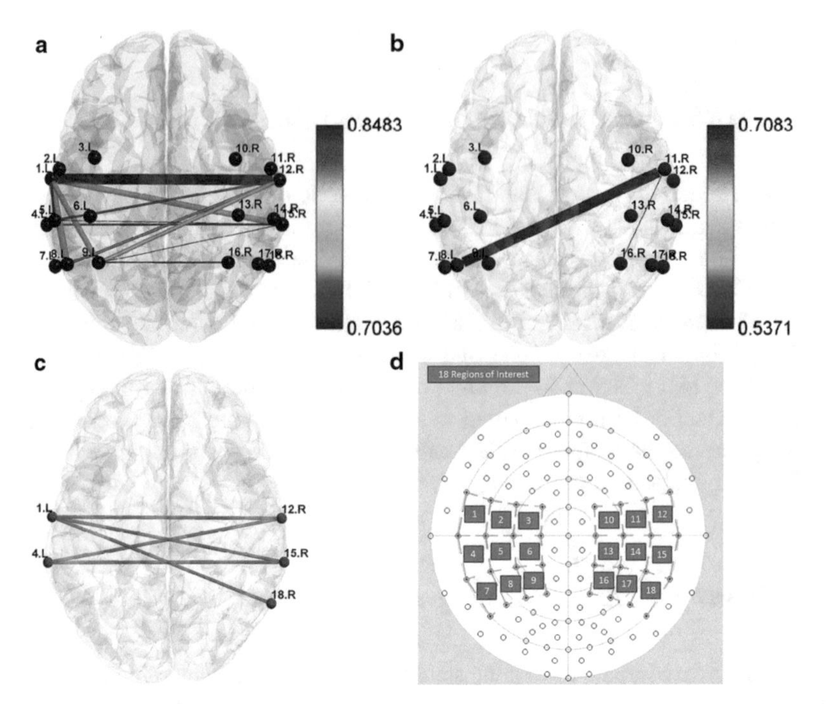

Fig. 1 Mean resting state connectivity in groups of 5 normal-hearing (*NH*) listeners (**a**) and 5 CI users (**b**). The *line colour* and *thickness* denote the strength of connectivity (r) in the different brain regions of interest (*colour bars* denotes r-values). Panel **c** shows the channels that are significantly more highly connected in NH compared to CI listeners. Panel **d** shows the montage used and the 18 regions of interest

2.3 Language Networks in CI Users Using fNIRS

In the second study, we aimed to measure the activity in regions in the brain involved in language processing in response to auditory and visual speech stimuli. We hypothesised that, compared to normal-hearing listeners, CI users would show altered functional organisation, and furthermore that this difference would be correlated with lip-reading ability and with their auditory speech understanding.

The optode montage for this experiment is shown in Fig. 2a. fNIRS data were collected using a block design (12-s stimulus blocks separated by 15–25-s silent gaps and preceded by resting state baseline) in two sessions. In the first session, stimuli were visual and auditory words: in the second session stimuli were auditory and audiovisual sentences. The auditory stimuli were presented via EAR-4 insert ear phones in normally-hearing listeners, or via direct audio input for the CI users. Sounds were only presented to the right ear of subjects (the CI users were selected to have right-ear implants).

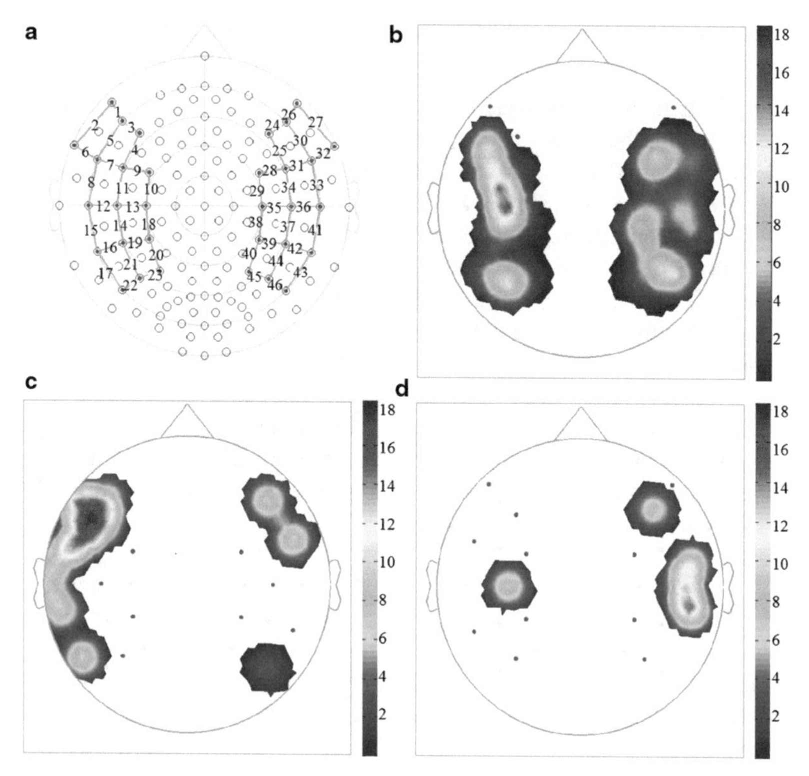

Fig. 2 Task-related activation due to audiovisual sentences (sound in *right* ear). **a** optode montage. **b** mean activation pattern from 10 NH listeners. **c** and **d** activation patterns from individual CI users with good and poor speech understanding, respectively. The colour bar scale denotes the significance (t-value) of the activation compared to resting state

3 Results

3.1 Preliminary fNIRS Data for Resting-State Connectivity

Figure 1 shows data for mean resting state connectivity in groups of (A) 5 normal-hearing (NH) listeners, and (B) 5 CI users: panel C shows channels that are more highly connected in the NH listeners than in the CI users ($p<0.001$), using Network Based Statistics (Zalesky et al. 2010, pp. 1197–1207). All NH listeners tested to date have shown particularly strong connectivity between analogous regions in left and right hemispheres, consistent with other studies using fNIRS (Medvedev 2014). In contrast, the CI users in this group have consistently shown lower inter-hemispheric connectivity than the NH listeners.

3.2 Preliminary fNIRS Data for Language Networks

Figure 2 illustrates the task-related activation pattern evoked by audiovisual sentences, comparing the mean pattern from a group of 10 NH listeners to example patterns from two individual CI users with good and poor speech understanding. All NH listeners showed a similar activation pattern in wide language-associated regions in both hemispheres with asymmetry favouring the left hemisphere. The CI user with good speech understanding (100 % correct sentences in quiet) shows a generally similar pattern to that of the NH listeners, with left hemisphere dominance but greater activation in frontal areas (Broca's). In contrast, the CI user with poor speech understanding (<50 % correct) has very little significant left-hemisphere activity, and little activation out of primary and associated auditory areas. Further work is being undertaken to compare activation patterns and connectivity in language pathways between CI users and NH listeners, and to correlate relevant differences to individual behavioural measures of speech understanding and lip-reading ability.

4 Discussion

Our preliminary data show that fNIRS has the potential to provide insight into the way that brain functional organisation is altered by deafness and subsequent cochlear implantation. The fNIRS tool may be particularly useful for longitudinal studies that track changes over time, so that the changes in brain function can be correlated with simultaneous changes in behavioural performance. In this way, new knowledge about functional organisation and how it relates to an individual's ability to process language can be gained. For example, an fNIRS pre-implant test may, in the future, provide valuable prognostic information for clinicians and patients, and provide guidance for individual post-implant therapies designed to optimise outcomes. Routine clinical use of fNIRS is feasible due to its low-cost. It will be particularly useful in studying language development in deaf children due to its patient-friendly nature.

Acknowledgments This research was supported by a **veski** fellowship to CMM, an Australian Research Council Grant (FT130101394) to AKS, a Melbourne University PhD scholarship to ZX, the Australian Fulbright Commission for a fellowship to RL, the Lions Foundation, and the Melbourne Neuroscience Institute. The Bionics Institute acknowledges the support it receives from the Victorian Government through its Operational Infrastructure Support Program.

References

Coez A, Zilbovicius M, Ferrary E, Bouccara D, Mosnier I, Ambert-Dahan E, Bizaguet E, Syrota A, Samson Y, Sterkers O (2008) Cochlear implant benefits in deafness rehabilitation: PET study of temporal voice activations. J Nucl Med 49(1):60–67. doi:10.2967/jnumed.107.044545

Cooper RJ, Magee E, Everdell N, Magazov S, Varela M, Airantzis D, Gibson AP, Hebden JC (2014) MONSTIR II: a 32-channel, multispectral, time-resolved optical tomography system for neonatal brain imaging. Rev Sci Instrum 85(5):053105. doi:10.1063/1.4875593

Cope M, Delpy DT, Reynolds EO, Wray S, Wyatt J, van der Zee P (1988) Methods of quantitating cerebral near infrared spectroscopy data. Adv Exp Med Biol 222:183–189

Dewey RS, Hartley DE (2015) Cortical cross-modal plasticity following deafness measured using functional near-infrared spectroscopy. Hear Res. doi:10.1016/j.heares.2015.03.007

Giraud AL, Price CJ, Graham JM, Frackowiak RS (2001a) Functional plasticity of language-related brain areas after cochlear implantation. Brain 124(Pt 7):1307–1316

Giraud AL, Price CJ, Graham JM, Truy E, Frackowiak RS (2001b) Cross-modal plasticity underpins language recovery after cochlear implantation. Neuron 30(3):657–663

Lazard DS, Lee HJ, Gaebler M, Kell CA, Truy E, Giraud AL (2010) Phonological processing in post-lingual deafness and cochlear implant outcome. Neuroimage 49(4):3443–3451. doi:10.1016/j.neuroimage.2009.11.013

Lazard DS, Lee HJ, Truy E, Giraud AL (2013) Bilateral reorganization of posterior temporal cortices in post-lingual deafness and its relation to cochlear implant outcome. Hum Brain Mapp 34(5):1208–1219. doi:10.1002/hbm.21504

Lee JS, Lee DS, Oh SH, Kim CS, Kim JW, Hwang CH, Koo J, Kang E, Chung JK, Lee MC (2003) PET evidence of neuroplasticity in adult auditory cortex of postlingual deafness. J Nucl Med 44(9):1435–1439

Lee HJ, Giraud AL, Kang E, Oh SH, Kang H, Kim CS, Lee DS (2007) Cortical activity at rest predicts cochlear implantation outcome. Cereb Cortex 17(4):909–917

Medvedev AV (2014) Does the resting state connectivity have hemispheric asymmetry? A near-infrared spectroscopy study. Neuroimage 85(Pt 1):400–407. doi:10.1016/j.neuroimage.2013.05.092

Quaresima V, Bisconti S, Ferrari M (2012) A brief review on the use of functional near-infrared spectroscopy (fNIRS) for language imaging studies in human newborns and adults. Brain Lang 121(2):79–89. doi:10.1016/j.bandl.2011.03.009

Raichle ME, Snyder AZ (2007) A default mode of brain function: a brief history of an evolving idea. Neuroimage 37(4):1083–1090; discussion 1097–1089. doi:10.1016/j.neuroimage.2007.02.041

Rouger J, Lagleyre S, Demonet JF, Fraysse B, Deguine O, Barone P (2012) Evolution of cross-modal reorganization of the voice area in cochlear-implanted deaf patients. Hum Brain Mapp 33(8):1929–1940. doi:10.1002/hbm.21331

Shah A, Seghouane AK (2014) An integrated framework for joint HRF and drift estimation and HbO/HbR signal improvement in fNIRS data. IEEE Trans Med Imaging 33(11):2086–2097. doi:10.1109/TMI.2014.2331363

Strelnikov K, Rouger J, Demonet JF, Lagleyre S, Fraysse B, Deguine O, Barone P (2013) Visual activity predicts auditory recovery from deafness after adult cochlear implantation. Brain 136(Pt 12):3682–3695. doi:10.1093/brain/awt274

Vigneau M, Beaucousin V, Herve PY, Jobard G, Petit L, Crivello F, Mellet E, Zago L, Mazoyer B, Tzourio-Mazoyer N (2011) What is right-hemisphere contribution to phonological, lexico-semantic, and sentence processing? Insights from a meta-analysis. Neuroimage 54(1):577–593. doi:10.1016/j.neuroimage.2010.07.036

Zalesky A, Fornito A, Bullmore ET (2010) Network-based statistic: identifying differences in brain networks. Neuroimage 53(4):1197–1207

Isolating Neural Indices of Continuous Speech Processing at the Phonetic Level

Giovanni M. Di Liberto and Edmund C. Lalor

Abstract The human ability to understand speech across an enormous range of listening conditions is underpinned by a hierarchical auditory processing system whose successive stages process increasingly complex attributes of the acoustic input. In order to produce a categorical perception of words and phonemes, it has been suggested that, while earlier areas of the auditory system undoubtedly respond to acoustic differences in speech tokens, later areas must exhibit consistent neural responses to those tokens. Neural indices of such hierarchical processing in the context of continuous speech have been identified using low-frequency scalp-recorded electroencephalography (EEG) data. The relationship between continuous speech and its associated neural responses has been shown to be best described when that speech is represented using both its low-level spectrotemporal information and also the categorical labelling of its phonetic features (Di Liberto et al., Curr Biol 25(19):2457–2465, 2015). While the phonetic features have been proven to carry extra-information not captured by the speech spectrotemporal representation, the causes of this EEG activity remain unclear. This study aims to demonstrate a framework for examining speech-specific processing and for disentangling high-level neural activity related to intelligibility from low-level activity in response to spectrotemporal fluctuations of speech. Preliminary results suggest that neural measure of processing at the phonetic level can be isolated.

Keywords Hierarchical · Intelligibility · EEG · Noise vocoding · Priming · Natural speech

G. M. Di Liberto (✉) · E. C. Lalor
Trinity College Institute of Neuroscience, Trinity College Dublin, 152–160 Pearse Street, Dublin, Ireland
e-mail: diliberg@tcd.ie

E. C. Lalor
e-mail: edlalor@tcd.ie

G. M. Di Liberto · E. C. Lalor
School of Engineering, Trinity College Dublin, Dublin, Ireland
Trinity Centre for Bioengineering, Trinity College Dublin, Dublin, Ireland

P. van Dijk et al. (eds.), *Physiology, Psychoacoustics and Cognition in Normal and Impaired Hearing,* Advances in Experimental Medicine and Biology 894,
DOI 10.1007/978-3-319-25474-6_35

1 Introduction

Speech processing is an active cognitive activity underpinned by a complex hierarchical system (Chang et al. 2010; Okada et al. 2010; Peelle et al. 2010; DeWitt and Rauschecker 2012; Hickok 2012). In particular, evidence for hierarchical speech processing emerged from functional magnetic resonance imaging (fMRI) and electrocorticography (ECoG) studies. Although these methodologies have provided important scientific insights, they have many limitations. For example, fMRI does not allow for the study of the fast temporal dynamics typical of continuous speech, and ECoG studies are limited to patients suffering from severe cases of epilepsy.

Electro- and magnetoencephalography (EEG/MEG), as macroscopic non-invasive technologies, have the potential for further progress on this topic. Traditionally, the low signal-to-noise ratio (SNR) has hampered the experiments conducted with these instruments, limiting the study of speech processing to repeated presentations of simple short sounds, which elicit neural responses known as event related potentials (ERPs). Although studies based on ERPs have provided important insights on this topic, the human auditory processing mechanisms are tuned to process continuous speech and respond differently to discrete sounds (Bonte et al. 2006). Recent studies have provided encouraging evidence on the capability of EEG and MEG to track neural correlates of the low-frequency amplitude envelope of continuous speech (Aiken and Picton 2008; Lalor and Foxe 2010; Ding and Simon 2012), a finding that has proven useful for investigating the mechanisms underlying speech processing (Luo and Poeppel 2007) and its dependence on attention and multisensory integration (Power et al. 2012; Zion Golumbic et al. 2013). However, the causes of this envelope-tracking phenomenon remain unclear.

Our recent research effort has focused on the identification of neural indices of continuous speech processing at different levels of this hierarchical system. In particular, phonetic-features of speech have been shown to capture information that does not emerge when using models based on the envelope, suggesting that EEG is indexing higher-level speech processing (Di Liberto et al. 2015). Here we seek to build on this work by disentangling the higher-level phonemic processing contributions from those related to the low-level acoustic properties of the speech signal. By utilising priming, we can control how much a degraded speech stimulus is recognised without changing the stimulus itself, exploiting the fact that speech recognition is an active process (Hickok 2012). In doing so, we demonstrate a framework for examining speech-specific processing using EEG.

2 Methods

2.1 Subjects and Data Acquisition

Ten healthy subjects (5 male, aged between 21 and 31) participated in this study. Electroencephalographic (EEG) data were recorded from 128 electrode positions, digitised at 512 Hz using a BioSemi Active Two system. Monophonic audio stimuli were presented at a sampling rate of 44,100 Hz using Sennheiser HD650 headphones and Presentation software from Neurobehavioral Systems (http://www.neurobs.com). Testing was carried out in a dark room and subjects were instructed to maintain visual fixation on a crosshair centred on the screen, and to minimize motor activities for the duration of each trial.

2.2 Stimuli and Experimental Procedure

Audio-book versions of two classic works of fiction read by the same American English speaker were partitioned into speech snippets, each with a duration of 10s. 120 snippets were randomly selected for the experiment. In order to alter the intelligibility of the speech, a method known as noise-vocoding was implemented (Shannon et al. 1995). This method filters the speech into a number of frequency-bands, and uses the amplitude envelope of each band to modulate band-limited noise. In our experiment, frequency-bands were logarithmically spaced between 70 and 5000 Hz. To determine the number of frequency-bands to use, an offline intelligibility test was performed for each subject using a separate set of 40 speech snippets. The number of frequency-bands was chosen such that subjects could understand an average of two words in each snippet. Each EEG trial consisted of the presentation of 3 speech snippets (Fig. 1). A standard trial consisted of a first presentation (snippet *a*) in which the speech was degraded using noise-vocoding as previously described. Therefore it was largely unintelligible. The second snippet (snippet *b*) contained the same speech, but in its original clear form. The noise-vocoded version was then played again (snippet *c*). As such, this third condition was affected by priming because the original clear speech was played immediately before and because the speech was vocoded at the limit of intelligibility. Importantly, the speech snippets used to determine the number of vocoding frequency-bands served also to reduce the adaptation to the noise-vocoded speech during the experiment (Davis et al. 2005).

In order to measure the effect of priming, we also introduced deviant trials. These trials consisted of a modified version of (*a*) and/or (*c*), where a random chunk of ~5s was replaced with words from a different trial. In both cases, the deviant probabilities were set to 10%. The participants were asked to identify the first and the second speech vocoded streams as a standard or a deviant presentation (after the presentations (*b*) and (*c*), respectively), using a level of confidence from 1 to 5 (from 'definitely a deviant' to 'definitely a standard').

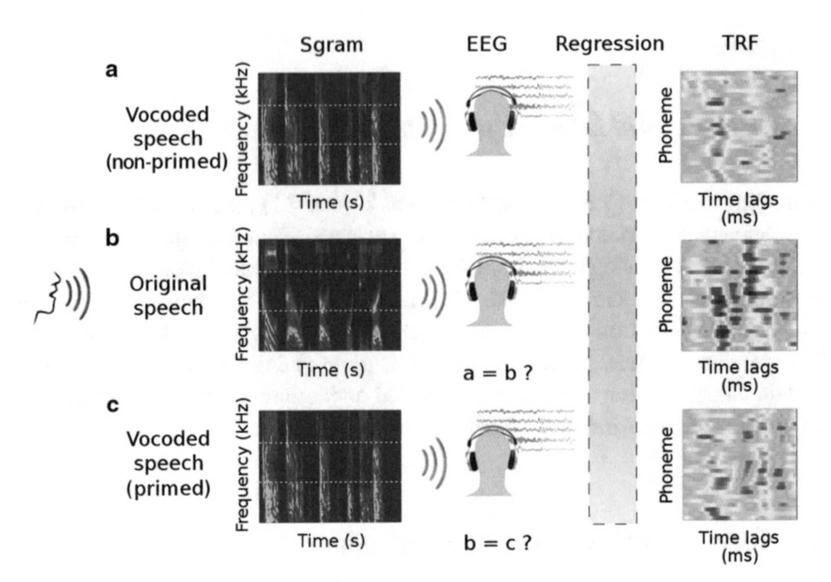

Fig. 1 EEG data were recorded while subjects listened to groups of three 10s long speech snippets. In the standard trial, the first (**a**) and third (**c**) speech streams of each group were the noise-vocoded version of the original one (**b**). The deviant presentations consisted of a modified version of (**a**) and/or (**c**). After both (**b**) and (**c**), the participants were asked to identify the first and the second speech vocoded streams as a standard or a deviant presentation. Linear regression was used to fit multivariate temporal response functions (TRFs) between the low-frequency EEG and each representation of the speech stimulus

2.3 EEG Data Analysis

The EEG data were inspected to identify channels with excessive noise or motor artefacts. Data were then digitally filtered using a Chebyshev type-2 band-pass filter with pass-band between 1 and 15 Hz and down-sampled to 128 Hz. Also, data was referenced to the average of the two mastoid channels.

Linear regression (http://sourceforge.net/projects/aespa) was used to create a mapping between the EEG and five different representations of the speech stimuli:

1. The broadband amplitude *envelope (Env)* of the speech signal, which was calculated as $Env = \left(x_a(t)\right)$, $x_a(t) = x(t) + jx(t)$, where $x_a(t)$ is the complex analytic signal obtained by the sum of the original speech $x(t)$ and its Hilbert transform $x(t)$.
2. The *spectrogram (Sgram)* was obtained by partitioning the speech signal into three frequency-bands logarithmically spaced between 70 and 5000 Hz according to Greenwood's equation (70–494–1680–5000 Hz, the same used for the vocoder), and computing the amplitude envelope for each band.

3. The *phonemic (Ph)* representation was computed using forced alignment (Yuan and Liberman 2008), given a speech file and the correspondent orthographic transcription broken into 26 phonemes in the International Phonetic Alphabet (IPA). A multivariate time-series composed of 26 indicator variables was then obtained.
4. The *phonetic features (Fea)* encoding is a linear mapping of the phonetic representation into a space of 18 features (Mesgarani et al. 2014), which describe specific articulatory and acoustic properties of the speech phonetic content (Chomsky and Halle 1968). In particular, the chosen features are related to the manner and place of articulation, to the voicing of a consonant, and to the backness of a vowel.
5. Finally, we propose a model that combines *Fea* and *Sgram* (*FS*) by applying a concatenation of the two representations.

For each representation of the speech, the result is a set of regression weights referred to as multivariate temporal response functions (TRFs). k-fold cross-validation ($k=10$) was employed to build predictions of the EEG signal from the TRFs of each distinct speech representation model. The prediction accuracies were obtained as an average of the correlation values over a set of 12 best predictive electrodes (6 on the left side of the scalp and the symmetrical counterparts on the right), on a time-lag window from 0 to 250 ms (Lalor et al. 2006; Di Liberto et al. 2015).

The study was undertaken in accordance with the Declaration of Helsinki and was approved by the Ethics Committee of the School of Psychology at Trinity College Dublin. Each subject provided written informed consent. Subjects reported no history of hearing impairment or neurological disorder.

3 Results

The main idea underlying the experimental paradigm was to take advantage of the priming effect to create two conditions in which the same speech stimuli were unintelligible and intelligible (condition *a* and *c* respectively; the prime is referred to as condition *b*). Such an experiment has the potential to disentangle acoustic and phonetic models and to allow us to address the non-overlapping cortical activity encoded by their TRFs.

3.1 Behavioural Results

The behavioural results (Fig. 2a) confirm that the experiment is creating these two conditions. Specifically, standard trials are detected with a higher confidence level (paired t-test, $p=3.1\times10^{-5}$) while no difference emerged for deviant trials ($p=0.41$). The distribution of responses (Fig. 2b, *top*) confirms that the participants

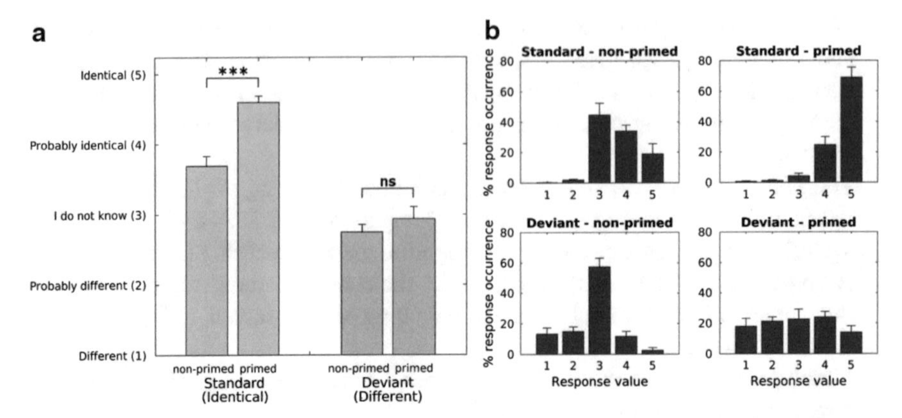

Fig. 2 Subjects were presented with sequences of vocoded-original-vocoded speech snippets and were asked to identify the two noise-vocoded streams (non-primed and primed conditions) as standard or deviant presentations by comparing them with the original speech snippet. Responses consisted of a level of confidence from 1 ('different') to 5 ('identical'). (**a**), the confidence in the identification of the standard trials was higher in the primed case for standard trials ($p<0.0005$), while no significant difference emerged for the deviants ($p>0.05$). (**b**), compares the distribution of response values across subjects in the primed and non-primed conditions, for standard and deviant trials

were more confident in the detection of standard trials; therefore they could better understand the vocoded speech in the primed condition. Interestingly, the deviant trials were not confidently detected (Fig. 2b, *bottom*), in fact they consisted of a vocoded speech snippet with a ~5s portion inconsistent with the original one, for which the prime has no effect. The hypothesis that subjects understand more the noise-vocoded speech presented after the prime is met for standard trials, therefore the analysis is conducted on them.

3.2 EEG Predictions and Model Comparison

128-channel EEG data were recorded from 10 subjects as they performed the task (Fig. 1). In this experiment, the original speech (condition *b*) was used as a prime for the second presentation of the noise-vocoded speech (condition *c*). Also, the results obtained in condition *b* provide a baseline with which to compare with previous research (Di Liberto et al. 2015). Indeed, the main interest relies on the comparison between vocoded speech in the primed (*a*) and in the non-primed (*c*) conditions. In particular, we studied the standard trials, in which the difference between the primed and non-primed conditions can originate only in the EEG signal and it is related to an improved intelligibility (Fig. 2a).

TRF models were fit using linear regression on the standard for each speech representation (*Env*, *Sgram*, *Ph*, *Fea*, and *FS*) in the primed and non-primed conditions. These models were then compared in terms of their ability of predicting the EEG data (see *Methods*). Interestingly, the phonetic models show an improvement

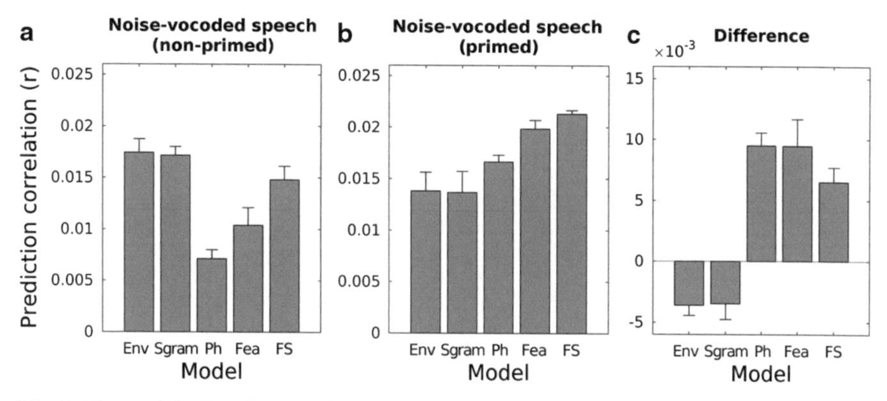

Fig. 3 The models *Env*, *Sgram*, *Ph*, *Fea*, and *FS* were compared using the Pearson correlation index between the EEG signal and its prediction as a quality indicator. The results for the non-primed and primed presentations of the vocoded speech are reported in (**a**) and (**b**) respectively, while (**c**) shows their difference (b–a). The error-bars represent the standard deviation of the Jack-knife distribution

in their EEG predictions unseen for the ones based only on acoustical information (Fig. 3). Importantly, this effect is driven by the whole distribution and not only by single subjects (Jack-knife analysis, two-way repeated measures ANOVA, Greenhouse-Geisser sphericity correction, $df = 1.8$, $F = 66.6$, $p < 0.0005$).

Further analysis can be performed on the model weights returned by the linear regression, which can be interpreted as the activation of macroscopic areas of the cerebral cortex in response to continuous speech at specific delays (time-lags). In the case of the current dataset, the three conditions can be compared in terms of response to *Env* (not shown) and to *Ph* and *Fea* (Fig. 4a and b).

4 Discussion

The ability of processing natural speech correctly is of extreme importance for humans as it represents the main method of direct communication between individuals. Although social interaction can be performed in other ways, deficits in any of its aspects can be the cause of discomfort, isolation, and depression. The awareness of this issue is essential for avoiding such consequences, however the diagnosis can be problematic, especially in the less severe cases. Certainly, methodologies for studying this articulated hierarchical system have the potential to drive the understanding of new insights into its mechanisms and to serve as an instrument of diagnosis for neural disorders related to speech. In particular, we are seeking the consolidation of a dependent measure capable of quantifying the higher-level processing of speech contribution to the scalp recorded neural activity.

EEG has been shown to be sensitive to phonetic-features of speech (Di Liberto et al. 2015). These categorical features capture information to which the envelope of speech is not sensitive, however the actual contribution of higher-level active

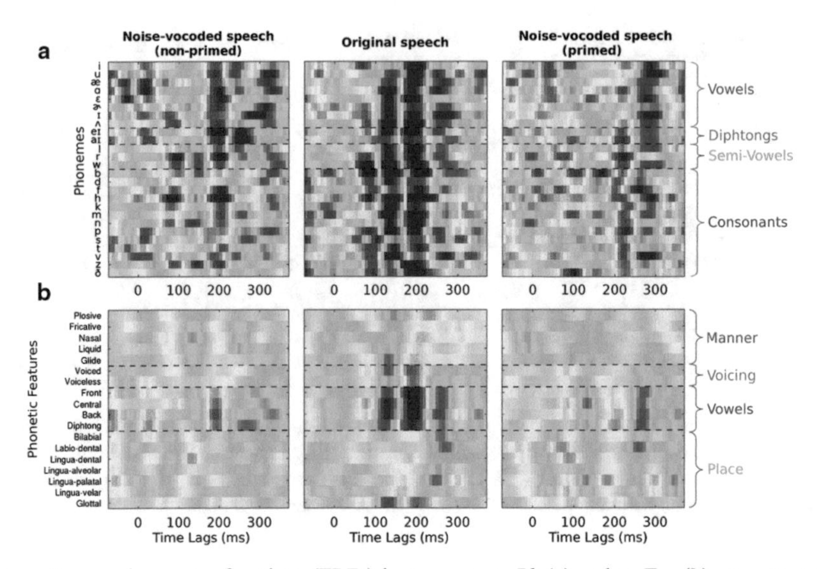

Fig. 4 Temporal response functions (TRFs) in response to *Ph* (**a**) and to *Fea* (**b**) are compared across conditions. The colour-scale relates to the neural activation in response to continuous speech for a specific phoneme or phonetic-feature. The *y-axis* labels are grouped in macroscopic categories which exhibit similar patterns of activations

processes related to intelligibility remains unclear. The framework we have defined here has the potential to disentangle these contributions from low-level EEG responses to spectrogram fluctuations. The results presented here support this hypothesis and motivate further studies.

The definition of a neural dependent measure of speech processing related to intelligibility could find application in clinical areas. For example, previous research (Pisoni 2000) suggests that the inability of deaf children with cochlear implants to discriminate fine phonetic differences in place of articulation and voicing is reflected in their perception of spoken words in terms of broad phonetic categories or functional equivalence classes. Indeed, such models are potentially useful in monitoring the development of patients in those situations and may provide new insights on the mechanisms underlying speech processing.

Acknowledgments Funding sources: Science Foundation Ireland; Irish Research Council.

References

Aiken SJ, Picton TW (2008) Envelope and spectral frequency-following responses to vowel sounds. Hear Res 245(1):35–47
Bonte M, Parviainen T, Hytönen K, Salmelin R (2006) Time course of top-down and bottom-up influences on syllable processing in the auditory cortex. Cereb Cortex 16(1):115–123
Chang EF, Rieger JW, Johnson K, Berger MS, Barbaro NM, Knight RT (2010) Categorical speech representation in human superior temporal gyrus. Nat Neurosci 13(11):1428–1432. [10.1038/nn.2641]. http://www.nature.com/neuro/journal/v13/n11/abs/nn.2641.html#supplementary-information
Chomsky N, Halle M (1968) The sound pattern of English. Harper & Row, New York
Davis MH, Johnsrude IS, Hervais-Adelman A, Taylor K, McGettigan C (2005) Lexical information drives perceptual learning of distorted speech: evidence from the comprehension of noise-vocoded sentences. J Exp Psychol Gen 134(2):222
DeWitt I, Rauschecker JP (2012) Phoneme and word recognition in the auditory ventral stream. Proc Natl Acad Sci U S A 109(8):E505–514. doi:10.1073/pnas.1113427109
Di Liberto GM, O'Sullivan JA, Lalor EC (2015) Low frequency cortical entrainment to speech reflects phonemic level processing. Curr Biol 25(19):2457–2465
Ding N, Simon JZ (2012) Emergence of neural encoding of auditory objects while listening to competing speakers. Proc Natl Acad Sci U S A 109(29):11854–11859. doi:10.1073/pnas.1205381109
Hickok G (2012) The cortical organization of speech processing: feedback control and predictive coding the context of a dual-stream model. J Commun Disord 45(6):393–402. doi:10.1016/j.jcomdis.2012.06.004
Lalor EC, Foxe JJ (2010) Neural responses to uninterrupted natural speech can be extracted with precise temporal resolution. Eur J Neurosci 31(1):189–193. doi:10.1111/j.1460-9568.2009.07055.x
Lalor EC, Pearlmutter BA, Reilly RB, McDarby G, Foxe JJ (2006) The VESPA: a method for the rapid estimation of a visual evoked potential. Neuroimage 32(4):1549–1561
Luo H, Poeppel D (2007) Phase patterns of neuronal responses reliably discriminate speech in human auditory cortex. Neuron 54(6):1001–1010
Mesgarani N, Cheung C, Johnson K, Chang EF (2014) Phonetic feature encoding in human superior temporal gyrus. Science 343(6174):1006–1010
Okada K, Rong F, Venezia J, Matchin W, Hsieh IH, Saberi K, Serences JT, Hickok G (2010) Hierarchical organization of human auditory cortex: evidence from acoustic invariance in the response to intelligible speech. Cereb Cortex 20(10):2486–2495. doi:10.1093/cercor/bhp318
Peelle JE, Johnsrude IS, Davis MH (2010) Hierarchical processing for speech in human auditory cortex and beyond. Front Hum Neurosci 4:51. doi:10.3389/fnhum.2010.00051
Pisoni DB (2000) Cognitive factors and cochlear implants: some thoughts on perception, learning, and memory in speech perception. Ear Hear 21(1):70–78
Power AJ, Foxe JJ, Forde EJ, Reilly RB, Lalor EC (2012) At what time is the cocktail party? A late locus of selective attention to natural speech. Eur J Neurosci 35(9):1497–1503. doi:10.1111/j.1460-9568.2012.08060.x
Shannon RV, Zeng FG, Kamath V, Wygonski J, Ekelid M (1995) Speech recognition with primarily temporal cues. Science 270(5234):303–304
Yuan J, Liberman M (2008) Speaker identification on the SCOTUS corpus. J Acoustical Soc Am 123(5):3878
Zion Golumbic EM, Cogan GB, Schroeder CE, Poeppel D (2013) Visual input enhances selective speech envelope tracking in auditory cortex at a "cocktail party". J Neurosci 33(4):1417–1426

Entracking as a Brain Stem Code for Pitch: The Butte Hypothesis

Philip X Joris

Abstract The basic nature of pitch is much debated. A robust code for pitch exists in the auditory nerve in the form of an across-fiber pooled interspike interval (ISI) distribution, which resembles the stimulus autocorrelation. An unsolved question is how this representation can be "read out" by the brain. A new view is proposed in which a known brain-stem property plays a key role in the coding of periodicity, which I refer to as "entracking", a contraction of "entrained phase-locking". It is proposed that a scalar rather than vector code of periodicity exists by virtue of coincidence detectors that code the dominant ISI directly into spike rate through entracking. Perfect entracking means that a neuron fires one spike per stimulus-waveform repetition period, so that firing rate equals the repetition frequency. Key properties are invariance with SPL and generalization across stimuli. The main limitation in this code is the upper limit of firing (~500 Hz). It is proposed that entracking provides a periodicity tag which is superimposed on a tonotopic analysis: at low SPLs and fundamental frequencies >500 Hz, a spectral or place mechanism codes for pitch. With increasing SPL the place code degrades but entracking improves and first occurs in neurons with low thresholds for the spectral components present. The prediction is that populations of entracking neurons, extended across characteristic frequency, form plateaus ("buttes") of firing rate tied to periodicity.

Keywords Autocorrelation · Brain stem · Phase-locking · Temporal

1 Introduction

> Pitch perception is a fascinating area because it appears so simple and yet is a process of considerable subtlety and complexity. (Green 1976)

P. X. Joris (✉)
Laboratory of Auditory Neurophysiology, University of Leuven, Herestraat 49 bus 1021, 3000 Leuven, Belgium
e-mail: Philip.Joris@med.kuleuven.be

P. van Dijk et al. (eds.), *Physiology, Psychoacoustics and Cognition in Normal and Impaired Hearing,* Advances in Experimental Medicine and Biology 894,
DOI 10.1007/978-3-319-25474-6_36

> Pitch perception is considered to represent the heart of hearing theory, and is, without doubt, the topic most discussed over the years. (Plomp 2002)

> Pitch may be the most important perceptual feature of sound. (Yost 2009)

Despite more than a century of study, there is no consensus regarding the basic nature of pitch, causing a palpable level of frustration among hearing researchers. The wealth of behavioral research on pitch perception contrasts with the paucity of physiological research into its neural basis beyond the level of the auditory nerve (AN). Processing in the central nervous system (CNS) is expected to be fundamentally different for the various temporal vs. spectral schemes that have been proposed, so physiological insights have the potential to reveal which (combination) of the two classes of schemes underlies human pitch perception.

A robust but implicit code for pitch exists in the AN in the form of an across-fiber pooled interspike interval (ISI) distribution (Cariani and Delgutte 1996a, 1996b; Meddis and Hewitt 1991a, b), which resembles the stimulus autocorrelation. An unsolved question is how and whether this implicit temporal representation is transformed into a more explicit representation. Following an early model (Licklider 1951), various autocorrelation-type schemes have been proposed, in which periodicity-tuning is generated by some combination of coincidence detection and a source of delay. It is generally assumed that such a computation is implemented at a brainstem level, where responses are temporally precise over a broad range of frequencies. However, recordings have not revealed convincing evidence for level-invariant, periodicity-tuned neurons or sources of delay that cover a sufficiently broad temporal range (Neuert et al. 2005; Sayles et al. 2013; Sayles and Winter 2008a, 2008b; Verhey and Winter 2006; Wang and Delgutte 2012).

Here, simple properties of the early central auditory system are brought into focus and it is argued that the relevant representation is fundamentally different from autocorrelation-type schemes *à la* Licklider (1951). The key proposal is that entrained phase-locking (contracted to "entracking") generates a scalar rate code for pitch early in the brainstem.

2 AN to Brain Stem: A Change in Orientation

In the AN, ISI histograms of responses to low-frequency pure tones are always multimodal (Kiang et al. 1965); i.e., after firing a spike, fibers often skip one or more cycles before firing another spike. Cycle skipping allows temporal and average rate behavior to be uncoupled. For example, neither the sigmoidal shape of rate-level functions (firing rate as a function of stimulus SPL), their dynamic range, nor maximum firing rate, are dependent on stimulus frequency *per se* but on stimulus frequency *relative* to characteristic frequency (CF). This behavior, combined with cochlear band-pass filtering, underlies rate-place coding: spectral components are translated into a firing rate profile of the population of AN fibers (Cedolin and Delgutte 2005; Larsen et al. 2008). This is a tonotopic or "vertical" view of the auditory system in which strength of response along the tonotopic axis is proportional to spectral energy. Experimental studies in which large populations of AN fibers are studied support this view (Delgutte and Kiang 1984; Kim and Molnar 1979; Sachs and Young 1979). Frequency *in an absolute sense* is inconsequential in these displays. For example, similar rate-place profiles would be expected for an animal with low-frequency hearing and one with high-frequency hearing, if the stimuli and filter widths and shapes could be appropriately scaled between the two species.

In contrast, in the CNS, skipping of cycles is often less prominent. At low stimulus frequencies, higher modes in the ISI distribution, at multiples of the stimulus period, can be strongly attenuated relative to the mode corresponding to the tone period (references: see Sect. 3). One corollary of this behavior is that average firing rate can approach the stimulus frequency. Perfect entracking means that a neuron fires one spike per cycle so that the firing rate equals the stimulus repetition frequency. A rate-place profile for neurons with perfect entracking would look quite different from that in the AN, and stimulus frequency *in an absolute sense* would now affect the display. For a low-frequency pure tone at some supra-threshold level, the rate-place profile would show a horizontal rather than a vertical pattern: all entracking neurons would be firing at the same firing rate, which would equal the stimulus frequency. Of course, only neurons with CF sufficiently close to the stimulus frequency would receive enough drive from their inputs to show entracking. The expected output is therefore a mixture of the vertical and horizontal pattern: a *butte* of activity in which all neurons would have the same average firing rate, and whose edges are formed by neurons whose CF is too far removed from the stimulus frequency for full entracking. A mixture of two low-frequency tones, if sufficiently separated in frequency, would be expected to generate two buttes with different height, corresponding to the two frequencies.

3 Entracking to Pure Tones

Pure tones are rare in nature but relevant in the context of pitch, not only to define and measure pitch, but also because of the dominant role of resolved harmonics (Plack and Oxenham 2005). Entracking to pure tones is visible in some of the early studies of the brainstem (Moushegian et al. 1967; Rose et al. 1974), and is extensively documented in studies from the Madison group (Joris et al. 1994b; Recio-Spinoso 2012; Rhode and Smith 1986). These data show that near-perfect entracking can be observed in low-CF neurons, or in the "low-frequency tail" of neurons with higher CFs. As expected from refractory behavior, there is an upper limit, which varies across neurons and across species. In rare instances neurons will fully entrack at 800 Hz or even higher, but more often the upper limit is near 500 Hz or lower. Over the frequency range of entracking, average firing rate increases linearly with frequency, and depends little on SPL except at low suprathreshold SPLs. One consequence is that neurons may fire at much higher rates to low-frequency stimuli in their "tail" than to tones near their CF (see Fig. 13 in Joris et al. 1994a).

Entracking is not a rare phenomenon. We and others have observed it in various cell types and nuclei, and in a number of species. In the ventral CN of the cat, most types of projection neuron display this behavior to some degree (spherical and globular bushy cells, octopus cells, commissural multipolar cells, stellate cells). We have observed it in the medial nucleus of the trapezoid body (Mc Laughlin et al. 2008) and in other neurons of the superior olivary complex (e.g. Joris and Smith 2008). We have observed entracking in different species (cat, chinchilla, gerbil; see also studies cited above), and have limited evidence in the CN of macaque monkey.

An important qualifier is that the entracking observed is not always the extreme form (exactly 1 spike/cycle), particularly at frequencies above a few 100 Hz. The defining property is that there is an effect of absolute frequency on average firing rate so that rate increases monotonically with frequency up to a certain maximum. Thus, in the brainstem, firing rate does not only depend on SPL and stimulus frequency relative to CF, as it does in the AN, but also depends on absolute stimulus frequency. For the extreme cases of this behavior a stronger statement can be made: SPL and stimulus frequency relative to CF have remarkably little effect, and absolute frequency is the overriding stimulus parameter determining response rate.

Besides being present in various cell types, nuclei, and species, entracking has inherent properties that make it an attractive coding mechanism. It is remarkably invariant with SPL, i.e. once perfect entracking is reached, further increases in SPL do not affect average response rate: the rate-level function shows a limited (20 dB) dynamic range, and at higher SPLs the firing rate remains clamped at the stimulus frequency. At the population level, increases in SPL cause an increase in the number of entracking neurons. A second striking property is the low variability in firing rate. In some cases there is no variability: exactly the same number of spikes is generated in response to the same stimulus presented at the same or other SPLs.

4 Yes We SAM

Data regarding entracking to pitch stimuli other than pure tones are limited. An early striking example is to click trains: octopus cells in the CN can fire one spike per click up to ~700 clicks/s (Godfrey et al. 1975; Oertel et al. 2000). This behavior occurs at high CFs, to which octopus neurons are biased. As mentioned, octopus cells that can be driven by low-frequency tones also show entracking to tones. In AN fibers, firing rate also shows some dependence on click train frequency, but much weaker than in octopus cells.

We have also observed entracking to sinusoidally amplitude modulated (SAM) tones. Figure 3G in Joris and Smith (2008) shows data at different SPLs over a range of modulation frequencies for one monaural neuron, recorded just laterally to the MSO, in an area which may correspond to the mLNTB (Spirou and Berrebi 1996). Similar to entracking to pure tones mentioned in the previous section, firing rate increased linearly with modulation frequency; was similar for the different SPLs tested; and declined steeply once an upper limit is reached. The CF of the neuron was 2.4 kHz, but the response to pure tones at CF was much lower than to low-frequency tones delivered in the tail of the tuning curve, to which the neuron entracked (Fig. 3C in Joris and Smith 2008).

5 Scalar *versus* Vector Code

Various schemes and neural mechanisms have been proposed for the encoding of pitch-related periodicity based on the temporal information carried by peripheral neurons. They have in common that they predict tuning in which a neuron is optimally responsive to a certain periodicity. Sounds that differ in pitch would activate different neurons; different sounds with the same pitch would activate the same neurons. Such schemes are referred to as vector codes (Churchland and Sejnowski 1992). Periodicity tuning is typically achieved by comparing spike trains with a delayed copy. The comparison involves multiplicative, subtractive, or additive neural interactions; and some source of delay from axons, cochlea, synapses, intrinsic membrane properties, etc. (reviewed by de Cheveigné 2005). From a physiological point of view, problems with this approach are that there is only limited evidence for such tuning in the brainstem and only over a limited range of periodicities; that no convincing delay mechanisms have been identified that cover a wide range of values (tens to tenths of ms); and that some of these schemes require elaborate (and biologically implausible) wiring.

The scheme proposed here dispenses with delays and suggests a scalar code, based on a property which is physiologically well documented. We surmise that entracking neurons at the brain stem level are not tuned to a particular periodicity but all code for a range of periodicities by their monotonic relationship between average firing rate and pitch-related period.

6 Buttes

To some extent (increasingly so with increasing SPL), the representation hypothesized here is "orthogonal" to the tonotopic representation. The rate-place profile in the AN to a low-frequency tone increasingly broadens with increasing SPL, and flattens due to rate saturation (Kim and Molnar 1979). If a population of central neurons shows entracking, this property causes a stratification in the rate-place profile. Instead of a vertical or "hilly" profile where resolved components cause local increases in firing rate and unresolved components lead to a broad mound of activity, entracking generates horizontally flattened profiles or buttes, for which firing rate is dictated by the dominant interval between spikes fed to neurons in these frequency channels. For multiple, truly resolved components, a staircase of buttes would result with increasing firing rates corresponding to the frequency of successive harmonics. For unresolved components, many factors come into play (filter shape, limits to fine-structure and envelope, component phase), but buttes could be formed by neurons entracking to the dominant stimulus period.

An obvious limitation in this scalar code is the upper limit of firing (usually near ~500 Hz). We surmise that at low SPLs and fundamental frequencies >500 Hz, a spectral or place mechanism codes for pitch (Cedolin and Delgutte 2005; Larsen et al. 2008). With increasing SPL the place code degrades but entracking improves and first occurs in the neurons with the lowest thresholds for the spectral components present.

For a given stimulus and time window, a neuron can only have one firing rate. One may wonder what distinguishes the low firing rate of an entracking neuron to a low frequency component vs. the low firing rate to a component above the entracking limit. The key would again be in the uniform firing rate, with low variability, across neurons in the case of a low frequency component. Above the entracking limit, firing rate is no longer clamped to the value dictated by the dominant ISI of a neuron's inputs and will vary across neurons.

7 Discussion

The main goal of this chapter is to introduce a new way of thinking about pitch coding, grounded in CNS physiology. If there is a robust representation of pitch in the dominant ISI distribution in the AN (Cariani and Delgutte 1996a, 1996b), and if some neurons convert ISI directly into a corresponding firing rate, then it seems possible that the dominant ISI interval is coded as predominant firing rate. Evidently, the scheme proposed here is incomplete. Questions arise how and where a butte profile would be read out; how such a representation would mesh with the spectral representation needed for F0 above ~500 Hz; how phase-invariant this representation would be; etc. We conclude with some interrelated issues.

An important issue is the effective bandwidth of central neurons. Several CN neuron types integrate over wide frequency regions (Godfrey et al. 1975; Winter and Palmer 1995): partials that are resolved at the level of the AN may be unresolved in these CN populations. The autocorrelogram-like display in the dominant ISI hypothesis sums across frequency channels (Cariani and Delgutte 1996a, 1996b): for such an operation a wide bandwidth would be beneficial. Another issue is whether there is a specific physiological subset of neurons or even a separate brainstem nucleus which codes periodicity via entracking. Entracking is observed in a diversity of structures and neuron types, suggesting a distributed mechanism, but this does not exclude the existence of a brainstem "pitch center" specialized in this form of encoding. For CN neurons showing entracking, convergence of multiple inputs from the AN is obviously required; the degree of entracking in responses beyond the CN suggests that there are multiple stages of such convergence. A strong form of the butte hypothesis is based on perfect entracking; a weaker form only requires a monotonic relationship between firing rate and pitch-related period without attaining equality. One of the most critical issues is phase invariance, which we see as an experimental issue. Some neurons in the CN show good envelope coding to quasi-frequency-modulated (QFM) stimuli (Rhode 1995). Possibly there are always some neurons entracking at the pitch-related period, no matter what the phase spectrum is. Finally, entracking is invariably accompanied by exquisite phase-locking. One could debate whether it fundamentally is a temporal or rate code. The temporal aspects of the response are disregarded here in the sense that it is the constancy in rate, within and across neurons and stimuli, that codes for pitch, while the phase of spiking is surmised to be irrelevant.

References

Cariani P, Delgutte B (1996a) Neural correlates of the pitch of complex tones. II. Pitch shift, pitch ambiguity, phase invariance, pitch circularity, rate pitch, and the dominance region for pitch. J Neurophysiol 76:1717–1734

Cariani P, Delgutte B (1996b) Neural correlates of the pitch of complex tones. I. Pitch and pitch salience. J Neurophysiol 76:1698–1716

Cedolin L, Delgutte B (2005) Pitch of complex tones: rate-place and interspike interval representations in the auditory nerve. J Neurophysiol 94(1):347–362

Churchland P, Sejnowski TJ (1992) The computational brain. MIT Press, Cambridge

De Cheveigné A (2005). Pitch perception models. In: Plack CJ, Fay RR, Oxenham AJ, Popper AN (eds) Pitch. Springer New York, New York, pp 169–233

Delgutte B, Kiang NYS (1984) Speech coding in the auditory nerve: i. Vowel-like sounds. J Acoust Soc Am 75:866–878

Godfrey DA, Kiang NYS, Norris BE (1975) Single unit activity in the posteroventral cochlear nucleus of the cat. J Comp Neurol 162:247–268

Green D (1976) An introduction to hearing. Lawrence Erlbaum Associates, Hillsdale

Joris PX, Smith PH (2008) The volley theory and the spherical cell puzzle. Neuroscience 154(1):65–76

Joris PX, Smith PH, Yin TC (1994a) Enhancement of neural synchronization in the anteroventral cochlear nucleus. II. Responses in the tuning curve tail. J Neurophysiol 71(3):1037–1051

Joris PX, Carney LH, Smith PH, Yin TC (1994b) Enhancement of neural synchronization in the anteroventral cochlear nucleus. I. Responses to tones at the characteristic frequency. J Neurophysiol 71(3):1022–1036

Kiang NYS, Watanabe T, Thomas EC, Clark LF (1965) Discharge patterns of single fibers in the cat's auditory nerve. MIT Press, Cambridge. Research Monograph No 35

Kim DO, Molnar CE (1979) A population study of cochlear nerve fibers: comparison of spatial distributions of average-rate and phase-locking measures of responses to single tones. J Neurophysiol 42:16–30

Larsen E, Cedolin L, Delgutte B (2008) Pitch representations in the auditory nerve: two concurrent complex tones. J Neurophysiol 100(3):1301–1319
Licklider JCR (1951) A duplex theory of pitch perception. Experientia 7(4):128–134
Mc Laughlin M, van der Heijden M, Joris PX (2008) How secure is in vivo synaptic transmission at the calyx of Held? J Neurosci 28(41):10206–10219
Meddis R, Hewitt MJ (1991a) Virtual pitch and phase sensitivity of a computer model of the auditory periphery. II: phase sensitivity. J Acoust Soc Am 89(6):2883–2894
Meddis R, Hewitt MJ (1991b) Virtual pitch and phase sensitivity of a computer model of the auditory periphery. I: pitch identification. J Acoust Soc Am 89(6):2866–2882
Moushegian G, Rupert AL, Langford TL (1967) Stimulus coding by medial superior olivary neurons. J Neurophysiol 30(5):1239–1261
Neuert V, Verhey JL, Winter IM (2005) Temporal representation of the delay of iterated rippled noise in the dorsal cochlear nucleus. J Neurophysiol 93(5):2766–2776
Oertel D, Bal R, Gardner SM, Smith PH, Joris PX (2000) Detection of synchrony in the activity of auditory nerve fibers by octopus cells of the mammalian cochlear nucleus. Proc Natl Acad Sci USA 97(22):11773–11779
Plack CJ, Oxenham AJ (2005) The psychophysics of pitch. In: Plack CJ, Oxenham AJ, Fay RR, Popper AN (eds) Pitch: neural coding and perception, vol 24. Spinger, New York, pp 7–55
Plomp R (2002) The intelligent ear: on the nature of sound perception. Lawrence Erlbaum Associates, Mahwah
Recio-Spinoso A (2012) Enhancement and distortion in the temporal representation of sounds in the ventral cochlear nucleus of chinchillas and cats. PloS ONE 7(9):e44286
Rhode WS (1995) Interspike intervals as a correlate of periodicity pitch in cat cochlear nucleus. J Acoust Soc Am 97(4):2414–2429
Rhode WS, Smith PH (1986) Encoding timing and intensity in the ventral cochlear nucleus of the cat. J Neurophysiol 56:261–286
Rose JE, Kitzes LM, Gibson MM, Hind JE (1974) Observations on phase-sensitive neurons of anteroventral cochlear nucleus of the cat: nonlinearity of cochlear output. J Neurophysiol 37:218–253
Sachs MB, Young ED (1979) Encoding of steady-state vowels in the auditory nerve: representation in terms of discharge rate. J Acoust Soc Am 66(2):470–479
Sayles M, Winter IM (2008a) Ambiguous pitch and the temporal representation of inharmonic iterated rippled noise in the ventral cochlear nucleus. J Neurosci 28(46):11925–11938
Sayles M, Winter IM (2008b) Reverberation challenges the temporal representation of the pitch of complex sounds. Neuron 58(5):789–801
Sayles M, Füllgrabe C, Winter IM (2013) Neurometric amplitude-modulation detection threshold in the guinea-pig ventral cochlear nucleus. J Physiol 591(Pt 13):3401–3419
Spirou GA, Berrebi AS (1996) Organization of ventrolateral periolivary cells of the cat superior olive as revealed by PEP-19 immunocytochemistry and Nissl stain. J Comp Neurol 368(1):100–120
Verhey JL, Winter IM (2006) The temporal representation of the delay of iterated rippled noise with positive or negative gain by chopper units in the cochlear nucleus. Hear Res 216–217:43–51
Wang GI, Delgutte B (2012) Sensitivity of cochlear nucleus neurons to spatio-temporal changes in auditory nerve activity. J Neurophysiol 108(12):3172–3195
Winter IM, Palmer AR (1995) Level dependence of cochlear nucleus onset unit responses and facilitation by second tones or broadband noise. J Neurophysiol 73:141–159
Yost WA (2009) Pitch perception. Atten Percept Psychophys 71(8):1701–1715

Can Temporal Fine Structure and Temporal Envelope be Considered Independently for Pitch Perception?

Nicolas Grimault

Abstract In psychoacoustics, works on pitch perception attempt to distinguish between envelope and fine structure cues that are generally viewed as independent and separated using a Hilbert transform. To empirically distinguish between envelope and fine structure cues in pitch perception experiments, a dedicated signal has been proposed. This signal is an unresolved harmonic complex tones with all harmonics shifted by the same amount of Hz. As the frequency distance between adjacent components is regular and identical than in the original harmonic complex tone, such a signal has the same envelope but a different fine structure. So, any perceptual difference between these signals is interpreted as a fine structure based percept. Here, as illustrated by very basic simulations, I suggest that this orthogonal point of view that is generally admitted could be a conceptual error. In fact, neither the fine structure nor the envelope is required to be fully encoded to explain pitch perception. Sufficient information is conveyed by the peaks in the fine structure that are located nearby a maximum of the envelope. Envelope and fine structure could then be in perpetual interaction and the pitch would be conveyed by "the fine structure under envelope". Moreover, as the temporal delay between peaks of interest is rather longer than the delay between two adjacent peaks of the fine structure, such a mechanism would be much less constrained by the phase locking limitation of the auditory system. Several data from the literature are discussed from this new conceptual point of view.

Keywords Pitch · Envelope · Fine structure

N. Grimault (✉)
Centre de Recherche en Neurosciences de Lyon, CNRS UMR 5292, Université, Lyon 1, Lyon, France
e-mail: nicolas.grimault@cnrs.fr

P. van Dijk et al. (eds.), *Physiology, Psychoacoustics and Cognition in Normal and Impaired Hearing,* Advances in Experimental Medicine and Biology 894, DOI 10.1007/978-3-319-25474-6_37

1 Introduction

Numerous works about pitch perception attempt to identify and to distinguish between relevant temporal cues as envelope and fine structure. These cues are generally viewed as independent. In fact, mathematically, these cues are orthogonal and can be extracted and separated using a Hilbert transform. Using this mathematical decomposition, some works suggest that musical pitch would rely mostly on the fine structure and that speech perception would rely mostly on envelope (Smith et al. 2002). To empirically distinguish between envelope and fine structure cues in pitch perception experiments, a dedicated signal has been proposed as early as 1956 (De Boer 1956). This signal is equivalent of an unresolved harmonic complex tone with all harmonics shifted by the same amount of Hz. As the frequency distance between adjacent components of such a signal is regular and identical than in the original harmonic complex tone, such a signal has the same envelope but a different fine structure. As a consequence, any perceptual difference between the harmonic and the shifted complex is often interpreted as a pure fine structure based percept.

Figure 1 represents the temporal information that potentially convey some pitch related temporal information. These temporal cues can be classified in three categories of periodicities. First, the periodicities of the carrier related to the delay dT1 in Fig. 1. This periodicity provides some information about the frequency of the pure tone or about the frequency of the carrier. To extract dT1 periodicities, the auditory system is required to be able to phase lock on the instantaneous phase of the signal. dT1 periodicities will be called temporal fine structure periodicities (TSF_{period}) in the following. Second, the periodicities of the envelope (delay dT2 in Fig. 1) provides some information about the F0 for harmonic complex tones. dT2 periodicities

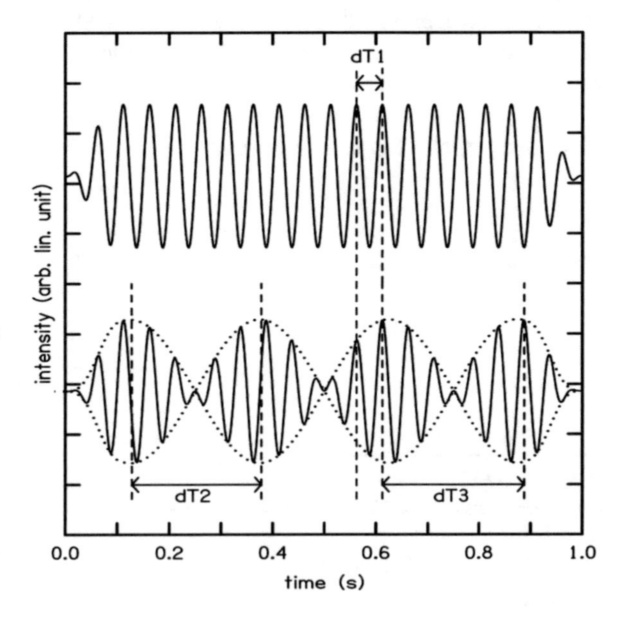

Fig. 1 Potential temporal periodicity cues usable for pith perception of a pure tone (*upper* waveform) or a shifted complex tone (*lower* waveform). dT1 is the time delay between two adjacent peaks in the temporal fine structure, dT2 is the time delay between two adjacent peaks in the temporal envelope and dT3 is the time delay between two maximums of the temporal fine structure located nearby two successive maximums of the temporal envelope. dT2 and dT3 are confounded for harmonic complex tones but different for shifted complex tones

will be called temporal envelope periodicities (TE_{period}) in the following. Third, the periodicities of the fine structure modulated by the envelope (delay dT3 in Fig. 1). This type of periodicities corresponds to the time delays between two energetic phases of the signal located nearby two successive maxima in the envelope. For harmonic complex tones, dT2 is equal to dT3 but for shifted complex tones, dT2 and dT3 are different. As dT3 is necessary a multiple of dT1, for a shifted complex, dT3 will be either equal to $n \times dT1$ or $(n+1) \times dT1$ choosing the integer n to verify the chain of inequations: $n \times dT1 \leq dT2 < (n+1) \times dT1$. As such, dT3 depends on both dT1 (fine structure information) and dT2 (envelope information) and can then be described as a periodicity related to an interaction between fine structure and envelope. dT3 periodicities will be called interaction periodicities ($TE \times TSF_{period}$) in the following.

However, most of the time, when a harmonic complex tone and a shifted complex tone with the same envelope frequency conduct to a different pitch percept, the pitch percept is supposed to be elicited by fine structure only and to be independent of envelope. This can appear as trivial but the aim of this study is simply to convince the reader that this assumption is not true.

2 Methods

2.1 Simulation

A very basic simulation has been performed to check for the potential use of various types of periodicity cues (i.e. TSF_{period}, TE_{period} or $TE \times TSF_{period}$) with various signals used in the literature. The basic idea was to pass the stimulus through a dynamic compressive gammachirp (the default auditory filter-bank of the Auditory Image Model) (Irino and Patterson 2006) and to transform the output waveform into a spiketrain by using a threshold dependent model. This model has a higher probability to spike each time the output is over a threshold value (see Eq. 1). Moreover, a reasonable refractory period of 2500 Hz is added to the simulation. The periodicities of the spiketrain are finally extracted with an autocorrelation. So, the spiketrain generation is based on the following formula which depends on both envelope and fine structure temporal information:

$$\text{spiketrain}(t) = \begin{cases} 1 \text{ if } U(t) \times TFS_{\text{signal}>0}(t) \times E(t) \times P_{\text{refract}}(t) > \text{Thres} \\ 0 \text{ if } U(t) \times TFS_{\text{signal}>0}(t) \times E(t) \times P_{\text{refract}}(t) > \text{Thres} \end{cases} \quad (1)$$

Where

U(t) is an uniform intern noise between 0 and 1,

$TSF_{signal>0}(t)$ is the positive part of the fine structure at the output of an auditory filter,

E(t) is the envelop at the output of an auditory filter,

$P_{refract}(t)$ is the refractory period. This function is either equal to 0 or 1 related to a refractory period equal to 1/2500. So, *if* $spiketrain(t_0) = 1$, $P_{refract}(t) = 0$ *for* $t_0 < t < t_0 + 1/2500$ *and* $P_{refract}(t_0 + 1/2500) = 1$,

Thres is the discharge threshold sets here to 0.5,

Which is exactly equivalent to:

$$\text{spiketrain}(t) = \begin{Bmatrix} 1 \text{ if } \mathrm{U}(t) \times \text{Signal}_{\text{signal}>0}(t) \times P_{\text{refract}}(t) > \text{Thres} \\ 0 \text{ if } \mathrm{U}(t) \times \text{Signal}_{\text{signal}>0}(t) \times P_{\text{refract}}(t) > \text{Thres} \end{Bmatrix} \quad (2)$$

where

$Signal_{signal>0}(t)$ is the positive part of the input signal at the output of an auditory filter.

As an intern noise U has been added, each signal is passed 300 times in the simulation model to estimate the distribution of the periodicities of the spiketrain. Moreover, for each signal, a single auditory filter output, located in the passband of the input signal and centred on the carrier frequency (f_c), has been used.

2.2 Stimuli

The stimuli used by Santurette and Dau (2011) and by Oxenham et al. (2011) have been generated and processed through the simulation. In Santurette and Dau (2011), the signals were generated by multiplying a pure tone carrier with frequency f_c with a half-wave rectified sine wave modulator with modulation frequency f_{env} and low-pass filtered by a 4th order Butterworth filter with cut-off frequency of $0.2 \times f_c$. All signals were generated at 50 dB SPL and mixed with a TEN noise at 34 dB SPL per ERB. All f_c and f_{env} values are indicated in Fig. 2. When f_c and f_{env} are not multiple from each other, this manipulation produce a shifted complex. In Oxenham et al. (2011), harmonic complex tones at various F0 values (indicated in Fig. 3) were generated by adding in random phase up to 12 consecutive harmonics, beginning on the sixth. Harmonics above 20 kHz were not generated. All harmonics were generated at 55 dB SPL per component and all signals were embedded in a broadband TEN noise at 45 dB per ERB. A shifted version of each harmonic complex tone was also generated by shifting all components of the complex tone by an amount of $0.5 \times$ F0.

3 Results

The outputs of the simulation provide the distributions of the temporal periodicities of the spiketrain. This is supposed to predict the perceived pitch evoked by the signal.

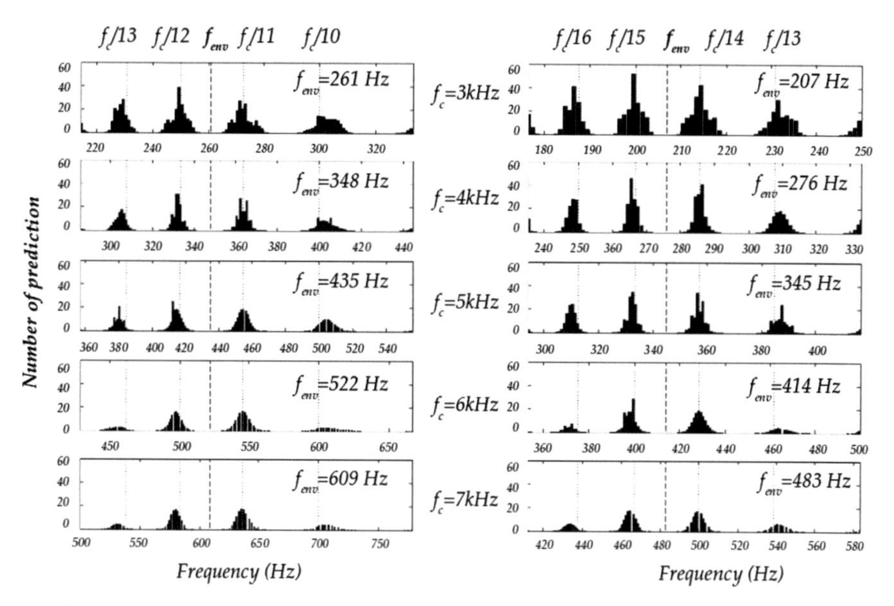

Fig. 2 Outputs of the simulation fed with the signals used in Santurette and Dau (2011). The distributions of pitch estimation are always related to TE × TSF_{period} and never on TE_{period}. This is closely consistent with the data reported in Figs. 4 and 6 in Santurette and Dau. (2011)

The results of the simulation plotted in Fig. 2 evidence that, as in Santurette and Dau (2011), the predicted pitch values are always related to TE × TSF_{period} and never related to TE_{period}. This is strongly consistent with the results reported by the authors.

The results of the simulation plotted in Fig. 3 are strongly consistent with Oxenham et al. (2011). Using a plausible refractory period of 2500 Hz, the simulation is able to extract the TE × TSF_{period} even if the TSF_{period} are too fast to be correctly encoded. The decrease in performances reported in Oxenham et al. (2011) when

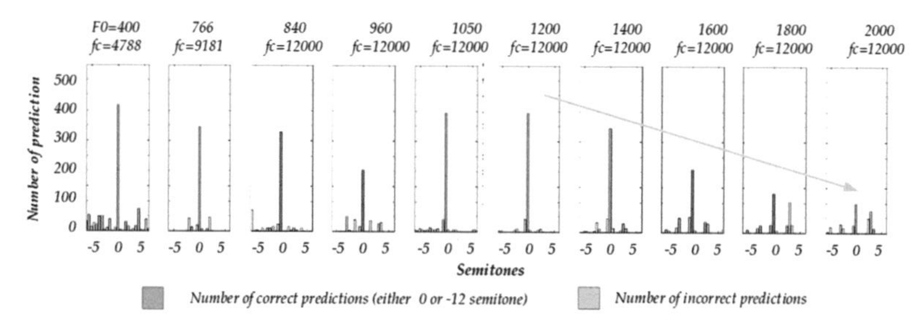

Fig. 3 Outputs of the simulation at the auditory filters centred on fc, fed with the signals used in Oxenham et al. (2011). The distributions of pitch estimation are closely consistent with the data reported in Fig. 2B in Oxenham et al. (2011). The *gray arrow* indicates the decrease in the number of predictions when increasing the resolvability of the complex by increasing the F0

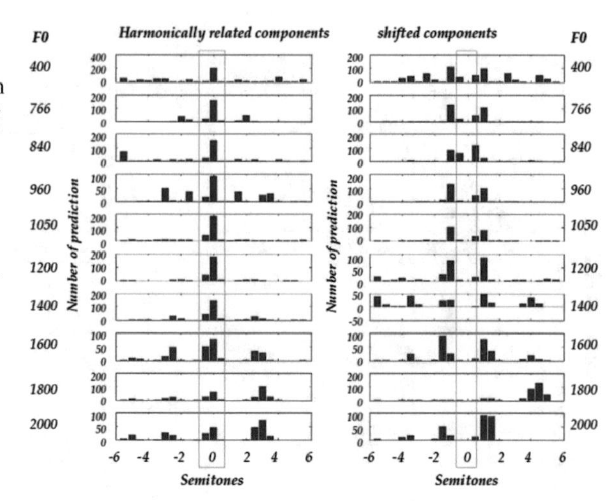

Fig. 4 Output of the simulation fed with the signals used in Oxenham et al. (2011) with harmonic complex tones (*left* column) and shifted complex tones (*right* column). The distributions of pitch estimation are always related to TE $\times$ TSF_{period} which predict a peak centred in the *gray rectangle* on the *left* and peaks on either side of the *gray rectangle* on the *right*. This is closely consistent with the data reported in experiment 1 and 2 in Oxenham et al. (2011)

increasing the F0 from 1200 Hz to up to 2000 Hz is also simulated. This decrease is probably explained by a decrease in resolvability (less and less interactions between harmonic components) when increasing the F0.

Finally, the results of the simulation plotted on the right column of Fig. 4 are strongly consistent with the results found with the signals used by Santurette and Dau (2011), and also evidence that the predicted pitch values are always related to TE $\times$ TSF_{period} and never related to TE_{period}.

As a control, the effect of threshold value used in the simulation has been tested with one complex tone having a F0 equal to 1400 Hz (Fig. 5). Varying the threshold value have some important incidence on the predictions. Using a threshold below 0.3 does not provide reliable periodicities estimations. Using a threshold from 0.4 to 0.8 provides reliable and consistent estimations. Using a high threshold value (over 0.9) prevents to report any periodicities. Using 0.5 as in the previous simulations appears then to be a good compromise. It is worth noting that such a threshold model is physiologically plausible and could be related to the thresholds of the auditory nerve fibres previously described in the literature (Sachs and Abbas 1974).

4 Discussion

4.1 Conclusions

These stimulations have a double interest.

First, this evidence that any perceptual effect that is empirically evidenced between harmonic complex tones and shifted complex tones should not been interpreted as a pure effect of fine structure. In fact, the pitch evoked by a shifted complex is based on interaction cues between envelope and fine structure (TE $\times$ TSF_{period}).

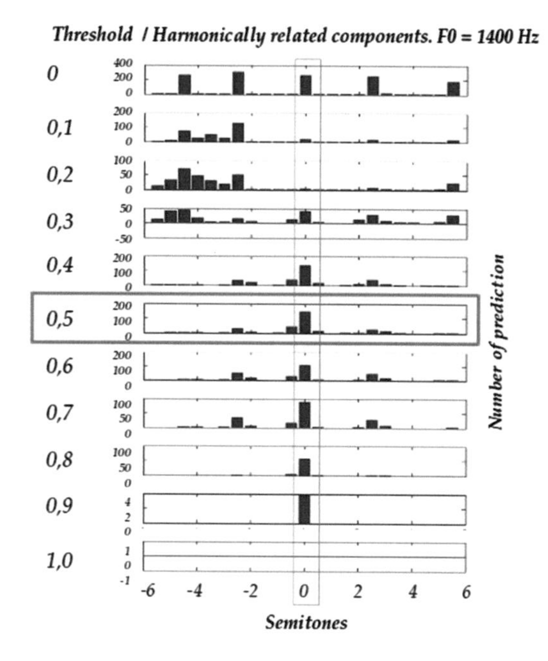

Fig. 5 Output of the simulation fed with a single harmonic complex tone (F0 = 1400 Hz) used in Oxenham et al. (2011). Effect of threshold value from 0 to 1 on periodicity estimations

Using these signals to tease apart temporal envelope cues from temporal fine structure cues is then a conceptual error. This impaired the conclusion that the pitch of unresolved complex tones is based *only* on fine structure information.

Second, when thinking about pitch perception of unresolved complex tones in terms of interaction between envelope and fine structure, it appears that the limitation of phase locking is probably much less critical than when thinking in terms of fine structure only. In fact, it seems clear that there is no need to encode every phases of the signal to encode the most intense phases located nearby an envelope maximum. This explains that the simulation can extract a periodicity related to pitch when the carrier frequency is over 10 kHz (Fig. 3).

4.2 Limitations

First, the current simulations are not a physiologically-based model of pitch perception and the refractory period which is used here does not accurately describe the physiological constraints for the phase locking. Some further works that would use realistic models of the auditory periphery should be used for further explorations.

Second, this simulation does not explain all the data reported in the literature about pitch perception. For example, experiment 2c in Oxenham et al. (2011) reports some pitch perception using dichotic stimulations with even-numbered harmonics presented on the right ear and odd-numbered harmonics presented to the left ear. This experimental manipulation increases the resolvability of the signals and prevents our simulation to extract any temporal periodicities and then to predict some pitch perception.

Acknowledgments This work was supported by institutional grants ("Centre Lyonnais d'Acoustique," ANR-10-LABX-60) and (Contint—Aïda, ANR-13-CORD-0001). I thank Andy Oxenham and Etienne Gaudrain for constructive and interesting comments about this work.

References

De Boer E (1956) Pitch of inharmonic signals. Nature 178(4532):535–536

Irino T, Patterson RD (2006) A dynamic compressive gammachirp auditory filterbank, IEEE Trans Audio, Speech, Lang process 14(6):2222–2232

Oxenham AJ, Micheyl C, Keebler MV, Loper A, Santurette S (2011). Pitch perception beyond the traditional existence region of pitch. Proc Nat Acad Sci U S A 108(18):7629–7634. http://doi.org/10.1073/pnas.1015291108

Sachs MB, Abbas PJ (1974) Rate versus level functions for auditory-nerve fibers in cats: tone-burst stimuli. J Acoust Soc Am 56(6):1835–1847

Santurette S, Dau T (2011). The role of temporal fine structure information for the low pitch of high-frequency complex tones. J Acoust Soc Am 129(1):282–292. http://doi.org/10.1121/1.3518718

Smith ZM, Delgutte B, Oxenham AJ (2002). Chimaeric sounds reveal dichotomies in auditory perception. Nature 416(6876):87–90. http://doi.org/10.1038/416087a

Locating Melody Processing Activity in Auditory Cortex with Magnetoencephalography

Roy D. Patterson, Martin Andermann, Stefan Uppenkamp and André Rupp

Abstract This paper describes a technique for isolating the brain activity associated with melodic pitch processing. The magnetoencephalograhic (MEG) response to a four note, diatonic melody built of French horn notes, is contrasted with the response to a control sequence containing four identical, "tonic" notes. The transient response (TR) to the first note of each bar is dominated by energy-onset activity; the melody processing is observed by contrasting the TRs to the remaining melodic and tonic notes of the bar (2–4). They have uniform shape within a tonic or melodic sequence which makes it possible to fit a 4-dipole model and show that there are two sources in each hemisphere—a melody source in the anterior part of Heschl's gyrus (HG) and an onset source about 10 mm posterior to it, in planum temporale (PT). The N1m to the initial note has a short latency and the same magnitude for the tonic and the melodic sequences. The melody activity is distinguished by the relative sizes of the N1m and P2m components of the TRs to notes 2–4. In the anterior source a given note elicits a much larger N1m-P2m complex with a shorter latency when it is part of a melodic sequence. This study shows how to isolate the N1m, energy-onset response in PT, and produce a clean melody response in the anterior part of auditory cortex (HG).

R. D. Patterson (✉)
Department of Physiology, Development and Neuroscience, University of Cambridge, Cambridge, UK
e-mail: rdp1@cam.ac.uk

M. Andermann · A. Rupp
Section of Biomagnetism, Department of Neurology, University of Heidelberg, Heidelberg, Germany
e-mail: andermann@uni-heidelberg.de

A. Rupp
e-mail: andre.rupp@uni-heidelberg.de

S. Uppenkamp
Medizinische Physik and Cluster of Excellence Hearing4All, Carl von Ossietzky Universität Oldenburg, 26111 Oldenburg, Germany
e-mail: stefan.uppenkamp@uni-oldenburg.de

P. van Dijk et al. (eds.), *Physiology, Psychoacoustics and Cognition in Normal and Impaired Hearing,* Advances in Experimental Medicine and Biology 894,
DOI 10.1007/978-3-319-25474-6_38

Keywords Auditory cortex · Melodic pitch · Magnetoencephalography

1 Introduction

Recently, Patterson et al. (2015) extended earlier imaging research on melody processing (Patterson et al. 2002; Rupp and Uppenkamp 2005; Uppenkamp and Rupp 2005) by presenting listeners with four-bar musical phrases played by a French horn. There were four notes per bar; they had the same amplitude and the inter-note gaps where minimized to minimize the energy onset component of the responses to the second and succeeding notes 2–4 (Krumbholz et al. 2003). In the first and third bars, the pitch was restricted to the tonic of the four-bar phrase; the second bar contained a novel, random tonal melody and the fourth bar contained a variation of that melody. Magnetoencephalographic recordings revealed that each note produced a prominent transient response (TR) in auditory cortex with a P1m, an N1m and a P2m. The structure of the four-bar phrase was reflected in the magnitude of the TR. In all bars, the TR to the first note was somewhat larger than that to the remaining three notes of that bar. In the melody bars, the P2m component of the TR was larger than in bars where the pitch was fixed. The stability of the TR magnitude to notes 2–4 of each bar made it possible to do a four-dipole analysis and show that the activity is distributed between two sources in each hemisphere, one in the medial part of Heschl's gyrus (HG) and the other slightly more lateral and posterior on the front edge of planum temporale (PT). The source waves showed that melody processing was largely confined to the P2m component of the TR in the medial part of HG.

This paper describes a shorter version of the melody experiment with alternating tonic and melodic bars and an inter-sequence interval of about 1325 ms—modifications that make it possible to measure the sustained field (SF) generated during the tonic and melodic bars. As in the previous experiment, the magnitude of the TR to the first note of any sequence is enhanced by the abrupt onset of energy. The purpose of this second experiment was to determine (1) whether the TR to the first note of each sequence emanates from a different source than the TRs from the remaining notes of that sequence, and (2) whether the source of the SF is in auditory cortex or beyond.

2 Methods

A natural French horn tone (Goto et al. 2003) served as the basic stimulus for the experiment. Four-note sequences were constructed by varying the f_0 of the original note within the original envelope using the vocoder STRAIGHT (Kawahara et al. 1999; Kawahara and Irino 2004). There were two different types of four-note sequences, "tonic" and "melodic": The f_0 was fixed in the tonic sequences at either 100 Hz or 141.5 Hz. In the melodic sequences, the f_0 value of each note was randomly drawn, without replacement, to be −1, 0, 2, 4, 5 or 7 semitones relative to the respective tonic value. Tonic and melodic sequences were alternated, and there were 400 of each (200 per tonic). Within a given sequence, each note had a length of 650 ms and was ramped on and off with 5- and 10-ms Hanning windows, respectively. There were no pauses between adjacent notes within a sequence; the inter-sequence interval varied randomly between 1300 and 1350 ms. The total duration of the experiment was 55 min and the overall stimulation level was 70 dB SPL.

The neuromagnetic activity in response to the four note sequences was measured with a Neuromag-122 (Elekta Neuromag Oy) whole-head MEG system, at a 1000 Hz sampling rate, having been lowpass filtered 330 Hz. Ten normal-hearing listeners (mean age: 33.9 ± 12.01 years) participated in the study. Subjects watched a silent movie, while their AEFs were recorded; they were instructed to direct their attention to the movie. Spatio-temporal source analyses (Scherg 1990) were performed offline using the BESA 5.2 software package (BESA Software GmbH). Separate source models with one equivalent dipole in each hemisphere were created to analyze the P50m (using unfiltered data) and the sustained field (SF) (using data lowpass filtered at 1 Hz). Then, a model with two dipoles per hemisphere was built to localize the N1m/P2m complex associated with energy onset at the start of each sequence, and to localize changes between adjacent notes within a given sequence (pooled notes 2–4). A 1–30 Hz bandpass filter was applied to the data for these four-dipole fits. Afterwards, unfiltered source waveforms were derived for each model for each participant to compute grand-average source waveforms for both the tonic and melodic sequences. A principal component analysis was performed on the last few milliseconds of the epoch to compensate for drift. Statistical evaluation of specific component differences was accomplished in MATLAB using the multivariate statistic t_{sum} from the permutation test of Blair and Karniski (1993). Latencies and amplitudes of the additional data set described in the discussion section were evaluated statistically using a bootstrap technique to derive critical t-values (Efron and Tibshirani 1993).

3 Results

The data were fitted using a 4-dipole model with 2 dipoles in each hemisphere. One pair focused on the N1m temporal region of the first note in each bar to locate the "onset generator"; the other focused on the N1m temporal region of the three adapted melodic notes of each trial (notes 2–4 in the melody bar) to locate the "melody generator"– a technique developed for analysing melody processing in Patterson et al. (2015). The melody generator was located in the medial part of Heschl's gyrus (HG); the onset generator about 10 mm more posterior in PT—a common location for the onset N1m. The source waves for the left and right hemispheres were very similar so the waves from the two hemispheres were averaged, in each case. The anterior-posterior difference between the average source locations was significant [left hemisphere: $t(9) = 4.21$, $p = 0.0023$; right hemisphere: $t(9) = 4.02$, $p = 0.003$)].

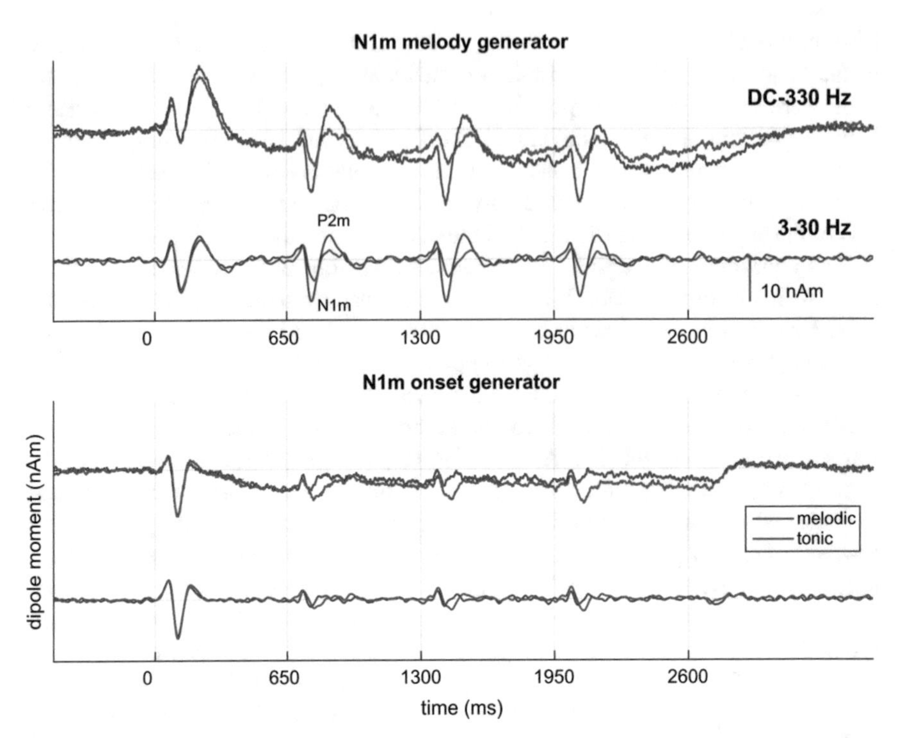

Fig. 1 Average source waves extracted from the data of the tonic (*red*) and melodic (*blue*) sequences using the melody generator (*upper* panel) and the onset generator (*lower* panel). The upper trace in each panel shows the unfiltered, grand average waveforms for the tonic (*red*) and melodic (*blue*) sequences. The sustained field is removed from the data in the lower trace of each panel by band pass filtering (3–30 Hz)

The upper panel of Fig. 1 shows the sources waves extracted with the melody generator separately for the tonic (red) and melodic sequences (blue) of each trial. The bottom panel shows the source waves extracted with the onset generator. The upper pair of curves are based on unfiltered data; the lower pair on data high-pass filtered at 3 Hz to remove the SF. The response to the first note of the tonic sequence is very similar to the first note of the melodic sequence. Thereafter, however, the adapted responses to notes 2–4 in the melodic sequence (blue) are substantially larger than their counterparts in the tonic sequence (red), both with regard to the N1m component and the P2m component (N100m: $t_{max\_note2}=-204$, $p=0.002$; $t_{max\_note3}=-229.35$, $p=0.002$; $t_{max\_note4}=-239.27$, $p=0.002$; P200m: $t_{max\_note2}=416.04$, $p=0.0001$; $t_{max\_note3}=427.91$, $p=0.002$; $t_{max\_note4}=443.30$, $p=0.002$). The level, duration and timbre of the notes in the tonic and melodic sequences are the same, so the differences in the source waves reflect the processing of the sequence of pitch changes that together constitute the melody. The lower panel shows the source waves extracted with the posterior source (the onset generator). The response to the first note of the tonic sequence is once again very similar to the first note of the melodic sequence. The responses to succeeding notes are greatly attenuated, illustrating the value of the 4-dipole analysis in the isolation of the energy-onset response.

In the melodic source generator (upper panel), the SF builds up (negatively) over about 500 ms following the P2m to the first note, from about 250 to 750 ms after the onset of the sequence. The magnitude of the SF decreases down to about −20 nAm below baseline for both the tonic and melodic sequences; it appears to get a little more negative towards the end of the melodic sequence, but not a great deal. Then, following the P2m of the last note, the SF decays back up to the baseline level over about 500 ms in both the tonic and melodic sequences. Statistically, over the course of the entire sequence, the SFs for the tonic and melodic conditions are not different, so it appears that the process that produces the SF does not distinguish the condition in which the notes produce a melody (SF differences based on melodic sources: -1973.99, $p=0.29$; SF based on the 2-dipole SF-fit (0–1 Hz): $t_{max}=1236.12$, $p=0.545$).

The lower panel of Fig. 1 shows the sources waves extracted with the onset generator. In this posterior source, the responses to the initial note of the tonic and melodic sequences is virtually identical ($t_{max}=21.81$, $p=0.688$). Small differences are observed between the tonic and melodic responses to the second and succeeding notes in the unfiltered data. However, they are mainly due to a small difference in the SF which is not itself significant. When the SF is removed by filtering, the differences between the source waves are minimal. Thus, the SF in the posterior source does not differ for melodic sequences.

4 Discussion

The transients produced by the adapted notes (2–4) in the current experiment have consistently larger magnitudes in the melodic source generator. This prompted us to return to the melody experiment of Patterson et al. (2015) and do a more detailed analysis of the transient responses in that four-bar experiment, where there were two tonic bars and two melodic bars in each trial. The six adapted responses from the bars with the same sequence type (tonic or melodic) were shown to be very similar in that paper, and this made it possible to do a powerful, four-dipole fit with two dipoles focusing on the N1m temporal window and two focusing on the P2m temporal window. The fit isolated melodic and tonic generators in HG in both hemispheres, near the position of the melodic generator in the current experiment.

Figure 2 presents a comparison of the average, adapted responses in the data of Patterson et al. (2015) to the tonic notes from bars 1 and 3 (red lines) and the melodic notes of bars 2 and 4 (blue lines). The left and right panels show the TRs from the source waves obtained with the N1m and P2m temporal windows respectively. The figure shows that the technique reveals a double dissociation; the N1m is prominent in the N1m temporal window (left panel) but the P2m is not; whereas the P2m is prominent in the P2m temporal window (right panel), but N1m is not.

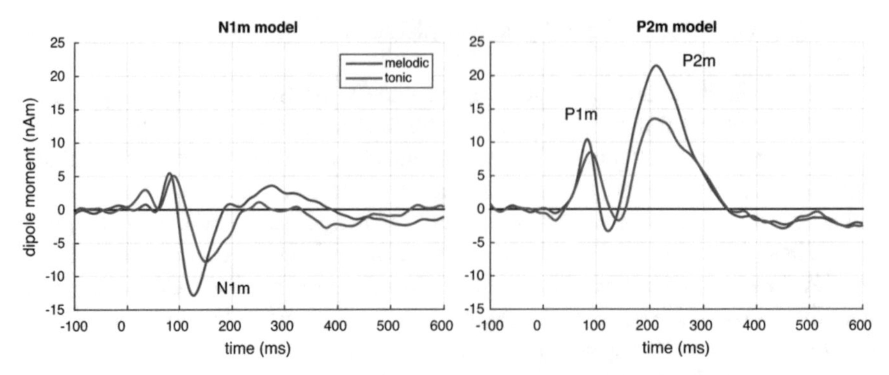

Fig. 2 Average transient responses from the 4-dipole model of Patterson et al. (2015) derived with an N1m temporal window (*left*) and a P2m temporal window (*right*)

Moreover, the adapted melodic notes (blue) produce a larger P2m (right panel) *and* a larger N1m (left panel) than the adapted tonic notes (red).

The P1m is larger with the P2m temporal window (right panel) but this is probably just because there is no N1m to reduce the apparent magnitude of the P1m with the P2m temporal window.

Both panels of Fig. 2 also show that the N1m peak occurs earlier in the *melodic* TR (blue), whereas one might have thought that it would take longer to produce a melodic response than a tonic response (latency $\text{diff}_{\text{passive}} = 25.63$ ms, crit-$t = 14.15$, $p < 0.05$; latency $\text{diff}_{\text{active}} = 14.22$ ms, crit-$t = 8.21$, $p < 0.05$). Although the peak delay differences might seem small, the timing of the N1m peak is one of the most precise statistics in MEG data, and these differences are highly significant—more so than the N1m amplitude differences in these panels (amplitude $\text{diff}_{\text{passive}} = 2.74$ nAm, crit-$t = 2.6$, n.s.; amplitude $\text{diff}_{\text{active}} = 5.52$ nAm, crit-$t = 5.35$, $p < 0.05$).

5 Conclusions

The shorter melody experiment presented at the start of this paper shows how the classic energy-onset generator in PT can be separated from the melody generator in the anterior part of HG. The transients obtained from the melody generator contain both N1m and P2m components, and both components are larger in melodic sequences than they are in tonic sequences. Moreover, the N1m peak delay is shorter in melodic sequences. Reanalysis of the TR data from the four-bar melody experiment suggests that the N1m and P2m components of the response in HG can be separated with a 4-dipole source model, and that melody processing is more readily observed in the shape and magnitude of the P2m response. Finally, the sustained field originates from a generator in HG not far from the melody generator; however, the SF itself does not distinguish between the tonic and melodic sequences.

References

Blair RC, Karniski W (1993) An alternative method for significance testing of waveform difference potentials. Psychophysiology 30:518–524

Efron B, Tibshirani RJ (1993) An introduction into the bootstrap. Chapman & Hall, New York

Goto M, Hashiguchi H, Nishimura T, Oka R (2003). RWC music database: music genre database and musical instrument sound database. In: Hoos HH, Bainbridge D (eds) Proceedings of the 4th International Conference on Music Information Retrieval, pp 229–230

Kawahara H, Irino T (2004) Underlying principles of a high-quality speech manipulation system STRAIGHT and its application to speech segregation. In: Divenyi PL (ed) Speech separation by humans and machines. Kluwer Academic, Dordrecht, pp 167–180

Kawahara H, Masuda-Kasuse I, de Cheveigne A (1999) Restructuring speech representations using pitch-adaptive time-frequency smoothing and instantaneous-frequency based F0 extraction: possible role of repetitive structure in sounds. Speech Comm 27:187–204

Krumbholz K, Patterson RD, Seither-Preisler A, Lammertmann C, Lütkenhöner B (2003) Neuromagnetic evidence for a pitch processing centre in Heschl's gyrus. Cereb Cortex 13:765–772

Patterson RD, Uppenkamp S, Johnsrude I, Griffiths TD (2002) The processing of temporal pitch and melody information in auditory cortex. Neuron 36:767–776

Patterson RD, Uppenkamp S, Andermann M, Rupp A (2015) Brain imaging the activity associated with pitch intervals in a melody. Proc Meet Acoust 21:050009

Rupp A, Uppenkamp S (2005). Neuromagnetic representation of short melodies in the auditory cortex. In: Fastl H, Fruhmann M (eds) Fortschritte der Akustik-DAGA 2005. DEGA e.V., Berlin, pp 473–474

Scherg M (1990) Fundamentals of dipole source potential analysis. In: Grandori F, Hoke M, Romani GL (eds) Auditory evoked magnetic fields and potentials. Advances in audiology 5. Karger, Basel, pp 40–69

Uppenkamp S, Rupp A (2005). Functional MR imaging of the processing of pitch changes in human listeners. In: Fastl H, Fruhmann M (eds) Fortschritte der Akustik-DAGA 2005. Deutsche Gesellschaft für Akustik e.V., Berlin, pp 471–472

Studying Effects of Transcranial Alternating Current Stimulation on Hearing and Auditory Scene Analysis

Lars Riecke

Abstract Recent studies have shown that perceptual detection of near-threshold auditory events may depend on the relative timing of the event and ongoing brain oscillations. Furthermore, transcranial alternating current stimulation (tACS), a non-invasive and silent brain stimulation technique, can entrain cortical alpha oscillations and thereby provide some experimental control over their timing. The present research investigates the potential of delta/theta-tACS to modulate hearing and auditory scene analysis. Detection of near-threshold auditory stimuli, which are modulated at 4 Hz and presented at various moments (phase lags) during ongoing tACS (two synchronous 4-Hz alternating currents applied transcranially to the two cerebral hemispheres), is measured in silence or in a masker. Results indicate that performance fluctuates as a function of phase lag and these fluctuations can be explained best by a sinusoid at the tACS frequency. This suggests that tACS may amplify/attenuate sounds that are temporally coherent/anticoherent with tACS-entrained cortical oscillations.

Keywords Brain stimulation · Neural oscillation · Entrainment · Auditory cortex · Phase alignment

1 Introduction

1.1 What Is TACS?

TACS is a non-invasive brain stimulation technique that can be applied safely and silently using portable and relatively inexpensive equipment. Its application involves inducing a weak electric current between electrodes at the scalp (Nitsche et al. 2008; Zaghi et al. 2010; Paulus 2011). The current partially penetrates the brain, where it propagates widely and forces neuronal excitability to oscillate along with its alternating waveform (Frohlich and McCormick 2010; Ozen et al. 2010).

L. Riecke (✉)
Department of Cognitive Neuroscience, Maastricht University, Maastricht, The Netherlands
e-mail: L.Riecke@MaastrichtUniversity.NL

P. van Dijk et al. (eds.), *Physiology, Psychoacoustics and Cognition in Normal and Impaired Hearing,* Advances in Experimental Medicine and Biology 894,
DOI 10.1007/978-3-319-25474-6_39

In this way, tACS may entrain ongoing cortical oscillations at corresponding frequencies and impact on different aspects of sensory-perceptual-motor processing and higher-order cognition, depending on stimulation parameters such as electrode positions (reviewed in Herrmann et al. 2013).

1.2 Cortical Oscillatory Phase Influences Auditory Perception

Cortical oscillations may also entrain with periodic auditory stimuli, so that high neuronal excitability phases become aligned with peaks in the sound envelope. This sound-brain entrainment is thought to facilitate cortical processing of upcoming acoustic events that concur with sound envelope peaks (Schroeder and Lakatos 2009; Arnal and Giraud 2012). While correlational studies showed that sound-brain entrainment is involved in the neural processing and perception of auditory stimuli (Stefanics et al. 2010; Besle et al. 2011; Gomez-Ramirez et al. 2011), only three published human studies sought to directly control this entrainment with periodic external stimulation and assess the perceptual effects. Neuling et al. (2012) presented tone pips in noise and simultaneously applied 10-Hz tACS above the auditory cortices while manipulating the relative timing of the tone and tACS (i.e., the phase lag). They found that detection performance varied as a function of phase lag and that this function resembled well the tACS waveform. Henry (2012; Henry et al. 2014) used a periodic 3- or 5-Hz modulation applied to an interrupted tone carrier as entraining stimulus. They manipulated the relative timing of the gap and the ongoing modulation and found that this influences gap detection in the phase-dependent manner described above for the tACS study. Based on electrophysiological recordings (Lakatos et al. 2005; Frohlich and McCormick 2010; Ozen et al. 2010; Ali et al. 2013), it appears plausible that the reported effects on auditory perception were mediated by auditory cortical oscillations that entrained to the periodic external stimulation and thereby caused the auditory targets to arrive at different neural excitability phases.

1.3 Can TACS Influence Auditory Perception?

Compared with periodic acoustic stimulation, tACS arguably provides the preferred entrainment stimulus as it can be applied both in silence and at selected scalp position, thus allowing to bypass the peripheral auditory system and reduce possible masking of the target sound. However, important issues remain unresolved: How reliable are auditory effects of tACS? Do they also apply to oscillations outside the alpha range? Do they also occur in more naturalistic situations where target sounds need to be tracked amidst other sounds?

This manuscript provides preliminary answers to these questions based on ongoing tACS research. A first experiment (Riecke et al. 2015) shows that auditory effects of alpha-tACS phase (Neuling et al. 2012) can be reproduced for the delta/theta range using a more rigorous methodological approach. An ongoing follow-up experiment is using an informational masking paradigm to explore whether these results can be extended to sustained sounds in scenes.

2 Methods

2.1 Stimuli

2.1.1 TACS

To facilitate homophasic stimulation in both auditory cortices (target regions), two approximately equivalent circuits were generated using two stimulator systems (Neuroconn, Ilmenau, Germany): Prior electric field simulations suggested that placing the stimulation electrodes at position T7 and T8, and the return electrodes close to position Cz, would produce relatively strong currents in the target regions. The same sinusoidal current (frequency: 4 Hz, starting phase: 0°) was applied to each lateralized circuit, the return electrodes were coupled, and the skin was prepared so that the left- and right-lateralized circuit impedances were matched, while keeping the net impedance below 10kΩ. Current strength was set individually to the point where participants reported being comfortable with receiving tACS under all electrodes.

2.1.2 Auditory Stimuli

Target sounds were modulated at the tACS frequency (4 Hz) to give listeners the opportunity to sample the target at fixed phase angle on consecutive tACS cycles. The relative level of target and background was set individually to a value near detection threshold determined beforehand.

In experiment 1, the target was a click train. It was generated by summing all harmonics (sinusoids with fixed amplitude and starting phase) of a 4-Hz fundamental within the range from 112 to 3976 Hz and bandpass-filtering the resulting waveform between 224 and 1988 Hz. The portion from 125 to 1125 ms was extracted to obtain four clicks centered on a 1-s interval.

In experiment 2, the target was a pulsed complex tone embedded in a masker. The target was generated by summing three sinusoidal harmonics of a 110-Hz fundamental that were chosen randomly from the range from 220 to 6270 Hz with the constraint that they differed by at least one critical band. The target was modulated with a square wave function (frequency: 4 Hz, duty cycle: 41.7 ms) and its duration was chosen randomly from the interval from 2 to 6 s. The masker was a continuous sequence of random complex tones, each of which was generated by summing three to five sinusoids (depending on the listener's target detection threshold) with random starting phases and the same amplitude and duration as the target components. The masker components were chosen randomly from the range from 200 to 6297 Hz with the constraint that they differed by at least one critical band from each other and the target components when the latter were present. For intervals when the target was present, overall amplitude cues were eliminated by reducing the number of masker components by the number of target components.

2.2 Design and Task

The critical parameter was the relative timing of acoustic and electric stimulation (Fig. 1a). This phase lag was manipulated across six conditions by varying the onset of the target in steps of 41.7 ms (30°) across the 4-Hz tACS cycles. Each experiment lasted 10 min. It was presented four times during continuous tACS and once without tACS. The latter sham stimulation was identical to tACS, except that the current was ramped down/up shortly after/before the experiment began/ended (Herrmann et al. 2013). Debriefings revealed that participants were unaware of whether they received tACS or sham stimulation. The order of trials and stimulation type (tACS, sham) was randomized individually.

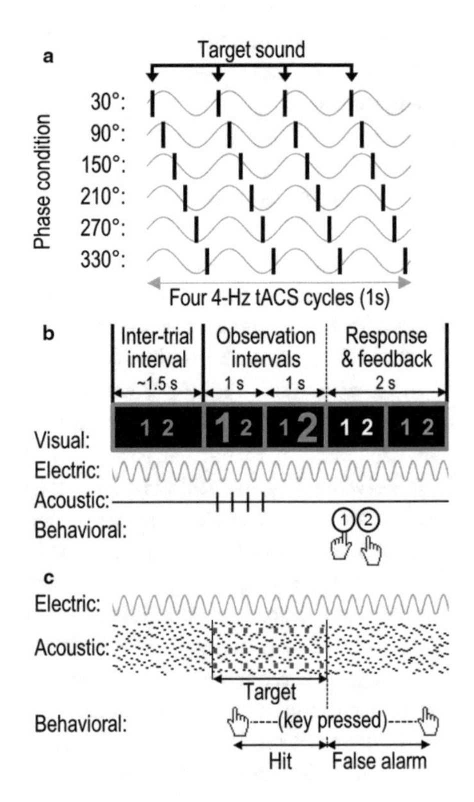

Fig. 1 Experimental designs. Panel A illustrates the six tACS phase conditions, each of which was characterized by a specific phase lag between the acoustic and electric stimuli. *Gray* waves schematize four consecutive 4-Hz tACS cycles. *Black dashes* sketch the target sound, which was either a 4-Hz click train (experiment 1) or a harmonic complex pulsed at 4 Hz (experiment 2). Panel B exemplifies the visual, electric, and auditory stimulation, and a behavioral response from a trial in experiment 1. Listeners detected clicks in a 2I2AFC task. Analogous to panel B, panel C exemplifies a portion of a trial from experiment 2. Listeners held a bar whenever they heard a repetitive tone, which occurred at unpredictable times within an ongoing multi-tone masker

In experiment 1, click detection performance was assessed using a two-interval two-alternative forced-choice task (2I2AFC); see Fig. 1b. The click train was presented in a randomly selected interval, whereas the other interval contained only silence. Listeners judged which interval contained the clicks and received visual feedback. Visual stimulus timing was jittered to provide no valid cue of the click timing.

In experiment 2 (Fig. 1c), target detection was assessed using a continuous yes/no task. On each trial, the masker was presented for 45 s during which several targets were presented. Consecutive targets were separated by non-target intervals whose duration varied randomly between 2 and 7 s, with the constraint that their overall duration matched that of the target intervals. Trials began and ended with non-target intervals. Listeners were instructed to press and hold a bar as soon and as long as they heard a repetitive tone.

2.3 Data Analysis

A performance measure was extracted for each tACS phase condition: For experiment 1, accuracy was extracted, defined as the number of correct responses divided by the number of trials. In experiment 2, the sensitivity index d' was computed based on the overall time during which the participant correctly or erroneously reported the target, divided by the overall target or non-target time (respectively).

Next, a 'time series' of the performance measure was reconstructed by concatenating the six phase conditions. To compensate for possible inter-individual variations in the phase for which performance was best, the maximum of the series was aligned to the 90°-point and the series was phase-wrapped (e.g., Ng et al. 2012). Following this 'best-phase alignment'—under the main hypothesis that 4-Hz tACS phase modulates detection—phases 30–150° should delimit a positive 4-Hz half-cycle, whereas phases 210–330° should delimit the opposite (negative) half-cycle. To verify this prediction, performance was averaged across the hypothesized positive half-cycle (phases 30 and 150°) and the opposite half-cycle (phases 210, 270, and 330°), and the two resulting averages were compared. The alignment point (90°) was excluded to preserve statistical independency. To verify whether the hypothesized phase effect cycled specifically at 4 Hz, spectral density was computed from the series and the resulting frequency components were compared. Only three bins centered on 4, 8, and 12 Hz could be resolved, due to the limited number of data points and sampling rate (six points spanning a 4-Hz period). Baseline was defined as the overall performance during the sham run, excluding trials presented during the tACS ramps.

2.4 Participants

Fourteen and eleven paid volunteers (ten females, ages: 20–38 years) participated in experiment 1 and experiment 2 so far. They reported no history of neurological, psychiatric, or hearing disorders, were eligible to undergo tACS as assessed by prior screening, and gave their written informed consent before taking part. They had normal audiometric thresholds, except for one participant with mild high-frequency hearing loss.

3 Results

Figure 2a shows listeners' average performance in experiment 1 as a function of phase condition. Performance was above baseline in the 30 and 150° conditions, whereas it was below baseline in the 210, 270, and 330° conditions. Furthermore, performance was similar at 30 and 150°, and at 210 and 330° respectively, i.e.,

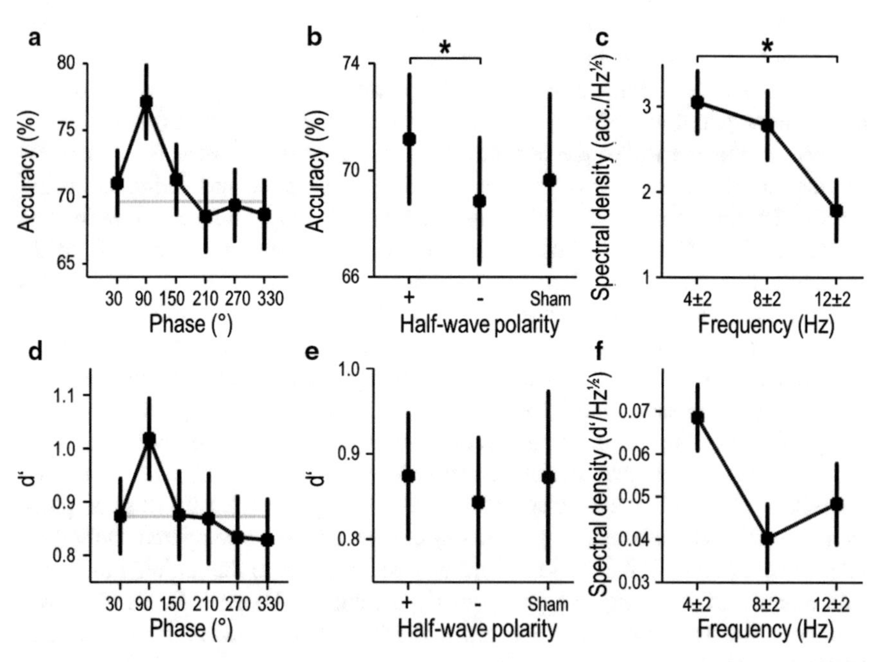

Fig. 2 Results. Panel A shows detection performance (mean ± sem across listeners) as a function of phase condition in experiment 1. The *gray* horizontal line shows performance under sham stimulation. Panel B shows averages of these data, i.e., the hypothesized positive half-cycle (average of conditions 30 and 150°), the opposite half-cycle (average of conditions 210, 270, and 330°), and sham stimulation. Panel C shows the magnitude spectrum of the performance waveform (mean ± sem across listeners' individual magnitude spectra). Analogously, panels D, E, and F show preliminary data from experiment 2

it was approximately symmetric with respect to the presumed best phase (90°). Although on average the best and worst performances were not associated with exactly opposite phases, possibly due to the contribution of higher harmonics, these observations match well the characteristics of the hypothesized 4-Hz cycle.

Figure 2b shows performance averaged across phase conditions to facilitate comparison of the hypothesized positive half-cycle (30, 150°) with the opposite half-cycle (210, 270, 330°) and baseline. Paired t-tests revealed an effect of half-cycle polarity on performance (positive vs. negative half-cycle: $t(13)=2.48$, corrected $p=0.042$); thus tACS phase modulated click detection. Comparison with baseline revealed no effect.

Figure 2c shows the magnitude spectrum of the series. While frequency bins centered on the higher harmonics (8 and 12 Hz) could explain a significant proportion of the overall variance in the series, this variance was explained best by a 4-Hz sinusoid; thus the phase-induced differences cycled predominantly at the tACS frequency. This observation was supported by a random-effects ANOVA including frequency as factor, which revealed a main effect of frequency on spectral magnitude ($F(2,26)=3.93$, $p=0.032$). Post-hoc paired t-tests showed that the 4-Hz bin was significantly stronger than the 12-Hz bin ($t(13)=2.65$, corrected $p=0.020$) but not the 8-Hz bin.

Analogously, Fig. 2d, f show preliminary results from experiment 2 based on d'. These data are shown at the descriptive level given the limited sample size. As for experiment 1, performance appears to be on average slightly better during the hypothesized positive 4-Hz half-cycle than during the opposite half-cycle (Fig. 2d, e) and the magnitude spectrum shows a peak at the tACS frequency (Fig. 2f). Statistical analyses based on a larger sample, which is being acquired at the moment, will assess enable to assess the significance of these preliminary observations.

4 Discussion

Experiment 1 revealed that click detection under 4-Hz tACS is modulated by tACS phase: performance was significantly better during one half of the 4-Hz tACS cycle than during the opposite half. Performance without tACS was intermediate, but did not differ significantly. The tACS phase effect was strongest in the vicinity of the tACS frequency but not strictly frequency-specific as it extended to higher frequencies. Overall, these data replicate the findings by Neuling et al. (2012), indicating that auditory effects of tACS phase are reproducible. They extend these previous findings, which were based on single tone pips and alpha-tACS, to new tasks and stimuli including periodic clicks and dual-channel delta/theta-tACS. Considering that clicks were presented at the tACS frequency and their detectability was found to depend on tACS phase, a possible interpretation is that tACS essentially enhanced the loudness of those click trains that were temporally coherent (rather than anti-coherent) with tACS-entrained oscillations. Such an 'envelope enhancement' would have interesting implications for auditory scene analysis problems that benefit from tracking the envelope of the target sound.

This idea is being addressed in experiment 2, where an otherwise audible modulated target sound is rendered inaudible by spectrally non-overlapping distractor sounds. Based on magnetoencephalography findings from a similar paradigm, Elhilali et al. (2009) proposed that the perceptual segregation of the modulated target from the distractors arises from temporal coherence of cortical activity patterns with the target. Whether tACS phase modulates this temporal coherence and therewith the segregability of the target remains to be shown; the preliminary data presented here indicate a trend toward a possible effect of 4-Hz tACS phase, as in experiment 1.

Observations of actual benefits of tACS compared with sham stimulation are rather modest so far. In experiment 1, click detection improved on average by 7% (at best phase), a difference that did not reach statistical significance. Neuling et al. (2012) observed a 0.3-dB improvement in tone detection. Future studies may identify which tACS parameters (e.g., single or dual channel tACS, 4 or 10 Hz frequency) are most effective and systematically optimize them on an individual basis in order to maximize possible perceptual benefits. For example, the tACS frequency could be chosen to match the individual's dominant ongoing cortical frequency as identified beforehand with EEG (Zaehle et al. 2010; Ali et al. 2013). Similarly, tACS electrode positions could be chosen based on more realistic current flow simulations using as head model the individual's anatomy obtained beforehand with MRI (e.g., (Wagner et al. 2014)). Due to the limited spatial specificity of transcranial stimulation, individual tonotopic locations cannot be selectively targeted. Nevertheless, the possibility to transmit sound envelope information non-invasively and bypass (possible deficits in) the auditory pathway up to the cortex might make transcranial stimulation a valuable tool.

Acknowledgments I would like to thank Alexander Sack, Daniel Hauke, and Marijn van Waardenburg for help with obtaining the data. This work was supported by NWO Veni grant 451-11-014. Some of the data presented here have been published elsewhere (Riecke et al. 2015).

References

Ali MM, Sellers KK, Frohlich F (2013) Transcranial alternating current stimulation modulates large-scale cortical network activity by network resonance. J Neurosci 33(27):11262–11275

Arnal LH, Giraud AL (2012) Cortical oscillations and sensory predictions. Trends Cogn Sci 16(7):390–398

Besle J, Schevon CA, Mehta AD, Lakatos P, Goodman RR, McKhann GM, Emerson RG, Schroeder CE (2011) Tuning of the human neocortex to the temporal dynamics of attended events. J Neurosci 31(9):3176–3185

Elhilali M, Xiang J, Shamma SA, Simon JZ (2009) Interaction between attention and bottom-up saliency mediates the representation of foreground and background in an auditory scene. PLoS Biol 7(6):e1000129

Frohlich F, McCormick DA (2010) Endogenous electric fields may guide neocortical network activity. Neuron 67(1):129–143

Gomez-Ramirez M, Kelly SP, Molholm S, Sehatpour P, Schwartz TH, Foxe JJ (2011) Oscillatory sensory selection mechanisms during intersensory attention to rhythmic auditory and visual inputs: a human electrocorticographic investigation. J Neurosci 31(50):18556–18567

Henry MJ, Obleser J (2012) Frequency modulation entrains slow neural oscillations and optimizes human listening behavior. Proc Natl Acad Sci U S A 109(49):20095–20100

Henry MJ, Herrmann B, Obleser J (2014) Entrained neural oscillations in multiple frequency bands comodulate behavior. Proc Natl Acad Sci U S A 111(41):14935–14940

Herrmann CS, Rach S, Neuling T, Struber D (2013) Transcranial alternating current stimulation: a review of the underlying mechanisms and modulation of cognitive processes. Front Hum Neurosci 7:279

Lakatos P, Shah AS, Knuth KH, Ulbert I, Karmos G, Schroeder CE (2005) An oscillatory hierarchy controlling neuronal excitability and stimulus processing in the auditory cortex. J Neurophysiol 94(3):1904–1911

Neuling T, Rach S, Wagner S, Wolters CH, Herrmann CS (2012) Good vibrations: oscillatory phase shapes perception. Neuroimage 63(2):771–778

Ng BS, Schroeder T, Kayser C (2012) A precluding but not ensuring role of entrained low-frequency oscillations for auditory perception. J Neurosci 32(35):12268–12276

Nitsche MA, Cohen LG, Wassermann EM, Priori A, Lang N, Antal A, Paulus W, Hummel F, Boggio PS, Fregni F, Pascual-Leone A (2008) Transcranial direct current stimulation: state of the art 2008. Brain Stimul 1(3):206–223

Ozen S, Sirota A, Belluscio MA, Anastassiou CA, Stark E, Koch C, Buzsaki G (2010) Transcranial electric stimulation entrains cortical neuronal populations in rats. J Neurosci 30(34):11476–11485

Paulus W (2011) Transcranial electrical stimulation (tES—tDCS; tRNS, tACS) methods. Neuropsychol Rehabil 21(5):602–617

Riecke L, Formisano E, Herrmann CS, Sack AT (2015). 4-Hz transcranial alternating current stimulation phase modulates hearing. Brain Stimul 8:777–783

Schroeder CE, Lakatos P (2009) Low-frequency neuronal oscillations as instruments of sensory selection. Trends Neurosci 32(1):9–18

Stefanics G, Hangya B, Hernadi I, Winkler I, Lakatos P, Ulbert I (2010) Phase entrainment of human delta oscillations can mediate the effects of expectation on reaction speed. J Neurosci 30(41):13578–13585

Wagner S, Rampersad SM, Aydin U, Vorwerk J, Oostendorp TF, Neuling T, Herrmann CS, Stegeman DF, Wolters CH (2014) Investigation of tDCS volume conduction effects in a highly realistic head model. J Neural Eng 11(1):016002

Zaehle T, Rach S, Herrmann CS (2010) Transcranial alternating current stimulation enhances individual alpha activity in human EEG. PLoS ONE 5(11):e13766

Zaghi S, Acar M, Hultgren B, Boggio PS, Fregni F (2010) Noninvasive brain stimulation with low-intensity electrical currents: putative mechanisms of action for direct and alternating current stimulation. Neuroscientist 16(3):285–307

Functional Organization of the Ventral Auditory Pathway

Yale E. Cohen, Sharath Bennur, Kate Christison-Lagay, Adam Gifford and Joji Tsunada

Abstract The fundamental problem in audition is determining the mechanisms required by the brain to transform an unlabelled mixture of auditory stimuli into coherent perceptual representations. This process is called auditory-scene analysis. The perceptual representations that result from auditory-scene analysis are formed through a complex interaction of perceptual grouping, attention, categorization and decision-making. Despite a great deal of scientific energy devoted to understanding these aspects of hearing, we still do not understand (1) how sound perception arises from neural activity and (2) the causal relationship between neural activity and sound perception. Here, we review the role of the "ventral" auditory pathway in sound perception. We hypothesize that, in the early parts of the auditory cortex, neural activity reflects the auditory properties of a stimulus. However, in latter parts of the auditory cortex, neurons encode the sensory evidence that forms an auditory decision and are causally involved in the decision process. Finally, in the prefrontal cortex, which receives input from the auditory cortex, neural activity reflects the actual perceptual decision. Together, these studies indicate that the ventral pathway contains hierarchical circuits that are specialized for auditory perception and scene analysis.

Keywords Auditory perception · Decision-making · Categorization · Perceptual grouping · Ventral auditory pathway

Y. E. Cohen (✉) · S. Bennur · J. Tsunada
Department of Otorhinolaryngology, University of Pennsylvania, Philadelphia, USA
e-mail: ycohen@mail.med.upenn.edu

Y. E. Cohen
Department of Neuroscience, University of Pennsylvania, Philadelphia, USA
e-mail: ycohen@mail.med.upenn.edu

Department of Bioengineering, University of Pennsylvania, Philadelphia, USA

K. Christison-Lagay · A. M. Gifford
Neuroscience Graduate Group, University of Pennsylvania, Philadelphia, USA

P. van Dijk et al. (eds.), *Physiology, Psychoacoustics and Cognition in Normal and Impaired Hearing,* Advances in Experimental Medicine and Biology 894,
DOI 10.1007/978-3-319-25474-6_40

1 Introduction

Hearing and communication present a variety of challenges to the auditory system. To be heard and understood, the auditory brain must transform a time-varying acoustic stimulus into a perceptual representation; that is, a sound.

Auditory perception is associated with a number of computational processes, which may act in parallel or serial, including: perceptual grouping, decision-making, attention, and categorization. (1) Perceptual grouping is a form of feature-based stimulus segmentation that determines whether acoustic events will be grouped into a single sound or be segregated into distinct sounds (Bregman 1990). (2) Auditory decision-making is a computational process in which the brain interprets sensory information in order to detect, discriminate, or identify the source or content of auditory stimuli (Gold and Shadlen 2007). Did I hear the stimulus? From where and whom did it come? What does it tell me? How can I use this information to plan an action? (3) Although attention is not always necessary, our awareness of a sound can be influenced by attention (Alain and Arnott 2000; Micheyl et al. 2003; Fritz et al. 2005; Shinn-Cunningham 2008; Snyder et al. 2012; Gutschalk et al. 2015). For example, we can choose whether to listen to—or ignore—the first violin, the string section, or even the whole orchestra. Likewise, we can selectively attend to the particular features in a person's voice that allow a listener to identify the speaker. (4) In auditory categorization, sounds are classified based on their acoustic features or more processed forms of information (e.g., semantic knowledge), providing an efficient means to interact with stimuli in our environment (Ashby and Berretty 1997; Gifford et al. 2014). For example, when we hear the word "Groningen" from different speakers, we can categorize the gender of each speaker based on the pitch of the speaker's voice. On the other hand, to analyze the linguistic content transmitted by a speech sound, we can ignore the unique pitch, timbre etc. of each speaker and categorize the sound into the distinct word category "Groningen".

It is thought that the neural computations and processes that mediate auditory perception are found in the ventral auditory pathway (Rauschecker and Scott 2009; Romanski and Averbeck 2009; Hackett 2011; Bizley and Cohen 2013). In rhesus monkeys, this pathway begins in core auditory cortex—specifically, primary auditory cortex (A1) and the rostral field. These core areas project to the middle lateral (ML) and anterolateral belt (AL) regions of auditory cortex. In turn, these belt regions project directly and indirectly to the ventrolateral prefrontal cortex (vlPFC).

It is important to briefly comment on the contribution of the dorsal ("spatial") pathway to auditory perception (Rauschecker 2012; Cloutman 2013). Spatial information can act as a grouping cue to assist the segregation of an acoustic stimulus into discrete sounds. For example, when a rhythmic sequence of identical sound bursts is presented from a single location, it is often perceived as one source. But, when the sound sequence is presented from two different locations, it can be perceived as two sounds (Middlebrooks and Onsan 2012; Middlebrooks and Bremen 2013). Such findings suggest that a mixture of spatial and non-spatial auditory cues from both the dorsal and ventral pathways may be needed in order to create a coherent auditory-perceptual representation that guides behavior.

Nonetheless, in this review, we focus on the hierarchical processing that occurs at different stages in the ventral auditory pathway. In particular, we identify—or, at least, suggest—the unique contributions of these different processing stages to auditory perception and categorization, with an acknowledgment that associating any single brain region with a particular computation oversimplifies the complexity of the auditory brain. Indeed, it is well known that neurons become increasingly sensitive to more complex stimuli along the beginning stages of the ventral auditory pathway (e.g., between the core and belt regions of the auditory cortex). For example, core neurons are more sharply tuned for tone bursts than neurons in the lateral belt, whereas lateral-belt neurons are more selective for particular spectro-temporal features of complex sounds, such as vocalizations (Rauschecker and Tian 2000). Here, though, we review hierarchical processing in the ventral pathway by focusing on those studies in which neural activity was collected simultaneously while a listener was engaged in an auditory task.

2 Neural Correlates of Auditory Perception Along the Ventral Auditory Pathway

A1's role in auditory perception is controversial. Part of that controversy stems from the putative role of A1 in processing auditory "objects" (Nelken 2008). We will take the position that auditory objects are analogous to perceptual representations (i.e., sounds) (Bizley and Cohen 2013). As a consequence of this definition, if a neuron encodes an auditory object, it should be modulated by a listener's perceptual reports. That is, by holding a stimulus constant and testing whether neural activity is modulated by a listener's reports, neural activity that is associated with the acoustic features of the stimulus can be dissociated from neural activity associated with the perceptual report. Thus, neurons with complex tuning properties or even those modulated by components of a task (Brosch et al. 2005; Fritz et al. 2005) may contribute to the construction of a perceptual representation; but by themselves do not offer direct evidence of a perceptual representation.

There has been a recent set of literature implicating A1 in auditory perceptual decision-making (Riecke et al. 2009; Kilian-Hutten et al. 2011; Niwa et al. 2012; Riecke et al. 2012; Bizley et al. 2013). In one study, ferrets were asked to report changes in a sound's pitch, and it was found that both local-field potentials and spiking activity in A1 were modulated by the ferrets' pitch judgments. Niwa and colleagues have also shown that A1 single-unit activity is modulated by monkeys' reports during a task in which monkeys reported whether or not a sound was amplitude modulated. Finally, human-imaging studies have revealed that regions of core auditory cortex are modulated by listener's reports of the identity of an ambiguous speech sound. However, a different body of work suggests that A1 does not encode auditory decisions. For example, when monkeys discriminate between two types of acoustic flutter, A1 activity is not modulated by the monkeys' choices (Lemus et al. 2009b).

What could be the bases for these apparent divergent sets of findings? We posit that these differences can be attributed to several non-exclusive possibilities. One possibility may be due to the relative perceptual and cognitive demands of the behavioural task: tasks with different demands might differentially engage neurons in core auditory cortex (Bizley and Cohen 2013; Nienborg and Cumming 2014). A second possibility focuses on how choice-related activity itself is analyzed. In choice analyses, it is imperative to restrict the analysis to those trials in which neural modulation related to choice can be clearly disassociated from the auditory stimulus. If this analysis is not carefully conducted, apparent choice activity may be conflated with stimulus-related activity. Finally, choice-related activity may not reflect a casual contribution of the auditory cortex to decision-making but may simply reflect feedback from higher choice-sensitive areas (Nienborg and Cumming 2009) or the structure of the correlated noise (Nienborg et al. 2012).

In the belt regions (ML and AL) of the ventral pathway, several lines of study from our laboratory suggest that this is not the case; but also see Niwa et al. 2013. While monkeys categorized speech sounds, we tested whether neural activity was modulated by monkeys' categorical judgements. We found that AL neurons were not modulated by the monkeys' reports (Tsunada et al. 2011, 2012). In a separate line of studies, we asked monkeys to listen to a sequence of tone bursts and report whether the sequence had a "low" or "high" pitch. We found that neither ML nor AL activity was modulated by the monkeys' choices (Tsunada et al., in press).

Do neurons in the different belt regions contribute differentially to auditory perception? The preliminary findings from our low-high pitch study suggest that AL neurons, but not ML neurons, might represent the sensory information (evidence) used to inform the monkeys' perceptual decisions. Specifically, AL neurometric sensitivity appeared to be correlated with both psychometric sensitivity and the monkeys' choices. Consistent with these findings, AL may play a causal role in these auditory judgments: microstimulation of an AL site tends to shift the monkeys' reports toward the pitch associated with the site's frequency tuning.

Whereas a single cortical locus of decision-making has yet to materialize, decision-related activity is seen throughout the frontal lobe. vlPFC neurons are strongly modulated by monkeys' choices (Russ et al. 2008; Lee et al. 2009). Neural activity in the inferior frontal lobe of the human cortex is also modulated by choice when listeners judge the content of ambiguous speech sounds (Binder et al. 2004). Neural correlates relating to a listeners' decision on auditory-flutter stimuli have also been observed in the ventral premotor cortex (Lemus et al. 2009a). Interestingly, as noted above, the dorsal pathway also contributes to auditory perception; consistent with that notion, activity in the human parietal lobe is modulated by listeners' choices (Cusack 2005).

In summary, we propose a model in which auditory information is hierarchically organized and processed in the ventral pathway. In early parts of the auditory cortex, neural activity encodes the acoustic features of an auditory stimulus and become increasingly sensitive to complex spectrotemporal properties (Rauschecker and Tian 2000). In later regions of the auditory cortex, this information informs perceptual judgments. However, neural activity that reflects a listener's perceptual judgments does not become apparent until the frontal lobe.

3 Category Representation in the Ventral Auditory Pathway

Next, we review the manner in which auditory-category information is hierarchically organized in the ventral auditory pathway. In brief, we will highlight how feature-based categories are represented early; whereas in later parts of the pathway, we find representations of "abstract" categories, which combine acoustic information with mnemonic, emotional, and other information sources.

In core auditory cortex, neural activity codes the category membership of simple feature conjunctions. For example, categorical representations of frequency-contours have been identified (Ohl et al. 2001; Selezneva et al. 2006). These neurons encode the direction of a frequency contour (increasing or decreasing), independent of its specific frequency content. These categories may be found in the firing rates of individual neurons or may be observed as a result of population-level computations.

Categories for more complex stimuli, such as speech sounds and vocalizations, can be found in the lateral belt (Chang et al. 2010; Steinschneider et al. 2011; Tsunada et al. 2011; Steinschneider 2013). For example, AL neurons respond categorically, and in a manner consistent with listeners' behavioral reports, to morphed versions of two speech sounds ("bad" and "dad"). AL neurons also respond categorically to species-specific vocalizations; however, the degree to which AL (and listeners) can categorize these vocalizations is constrained by the vocalizations' acoustic variability (Christison-Lagay et al. 2014). In humans, the superior temporal gyrus is categorically and hierarchically organized by speech sounds (Binder et al. 2000; Chang et al. 2010; Leaver and Rauschecker 2010): phoneme categories are found in the middle aspect; word categories in the anterior-superior aspect; and phrases in the most anterior aspect (DeWitt and Rauschecker 2012; Rauschecker 2012).

Beyond the auditory cortex, neurons represent categories that are formed based on the abstract information that is transmitted by sounds. vlPFC neurons represent the valence of food-related calls (e.g., high quality food vs. low quality food) (Gifford et al. 2005). That is, vlPFC neurons encode the "referential" information that is transmitted by vocalizations, independent of differences in their acoustic properties. Prefrontal activity also contributes to the formation of categories that reflect the emotional valence of a speaker's voice (Fecteau et al. 2005) as well as the semantic information transmitted by multisensory stimuli (Joassin et al. 2011; Werner and Noppeney 2011; Hu et al. 2012). Together, these studies are consistent with the idea that ventral auditory pathway is an information-processing pathway that more complex stimuli and categories are processed in a hierarchically organized manner.

4 Future Questions

Of course, several fundamental questions remain. First, as alluded to above, understanding how feedforward versus feedback information contributes to neural correlates of perceptual judgments remains an open question. Second, the degree to which different types of auditory judgements differentially engage different components of the ventral pathway has yet to be fully articulated. A third question is to identify how the different computational processes (e.g., perceptual grouping, attention, decision-making, and categorization) that underlie auditory perception interact with one another. For example, it remains an open issue as to whether and how attention differentially modulates neural correlates of auditory perception at different hierarchical levels of the ventral auditory pathway (e.g., A1 versus AL) (Atiani et al. 2014). Another example is to identify the potential interactions between auditory perceptual grouping and decision-making. Finally, it is important to identify the manner by which the dorsal and ventral auditory pathways interact in order to form a consistent and coherent representation of the auditory scene.

References

Alain C, Arnott SR (2000) Selectively attending to auditory objects. Front Biosci 5:d202–212

Ashby FG, Berretty PM (1997) Categorization as a special case of decision-making or choice. In: Marley AAJ (ed) Choice, decision, and measurement: essays in honor of R. Duncan Luce. Lawrence Erlbaum Associates, Hillsdale, pp 367–388

Atiani S, David SV, Elgueda D, Locastro M, Radtke-Schuller S, Shamma SA, Fritz JB (2014) Emergent selectivity for task-relevant stimuli in higher-order auditory cortex. Neuron 82(2):486–499. doi:10.1016/j.neuron.2014.02.029

Binder JR, Frost JA, Hammeke TA, Bellgowan PS, Springer JA, Kaufman JN, Possing ET (2000) Human temporal lobe activation by speech and nonspeech sounds. Cereb Cortex 10(5):512–528

Binder JR, Liebenthal E, Possing ET, Medler DA, Ward BD (2004) Neural correlates of sensory and decision processes in auditory object identification. Nat Neurosci 7:295–301

Bizley JK, Cohen YE (2013) The what, where, and how of auditory-object perception. Nat Rev Neurosci 14:693–707

Bizley JK, Walker KM, Nodal FR, King AJ, Schnupp JW (2013) Auditory cortex represents both pitch judgments and the corresponding acoustic cues. Curr Biol 23(7):620–625 (Research Support, Non-U.S. Gov't)

Bregman, AS (1990) Auditory scene analysis. MIT Press, Boston, MA

Brosch M, Selezneva E, Scheich H (2005) Nonauditory events of a behavioral procedure activate auditory cortex of highly trained monkeys. J Neurosci 25(29):6797–6806

Chang EF, Rieger JW, Johnson K, Berger MS, Barbaro NM, Knight RT (2010) Categorical speech representation in human superior temporal gyrus. Nat Neurosci 13:1428–1432

Christison-Lagay KL, Bennur S, Lee JH, Blackwell J, Schröder T, Cohen YE (2014) Natural variability in species-specific vocalizations constrains behavior and neural activity. Hear Res 312:128–142

Cloutman LL (2013) Interaction between dorsal and ventral processing streams: where, when and how? Brain Lang 127(2):251–263. doi:10.1016/j.bandl.2012.08.003
Cusack R (2005) The intraparietal sulcus and perceptual organization. J Cogn Neurosci 17:641–651
DeWitt I, Rauschecker JP (2012) Phoneme and word recognition in the auditory ventral stream. Proc Natl Acad Sci U S A 109(8):E505–E514
Fecteau S, Armony JL, Joanette Y, Belin P (2005) Sensitivity to voice in human prefrontal cortex. J Neurophysiol 94(3):2251–2254. doi:10.1152/jn.00329.2005
Fritz J, Elhilali M, Shamma S (2005) Active listening: task-dependent plasticity of spectrotemporal receptive fields in primary auditory cortex. Hear Res 206(1–2):159–176. doi: S0378-5955(05)00083-3 [pii] 10.1016/j.heares.2005.01.015
Gifford GW, 3rd, MacLean KA, Hauser MD, Cohen YE (2005) The neurophysiology of functionally meaningful categories: macaque ventrolateral prefrontal cortex plays a critical role in spontaneous categorization of species-specific vocalizations. J Cogn Neurosci 17(9):1471–1482
Gifford AM, Cohen YE, Stocker AA (2014) Characterizing the impact of category uncertainty on human auditory categorization behavior. PloS Comp Biol 10:e1003715
Gold JI, Shadlen MN (2007) The neural basis of decision making. Annu Rev Neurosci 30:535–574
Gutschalk A, Rupp A, Dykstra AR (2015) Interaction of streaming and attention in human auditory cortex. PLoS One 10(3):e0118962
Hackett TA (2011) Information flow in the auditory cortical network. Hear Res 271:133–146
Hu Z, Zhang R, Zhang Q, Liu Q, Li H (2012) Neural correlates of audiovisual integration of semantic category information. Brain Lang 121(1):70–75. doi:10.1016/j.bandl.2012.01.002
Joassin F, Maurage P, Campanella S (2011) The neural network sustaining the crossmodal processing of human gender from faces and voices: an fMRI study. Neuroimage 54(2):1654–1661. doi:10.1016/j.neuroimage.2010.08.073
Kilian-Hutten N, Valente G, Vroomen J, Formisano E (2011) Auditory cortex encodes the perceptual interpretation of ambiguous sound. J Neurosci 31(5):1715–1720. doi:10.1523/jneurosci.4572-10.2011
Leaver AM, Rauschecker JP (2010) Cortical representation of natural complex sounds: effects of acoustic features and auditory object category. J Neurosci 30:7604–7612
Lee JH, Russ BE, Orr LE, Cohen YE (2009) Prefrontal activity predicts monkeys' decisions during an auditory category task. Fron Integr Neurosci, 3(16)
Lemus L, Hernandez A, Romo R (2009a) Neural encoding of auditory discrimination in ventral premotor cortex. Proc Natl Acad Sci U S A 106(34):14640–14645. doi:10.1073/pnas.0907505106
Lemus L, Hernandez A, Romo R (2009b) Neural codes for perceptual discrimination of acoustic flutter in the primate auditory cortex. Proc Natl Acad Sci U S A 106:9471–9476
Micheyl C, Carlyon RP, Shtyrov Y, Hauk O, Dodson T, Pullvermuller F (2003) The neurophysiological basis of the auditory continuity illusion: a mismatch negativity study. J Cogn Neurosci 15(5):747–758. doi:10.1162/089892903322307456
Middlebrooks JC, Bremen P (2013) Spatial stream segregation by auditory cortical neurons. J Neurosci 33:10986–11001
Middlebrooks JC, Onsan ZA (2012) Stream segregation with high spatial acuity. J Acoust Soc Am 132:3896–3911
Nelken I (2008) Processing of complex sounds in the auditory system. Curr Opin Neurobiol 18:413–417
Nienborg H, Cumming BG (2009) Decision-related activity in sensory neurons reflects more than a neuron's causal effect. Nature 459(7243):89–92. doi:10.1038/nature07821
Nienborg H, Cumming BG (2014) Decision-related activity in sensory neurons may depend on the columnar architecture of cerebral cortex. J Neurosci 34(10):3579–3585. doi:10.1523/jneurosci.2340-13.2014
Nienborg H, Cohen MR, Cumming BG (2012) Decision-related activity in sensory neurons: correlations among neurons and with behavior. Annu Rev Neurosci 35:463–483 (Review)
Niwa M, Johnson JS, O'Connor KN, Sutter ML (2012) Active engagement improves primary auditory cortical neurons' ability to discriminate temporal modulation. J Neurosci 32(27):9323–9334

Niwa M, Johnson JS, O'Connor KN, Sutter ML (2013) Differences between primary auditory cortex and auditory belt related to encoding and choice for AM sounds. J Neurosci 33(19):8378–8395

Ohl FW, Scheich H, Freeman WJ (2001) Change in pattern of ongoing cortical activity with auditory category learning. Nature 412(6848):733–736

Rauschecker JP (2012) Ventral and dorsal streams in the evolution of speech and language. Front Evol Neurosci 4:7

Rauschecker JP, Scott SK (2009) Maps and streams in the auditory cortex: nonhuman primates illuminate human speech processing. Nat Neurosci 12(6):718–724

Rauschecker JP, Tian B (2000) Mechanisms and streams for processing of "what" and "where" in auditory cortex. Proc Natl Acad Sci U S A 97(22):11800–11806

Riecke L, Mendelsohn D, Schreiner C, Formisano E (2009) The continuity illusion adapts to the auditory scene. Hear Res 247(1):71–77. doi:10.1016/j.heares.2008.10.006

Riecke L, Vanbussel M, Hausfeld L, Baskent D, Formisano E, Esposito F (2012) Hearing an illusory vowel in noise: suppression of auditory cortical activity. J Neurosci 32(23):8024–8034. doi:10.1523/jneurosci.0440-12.2012

Romanski LM, Averbeck BB (2009) The primate cortical auditory system and neural representation of conspecific vocalizations. Annu Rev Neurosci 32:315–346

Russ BE, Orr LE, Cohen YE (2008) Prefrontal neurons predict choices during an auditory same-different task. Curr Biol 18(19):1483–1488

Selezneva E, Scheich H, Brosch M (2006) Dual time scales for categorical decision making in auditory cortex. Curr Biol 16(24):2428–2433

Shinn-Cunningham BG (2008) Object-based auditory and visual attention. Trends Cogn Sci 12:182–186

Snyder JS, Gregg MK, Weintraub DM, Alain C (2012) Attention, awareness, and the perception of auditory scenes. Front Psychol 3:15

Steinschneider M (2013) Phonemic representations and categories. In Cohen YE, Popper AN, Fay RR (eds), Neural correlates of auditory cognition, vol 45. Springer-Verlag, New York, pp 151–161

Steinschneider M, Nourski KV, Kawasaki H, Oya H, Brugge JF, Howard MA (2011) Intracranial study of speech-elicited activity on the human posterolateral superior temporal gyrus. Cereb Cortex 10:2332–2347

Tsunada J, Lee JH, Cohen YE (2011) Representation of speech categories in the primate auditory cortex. J Neurophysiol 105:2634–2646

Tsunada J, Lee JH, Cohen YE (2012) Differential representation of auditory categories between cell classes in primate auditory cortex. J Physiol 590:3129–3139

Tsunada, J, Liu, AS, Gold, JI, Cohen, YE (?) Causal role of primate auditory cortex in auditory perceptual decision-making. Nature neuroscience (in press)

Werner S, Noppeney U (2011) The contributions of transient and sustained response codes to audiovisual integration. Cereb Cortex 21(4):920–931. doi:10.1093/cercor/bhq161

Neural Segregation of Concurrent Speech: Effects of Background Noise and Reverberation on Auditory Scene Analysis in the Ventral Cochlear Nucleus

Mark Sayles, Arkadiusz Stasiak and Ian M. Winter

Abstract Concurrent complex sounds (e.g., two voices speaking at once) are perceptually disentangled into separate "auditory objects". This neural processing often occurs in the presence of acoustic-signal distortions from noise and reverberation (e.g., in a busy restaurant). A difference in periodicity between sounds is a strong segregation cue under quiet, anechoic conditions. However, noise and reverberation exert differential effects on speech intelligibility under "cocktail-party" listening conditions. Previous neurophysiological studies have concentrated on understanding auditory scene analysis under ideal listening conditions. Here, we examine the effects of noise and reverberation on periodicity-based neural segregation of concurrent vowels /a/ and /i/, in the responses of single units in the guinea-pig ventral cochlear nucleus (VCN): the first processing station of the auditory brain stem. In line with human psychoacoustic data, we find reverberation significantly impairs segregation when vowels have an intonated pitch contour, but not when they are spoken on a monotone. In contrast, noise impairs segregation independent of intonation pattern. These results are informative for models of speech processing under ecologically valid listening conditions, where noise and reverberation abound.

Keywords Vowels · Double vowels · Intonation · Reverberation · Background noise · Auditory scene analysis · Concurrent speech · Perceptual segregation · Inter-spike intervals · Temporal envelope · Brain stem · Cochlear nucleus

M. Sayles (✉) · A. Stasiak · I. M. Winter
Centre for the Neural Basis of Hearing, The Physiological Laboratory, Department of Physiology, Development and Neuroscience, Downing Street, Cambridge CB2 3EG, UK
e-mail: sayles.m@gmail.com

M. Sayles
Laboratory of Auditory Neurophysiology, Campus Gasthuisberg, O&N II, Herestraat 49—bus 1021, 3000 Leuven, Belgium

P. van Dijk et al. (eds.), *Physiology, Psychoacoustics and Cognition in Normal and Impaired Hearing,* Advances in Experimental Medicine and Biology 894,
DOI 10.1007/978-3-319-25474-6_41

1 Introduction

A difference in periodicity between simultaneous complex sounds is a strong segregation cue under quiet, anechoic conditions (Brokx and Nooteboom 1982). However, noise and reverberation can both degrade speech intelligibility under realistic "cocktail-party" listening conditions (Nabelek 1993; Culling et al. 1994, 2003). Neurophysiological studies of concurrent-sound segregation have concentrated on harmonic complex sounds, with a fundamental-frequency difference (ΔF0), heard under idealized (quiet, anechoic) conditions (e.g., Palmer 1990; Keilson et al. 1997; Larsen et al. 2008). We examine the effects of noise and reverberation, separately, on periodicity-based neural segregation of ΔF0 concurrent vowels, with and without simulated intonation, in the ventral cochlear nucleus (VCN); the first processing station in the auditory brain stem.

2 Methods

2.1 Animal Model

Experiments were carried out in accordance with the United Kingdom Animals (Scientific Procedures) Act (1986), with approval of the University of Cambridge Animal Welfare Ethical Review Board. Details of our recording techniques are available elsewhere (Sayles and Winter 2008; Sayles et al. 2015). Adult guinea pigs (*Cavia porcellus*) were anesthetized with urethane and hypnorm (fentanyl/fluanisone). The cochlear nucleus was exposed via a posterior-fossa craniotomy and unilateral cerebellotomy. A glass-insulated tungsten microelectrode was advanced in the sagittal plane through the VCN, using a hydraulic microdrive. Upon isolation of a single unit, best frequency (BF) and threshold were determined. Units were classified on their responses to BF-tones.

2.2 Complex Stimuli

Stimuli were synthetic vowels /a/ and /i/, generated using a MATLAB implementation of the Klatt formant synthesizer. Formant frequencies were {0.7, 1.0, 2.4, 3.3} kHz for /a/, and {0.3, 2.2, 3.0, 3.7} kHz for /i/. Stimuli were presented monaurally; either alone, or as "double vowels" /a, i/ (Fig. 1). F0 was either static (125 Hz, or 250 Hz for /a/, and 100 Hz, or 200 Hz for /i/), or sinusoidally modulated at 5 Hz, by ±2 semitones. Reverberation was added by convolution with real-room impulse responses recorded in a long corridor, at source-receiver distances of 0.32, 2.5, and 10 m (Tony Watkins; University of Reading, UK). We refer to these as "mild", "moderate" and "strong" reverberation. For noise-masked vowels, Gaussian noise (5-kHz bandwidth) was added at signal-to-noise ratios of {10, 3, 0} dB.

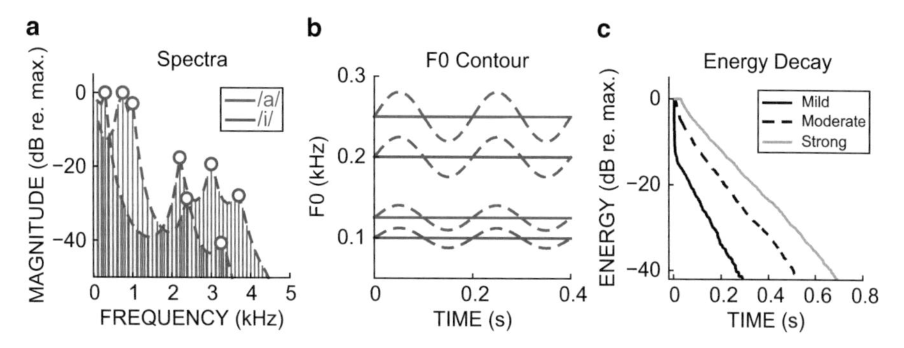

Fig. 1 Synthetic-vowel stimuli. **a** Magnitude spectra. **b** Fundamental-frequency contours: *solid lines*, static-F0 vowels; *dashed lines*, modulated-F0 vowels. **c**, Impulse-response energy-decay curves

2.3 Analyses

2.3.1 Correlograms

Analyses are based on normalized shuffled inter-spike-interval distributions derived from the VCN-unit spike trains (Fig. 2; Joris et al. 2006; Sayles et al. 2015). We computed the *across*-spike-train shuffled inter-spike-interval distribution in a short (30-ms duration) rectangular time window, slid in 5-ms steps through the 400-ms duration response. We refer to these time-varying inter-spike-interval distributions as *correlograms*.

2.3.2 Periodic Templates

We applied a "periodicity-sieve" analysis to the correlograms to estimate the dominant period(s) in the inter-spike-interval statistics (e.g., Larsen et al. 2008); yielding *template-contrast functions*. To assess the statistical significance of peaks in the template-contrast function we used a bootstrap technique (permutation analysis; 1000 replications: $p<0.01$ considered significant). Based on the template-contrast functions in response to double vowels, we computed the "periodicity-tagged" firing rate for each vowel of the mixture (similar to that proposed by Keilson et al. 1997); e.g., for the /a/ component:

$$\bar{R}_{/a/} = \max\left(\left(\frac{\bar{R}\cdot C_{/a/}}{C_{/a/}+C_{/i/}}\right) - (0.5\cdot\bar{R}),\ 0\right)$$

Where $\bar{R}$ is the mean firing rate of that single unit to the double vowel /a, i/, and $C_{/a/}$ and $C_{/i/}$ are template-contrast values for the two double-vowel components, respectively.

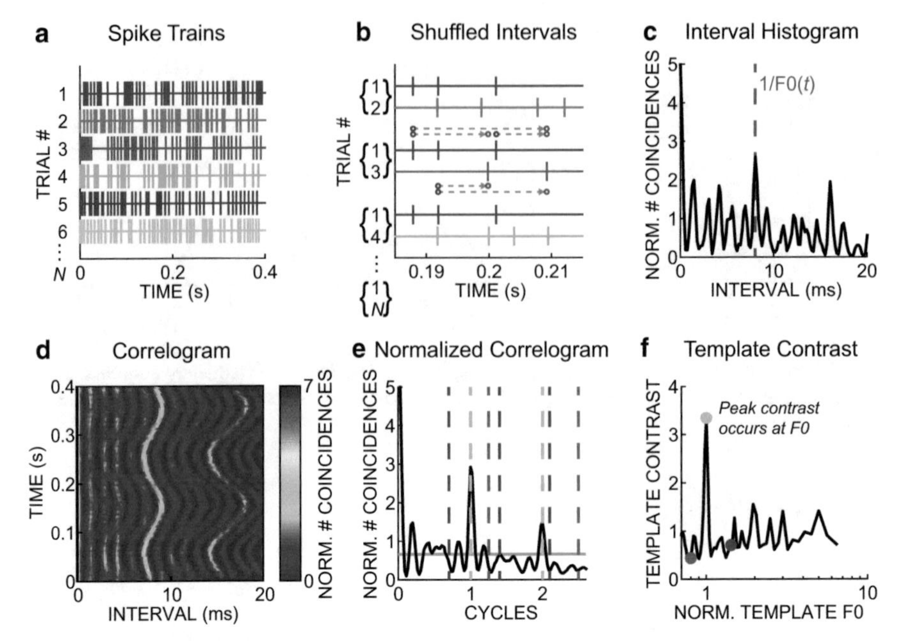

Fig. 2 Example analyses for a single unit (Primary-like, BF =2.1 kHz), responding to a single vowel /a/ with F0 modulated at 5 Hz (±2 semitones) around a mean F0 of 125 Hz. **a** Spike times were collected in response to *N* (typically, 50) repetitions of each stimulus. **b** Forward inter-spike intervals were calculated between all non-identical spike-train pairs, in 30-ms time windows. **c** Intervals were tallied in a histogram, and the analysis window slid in 5-ms steps to give the interval distribution as a function of time **d**. **e** Time-varying interval distributions were normalized for instantaneous stimulus F0, and averaged over time. **f** Harmonic periodicity sieves were applied to compute the template-contrast function

3 Results

We recorded responses to single and double vowels, in anechoic and reverberant conditions, from 129 units with BFs between 0.1 and 6 kHz (36 primary-like [PL/PN], 47 chopper [CT/CS], 24 onset [OC/OL], 19 low-frequency [LF], 3 unusual [UN]). From 52 of these, we also recorded responses to vowels in noise. The effects of noise and reverberation on double-vowel segregation are not equivalent. The results can be summarized as: (1), There is a strong interaction between F0 modulation (simulating intonated speech) and reverberation to reduce template contrast. (2), There is no interaction between F0 modulation and signal-to-noise ratio. (3), Noise-induced deficits in neural periodicity-based double-vowel segregation are strongly BF-dependent, due to more total (masking) noise power passed by higher-BF filters. (4), Reverberation impairs neural segregation of intonated double vowels independent of BF, but has only marginally detrimental effects on segregation of double vowels with steady F0s.

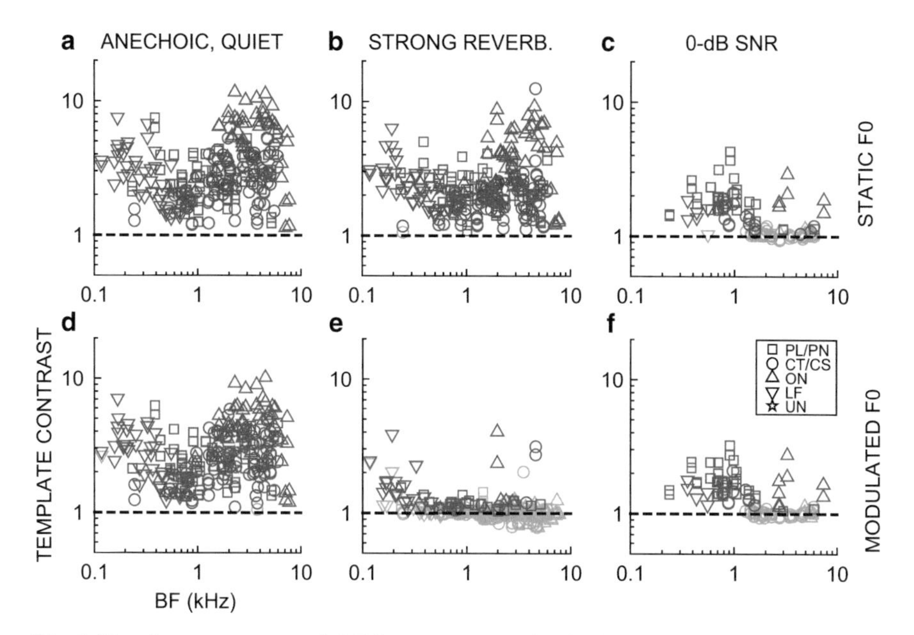

Fig. 3 Template contrast *vs.* unit BF for responses to the single vowel /a/. *Top row*, responses to static-F0 vowels; *Bottom row*, responses to modulated-F0 vowels. Different symbols indicate unit type (legend in F). *Red symbols* indicate significant values ($p < 0.01$), *gray symbols* are non-significant. *Dashed line* indicates contrast of 1, the expected value for a flat correlogram

Figure 3 shows the template contrast calculated from responses to the single vowel /a/ under ideal (anechoic, quiet) conditions, in the presence of "strong" reverberation, and in the presence of background noise at 0-dB SNR. There is a small reduction in template contrast due to reverberation alone (Fig. 3a, b), and a similar small reduction due to F0 modulation alone (Fig. 3a, d). However, the major detrimental effect on template contrast is an interaction between reverberation and F0 modulation (Fig. 3a, e). In the combined presence of reverberation and F0 modulation many units do not have a significant representation of the periodicity corresponding to this single vowel in their inter-spike-interval distributions; the many non-significant gray symbols in Fig. 3e. In contrast, the presence of masking noise reduces template contrast at the F0 period independent of F0 modulation, but in a BF-dependent manner with higher-BF units affected most strongly (Fig. 3c, f).

The interaction between reverberation and F0 modulation is further illustrated by plotting template contrast in response to modulated-F0 vowels against that in response to static-F0 vowels (Fig. 4). For both single- and double-vowel responses, the reduction in template contrast in the presence of reverberation is much greater in the modulated-F0 condition relative to the static-F0 condition (Fig. 4b, e). In the presence of noise, the maximum template contrast is reduced equally for modulated- and static-F0 vowels (Fig. 4c, f).

Based on template contrast calculated from double-vowel responses, we computed a "periodicity-tagged" discharge rate for each vowel of the concurrent-speech

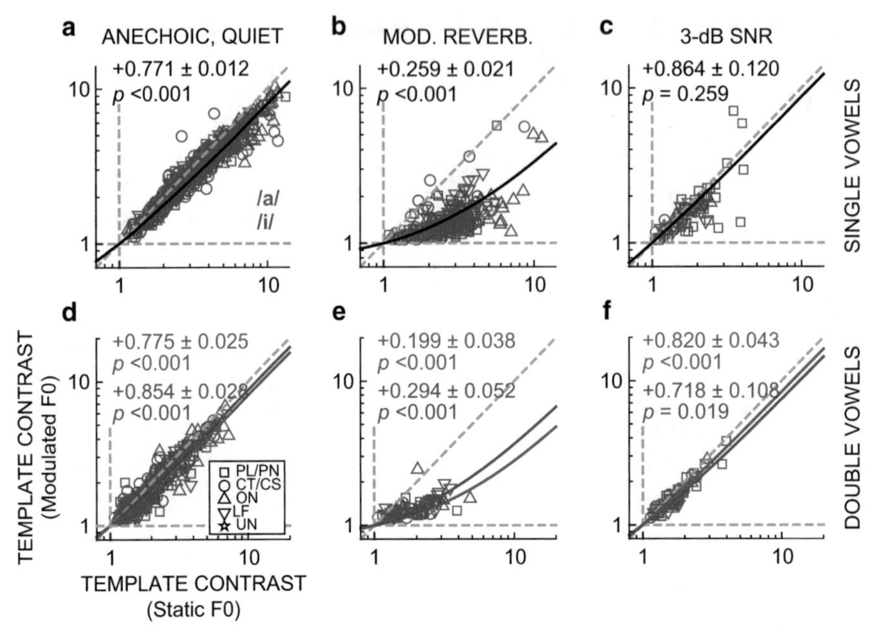

Fig. 4 Template contrast: modulated-F0 *vs.* static-F0 vowels. *Top row*, responses to single vowels; *Bottom row*, responses to double vowels. Red symbols indicate responses to /a/, and blue symbols /i/. Horizontal and vertical gray dashed lines indicate template contrast of 1. Diagonal gray dashed line indicates equality. Solid lines indicate least-squares linear fits to the significant data, with the fit constrained to (1,1). Text in each panel indicates the linear-regression slope (β_1)±S.E., and the p-value for a two-tailed t-test with the null hypothesis H_0: β_1=1. For clarity, only those data from responses with significant contrast ($p<0.01$) for both static *and* modulated vowels are displayed

mixture (Fig. 5). In the across-BF profiles of periodicity-tagged rate, there are peaks and troughs corresponding to the formant structure of each vowel under quiet, anechoic listening conditions. This is the case with both static- and modulated-F0 double vowels (Fig. 5a, d). With reverberation, there are no formant related peaks and troughs remaining in the modulated-F0 case (Fig. 5e), although a clear formant-related pattern of periodicity-tagged discharge rate remains in the static-F0 case in reverberation (Fig. 5b). Information about higher formants is degraded in the presence of noise; however, formant-related peaks in periodicity-tagged discharge rate remain in the first-formant region for both static- and modulated-F0 vowels (Fig. 5c, f).

To quantify the "segregation" of the two vowels of the double-vowel mixture by the periodicity-tagged firing-rate profile across BF, we computed the normalized Euclidean distance between the /a/ and /i/ profiles in response to /a, i/ as (Fig. 6),

$$d_{(/a/,/i/)} = \frac{\sum_{j=1}^{n}\left(\overline{R}_{/a/,j} - \overline{R}_{/i/,j}\right)^2}{\sqrt{n}}$$

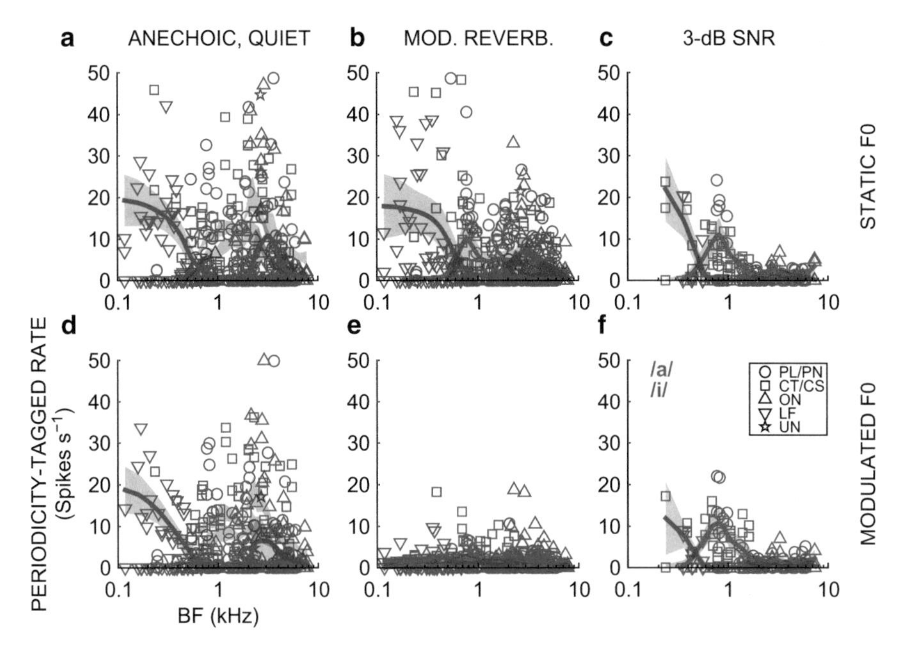

Fig. 5 Periodicity-tagged firing rate *vs.* BF. *Top row*, responses to static-F0 vowels; *Bottom row*, responses to modulated-F0 vowels. *Solid lines*, *lowess* smoothing fits; *shaded areas*, 95 % confidence intervals

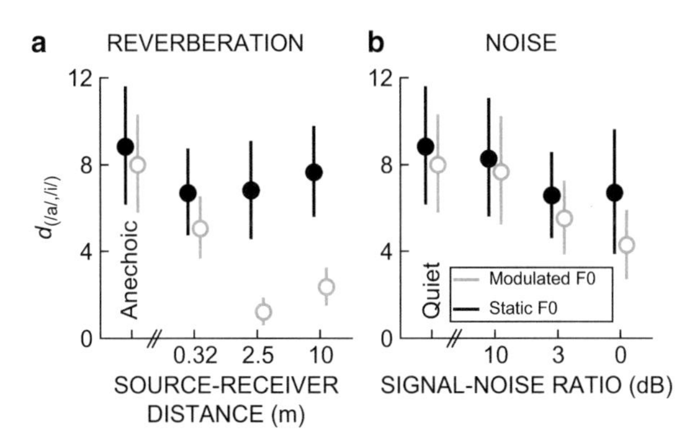

Fig. 6 Euclidean distance between /a/ and /i/ periodicity-tagged spike-rate *vs.* BF profiles, calculated from the double-vowel responses

The double-vowel sounds are well segregated on the basis of the periodicity-tagged discharge rate statistic, except when F0 is modulated and the sounds are heard in "moderate" or "strong" reverberation (Fig. 6a).

4 Discussion

The physical effects of noise and reverberation have some similarities: small changes to the magnitude and phase of each component of a complex sound. The impulse response of a room is a linear filter. Therefore, for a steady-state source no new frequency components are added to a reverberant acoustic signal. Contrast this with additive noise. Each indirect sound component in a reverberant space adds to the direct sound at the receiver with essentially random phase, reducing the depth of temporal-envelope modulation at the output of cochlear band-pass filters (Sabine 1922; Sayles and Winter 2008; Sayles et al. 2015; Slama and Delgutte 2015). Perceptually, noise and reverberation can both decrease speech intelligibility (e.g., Nabelek 1993; Payton et al. 1994), and can be particularly troublesome for cochlear-implant listeners (e.g., Qin and Oxenham 2005). However, error patterns differ substantially between noisy and reverberant spaces (Nabelek 1993; Assmann and Summerfield 2004). Moreover, normal-hearing listeners have remarkably good speech understanding in moderately reverberant spaces (e.g., Poissant et al. 2006).

Our results can be understood in terms of the effects of reverberation on the acoustic temporal envelope. For static-F0 vowels, the reduction in envelope modulation due to reverberant energy is evident in higher-BF units (Fig. 3a, b; Fig. 5a, b). The addition of frequency modulation to the vowels has a dramatic effect on template contrast in reverberation, due to spectral smearing: i.e., monaural decorrelation. Noise impairs neural segregation of concurrent vowels independent of intonation pattern, but in a BF-dependent manner. Similar to other studies of noise-masked single-vowel coding (e.g., Delgutte and Kiang 1984; May et al. 1998), this is the consequence of higher-BF filters passing more total noise power than low-BF filters. The within-band signal-to-noise ratio is therefore much lower for high-BF units. The negative spectral tilt of the vowels, masked by flat-spectrum noise, imposes a similar frequency-dependency on signal-to-noise ratio after cochlear filtering.

The differential effects of noise and reverberation on speech representations in the early stages of brain-stem neural processing are in clear contrast to noise- and reverberation-invariant representations of speech in auditory cortex (Mesgarani et al. 2014). Perhaps one clue to the neural underpinnings of robust speech understanding in challenging acoustic environments is compensation for the effects of reverberation on coding of temporal-envelope modulation in the inferior colliculus, based on sensitivity to inter-aural correlation (Slama and Delgutte 2015). What is clear from our data is that neurophysiological effects of room reverberation cannot simply be assimilated to those of broadband noise for signal detection and discrimination.

Acknowledgements Funded by a grant from the BBSRC (IMW), and a University of Cambridge MD/PhD scholarship (MS).

References

Assmann PF, Summerfield AQ (2004) The perception of speech under adverse conditions. In: Greenberg S, Ainsworth WA, Fay RR, Popper A (eds) Speech processing in the auditory system. Springer, New York

Brokx JPL, Nooteboom SG (1982) Intonation and the perceptual separation of simultaneous voices. J Phon 10(1):23–36

Culling JF, Summerfield Q, Marshall DH (1994) Effects of simulated reverberation on the use of binaural cues and fundamental-frequency differences for separating concurrent vowels. Speech Comm 14(1):71–95

Culling JF, Hodder KI, Toh CY (2003) Effects of reverberation on perceptual segregation of competing voices. J Acoust Soc Am 114(5):2871–2876

Delgutte B, Kiang NYS (1984) Speech coding in the auditory-nerve. V. vowels in background-noise. J Acoust Soc Am 75(3):908–918

Joris PX, Louage DH, Cardoen L, van der Heijden M (2006). Correlation index: a new metric to quantify temporal coding. Hear Res, 216–217(1), 19–30

Keilson SE, Richards VM, Wyman BE, Young ED (1997) The representation of concurrent vowels in the cat anaesthetized ventral cochlear nucleus: evidence for a periodicity-tagged spectral representation. J Acoust Soc Am 102(2):1056–1070

Larsen E, Cedolin L, Delgutte B (2008) Pitch representations in the auditory nerve: two concurrent complex tones. J Neurophysiol 100(3):1301–1319

May BJ, Prell GS, Sachs MB (1998) Vowel representations in the ventral cochlear nucleus of the cat: effects of level, background noise, and behavioral state. J Neurophysiol 79(4):1755–1767

Mesgarani N, David SV, Fritz JB, Shamma SA (2014) Mechanisms of noise robust representation of speech in primary auditory cortex. Proc Natl Acad Sci USA 111(18):6792–6797

Nabelek AK (1993) Communication in noisy and reverberant environments. In: Stubebaker GA, Hochberg I (eds) Acoustical factors affecting hearing aid performance. Allyn and Bacon, Needham Heights

Palmer AR (1990) The representation of the spectra and fundamental frequencies of steady-state single- and double-vowel sounds in the temporal discharge patterns of guinea pig cochlear-nerve fibers. J Acoust Soc Am 88(3):1412–1426

Payton KL, Uchanski RM, Braida LD (1994) Intelligibility of conversational and clear speech in noise and reverberation for listeners with normal and impaired hearing. J Acoust Soc Am 95(3):1581–1592

Poissant SF, Whitmal NA 3rd, Freyman RL (2006) Effects of reverberation and masking on speech intelligibility in cochlear implant simulations. J Acoust Soc Am 119(3):1606–1615

Qin MK, Oxenham AJ (2005) Effects of envelope-vocoder processing on F0 discrimination and concurrent-vowel identification. Ear Hear 26:451–460

Sabine WC (1922) Collected papers on acoustics. Harvard University Press, Cambridge

Sayles M, Winter IM (2008) Reverberation challenges the temporal representation of the pitch of complex sounds. Neuron 58:789–801

Sayles M, Stasiak A, Winter IM (2015) Reverberation impairs brainstem temporal representations of voiced vowel sounds: challenging "periodicity-tagged" segregation of competing speech in rooms. Front Syst Neurosci 8:248

Slama MC, Delgutte B (2015) Neural coding of sound envelope in reverberant environments. J Neurosci 35(10):4452–4468

Audio Visual Integration with Competing Sources in the Framework of Audio Visual Speech Scene Analysis

Attigodu Chandrashekara Ganesh, Frédéric Berthommier and Jean-Luc Schwartz

Abstract We introduce "Audio-Visual Speech Scene Analysis" (AVSSA) as an extension of the two-stage Auditory Scene Analysis model towards audiovisual scenes made of mixtures of speakers. AVSSA assumes that a coherence index between the auditory and the visual input is computed prior to audiovisual fusion, enabling to determine whether the sensory inputs should be bound together. Previous experiments on the modulation of the McGurk effect by audiovisual coherent vs. incoherent contexts presented before the McGurk target have provided experimental evidence supporting AVSSA. Indeed, incoherent contexts appear to decrease the McGurk effect, suggesting that they produce lower audiovisual coherence hence less audiovisual fusion. The present experiments extend the AVSSA paradigm by creating contexts made of competing audiovisual sources and measuring their effect on McGurk targets. The competing audiovisual sources have respectively a high and a low audiovisual coherence (that is, large vs. small audiovisual comodulations in time). The first experiment involves contexts made of two auditory sources and one video source associated to either the first or the second audio source. It appears that the McGurk effect is smaller after the context made of the visual source associated to the auditory source with less audiovisual coherence. In the second experiment with the same stimuli, the participants are asked to attend to either one or the other source. The data show that the modulation of fusion depends on the attentional focus. Altogether, these two experiments shed light on audiovisual binding, the AVSSA process and the role of attention.

Keywords Audio visual binding · Auditory speech analysis · McGurk effect · Attention

---

A. C. Ganesh (✉) · F. Berthommier · J.-L. Schwartz
Grenoble Images Parole Signal Automatique-Lab, Speech and Cognition Department, CNRS, Grenoble University, UMR 5216, Grenoble, France
e-mail: ganesh.attigodu@gipsa-lab.grenoble-inp.fr

F. Berthommier
e-mail: frederic.berthommier@gipsa-lab.grenoble-inp.fr

J.-L. Schwartz
e-mail: jean-luc.schwartz@gipsa-lab.grenoble-inp.fr

P. van Dijk et al. (eds.), *Physiology, Psychoacoustics and Cognition in Normal and Impaired Hearing,* Advances in Experimental Medicine and Biology 894,
DOI 10.1007/978-3-319-25474-6_42

1 Introduction

This paper is focused on a tentative fusion between two separate concepts: Auditory Scene Analysis and Audio-Visual fusion in speech perception.

Auditory Scene Analysis (ASA) introduced the principle of a two-stage process in the auditory processing of complex auditory scenes with competing sources (Bregman 1990). A first stage would involve segmenting the scene into auditory elements, which would be segregated or grouped in respect to their common source, either by bottom-up innate primitives or by learnt top-down schemas. Decision and formation of a final percept would be done at a second later stage.

Audio-Visual fusion in speech perception refers to the well-known fact that speech perception involves and integrates auditory and visual cues, as shown in various paradigms such as speech in noise (Sumby and Pollack 1954; Erber 1969) or the perception of conflicting stimuli (the so-called McGurk effect, McGurk and MacDonald 1976; also see Tiippana 2014).

Since a pioneer proposal by Berthommier (2004), our group proposed that auditory scene analysis and multisensory interactions in speech perception should be combined into a single "Audio-Visual Speech Scene Analysis" (AVSSA) process. The basic claim is that the two-stage analysis-and-decision process at work in ASA should be extended to audiovisual speech scenes made of mixtures of auditory and visual speech sources. A first audiovisual binding stage would involve segmenting the scene into audiovisual elements, which should be segregated or grouped in respect to their common multisensory speech source, either by bottom-up *audiovisual primitives* or by learnt top-down *audiovisual schemas*. This audiovisual binding stage would control the output of the later decision stage, and hence intervene on the output of the speech-in-noise or McGurk paradigms.

To provide evidence for this "binding and fusion" AVSSA process, Nahorna et al. (2012, 2015) showed that the McGurk effect can be significantly and strongly reduced by an audiovisual context made of a few seconds of incoherent material (sounds and images coming from different speech sources) presented before the McGurk target (audio "ba" plus video "ga"): the target, classically perceived as "da" in the McGurk effect, was more often perceived as "ba", suggesting a decreased weight of the visual input in the fusion process. The interpretation was that the incoherent context resulted in an "unbinding" effect decreasing the visual weight and hence diminishing the McGurk effect. This modulation of the McGurk effect through incoherent contexts was further extended to speech in noise (Ganesh et al. 2013), and a possible neurophysiological correlate of the binding/unbinding process was provided in an EEG experiment (Ganesh et al. 2014).

However, these studies were based on audiovisual scenes that never implied competing sources. The objective of the present study was to test the "binding and fusion" AVSSA process in scenes including competition between audiovisual sources. For this aim, we generated two audiovisual sources, one made of a sequence of isolated syllables, and the other one made of a sequence of sentences. We prepared two kinds of combinations, with the same auditory content (mixing the two audio sources, syllables and sentences) and two different video contents, either the syllables or the sentences. These two combinations (“Video syllables” and “Video sentences”) were used as the context in a McGurk experiment. We hypothesized that since syllables correspond to stronger audiovisual modulations in time and hence stronger audiovisual coherence than sentences, the association between the visual input and the corresponding auditory input would be stronger for syllables than for sentences. Hence the coherence of the audiovisual context would be stronger for syllables, and it would lead to a larger visual weight and more McGurk effect than with visual sentences. Furthermore, the introduction of a competition between sources made it possible to introduce attention factors in the paradigm, and we tested whether the attentional focus put by the participants on either syllables or sentences would play a role in the AVSSA process.

2 Method

2.1 Participants

The study involved twenty-nine French participants without hearing or vision problems (22 women and 7 men; 27 right-handed and 2 left handed; mean age = 29.2 years; SD = 10. 4 years), who all gave informed consent to participate in the experiment.

2.2 Stimuli

The stimuli were similar to those of the previous experiment by (Nahorna et al. 2015) with suitable modification in the paradigm. They were prepared from two sets of audiovisual material, a “syllables” material and a “sentences” material, produced by a French male speaker, with lips painted in blue to allow precise video analysis of lip movements (Lallouache 1990). The whole experiment consisted of two types of contexts followed by a target.

2.2.1 Target

The target was either a congruent audiovisual “ba” syllable (“ba-target” in the following), serving as a control—or an incongruent McGurk stimulus with an audio “ba” mounted on a video “ga” (“McGurk target” in the following).

2.2.2 Context

There were two types of contexts i.e. “Video syllables” and “Video sentences”. In both contexts, the set of audio stimuli was the same. It consisted of a sequence of 2

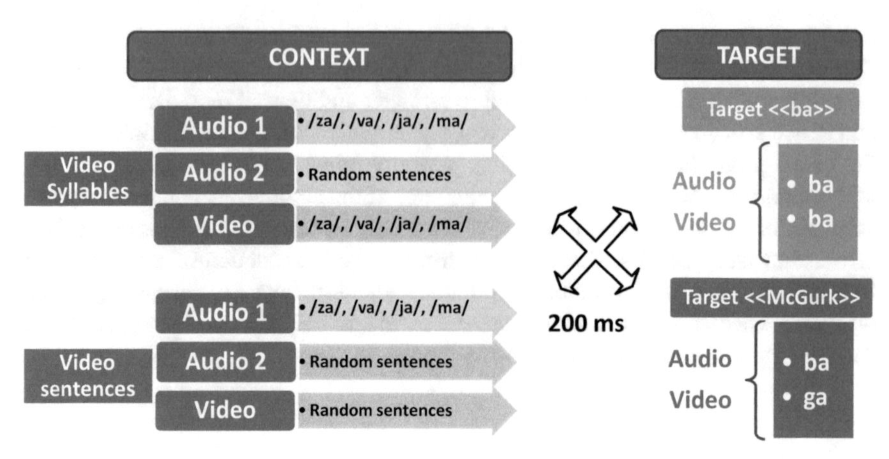

Fig. 1 Description of the audiovisual material

or 4 syllables (A-syl-2 or A-syl-4) randomly extracted from the set "pa," "ta," "va," "fa," "za," "sa," "ka," "ra," "la," "ja," "cha," "ma," or "na," mixed with a portion of a random set of sentences with the adequate duration (A-sent-2 or A-sent-4). The 2- vs. 4-syllable duration was selected from earlier experiments by Nahorna et al. (2015), showing that the effect of incoherent context was maximal (maximal reduction of the McGurk effect) for short 2-syllable contexts and slightly less for longer 4-syllable contexts. The visual components of the context were the visual stream associated with either the auditory syllables (V-syl-2 or V-syl-4) or the auditory sentences (V-sent-2 or V-sent-4). Therefore, in the "Video syllables" contexts, there was an audiovisual "syllables" source competing with an audio "sentences" source, while in the "Video sentences" contexts, there was an audiovisual "sentences" source competing with an audio "syllables" source (Fig. 1). A 200 ms fading transition stimulus (five images) was implemented between context and target to ensure continuity between images.

There were altogether 120 stimuli with four times more "McGurk" than "Ba" targets (serving as controls), and with the same number of occurrences of the V-syl-2, V-syl-4, V-sent-2 and V-sent-4 contexts (6 occurrences each for "Ba" targets, 24 occurrences each for McGurk targets). Exactly the same set of 30 targets was presented after the 4 types of contexts. The 120 stimuli were presented in a random order and concatenated into a single 7-min film.

2.3 Procedure

The study included two consecutive experiments, Exp. A. followed by Exp. B (always in this order). In Exp. A, the participants were involved in a monitoring paradigm in which they were asked to constantly look at the screen and monitor for possible "ba" or "da" targets by pressing an appropriate key, as in Nahorna et al. (2012, 2015). In Exp. B the monitoring "ba" vs. "da" task remained the same (with a different order of the 120 stimuli in the film), but in addition, specific instructions were given to participants, either to put more attention to syllables ("Attention syllables") or to put more attention to sentences ("Attention sentences"). The order of the "Attention syllables" and "Attention sentences" conditions was counterbalanced between the participants. To increase the efficiency of the attentional demand, participants were informed that they would be questioned on the content of either the "syllables" or the "sentences" material at the end of the experiment. A practice session was provided for all of them and most of the participants were indeed able to recall specific syllables or words. The films were presented on a computer monitor with high-fidelity headphones set at a comfortable fixed level.

2.4 Processing of Responses

Response time was computed in reference to the acoustic onset of the burst of the "b" in the target syllable, discarding values higher than 1200 ms or lower than 200 ms. "ba" and "da" responses were taken into account only when they occurred within this time window (200–1200 ms) and in case of two different responses inside the time window, both responses were also discarded. Finally, for each participant and each condition of context and target (and attention in Exp. B), a global score of "ba" responses was calculated as the percentage of "ba" responses divided by the sum of "ba" and "da" responses to the target, and a mean response time was calculated as the average of response times for all the responses to the target.

3 Results

First of all, the mean percentage of "ba" scores for McGurk targets over all conditions in Exp. A was computed for each subject, and participants providing mean scores larger than 95 % or less than 5 % were discarded, considering that these subjects provided either too strong or too low McGurk effects to enable binding modulations to be displayed. This resulted in discarding 8 out of 29 participants. All further analyses for both Exp. A and B will hence concern only the 21 remaining subjects. As expected, the global score (percentage of "ba" responses relative to "ba" + "da" responses) for all control "ba" targets was close to 100 % in all conditions in both experiments. Therefore, from now on we will concentrate on McGurk targets.

3.1 On the Role of Context Type Without Explicit Attention Focus (Exp. A)

Percentages of "ba" responses to McGurk targets in Exp. A (without explicit attentional focus) are displayed on the left part of Fig. 2. A two-factor repeated measures ANOVA with *context type* ("Video syllables" vs. "Video sentences") and *context*

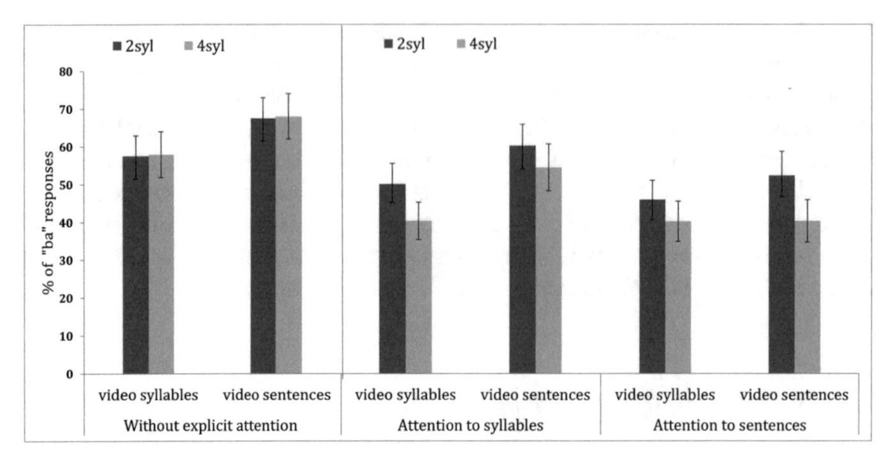

Fig. 2 The percentage of "ba" responses (relative to the total number of "ba" or "da" responses) for "McGurk" targets, in the "Video syllables" vs. "Video sentences" contexts in Experiment A and Experiment B

duration (2- vs. 4-syllables) as the independent variables was administered on these percentages (applying Greenhouse-Geisser correction when applicable). The effect of *context type* is significant [$F(1, 20) = 34.65$, $p < 0.001$], with a higher McGurk effect (10 % less "ba" responses) with the "Video syllables" context. This is in line with our prediction that audiovisual coherence is higher in the "Video syllables" condition, leading to a higher binding level, a larger visual weight and hence a larger number of McGurk fusion ("da" responses). *Context duration* displayed no significant effect on "ba" scores, either in isolation or in interaction with *context type*.

3.2 On the Interaction Between Context Type and Attention Focus (Exp. B)

Percentages of "ba" responses to McGurk targets in Exp. B (involving explicit attention towards one or the other source) are displayed on the right part of Fig. 2. A repeated-measures ANOVA was administered on these percentages with three factors, *context type* ("Video syllables" vs. "Video sentences"), *context duration* (2- vs 4-syllables) and *attention* ("Attention syllables" vs. "Attention sentences") by applying Greenhouse-Geisser correction when applicable.

The effect of context type [$F(1, 20) = 11.91$, $p < 0.001$] is significant, as in Exp. A: video syllables produce more McGurk than video sentences. Contrary to Exp. A, the effect of context duration [$F(1, 20) = 33.86$, $p < 0.001$] is also significant, with no interaction with context type. The attention factor alone is not significant, but its interaction with context type is significant [$F(1, 20) = 11.07$, $p < 0.005$]. Post-hoc analyses with Bonferroni corrections show that while there is no significant difference between the two attention conditions for the "Video syllables" context type, there is a difference for the “Video sentences” condition, with a lower “ba” percentage (a higher McGurk effect) in the “Attention sentence” condition. Interestingly, while the “ba” percentage is higher for the “Video sentences” than for the “Video syllables” condition when attention is put in syllables, there is no more significant difference when attention is put on sentences.

Finally, the three-way interaction between context type, context duration and attention is significant [$F(1, 20)=6.51$, $p<0.05$], with a larger difference between durations from the “Video syllables” to the “Video sentences” condition in the “Attention sentences” than in the “Attention syllables” condition.

3.3 Response Time

The results are consistent with the previous findings (Nahorna et al. 2012) in which response times were larger for McGurk targets, independently on context. In both experiments and in all contexts, the processing of “ba” responses was indeed quicker compared to McGurk responses. Two-way repeated-measures ANOVA on *target* and *context type* in Exp. A displays an effect of target [70 ms quicker response for “ba” targets, $F(1, 20)=14.25$, $p<0.005$] and no effect of context or any interaction effect. A three-way repeated-measures ANOVA on *condition type*, *attention* and *target* in Exp. B displays once again an effect of target [80 ms quicker response for “ba” targets, $F(1, 20)=4.47$, $p<0.05$] and no other significant effect of other factors, alone or in interaction.

4 Discussion

Audiovisual fusion has long been considered as an automatic fusion process (e.g. Massaro 1987). However Exp. A. confirms that the contextual stimulus may modulate fusion as in our previous experiments (Nahorna et al. 2012; Ganesh et al. 2013; Nahorna et al. 2015) and extends the concept to the case of competing sources.

This supports the AVSSA hypothesis, in which a first speech scene analysis process would group together the adequate audiovisual pieces of information and estimate the degree of audiovisual coherence. The effect of context type (larger McGurk effect in the “Video syllables” condition) could be due to the differences in audiovisual correlations for syllables and sentences. Indeed, correlation analysis between audio (full band envelope) and video (mouth opening area) material for syllables and sentences provides a mean correlation value of 0.59 for “Video syllables” and 0.10 for “Video sentences” (Fig. 3). Another factor could increase binding with syllables, i.e. the presence of a streaming mechanism in which the syllabic target would be associated to the syllables stream rather than to the sentences stream.

A number of recent experiments pointed the role of general attentional mechanisms able to globally decrease the amount of fusion (Tiippana et al. 2004; Alsius

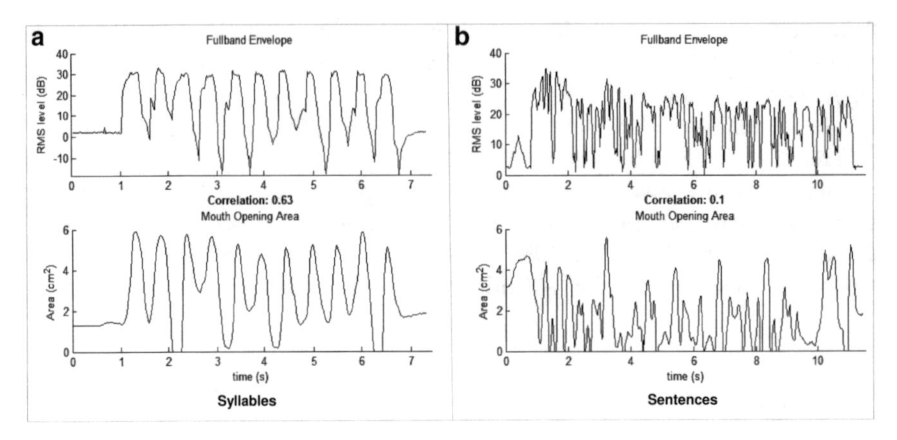

Fig. 3 Variations in time of the audio full band envelope (*top row*) and the video mouth opening area (*bottom row*) for syllables (**a**, *left*) and sentences (**b**, *right*). Notice that the fluctuations in time of the audio and video information are much more coherent between the audio and the video streams for syllables than for sentences

et al. 2007; Navarra et al. 2010; Alsius et al. 2014). Experiment B shows that attentional mechanisms may intervene at the level of single audiovisual sources in an audiovisual speech scene, selectively increasing or decreasing the amount of fusion depending on the coherence of the attended source. Interestingly, attention intervened only for "Video sentences". Our interpretation is that binding could be pre-attentive for syllables, because of their salient audiovisual comodulations making them pop out as strong bottom-up audiovisual primitives. In contrast, since the coherence of AV sentences is low, the attentional focus could enhance audiovisual top-down schemas increasing binding.

These two studies provide confirmation and development to the view that audiovisual fusion in speech perception includes a first stage of audiovisual speech scene analysis. A number of previous studies suggested that the presentation of a visual stream can enhance segregation by affecting primary auditory streaming (Rahne et al. 2007; Marozeau et al. 2010; Devergie et al. 2011) or that visual cues can improve speech detection and cue extraction (Grant and Seitz 2000; Kim and Davis 2004; Schwartz et al. 2004; Alsius and Munhall 2013); though some contradictory studies highlight cases where unimodal perceptual grouping precedes multisensory integration (Sanabria et al. 2005).

Altogether, the "binding stage" in the AVSSA process, in which the coherence between auditory and visual features would be evaluated in a complex scene, provides a mechanism in order to properly associate the adequate components inside a coherent audiovisual speech source. The present study confirms the importance of this mechanism in which the "binding stage" enables the listener to integrate "similar sources" or segregate "dissimilar sources" in Audio Visual fusion.

Acknowledgments This project has been supported by Academic Research Community "Quality of life and ageing" (ARC 2) of the Rhône-Alpes Region, which provided a doctoral funding for Ganesh Attigodu Chandrashekara.

References

Alsius A, Munhall KG (2013) Detection of audiovisual speech correspondences without visual awareness. Psychol Sci 24(4):423–431. doi:10.1177/0956797612457378

Alsius A, Navarra J, Soto-Faraco S (2007) Attention to touch weakens audiovisual speech integration. Exp Brain Res 183(3):399–404. doi:10.1007/s00221-007-1110-1

Alsius A, Mottonen R, Sams ME, Soto-Faraco S, Tiippana K (2014) Effect of attentional load on audiovisual speech perception: evidence from ERPs. Front Psychol 5:727. doi:10.3389/fpsyg.2014.00727

Berthommier F (2004) A phonetically neutral model of the low-level audio-visual interaction. Speech Comm 44(1–4):31–41. doi:10.1016/j.specom.2004.10.003

Bregman AS (1990) Auditory scene analysis. MIT Press, Cambridge

Devergie A, Grimault N, Gaudrain E, Healy EW, Berthommier F (2011) The effect of lip-reading on primary stream segregation. J Acoust Soc Am 130(1):283–291. doi:10.1121/1.3592223

Erber NP (1969) Interaction of audition and vision in the recognition of oral speech stimuli. J Speech Hear Res 12(2):423–425

Ganesh AC, Berthommier F, Schwartz J-L (2013). Effect of context, rebinding and noise on audiovisual speech fusion. Paper presented at the 14th Annual Conference of the International Speech Communication Association (Interspeech 2013), Lyon, France

Ganesh AC, Berthommier F, Vilain C, Sato M, Schwartz J-L (2014) A possible neurophysiological correlate of audiovisual binding and unbinding in speech perception. Front Psychol 5:1340. doi:10.3389/fpsyg.2014.01340

Grant KW, Seitz PF (2000) The use of visible speech cues for improving auditory detection of spoken sentences. J Acoust Soc Am 108(3):1197–1208. doi:10.1121/1.1288668

Kim J, Davis C (2004) Investigating the audio–visual speech detection advantage. Speech Comm 44(1–4):19–30. doi:10.1016/j.specom.2004.09.008

Lallouache MT (1990). Un poste 'visage-parole.' Acquisition et traitement de contours labiaux (A 'face-speech' workstation. Acquisition and processing of labial contours). Paper presented at the Proceedings of the eighteenth Journées d'Etudes sur la Parole, Montréal, QC

Marozeau J, Innes-Brown H, Grayden DB, Burkitt AN, Blamey PJ (2010) The effect of visual cues on auditory stream segregation in musicians and non-musicians. PLoS ONE 5(6):e11297. doi:10.1371/journal.pone.0011297

Massaro DW (1987) Speech perception by ear and eye. Lawrence Erlbaum Associates, Hillsdale

McGurk H, MacDonald J (1976) Hearing lips and seeing voices. Nature 264(5588):746–748 [10.1038/264746a0]

Nahorna O, Berthommier F, Schwartz JL (2012) Binding and unbinding the auditory and visual streams in the McGurk effect. J Acoust Soc Am 132(2):1061–1077. doi:10.1121/1.4728187

Nahorna O, Berthommier F, Schwartz JL (2015) Audio-visual speech scene analysis: characterization of the dynamics of unbinding and rebinding the McGurk effect. J Acoust Soc Am 137(1):362–377. doi:10.1121/1.4904536

Navarra J, Alsius A, Soto-Faraco S, Spence C (2010) Assessing the role of attention in the audiovisual integration of speech. Inf Fusion 11(1):4–11. doi:10.1016/j.inffus.2009.04.001

Rahne T, Bockmann M, von Specht H, Sussman ES (2007) Visual cues can modulate integration and segregation of objects in auditory scene analysis. Brain Res 1144:127–135. doi:10.1016/j.brainres.2007.01.074

Sanabria D, Soto-Faraco S, Chan J, Spence C (2005) Intramodal perceptual grouping modulates multisensory integration: evidence from the crossmodal dynamic capture task. Neurosci Lett 377(1):59–64. doi:10.1016/j.neulet.2004.11.069

Schwartz JL, Berthommier F, Savariaux C (2004) Seeing to hear better: evidence for early audio-visual interactions in speech identification. Cognition 93(2):B69–B78. doi:10.1016/j.cognition.2004.01.006

Sumby WH, Pollack I (1954) Visual contribution to speech intelligibility in noise. J Acoust Soc Am 26(2):212–215. doi:10.1121/1.1907309

Tiippana K (2014) What is the McGurk effect? [Opinion]. Front Psychol 5. doi:10.3389/fpsyg.2014.00725

Tiippana K, Andersen TS, Sams M (2004) Visual attention modulates audiovisual speech perception. Eur J Cog Psychol 16(3):457–472. doi:10.1080/09541440340000268

Relative Pitch Perception and the Detection of Deviant Tone Patterns

Susan L. Denham, Martin Coath, Gábor P. Háden, Fiona Murray and István Winkler

Abstract Most people are able to recognise familiar tunes even when played in a different key. It is assumed that this depends on a general capacity for relative pitch perception; the ability to recognise the pattern of inter-note intervals that characterises the tune. However, when healthy adults are required to detect rare deviant melodic patterns in a sequence of randomly transposed standard patterns they perform close to chance. Musically experienced participants perform better than naïve participants, but even they find the task difficult, despite the fact that musical education includes training in interval recognition.

To understand the source of this difficulty we designed an experiment to explore the relative influence of the size of within-pattern intervals and between-pattern transpositions on detecting deviant melodic patterns. We found that task difficulty increases when patterns contain large intervals (5–7 semitones) rather than small intervals (1–3 semitones). While task difficulty increases substantially when transpositions are introduced, the effect of transposition size (large vs small) is weaker. Increasing the range of permissible intervals to be used also makes the task more difficult. Furthermore, providing an initial exact repetition followed by subsequent

S. L. Denham (✉) · M. Coath
Cognition Institute and School of Psychology, Plymouth University, Plymouth, UK
e-mail: sdenham@plymouth.ac.uk

M. Coath
e-mail: mcoath@gmail.com

G. P. Háden · I. Winkler
Institute of Cognitive Neuroscience and Psychology, Research Centre for Natural Sciences, Hungarian Academy of Sciences, Budapest, Hungary
e-mail: haden.gabor@ttk.mta.hu

F. Murray
School of Psychology, Plymouth University, Plymouth, UK
e-mail: fi_murray@hotmail.co.uk

I. Winkler
Institute of Psychology, University of Szeged, Szeged, Hungary
e-mail: winkler.istvan@ttk.mta.hu

P. van Dijk et al. (eds.), *Physiology, Psychoacoustics and Cognition in Normal and Impaired Hearing,* Advances in Experimental Medicine and Biology 894,
DOI 10.1007/978-3-319-25474-6_43

transpositions does not improve performance. Although musical training correlates with task performance, we find no evidence that violations to musical intervals important in Western music (i.e. the perfect fifth or fourth) are more easily detected. In summary, relative pitch perception does not appear to be conducive to simple explanations based exclusively on invariant physical ratios.

Keywords Relative pitch perception · Musical intervals · Oddball paradigm · Pattern detection · Deviant detection · Translation-invariant perception

1 Introduction

Most people easily recognise well known melodies even when they are transposed to a different key. The invariant property of transposed melodies is the preserved pitch ratio relationship between notes of the melody; i.e. pitch intervals of the melody remain the same despite changes in absolute pitch. For this reason, it is assumed that the ability to recognise pitch relationships (relative pitch perception) is rather robust and commonly found in the population. Recognition of preserved pitch interval patterns irrespective of absolute pitch is an auditory example of translation-invariant object perception (Kubovy and Van Valkenburg 2001; Griffiths and Warren 2004; Winkler et al. 2009).

The robustness of the ability to recognise tone patterns has been supported by recent findings showing that listeners can detect random tone patterns very quickly (after ca. 1.5 repetitions) within rapidly presented tone sequences, even if the patterns are quite long (up to 20 tones in a pattern) (Barascud 2014). The human brain is also sensitive to pattern violations, with regular to random transitions (Chait et al. 2007) being detected within about 150 ms (~3 tones) from deviation onset (Barascud 2014). However, in these examples tone patterns were always repeated exactly, i.e. without transposition, so it is not clear whether listeners were remembering absolute pitch sequences or relative pitch relationships.

In support of the assumed generality of relative pitch perception, it has been shown that violations of transposed pitch patterns elicit discriminative brain responses in neonates (Stefanics et al. 2009) and young infants (Tew et al. 2009). So it is surprising that relative pitch perception can be rather poor (e.g. see (Foster and Zatorre 2010; McDermott et al. 2010)), especially if contour violations and tonal melodies are excluded (Dowling 1986). McDermott et al. (2010), commenting on the poor pitch interval discrimination threshold they found, suggested that the importance of pitch as an expressive musical feature may rest more on an ability to detect pitch differences between tones, rather than an ability to recognise complex patterns of pitch intervals.

Some years ago, in a pilot experiment we noticed that an oddball interval (e.g. a tone pair separated by 7 semitones) did not pop out as expected within a randomly transposed series of standard intervals (e.g. 3 semitones). We subsequently ran a series of experiments in which we maintained a standard pitch contour, but varied the number of repetitions of the standard phrase (2 or 3), the number of tones in a phrase (2–6), the size of the deviance (1–3 semitones), and the tonality of the short melodies (Coath 2008). Most listeners, including those with musical education, found it very difficult to detect an oddball melodic phrase in a sequence of randomly transposed standard phrases, performing close to chance. The source of the surprising difficulty of the task was not clarified by this experiment, as the variables tested only weakly influenced performance. Here we report another attempt to discover what makes this task so hard.

Consistent with Gestalt grouping principles (Köhler 1947), auditory streaming experiments show that featural separation (such as pitch differences) promote segregation and conversely that featural similarity promotes integration (Bregman 1990; Moore and Gockel 2012). It is also known that within-stream (within-pattern) comparisons are far easier to make than between stream comparisons; (e.g. (Bregman 1990; Micheyl and Oxenham 2010)). Therefore, we hypothesized that if the standard pattern satisfied Gestalt grouping principles and could thus be more easily grouped, this would facilitate pattern comparisons, and that deviations within such patterns would be easier to detect. Another possibility is that confusion between within-pattern intervals and between-pattern transpositions may make individual patterns less distinctive, and so increase the task difficulty. Therefore, we also investigated the effects of transposition size and interactions between transposition size and within-phase intervals. Finally, the predictive coding account of perception (Friston 2005) suggests that the precision with which perceptual discriminations can be made is inversely related to stimulus variance, suggesting that task difficulty would increase with variance of standard phrase pitch intervals.

Our specific hypotheses were:

1. Small within-pattern intervals will promote grouping and thus improve performance (Gestalt proximity/similarity);
2. Small transpositions, especially when within-pattern intervals are large, may make individual patterns less distinctive, and thus impair performance;
3. Exact repetitions with no transposition will result in very good performance;
4. One exact repeat (i.e. pattern 1 = pattern 2) before introducing transpositions may allow a better pattern representation to be built and used as a template for subsequent patterns, and so improve task performance;
5. Smaller variance in the intervals within a pattern (either only small or only large intervals) will increase the predictability of the pattern and allow the formation of a more precise representation of the pattern. Therefore, task performance will decrease with increasing interval variance.
6. Musical training and experience will facilitate task performance.

2 Methods

The study was approved by the ethical review board of Plymouth University. Participants either received credits in a university course for their participation, or volunteered to take part.

2.1 Participants

Data were collected from 54 participants in total (32 females, 22 males; age range 19–65 years, median 20.5 years). The majority were undergraduate Psychology students at Plymouth University. Additional participants recruited from a doctoral programme and the University orchestra. All participants confirmed they had normal hearing. Details of musical training (years of formal tuition) and playing experience (years playing) were recorded for each participant. Four participants' data were excluded from the analysis as they achieved less than 30 % in at least one experimental block (chance level being 50 %), suggesting that they may not have understood the task correctly.

2.2 Materials

The experiment was conducted using a bespoke Matlab programme. Participants listened to the stimuli using Sennheiser HD215 headphones, individually adjusted to a comfortable sound level during the initial practice trial. The absolute level selected by each participant was not recorded.

2.2.1 Stimuli

Each trial consisted of four patterns separated by 700 milliseconds (ms) silence, and each pattern consisted of six tones. Three of the patterns had the same sequence of pitch intervals (standard pattern); the last pitch interval of either the final or the penultimate pattern of the trial deviated from the other three. A different standard pattern was delivered on each trial and no pattern was used more than once in the experiment. Patterns were generated by randomly selecting a set of five intervals, with the restrictions that each interval should only occur once within a pattern, and two intervals with same magnitude but opposite sign should not follow each other immediately in the sequence (to prevent the occurrence of repeated tones in the pattern).

All tones making up the pitch sequences were harmonic complexes, consisting of the first four harmonics of the nominal pitch, exponentially decreasing in amplitude (1:1/2:1/4:1/8) to give an oboe-like timbre. Tone duration was 110 ms, with 5 ms onset and offset linear ramps and 40 ms silence between tones, giving a tone onset to onset interval of 150 ms. Deviant intervals were always four semitones. Since standard pattern intervals were chosen from the set {1, 2, 3, 5, 6, 7 semitones}, depending on the condition (see Table 1), the difference between the standard and the deviant pitch interval was always 1, 2 or 3 semitones. The first tone of the first pattern always had a pitch of 450 Hz. To avoid the use of pitches which may not be clearly audible to everyone despite reporting normal hearing, all pitches were restricted to lie between 100 and 3200 Hz.

Table 1 Details of the within-pattern and transposition intervals used and the number of trials in each test block

Block	Within-pattern intervals	Transposition intervals	Number of trials
1	Big	None	10
2	Big	One exact repeat, then two big transpositions	10
3	Small	Small	20
4	Small	Big	20
5	Big	Small	20
6	Big	Big	20
7	All: 1, 2,3,5,6,7 ST	Big	20

The experiment consisted of one practice block and seven test blocks, each distinguished by the set of intervals used, as detailed in Table 1. Intervals were nominally divided into two sets: *small* {1, 2, 3} semitones, and *big* {5, 6, 7} semitones.

The practice block consisted of 10 trials. The first four were very easy with no transpositions and small within pattern intervals. The next four were slightly harder with two exact repeats of the pattern before two transpositions, with small within-pattern intervals and small transpositions. The final two examples were similar to trials in block 3 with small within-pattern intervals and small transpositions. Participants were given feedback after each trial (the response button briefly turned green for correct and red for incorrect) and a final practice score.

2.2.2 Procedure

Participants were required to indicate using two on-screen response buttons (labelled '2nd Last' and 'Last') whether the penultimate or last pattern was different from the rest. They were told that any difference was in the last interval of the pattern.

Participants began by entering their personal details and then continued with the practice block. They were encouraged to repeat the practice block as many times as they needed to familiarize themselves with the task; 1–3 repetitions were judged to suffice in all cases.

Following the practice block, participants were presented with seven test blocks, with no feedback. Blocks as detailed in Table 1 were presented in random order. Once they had completed all the test blocks, participants were presented with a bar graph showing their score in each block. Each 20-trial block took 3–4 min to complete and the experiment lasted roughly 30 min.

2.2.3 Analysis

In all cases confidence was assessed at the .05 level. Score distributions in each test block were compared against chance using the t-test. The effect of block was assessed using a 1-way ANOVA with all test blocks. The effect of transposition was assessed by contrasting block 1 with the average of blocks 5 and 6. The effect of one exact repetition was assessed by contrasting block 2 with block 6. The effect of variance in interval range was assessed by contrasting block 7 with the average of blocks 4 and 6. The effects of within-phrase intervals and between-phrase transpositions on performance were assessed using a two way ANOVA on data from blocks 3–6. The effect of interval variance was also tested using correlation analysis on data from blocks 3–7. The effect of final interval size of performance was tested using correlation analysis on data from all test blocks. The influence of musical experience was tested using correlation analysis on data from all test blocks. Correlation analysis was carried out using Spearman's correlation coefficient as the data were not normally distributed.

3 Results

Figure 1 shows the score distributions for each block for all participants.

Performance in all blocks was found to be significantly different from chance (shown by dotted line in the figure; $p<0.05$).

There was a main effect of block ($F(6,294)=41.61$, $p<0.001$, $\varepsilon=0.790$, partial $\eta^2=0.459$). The effect of transposition (contrasting block 1 with the average of blocks 5 and 6) was significant ($t=-10.36$, $p<0.001$). The effect of one exact repetition (contrasting block 2 with block 6) was not significant ($t=0.59$, $p=0.559$). The effect of variance in interval range (contrasting block 7 with the average of blocks 4 and 6) was significant ($t=3.20$, $p=0.002$). The more detailed trial-level correlation analysis showed performance correlated negatively with the variance of the pattern intervals (correlation coefficient $=-0.336$, $p<0.001$). There was no significant correlation between the magnitude of the final interval and performance (correlation coefficient -0.298, $p=0.147$). Posthoc multiple comparison analysis showed performance for musically important final intervals (perfect fourth and fifth, 5 and 7 semitones, respectively) was significantly lower than that for 1 semitone.

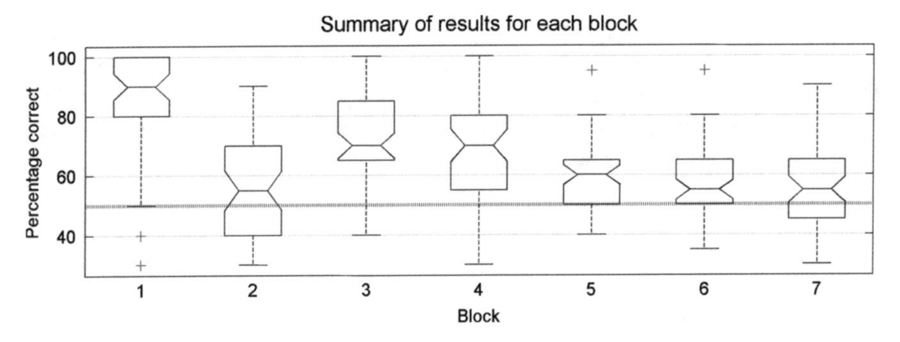

Fig. 1 Distribution of percentage correct scores in each block for all participants

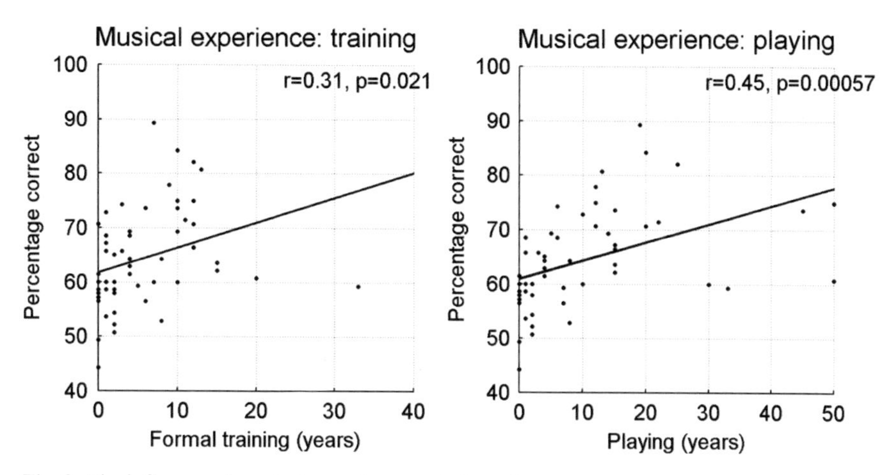

Fig. 2 The influence of musical experience on task performance

The two-way ANOVA assessing the effect of within-pattern and transposition intervals showed a significant main effect of within-pattern intervals ($F(1,49)=37.45$, $p<0.001$, partial $\eta^2=0.433$) and transposition size ($F(1,49)=12.16$, $p=0.001$, partial $\eta^2=0.199$) but their interaction was not significant ($F(1,49)=1.45$, $p=0.235$, partial $\eta^2=0.029$). A posthoc multiple comparison analysis showed that there was a tendency for large transpositions to impair performance more than small transpositions.

Performance correlated positively with musical experience; years of formal training (correlation coefficient $=0.342$, $p=0.015$), as well as years of playing (correlation coefficient $=0.435$, $p=0.002$).

The influence of musical training on task performance is illustrated in Fig. 2.

4 Discussion

In this study we investigated some of the potential sources of difficulty in detecting a pattern with a deviant pitch interval amongst transposed repetitions of a standard pattern, a task that is assumed to depend on relative pitch perception. Our results are consistent with a number of previous studies, e.g. (McDermott and Oxenham 2008), showing that relative pitch perception may be more limited than is commonly assumed. Performance is best when the standard phrase is repeated exactly with no transpositions (block 1), but falls substantially when transpositions are introduced (block 1 versus the average of blocks 3–7). Without transpositions, the task can be performed by direct comparisons between pitches, rather than using the interval relationships between successive pitches. Performance is not helped by one exact repetition of the standard pattern (block 2 versus block 6). This shows that although listeners may become sensitive to a repeating pattern after only 1.5

repetitions (Barascud 2014), they are unable to use this pattern for comparison with transposed versions of the pattern.

When patterns are transposed, then performance is best for standard patterns consisting of small intervals. This is consistent with the notion that grouping is promoted by featural similarity, and that representations of phrases consisting of small intervals are more easily formed, suggesting that comparisons between patterns may be facilitated by having a more coherent representation of the standard. With transpositions, large within-pattern intervals make the task very difficult. However, contrary to our hypothesis, large transpositions impaired performance more than small transpositions. This suggests that comparisons between pitch interval patterns are facilitated by proximity in pitch space. Increasing the variance in the pattern intervals, as predicted, impairs performance.

The idea that relative pitch perception depends solely on detecting a pattern of invariant pitch intervals is not supported by our results. Although the invariant property of the patterns in each trial is the sequence of pitch intervals defining the standard, listeners often could not use this information in the current experiment. Our results are compatible with the notion that in constructing object representations, the tolerance of the representation is a function of the variance in the pattern, i.e. increasing variance in object components, leads to more permissive representations. This makes sense when the general problem of perceptual categorisation is considered; e.g. the variability of the spoken word.

Relative pitch perception has been likened to translation invariant object recognition in vision (Kubovy and Van Valkenburg 2001). Interestingly the literature on visual perceptual learning has shown that learning can be surprisingly specific to the precise retinal location of the task stimulus (Fahle 2005). The most influential model of translation invariant object recognition is the so-called trace model (Stringer et al. 2006), which assumes that this ability actually depends on learning the activity caused by the same stimulus being shown at many different locations; invariant recognition then emerges at a higher level by learning that these different activations are caused by the same object. Perhaps this is what happens when we learn a tune. The categorisation of the tune depends on hearing it at many different pitch levels within a context that provides clear links between the various repetitions (e.g. within the same piece of music, or same social context).

Acknowledgments We would like to thank Dávid Farkas for help with the statistical analysis. IW was supported by the Hungarian Academy of Sciences (Lendület project, LP36/2012). GPH was supported by a post-doctoral research fellowship of the Hungarian Academy of Sciences.

References

Barascud N (2014). Auditory pattern detection. PhD thesis, University College London (University of London), London
Bregman AS (1990) Auditory scene analysis: the perceptual organization of sound. MIT Press, Cambridge
Chait M, Poeppel D, de Cheveigne A, Simon JZ (2007) Processing asymmetry of transitions between order and disorder in human auditory cortex. J Neurosci 27(19):5207–5214. doi:10.1523/JNEUROSCI.0318-07.2007
Coath M, Háden GP, Winkler I, Denham SL (2008). Are short tone sequences memorable? Paper presented at the British Society of Audiology Annual Conference, Leeds
Dowling WJ (1986) Context effects on melody recognition: scale-step versus interval representations. Music Percept 3(3):281–296
Fahle M (2005) Perceptual learning: specificity versus generalization. Curr Opin Neurobiol 15(2):154–160. doi:10.1016/j.conb.2005.03.010
Foster NE, Zatorre RJ (2010) A role for the intraparietal sulcus in transforming musical pitch information. Cereb Cortex 20(6):1350–1359. doi:10.1093/cercor/bhp199
Friston K (2005) A theory of cortical responses. Philos Trans R Soc Lond B Biol Sci 360(1456):815–836. doi:10.1098/rstb.2005.1622
Griffiths TD, Warren JD (2004) What is an auditory object? Nat Rev Neurosci 5(11):887–892. doi:10.1038/nrn1538
Köhler W (1947) Gestalt psychology: an introduction to new concepts in modern psychology. Liveright Publishing Corporation, New York
Kubovy M, Van Valkenburg D (2001) Auditory and visual objects. Cognition 80(1–2):97–126
McDermott JH, Oxenham AJ (2008) Music perception, pitch, and the auditory system. Curr Opin Neurobiol 18(4):452–463. doi:10.1016/j.conb.2008.09.005
McDermott JH, Keebler MV, Micheyl C, Oxenham AJ (2010) Musical intervals and relative pitch: frequency resolution, not interval resolution, is special. J Acoust Soc Am 128(4):1943–1951. doi:10.1121/1.3478785
Micheyl C, Oxenham AJ (2010) Objective and subjective psychophysical measures of auditory stream integration and segregation. J Assoc Res Otolaryngol 11(4):709–724. doi:10.1007/s10162-010-0227-2
Moore BC, Gockel HE (2012) Properties of auditory stream formation. Philos Trans R Soc Lond B Biol Sci 367(1591):919–931. doi:10.1098/rstb.2011.0355
Stefanics G, Haden GP, Sziller I, Balazs L, Beke A, Winkler I (2009) Newborn infants process pitch intervals. Clin Neurophysiol 120(2):304–308. doi:10.1016/j.clinph.2008.11.020
Stringer SM, Perry G, Rolls ET, Proske JH (2006) Learning invariant object recognition in the visual system with continuous transformations. Biol Cybern 94(2):128–142. doi:10.1007/s00422-005-0030-z
Tew S, Fujioka T, He C, Trainor L (2009) Neural representation of transposed melody in infants at 6 months of age. Ann N Y Acad Sci 1169:287–290. doi:10.1111/j.1749-6632.2009.04845.x
Winkler I, Denham SL, Nelken I (2009) Modeling the auditory scene: predictive regularity representations and perceptual objects. Trends Cogn Sci 13(12):532–540. doi:10.1016/j.tics.2009.09.003

Do Zwicker Tones Evoke a Musical Pitch?

Hedwig E. Gockel and Robert P. Carlyon

Abstract It has been argued that musical pitch, i.e. pitch in its strictest sense, requires phase locking at the level of the auditory nerve. The aim of the present study was to assess whether a musical pitch can be heard in the absence of peripheral phase locking, using Zwicker tones (ZTs). A ZT is a faint, decaying tonal percept that arises after listening to a band-stop (notched) broadband noise. The pitch is within the frequency range of the notch. Several findings indicate that ZTs are unlikely to be produced mechanically at the level of the cochlea and, therefore, there is unlikely to be phase locking to ZTs in the auditory periphery. In stage I of the experiment, musically trained subjects adjusted the frequency, level, and decay time of an exponentially decaying sinusoid so that it sounded similar to the ZT they perceived following a broadband noise, for various notch positions. In stage II, subjects adjusted the frequency of a sinusoid so that its pitch was a specified musical interval below that of either a preceding ZT or a preceding sinusoid (as determined in stage I). Subjects selected appropriate frequency ratios for ZTs, although the standard deviations of the adjustments were larger for the ZTs than for the equally salient sinusoids by a factor of 1.1–2.2. The results suggest that a musical pitch may exist in the absence of peripheral phase locking.

Keywords Auditory afterimage · Tonal percept · Band-stop noise · Notched noise · Phase locking · Pitch match · Frequency adjustment · Pitch salience · Musical interval

H. E. Gockel (✉) · R. P. Carlyon
MRC Cognition and Brain Sciences Unit, 15 Chaucer Road, Cambridge CB2 7EF, UK
e-mail: hedwig.gockel@mrc-cbu.cam.ac.uk

R. P. Carlyon
e-mail: bob.carlyon@mrc-cbu.cam.ac.uk

P. van Dijk et al. (eds.), *Physiology, Psychoacoustics and Cognition in Normal and Impaired Hearing,* Advances in Experimental Medicine and Biology 894,
DOI 10.1007/978-3-319-25474-6_44

1 Introduction

Periodic sounds produce periodic patterns of phase-locked activity in the auditory nerve. It has been argued that this temporal code is the basis for our sensation of pitch, and, specifically, that musical pitch i.e. pitch in its strictest sense, requires phase locking. In this vein, most current models of pitch perception rely on the precise timing of action potentials, or spikes, within the auditory nerve (Cariani and Delgutte 1996; Meddis and O'Mard 1997). Furthermore, the precision of phase locking weakens at high frequencies, and the putative upper limit of phase locking at frequencies of about 4–5 kHz (Johnson 1980; Palmer and Russell 1986) has been used to explain the finding that the ability to recognize melodies, the accuracy of musical interval judgements, and frequency discrimination are all severely degraded for pure tones above about 4–5 kHz (Ward 1954; Attneave and Olson 1971; Moore 1973; Sek and Moore 1995).

There exist, however, some observations that cast doubt on the generally accepted assumption that phase locking is necessary for musical pitch. For example, while most subjects in Ward's (1954) study were unable to make octave judgements when the reference frequency, f1, was above 2700 Hz, for two of his subjects the variability in octave judgements was essentially the same for f1 = 5 kHz (with the octave match at ~ 10 kHz) as for lower frequencies. However, octave judgements were more difficult and subjects took a greater time when f1 was 5 kHz. Ward suggested that experience might play a role, as these two subjects were the only ones with experience in judging the pitch of pure tones. Similarly, Burns and Feth (1983) asked musically trained subjects to adjust various musical intervals for reference frequencies of 1 kHz and 10 kHz. Even for the high frequency, all three subjects could do the task. For the 10-kHz reference, the standard deviations, SDs, averaged across all musical intervals, were about 3.5–5.5 times larger than for the 1-kHz reference. This increase was less than that observed for unison adjustments, which were taken as an estimate of the difference limen for frequency, DLF. Burns and Feth concluded that their results were not incompatible with a temporal basis for both frequency discrimination and musical interval adjustment, as phase locking information decreases with increasing frequency. More recently, Oxenham et al. (2011) showed that complex tones whose audible harmonics all fall above 6 kHz can evoke a robust sense of pitch and musical interval, but high-frequency pure tones did not, even though the just noticeable difference for the latter was less than 1.5 %. They concluded either that there is sufficient phase-locking present at high frequencies to derive complex musical pitch, or that a complex pitch can be derived in the absence of phase locking. Pertaining to this, Moore and Ernst (2012) reported DLFs for center frequencies from 2 to 14 kHz that were consistent with the idea that there is a transition from a temporal to a place mechanism at about 8 kHz, rather than at 4–5 kHz, as commonly assumed.

Here we assessed whether a musical pitch can be heard in the *absence* of phase locking, using Zwicker tones, ZTs (Zwicker 1964). A ZT is a faint, decaying tonal percept that can arise following the presentation of a notched broadband noise and that can last up to 5–6 s. The pitch is always within the frequency range of the notch, and depends on the level of the noise and the width of the notch. Several findings indicate that ZTs are unlikely to be produced mechanically at the level of the cochlea. Normally, no otoacoustic emissions (OAEs) were found at the frequency corresponding to the pitch of the ZT, except in the rare case when a subject had a spontaneous OAE which could be made temporarily audible by a preceding notched noise (without increasing its physical level), and no beating was observed between ZTs and a soft physical tone (Krump 1993). Also, while low frequency tones at moderate to high levels can affect the level of evoked OAEs, they did not affect the ZT (Wiegrebe et al. 1995). Thus, no mechanical activity correlated to the ZT exists in the cochlea, and there is unlikely to be phase locking to ZTs in the auditory periphery. The question addressed here is whether ZTs can evoke a musical pitch.

2 Methods

2.1 *Experimental Design*

Four young normal-hearing musically trained subjects took part. In stage I, subjects adjusted the frequency, level, and decay time of an exponentially decaying sinusoid so that it sounded similar to the ZT they perceived following a broadband noise, for various notch positions. In stage II, subjects adjusted the frequency (and level) of a sinusoid so that its pitch was a specified musical interval below that of either a preceding ZT or a preceding sinusoid (and to be equally loud). Importantly, for each subject, the reference sinusoids corresponded to those that were adjusted in stage I to sound similar to, i.e. have equal pitch, loudness and decay time as, the ZTs. The precision of the musical interval adjustments for the ZTs and the matched sinusoids (PT condition) was compared.

2.2 *Stimuli and General Procedure*

To evoke ZTs, 5-s (including 20-ms onset and offset ramps) notched broadband noises (30-16000 Hz) were presented diotically at an rms level of 51 dB SPL. From one match to the next, the lower edge frequency (LEF) of the notch could take one of eight values: 2000, 2144, 2297, 2460, 2633, 2818, 3014, and 3500 Hz. The higher edge frequency (HEF) of the notch was always 1.5 times the LEF. The adjustable sinusoid had a maximum duration of 5 s (including 20-ms onset and offset ramps) and followed the ZT-exciting noise with an inter-stimulus interval of 5.5 s. After cessation of the sinusoid, the subject indicated by button presses the desired direction of change for the frequency, level, and (stage I only) time constant of the sinusoid for the next presentation, and/or initiated the next sound presentation, i.e. the next trial. In each trial, the subject was allowed an unlimited number of button presses before s/he initiated the next trial. For each parameter that needed adjusting, three different step sizes were available for each direction of change. The smallest available step sizes were 1/16 semitone, 1 dB, and a factor of $2^{1/4}$ for the frequency, level, and time constant adjustments, respectively. Subjects could take as many trials as they liked to finish a match. The first session (2 h including breaks) was considered practice, and matches from this session were discarded. For the experiment proper, typically about ten matches were collected for each condition from each subject.

In stage I, in most cases, the adjusted frequency of the matched sinusoid increased with increase in LEF and was a factor of 1.1–1.2 above the LEF. Two of the subjects reliably matched higher frequencies (factor of 1.3–1.4 above the LEF) in two or three LEF conditions, leading to functions where the matched frequency sometimes did not increase with increase in LEF. For the present purposes, these cases were excluded from further analysis of stage II data, as they could be due to spontaneous OAEs that became audible after listening to the notched noise (Krump 1993; Wiegrebe et al. 1995). The initial level and time constant of the matched sinusoids had mean values that ranged from 5 to 23 dB above threshold in quiet and from 1.3 to 5.1 s, respectively, across subjects and LEFs. In the following, we consider the interval adjustments of stage II of the experiment only.

2.3 Musical Interval Adjustments

In stage II, subjects adjusted a sinusoid to be a specified musical interval below the preceding reference tone. The musical intervals that had to be adjusted were a minor third (3 semitones down) and a perfect fifth (7 semitones down). In different sessions, the reference tone was either a ZT or a physically presented pure tone (PT). The latter corresponded to the matched tones from stage I. From one match to the next, the reference tone could be any one of the set of eight. The general trial structure was the same as in stage I. Typically about 12 (at least 10) interval adjustments were collected for each condition from each subject.

3 Results

Figures 1 and 2 show, for each subject, several measures of the accuracy, repeatability, and ease of the musical interval adjustments in the ZT and PT conditions. Figure 1 shows the geometric mean (and SDs across the 8 reference frequencies or LEFs) of the ratio of the adjusted frequency of the variable pure tone to the expected frequency. The expected frequency was determined on the equal temperament scale. That is, the expected frequencies for the minor third and the perfect fifth were exactly 3 semitones (a factor of 1/1.189) and 7 semitones (a factor of 1/1.498) below the reference frequency. For all subjects, the adjusted frequencies were somewhat flat, i.e. slightly lower than expected, leading to somewhat larger musical intervals. However, this was true for both the PT and the ZT conditions (different coloured bars). Generally,

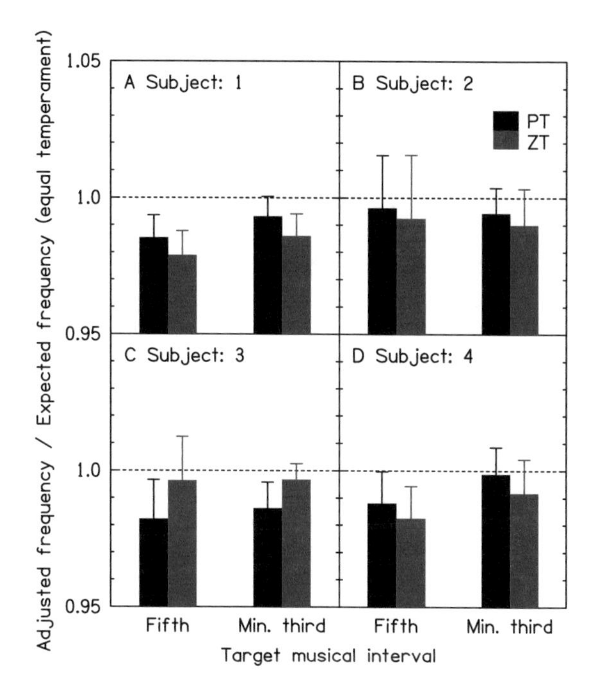

Fig. 1 Geometric mean (and SDs across the 8 reference frequencies or LEFs) of the ratio of the adjusted frequency of the variable pure tone to the expected frequency for musical intervals of a perfect fifth and a minor third

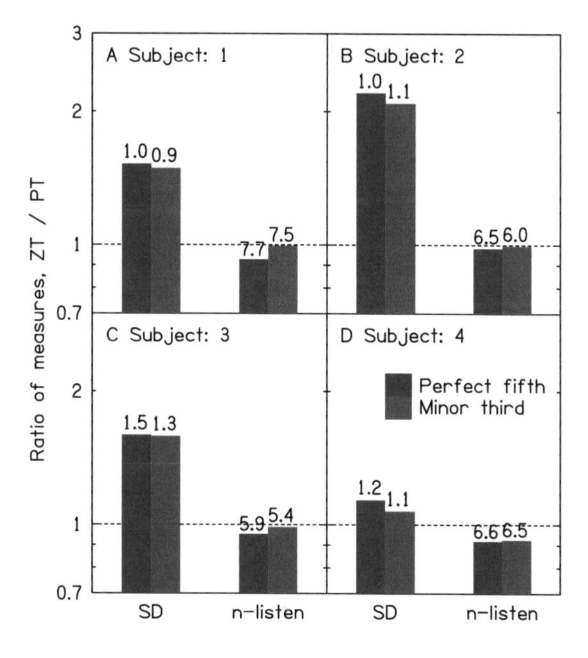

Fig. 2 Ratio of the geometric mean (across the 8 reference frequencies or LEFs) of the SDs (*left-hand side*) and of the average number of trials (*right-hand side*) between ZT and PT conditions, for musical intervals of a perfect fifth and a minor third. The value of each measure for the PT condition is given by the number above the corresponding bar; for the SD this is the representative SD (see text), expressed as a percentage of the geometric mean of the adjusted frequencies

the adjusted frequencies were within about 2% of the expected frequencies, with no systematic differences between the ZT and the PT conditions. Thus, subjects were able to match musical intervals with good accuracy for both conditions.

To compare the reliability of the musical interval adjustments in the ZT conditions with that in the matched PT conditions, we calculated the ratio of the respective SDs. This was done as follows: (1) Separately for each subject, each musical interval (minor third, perfect fifth), each reference condition (ZT, PT) and each LEF (or reference PT tone frequency) we determined the SD of the adjusted frequencies across the 12 interval adjustments expressed as a percentage of the geometric mean; (2) These were geometrically averaged across the eight LEFs (or PT frequencies) to obtain the representative SD; (3) The representative SD for the ZT was divided by the representative SD for the PT. This ratio is shown in the group of two bars on the left-hand side of each panel in Fig. 2. Musical interval adjustments for the ZTs were somewhat more variable than for the matched PTs; the ratio of the SDs was between 1.1 and 2.2. Four univariate analyses of variance conducted separately on the data for each subject (using the SD of the logarithms of the individual adjusted frequencies within a given LEF or reference frequency as input data) showed that for three of the subjects the SDs were significantly larger for the ZT than for the PT conditions (with $p<0.01$ for subjects 1 and 2, and $p<0.05$ for subject 3). The number above each bar in Fig. 2 indicates the size of the representative SD for the PT condition in percent (ranging from 0.9−1.5%). The representative SD for the ZT condition ranged from 1.2−2.5%. By comparison, the SDs observed by Burns and Feth (1983) for the same musical interval adjustments were on average 1.2% and 4.2% for 1-kHz and 10-kHz 70-dB SPL reference tones, respectively.

The reliability of adjustments can be increased by listening more often to the stimuli, and thus needs to be considered as well. The group of two bars on the right-hand side of each panel in Fig. 2 shows the ratio of the average number of listening times (trials) in the ZT and the PT conditions. The number above each bar indicates the average number of trials (n-listen) for the PT condition. There were inter-individual differences in how many trials a subject needed on average to make a match. For example, in the PT conditions, subject 1 needed on average 7.6 trials while subject 3 needed on average 5.7 trials. However, n-listen was very similar for the two conditions, i.e. the ratio was very close to one. If anything, n-listen was very slightly lower for the ZTs than for the PT. This means that the achieved accuracy and reliability in the ZT conditions did not come at the cost of listening more often in the ZT than the PT conditions.

4 Discussion

The results showed that, on average, subjects selected similar frequencies in a musical interval adjustment task irrespective of whether the reference tone was a ZT or an equally salient pure tone. The adjusted frequencies were slightly flat in both cases. This might partly result from subjects' bias towards a "just scale" with musical intervals corresponding to integer ratios of 6/5 and 3/2 for the minor third and the perfect fifth, respectively. If this was the only underlying reason, the "bias" should have been larger for the minor third than for the perfect fifth, which was not observed. The SDs of the musical interval adjustments were only slightly larger (a factor of 1.1–2.2) for the ZTs than for the PTs, with no increase in listening time. Thus, overall the results suggest that a weak musical pitch can exist in the absence of peripheral phase locking.

Burns and Feth (1983) had three musically trained subjects adjust various musical intervals for reference frequencies of 1 kHz and 10 kHz. For the same musical intervals as used in the present study, SDs in their data were a factor of 1.6–11.7 (mean of 4) times larger for the 10-kHz reference than for the 1-kHz reference. Thus, the increase in SDs for the 10-kHz frequency was larger than that observed here for the ZTs. Burns and Feth (1983) interpreted their results in terms of phase locking, and a decrease thereof with increasing frequency. The present results suggest that their relatively large increase in SDs at 10 kHz reflects the combined effect of lack of familiarity with high-frequency pure tones and the transition from a temporal code to a place code.

Acknowledgments Supported by intramural funding from the MRC. Thanks to Brian Moore for helpful comments.

References

Attneave F, Olson RK (1971) Pitch as a medium—new approach to psychophysical scaling. Am J Psychol 84:147–166

Burns EM, Feth LL (1983) Pitch of sinusoids and complex tones above 10 kHz. In: Klinke R, Hartmann R (eds) Hearing—physiological bases and psychophysics. Springer, Berlin, pp 327–333

Cariani PA, Delgutte B (1996) Neural correlates of the pitch of complex tones. II. Pitch shift, pitch ambiguity, phase invariance, pitch circularity, rate pitch, and the dominance region for pitch. J Neurophysiol 76:1717–1734

Johnson DH (1980) The relationship between spike rate and synchrony in responses of auditory-nerve fibers to single tones. J Acoust Soc Am 68:1115–1122

Krump G (1993) Beschreibung des akustischen Nachtones mit Hilfe von Mithörschwellenmustern. Ph.D. thesis. Technical University, München, Germany

Meddis R, O'Mard L (1997) A unitary model of pitch perception. J Acoust Soc Am 102:1811–1820

Moore BCJ (1973) Frequency difference limens for short-duration tones. J Acoust Soc Am 54:610–619

Moore BCJ, Ernst SMA (2012) Frequency difference limens at high frequencies: evidence for a transition from a temporal to a place code. J Acoust Soc Am 132:1542–1547
Oxenham AJ, Micheyl C, Keebler MV, Loper A, Santurette S (2011) Pitch perception beyond the traditional existence region of pitch. Proc Natl Acad Sci USA 108:7629–7634
Palmer AR, Russell IJ (1986) Phase-locking in the cochlear nerve of the guinea-pig and its relation to the receptor potential of inner hair-cells. Hear Res 24:1–15
Sek A, Moore BCJ (1995) Frequency discrimination as a function of frequency, measured in several ways. J Acoust Soc Am 97:2479–2486
Ward WD (1954) Subjective musical pitch. J Acoust Soc Am 26:369–380
Wiegrebe L, Kössl M, Schmidt S (1995) Auditory sensitization during the perception of acoustical negative afterimages—analogies to visual processing. Naturwissenschaften 82:387–389
Zwicker E (1964) 'Negative afterimage' in hearing. J Acoust Soc Am 36:2413–2415

Speech Coding in the Midbrain: Effects of Sensorineural Hearing Loss

Laurel H. Carney, Duck O. Kim and Shigeyuki Kuwada

Abstract In response to voiced speech sounds, auditory-nerve (AN) fibres phase-lock to harmonics near best frequency (BF) and to the fundamental frequency (F0) of voiced sounds. Due to nonlinearities in the healthy ear, phase-locking in each frequency channel is dominated either by a single harmonic, for channels tuned near formants, or by F0, for channels between formants. The alternating dominance of these factors sets up a robust pattern of F0-synchronized rate across best frequency (BF). This profile of a temporally coded measure is transformed into a mean rate profile in the midbrain (inferior colliculus, IC), where neurons are sensitive to low-frequency fluctuations. In the impaired ear, the F0-synchronized rate profile is affected by several factors: Reduced synchrony capture decreases the dominance of a single harmonic near BF on the response. Elevated thresholds also reduce the effect of rate saturation, resulting in increased F0-synchrony. Wider peripheral tuning results in a wider-band envelope with reduced F0 amplitude. In general, sensorineural hearing loss reduces the *contrast* in AN F0-synchronized rates across BF. Computational models for AN and IC neurons illustrate how hearing loss would affect the F0-synchronized rate profiles set up in response to voiced speech sounds.

Keywords Vowels · Modulation transfer function · Neural coding

1 Introduction

The impact of sensorineural hearing loss (SNHL) on speech communication is perhaps the most significant problem faced by listeners with hearing loss. A puzzle regarding the effects of SNHL is that relatively small amounts of loss have a surprisingly large impact on listeners, especially in noisy acoustic environments. Here

L. H. Carney (✉)
Departments of Biomedical Engineering, Neurobiology & Anatomy,
Electrical & Computer Engineering, University of Rochester, Rochester, NY, USA
e-mail: Laurel.Carney@Rochester.edu

D. O. Kim · S. Kuwada
Department f Neuroscience, University of Connecticut Health Center, Farmington, CT, USA

P. van Dijk et al. (eds.), *Physiology, Psychoacoustics and Cognition in Normal and Impaired Hearing,* Advances in Experimental Medicine and Biology 894,
DOI 10.1007/978-3-319-25474-6_45

we focus on the effects of SNHL on the representation of voiced speech sounds in the auditory periphery and in the midbrain (inferior colliculus, IC). Responses of models for auditory-nerve (AN) fibres and IC neurons provide insight into how changes in the temporal structure of AN responses impact representations at the level of the IC. Neurons in the IC are sensitive to low-frequency rate fluctuations, including rates synchronized to envelopes of amplitude-modulated (AM) stimuli.

Vowels are an essential component of all languages, and play an important role in carrying information, especially in running speech (e.g. Kewley-Port et al. 2007). Vowel sounds are characterized by voicing, which results in a spectrum with a fundamental frequency (F0), related to voice pitch, and harmonics of F0. The spectrum is shaped by vocal tract resonances, or formants. The locations of the lowest two frequency formants (F1 and F2), distinguish different vowel sounds (Fant 1960). Neural coding of these formants is essential for understanding how the brain processes speech.

Neural studies of vowel coding have focused on rate-place and fine-structure temporal coding of formants in AN responses (e.g. Sachs and Young 1979; Young and Sachs 1979; Delgutte and Kiang 1984; reviewed by Young 2008). Consideration of midbrain responses to vowels shifts the focus to the peripheral representation of low-frequency fluctuations ($\lesssim 250$ Hz) associated with F0. IC cells are particularly sensitive to amplitude modulation in the frequency range of F0 (Langner 1992). In the healthy ear, nonlinear AN response properties result in systematic patterns of neural fluctuations synchronized to F0 or higher harmonics near AN characteristic frequencies. In particular, synchrony capture results in the dominance of a single harmonic near a spectral peak on the fine-structure of AN responses to voiced sounds (Delgutte and Kiang 1984; Deng and Geisler 1987). After trauma-induced SNHL, synchrony capture is reduced, and AN fibres respond to multiple harmonics (Miller et al. 1997). These studies focused how synchrony capture effects fine-structure coding; however, this mechanism also sets up a pattern of low-frequency fluctuations across the AN population. In the healthy ear, the fibres that are "captured" by a single harmonic and have responses with relatively flat envelopes. In contrast, fibres tuned to frequencies between formants respond to multiple harmonics and strong rate fluctuations at F0. The amplitude of F0-related neural fluctuations provides a robust code for the formant frequencies. In the ear with SNHL, decreased synchrony capture results in AN fibres that respond to multiple harmonics, thus F0-related fluctuations prevail across all driven frequency channels. The profile of F0-related fluctuations that codes the formants, and that ultimately drives modulation-sensitive midbrain neurons, is diminished by SNHL.

In this study, computational models illustrate vowel responses of AN fibres with and without SNHL. Next, models for three types of modulation sensitivity in the IC illustrate the effect of SNHL on population responses in the midbrain.

2 Methods

The Zilany et al. (2014) model for AN responses has a provision for including impairment due to outer and inner hair cell dysfunction (Zilany and Bruce 2007), which reduces synchrony capture in a manner comparable to that observed in acoustically traumatized ears (Miller et al. 1997). Here SNHL was simulated by setting AN model parameters C_{OHC} and C_{IHC} to 0.3; these values simulate mild hearing loss, with threshold elevations ranging from about 15 dB at 500 Hz to 30 dB at 2000 Hz and higher.

Phenomenological models were used to simulate the three types of rate modulation transfer functions (MTFs) that comprise the bulk of IC neurons (Kim et al. 2015a): band-enhanced, band-suppressed, and hybrid (Fig. 1). The first two MTF types have discharge rates to AM stimuli that are enhanced or suppressed relative to responses to unmodulated stimuli (Kim et al. 2015a, 2015b). Hybrid MTFs have discharge rates that are enhanced at some modulation frequencies and suppressed at others (Krishna and Semple 2000; Kim et al. 2015a).

The band-enhanced model (Fig. 1, BE) was Nelson and Carney's (2004) same-frequency inhibitory-excitatory (SFIE) model. The key mechanism in this model is a dynamic interaction between relatively short-duration excitatory and relatively long-duration, delayed inhibitory synaptic potentials. The durations of the excitatory and inhibitory potentials and the delay of the inhibition with respect to the excitation determine the model's best modulation frequency (BMF).

An extension of the SFIE model was used for the band-suppressed model (Fig. 1, BS). This model receives the same ascending excitatory input as the band-enhanced model, but it is inhibited by the band-enhanced model (Carney et al. 2015; Kim et al. 2015b). The BMF of the band-enhanced model determines the worst modulation frequency (WMF) for the band-suppressed model.

Finally, the hybrid MTF model receives the same ascending excitatory input as the other model IC neurons, but is inhibited by both band-enhanced and band-suppressed cells (Fig. 1, Hybr). The relation between the BMF and WMF of the inhibitory inputs determines the shape of the hybrid model cell's MTF.

3 Results

The responses of AN models to the vowel/æ/(in "had") are illustrated in Fig. 2, with the vowel waveform (Fig. 2a) and spectrum (Fig. 2b). Discharge rate vs. BF is shown for healthy (Fig. 2c) and impaired (Fig. 2e) model populations. At conversational speech levels, these profiles encode the formant frequencies. However, in background noise or as sound level increases the profiles tend to saturate, though intelligibility does not decrease until high SPLs are reached (Studebaker et al. 1999).

Figure 2d, f illustrates features of the temporal responses of AN fibres to vowels shown using the dominant components, which are the largest spectral components

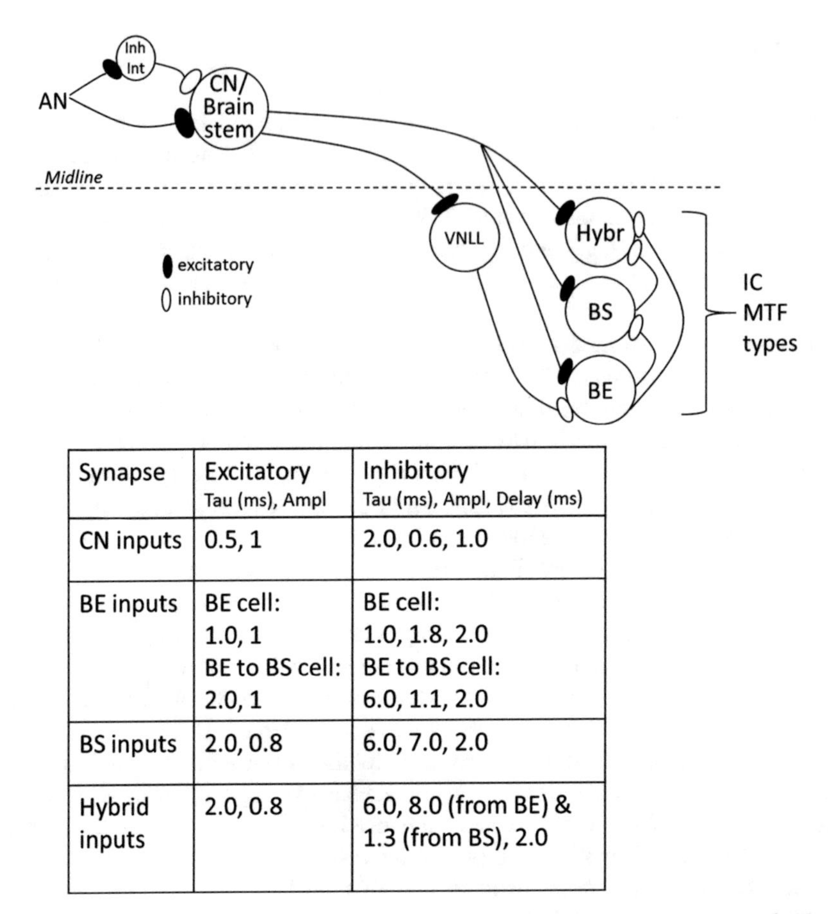

Synapse	Excitatory Tau (ms), Ampl	Inhibitory Tau (ms), Ampl, Delay (ms)
CN inputs	0.5, 1	2.0, 0.6, 1.0
BE inputs	BE cell: 1.0, 1 BE to BS cell: 2.0, 1	BE cell: 1.0, 1.8, 2.0 BE to BS cell: 6.0, 1.1, 2.0
BS inputs	2.0, 0.8	6.0, 7.0, 2.0
Hybrid inputs	2.0, 0.8	6.0, 8.0 (from BE) & 1.3 (from BS), 2.0

Fig. 1 Schematic diagram of the extended SFIE model for IC neurons with three types of MTFs: band-enhanced (*BE*), band-suppressed (*BS*) and hybrid. For simplicity, the simulations presented here were driven only by contralateral inputs. The AN excites and inhibits (via an interneuron) a model cochlear nucleus (*CN*) or other brainstem-level neuron, which excites all three IC cell types, and inhibits (via the ventral nucleus of the lateral lemniscus, *VNLL*) the BE model. The BE model inhibits the BS model, and both BE and BS models inhibit the hybrid model. In each model, inputs were convolved with the post-synaptic potential waveforms, summed, and half-wave rectified

of the post-stimulus time histograms (Delgutte and Kiang 1984; Miller et al. 1997). Dominant components for the healthy AN model (Fig. 2d) show synchrony capture, or dominance of the temporal responses by a single harmonic near the formant peak. The synchrony capture extends across a span of fibres with BFs near each formant frequency (red arrows). The dominance of these responses by a harmonic near the formants (orange arrows) results in absence of synchrony to F0 (black arrow) for BFs near formants. In contrast, fibres tuned between the formants are not dominated by a single harmonic. Because these fibres respond to multiple harmonics near their BFs (green arrow), their responses synchronize to the "beats" at F0 (black arrow).

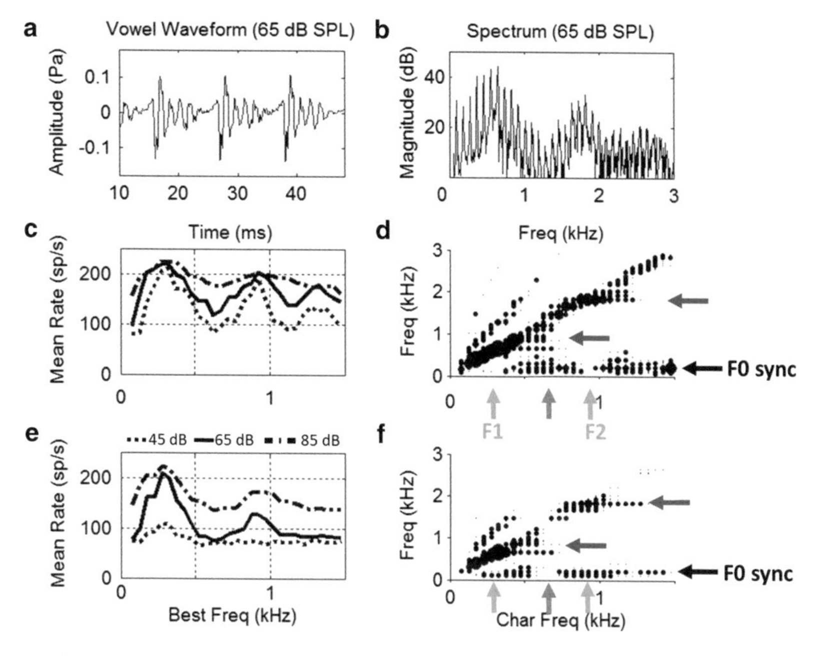

Fig. 2 Time waveform (**a**) and spectrum (**b**) for the vowel/æ/(in "had") at 65 dB SPL (from Hillenbrand et al. 1995; F0=95 Hz). (**c**) Healthy and (**e**) impaired AN model rate profiles for 3 sound levels. (**d**, **f**) Dominant components of synchronized rate for healthy and impaired responses. Symbol sizes are proportional to synchronized rate (components smaller than 15 sp/sec are not shown)

The dominant components for model AN fibres with SNHL simulated by outer and inner hair cell dysfunction are shown in Fig. 2f. Physiological studies have shown that synchrony capture is weaker in these fibres (Miller et al. 1997). The model response illustrates reduced synchrony capture, especially to F2 (Fig. 2f, red arrow), even for the mild hearing loss in this simulation. The reduced synchrony capture is more evident in the pattern of synchrony to F0, which extends across all regions where there is sufficient energy to drive the neurons (Fig. 2f, black arrow). Thus, the profile of F0-synchronized rates differs qualitatively between the impaired and the healthy AN populations. For BFs tuned between F1 and F2 (green arrow), the response to F0 (black arrow) is reduced in the impaired nerve, whereas in the BF regions of F1 and of F2 (orange arrows), the response to F0 is increased in the impaired nerve. The net effect of SNHL is a reduction in the contrast of the F0-synchronized rates across BF.

Figure 3 shows examples of IC neural (left) and model (right) band-enhanced, band-suppressed, and hybrid MTFs. Kim et al. (2015a) used multiple types of AM envelopes (e.g., raised-sine envelopes with exponents of 1, 8 and 32) and a 25% criterion for a change in rate with respect to the unmodulated response, and found that approximately 25% of IC neurons had band-enhanced, 50% band-suppressed, and 25% hybrid MTFs.

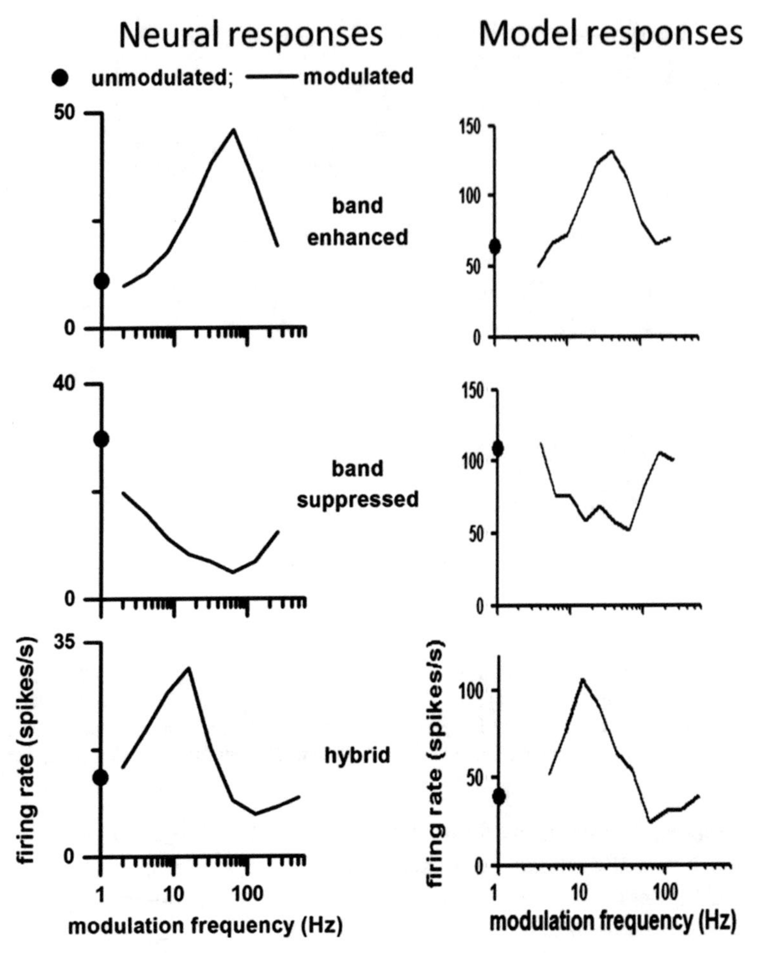

Fig. 3 *Left*: Three types of MTFs in the IC of awake rabbit (from Kim et al. 2015a). *Right*: IC model MTFs with BF = 2 kHz. All neural and model responses are to 1-octave bandwidth sinusoidal AM noise (100 % modulation depth) centred at BF. Filled dot is response rate to unmodulated noise

Figure 4 shows model responses to the vowel/æ/for the three MTF types for healthy and impaired model AN inputs. The healthy band-enhanced responses (Fig. 4a, blue) reflect the patterns of F0-synchronized rate shown in Fig. 2; responses are decreased near formant frequencies (orange arrows), because synchrony capture in the healthy AN reduces the low-frequency fluctuations that ultimately drive this IC type. Band-enhanced neurons tuned to frequencies between formants (Fig. 4a, green arrow) have strong responses due to the strong F0-synchronized rate in the healthy AN for these frequency channels (Fig. 2d, green arrow). The strong contrast in rate across frequency channels for the band-enhanced responses is robust across SPLs and in background noise (Carney et al. 2015).

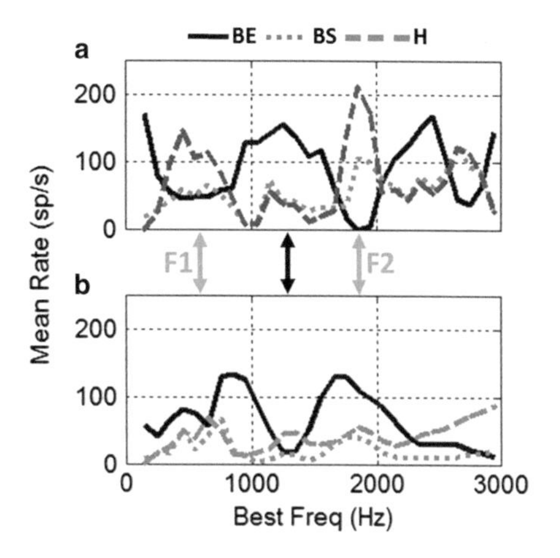

Fig. 4 Responses of model IC MTF types to the vowel/æ/at 65 dB SPL. IC models with (**a**) healthy AN model inputs, (**b**) impaired AN inputs. Model parameters provided in Fig. 1

The rate profile for healthy band-suppressed IC model neurons has peaks at the formant frequencies (Fig. 4a, magenta). These cells respond more strongly at frequencies where the low-frequency fluctuations are reduced (Fig. 2d, orange arrows). The peaks in the rate profile at the formant frequencies are robust in the model hybrid population (Fig. 4a, green). This model result motivates further physiological and computational investigation of these cells. The hybrid neuron and model in Fig. 3 are most strongly driven by a contrast in neural fluctuations (i.e., strong neural fluctuations in the 8–30 Hz range and reduced fluctuations in the 30–200 Hz range); other MTF shapes are also observed for hybrid neurons (Krishna and Semple 2000).

Model IC responses with impaired AN inputs (Fig. 4b) differ dramatically from the healthy responses. The band-enhanced rate profile (Fig. 4b, blue) has peaks approximately where the healthy model has valleys, and vice versa. The impaired model profile is explained by reduced synchrony capture: the impaired AN model synchronizes to F0 whenever the fibres are adequately driven by stimulus energy (see Fig. 2f). Thus the impaired band-enhanced rate profile reflects the energy spectrum of the vowel, unlike the healthy rate profile, which is inversely related to energy. The impaired band-suppressed (Fig. 4b, magenta) and hybrid (Fig. 4b, green) model responses also have peaks near F1 and F2. Unlike the healthy case, the inhibition from the shifted peaks in the impaired band-enhanced model suppresses the peaks in the band-suppressed and hybrid models. These results emphasize the fact that the strong rate profiles in the healthy IC model are not simply explained by the rate vs.BF profile of the AN. The healthy band-suppressed and hybrid profiles are created by a synchrony-to-rate transformation between the AN and IC and enhanced by disinhibition at the formant frequencies. In general, the rate profiles for all three response types are qualitatively different for this mildly impaired model due to differences in the F0-synchronized rate between the healthy and impaired AN models (Fig. 2d, f).

4 Discussion

Vowel coding studies have historically focussed on representation of spectral peaks, or formants, in AN rates or synchrony to stimulus components near BF. These representations are vulnerable to additive noise and vary with sound level. The sensitivity of midbrain neurons to low-frequency fluctuations inspired a shift of focus to the contrast in pitch-related fluctuations along the BF axis in vowel coding (Carney et al. 2015). In the healthy AN, there is a robust contrast in the profile of the F0-synchronized discharge rate across the BF axis (Fig. 2d). Midbrain neurons' sensitivities to these F0-synchronized rates transforms the temporally coded AN profile into a discharge rate profile (Fig. 4a).

The representation of F0-synchronized rates is affected by several mechanisms that are influenced by SNHL: synchrony capture, threshold (and thus rate saturation), and bandwidth of tuning. Bandwidth, in turn, affects the modulation spectrum of peripheral responses. Wider peripheral bandwidths associated with SNHL result in a wider modulation bandwidth and a reduction of the amplitude of low-frequency fluctuations. These factors have little influence on the mean rates of AN fibres; however, they have a large effect on F0-synchrony in the AN, which in turn affects the responses of IC neurons.

These IC model responses have implications for the design of signal-processing strategies for listeners with SNHL. Recreating stimulus spectra in the AN rate profiles will not elicit appropriate responses from central neurons, whereas restitution of the F0-synchronized rate profiles may (Rao and Carney 2014). This result suggests an experimental question: If the profile of peripheral F0-synchrony in response to voiced sounds can be restored to normal for these listeners, can intelligibility of speech, and especially speech in background noise, be improved? The results of such a test would depend on the status of central auditory pathways, which may undergo change following cochlear trauma (Suneja et al. 1998; Salvi et al. 2000).

Acknowledgments Supported by NIDCD-010813.

References

Carney LH, Li T, McDonough JM (2015) Speech coding in the brain: representation of formants by midbrain neurons tuned to sound fluctuations. eNeuro 2(4) e0004-15.2015 1–1. doi: 10.1523/ENEURO.0004-15.2015
Delgutte B, Kiang NY (1984) Speech coding in the auditory nerve: i. Vowel-like sounds. J Acoust Soc Am 75:866–878
Deng L, Geisler CD (1987) Responses of auditory-nerve fibers to nasal consonant–vowel syllables. J Acoust Soc Am 82:1977–1988
Deng L, Geisler CD, Greenberg S (1987) Responses of auditory-nerve fibers to multitone complexes. J Acoust Soc Am 82:1989–2000
Fant G (1960) Acoustic theory of speech production. Mouton, Hague
Hillenbrand J, Getty LA, Clark MJ, Wheeler K (1995) Acoustic characteristics of American English vowels. J Acoust Soc Am 97:3099–3111
Kewley-Port D, Burkle TZ, Lee JH (2007) Contribution of consonant versus vowel information to sentence intelligibility for young normal-hearing and elderly hearing-impaired listeners. J Acoust Soc Am 122:2365–2375
Kim DO, Bishop BB, Kuwada S, Carney LH (2015a). Band-Enhanced and Band-Suppressed Rate Modulation Transfer Functions of Inferior Colliculus Neurons and a Model: effects of Duty Cycle and Rise/Fall Rate, Association for Research in Otolaryngology. 38th Annual Midwinter Meeting, abstract #PS-423
Kim DO, Zahorik P, Carney LH, Bishop BB, Kuwada S (2015b) Auditory distance coding in rabbit midbrain neurons and human perception: monaural amplitude modulation depth as a cue. J Neurosci 35(13):5360–5372
Krishna BS, Semple MN (2000) Auditory temporal processing: responses to sinusoidally amplitude-modulated tones in the inferior colliculus. J Neurophysiol 84:255–273
Langner G (1992) A review: periodicity coding in the auditory system. Hear Res 60:115–142
Miller RL, Schilling JR, Franck KR, Young ED (1997) Effects of acoustic trauma on the representation of the vowel/e/in cat auditory-nerve fibers. J Acoust Soc Am 101:3602–3616
Nelson PC, Carney LH (2004) A phenomenological model of peripheral and central neural responses to amplitude-modulated tones. J Acoust Soc Am 116:2173–2186
Rao A, Carney LH (2014) Speech enhancement for listeners with hearing loss based on a model for vowel coding in the auditory midbrain. IEEE Trans Bio-med Eng 61:2081–2091
Sachs MB, Young ED (1979) Encoding of steady-state vowels in the auditory nerve: representation in terms of discharge rate. J Acoust Soc Am 66:470–479
Salvi RJ, Wang J, Ding D (2000) Auditory plasticity and hyperactivity following cochlear damage. Hear Res 147(1):261–274
Studebaker GA, Sherbecoe RL, McDaniel DM, Gwaltney CA (1999) Monosyllabic word recognition at higher-than-normal speech and noise levels. J Acoust Soc Am 105:2431–2444
Suneja SK, Potashner SJ, Benson CG (1998) Plastic changes in glycine and GABA release and uptake in adult brain stem auditory nuclei after unilateral middle ear ossicle removal and cochlear ablation. Exp Neurol 151(2):273–288
Young ED (2008) Neural representation of spectral and temporal information in speech. Philos Royal Soc London B: Biol Sci 363:923–945
Young ED, Sachs MB (1979) Representation of steady-state vowels in the temporal aspects of the discharge patterns of populations of auditory-nerve fibers. J Acoust Soc Am 66:1381–1403
Zilany MS, Bruce IC (2007) Representation of the vowel/in normal and impaired auditory nerve fibers: model predictions of responses in cats. J Acoust Soc Am 122(1):402–417
Zilany MSA, Bruce IC, Carney LH (2014) Updated parameters and expanded simulation options for a model of the auditory periphery. J Acoust Soc Am 135:283–286

Sources of Variability in Consonant Perception and Implications for Speech Perception Modeling

Johannes Zaar and Torsten Dau

Abstract The present study investigated the influence of various sources of response variability in consonant perception. A distinction was made between source-induced variability and receiver-related variability. The former refers to perceptual differences induced by differences in the speech tokens and/or the masking noise tokens; the latter describes perceptual differences caused by within- and across-listener uncertainty. Consonant-vowel combinations (CVs) were presented to normal-hearing listeners in white noise at six different signal-to-noise ratios. The obtained responses were analyzed with respect to the considered sources of variability using a measure of the perceptual distance between responses. The largest effect was found across different CVs. For stimuli of the same phonetic identity, the speech-induced variability across and within talkers and the across-listener variability were substantial and of similar magnitude. Even time-shifts in the waveforms of white masking noise produced a significant effect, which was well above the within-listener variability (the smallest effect). Two auditory-inspired models in combination with a template-matching back end were considered to predict the perceptual data. In particular, an energy-based and a modulation-based approach were compared. The suitability of the two models was evaluated with respect to the source-induced perceptual distance and in terms of consonant recognition rates and consonant confusions. Both models captured the source-induced perceptual distance remarkably well. However, the modulation-based approach showed a better agreement with the data in terms of consonant recognition and confusions. The results indicate that low-frequency modulations up to 16 Hz play a crucial role in consonant perception.

J. Zaar (✉)
Hearing Systems Group, Department of Electrical Engineering, Technical University of Denmark, Ørsteds Plads Building 352, room 108, 2800 Kongens Lyngby, Denmark
e-mail: jzaar@elektro.dtu.dk

T. Dau
Hearing Systems Group, Department of Electrical Engineering, Technical University of Denmark, Ørsteds Plads Building 352, room 120, 2800 Kongens Lyngby, Denmark
e-mail: tdau@elektro.dtu.dk

P. van Dijk et al. (eds.), *Physiology, Psychoacoustics and Cognition in Normal and Impaired Hearing,* Advances in Experimental Medicine and Biology 894,
DOI 10.1007/978-3-319-25474-6_46

Keywords Talker effects · Noise effects · Listener effects · Internal noise · Auditory modeling · Perceptual distance · Envelope domain · Microscopic · Perceptual distance · Modulation filters

1 Introduction

Speech perception is often studied from a *macroscopic* perspective, i.e., using meaningful long-term speech stimuli (e.g., in additive noise). To solely investigate the relation between the acoustic properties of the stimulus and the resulting speech percept (excluding lexical, semantic, and syntactic effects), an alternative is to take a *microscopic* perspective by investigating the perception of smaller units of speech such as consonants. Miller and Nicely (1955) measured the perception of consonant-vowel combinations (CVs, e.g.,/ba/,/ta/) in white noise and different band-pass filtering conditions and observed distinct consonant confusions. Wang and Bilger (1973) demonstrated that consonant perception also depends on the vowel context. Phatak and Allen (2007) measured consonant perception in speech-weighted noise and demonstrated noise-type induced perceptual differences to the Miller and Nicely (1955) data. In following studies, perceptual differences across different speech tokens of the same phonetic identity came more into focus (e.g., Phatak et al. 2008).

A few studies have attempted to simulate consonant perception. Li et al. (2010) successfully related consonant recognition data to the so-called AI Gram, which is related to the energy-based Articulation Index (ANSI 1969). Gallun and Souza (2008) considered noise-vocoded VCVs and demonstrated that the correlation of long-term modulation power representations was a strong predictor of consonant confusions. Jürgens and Brand (2009) used an auditory model with a modulation-frequency selective preprocessing stage in combination with a template-matching back end. The model showed convincing recognition predictions while the confusion predictions were inconclusive.

Motivated by the increasing evidence for a major variability in consonant perception that cannot be accounted for by the phonetic identity of the stimuli, the present study attempted to quantify some of the sources of variability that influence consonant perception. It was distinguished between *source-induced variability* and *receiver-related variability*. The former was subdivided into speech- and noise-induced variability; the latter was subdivided into across- and within-listener variability. Consonant perception data were collected and analyzed with respect to the considered sources of variability using a measure of the perceptual distance between responses. Predictions of the data were obtained using an energy- and a modulation-based model in combination with a template-matching back end. The model predictions were compared to the data (i) in terms of how well they reflected the source-induced variability measured in listeners and (ii) in terms of the agreement between perceptual and predicted consonant recognition and confusions.

2 Methods

2.1 *Experiment 1: Speech Variability*

CVs consisting of the 15 consonants /b, d, f, g, h, j, k, l, m, n, p, s, ʃ, t, v/ followed by the vowel /i/ (as in "feed") were used. Six recordings of each CV (three spoken by a male, three spoken by a female talker) were taken from the Danish nonsense syllable speech material collected by Christiansen and Henrichsen (2011). Six SNR conditions (12, 6, 0, −6, −12, and −15 dB) were created by fixing the noise sound pressure level (SPL) to 60 dB and adjusting the speech SPL. One particular white masking noise waveform with a duration of 1 s was generated for each speech token in each SNR condition and mixed with it such that the speech token onset was temporally positioned 400 ms after the noise onset.

2.2 *Experiment 2: Noise Variability*

Only one male-talker speech token of each CV was used. Three masking-noise conditions (frozen noise A, frozen noise B, and random noise) were considered. For each speech token, one particular white-noise waveform with a duration of 1 s was generated and labeled "frozen noise A"; the same noise token was then circularly shifted in time by 100 ms to obtain "frozen noise B". The noise waveforms for the random noise condition (added to prevent noise learning) were randomly generated during the experimental procedure. The noisy speech tokens were created as described in Sect. 2.1.

2.3 *Experimental Procedure*

Two different groups of eight normal-hearing native Danish listeners participated in the two experiments (average age: 26 and 24 years, respectively). The stimuli were monaurally presented to the listeners via headphones in experimental blocks ordered according to the SNR in descending order. An additional quiet condition (clean speech at 60 dB SPL) preceded the SNR conditions (noise level at 60 dB SPL). Each block started with a short training run. The order of presentation within each experimental block was randomized. In experiment 1, each stimulus was presented three times to each listener. In experiment 2, each stimulus was presented five times to each listener. Listeners had to choose one of the response alternatives displayed as 15 buttons labeled "b, d, f, g, h, j, k, l, m, n, p, s, Sj, t, v" and one button labeled "I don't know" on a graphical user interface (the Danish "Sj" corresponds to /ʃ/). Experiment 2 was repeated with four of the originally eight listeners to obtain test-retest data.

2.4 Data Analysis

For each stimulus and listener, the responses obtained in the experiments were converted to proportions of responses by distributing any "I don't know" response evenly across the 15 other response alternatives and dividing the occurrences of responses by the number of stimulus presentations. The response of a given listener obtained with a given stimulus was thus calculated as a vector $r=[p_b, p_d, \ldots, p_v]$, where p_x denotes the proportion of response "x". The perceptual distance between two response vectors r_1 and r_2 was defined as the normalized angular distance between them:

$$D(r_1, r_2) = \arccos\left(\frac{\langle r_1, r_2 \rangle}{\| r_1 \| \cdot \| r_2 \|}\right) \cdot \frac{100\%}{\pi/2}$$

The perceptual distance was calculated across six different factors: (i) across CVs, (ii) across talkers, (iii) within talkers, (iv) across masking-noise tokens, (v) across listeners, and (vi) within listeners. Apart from the across-CV factor, only responses obtained with stimuli of the same phonetic identity were compared. For each considered factor, the perceptual distance was calculated across all pairwise comparisons of response vectors representative of that factor. The calculations for all factors but (v) were performed for each listener and each SNR condition separately. The calculations for (v) were performed for each SNR condition separately, comparing responses across listeners. The individual distance values were then averaged across the considered response pairs and (where applicable) across listeners. As a result, the respective perceptual distances were obtained as a function of SNR.

3 Experimental Results

Figure 1 shows examples of perceptual *across-talker variability* and perceptual *across-noise variability* in terms of across-listener average confusion patterns: /pi/ spoken by talker A (panel a) was more recognizable (and less confusable) than /pi/ spoken by talker B (panel b); the perception of a given speech token /gi/ was very differently affected by a specific white masking-noise waveform "A" (panel c) than by a time-shifted version of that waveform ("B", panel d).

Figure 2 shows the perceptual distances derived from the experimental data for all SNR conditions (see Sect. 2.4). On the left, the across-SNR average is shown. The largest perceptual distance of 91 % was found across CVs (black bar). Regarding the source-induced perceptual distances across stimuli of the same phonetic identity, the largest perceptual distance of 51 % was obtained across talkers (blue bar), followed by the within-talker factor (47 %, green bar). A temporal shift in the masking-noise waveform induced a perceptual distance of 39 % (red bar). Regarding the receiver-related effects, a substantial perceptual distance of 46 % across

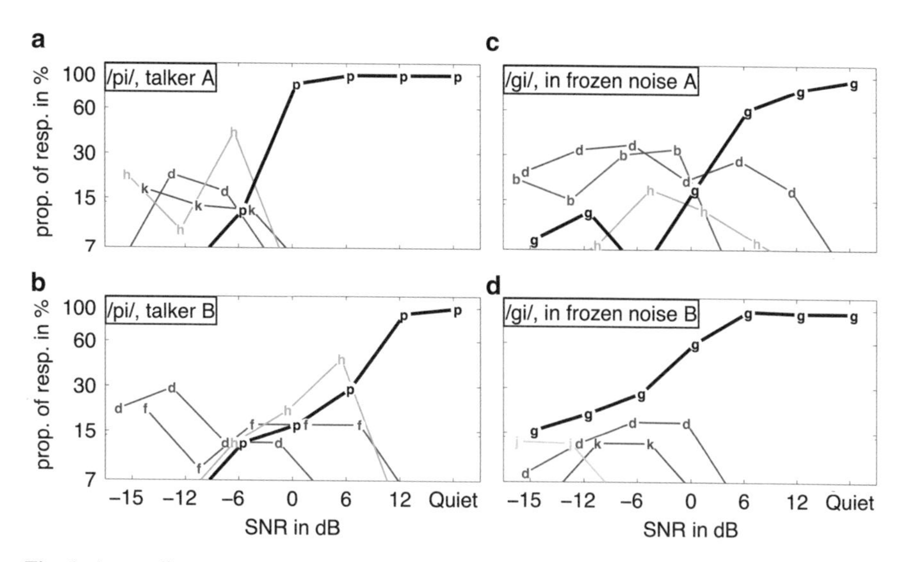

Fig. 1 Across-listener average example confusion patterns (CPs). *Left*: CPs obtained with two different speech tokens /pi/ spoken by male talker A (*top*) and female talker B (*bottom*). *Right*: CPs obtained with one specific speech token /gi/ mixed with frozen noise A (*top*) and frozen noise B (*bottom*)

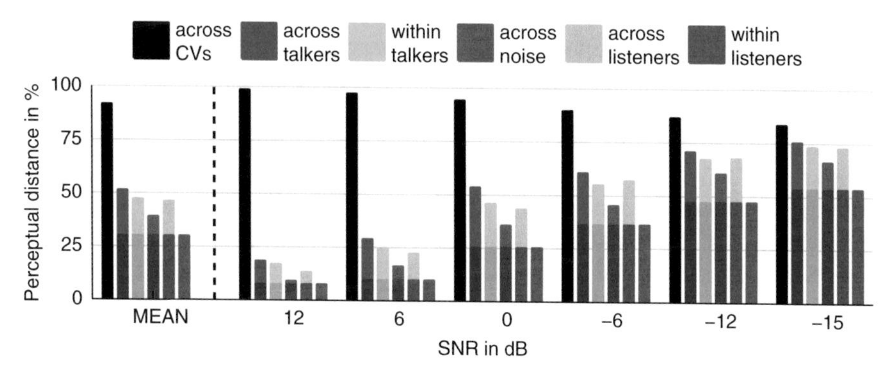

Fig. 2 Mean (*left*) and SNR-specific perceptual distances across CVs, across talkers, within talkers, across noise, across listeners, and within listeners. The *shaded* areas represent values below the within-listener distance, i.e., below the internal-noise baseline

listeners was found for physically identical stimuli (light gray bar). In contrast, the relatively low perceptual distance of 30 % within listeners (test vs. retest, dark gray bar) indicated that the individual listeners were able to reproduce their responses fairly reliably. Pairwise t-tests across all combinations of conditions (excluding the across-CV condition) demonstrated that all conditions were significantly different from each other ($p < 0.05$) except for the across-talker (blue), within-talker (green), and across-listener (light gray) conditions.

Regarding the trends across SNR in Fig. 2, it can be seen that the across-CV distance (black bars) was at ceiling for large SNRs and decreased with decreasing SNR, as listeners made more speech-token specific confusions. All other perceptual distance types showed low values for large SNRs and increased with decreasing SNR due to stimulus-specific confusions and listener uncertainty. The within-listener distance ("internal noise") represented the baseline and strongly increased with decreasing SNR as the task became more challenging.

4 Modeling

4.1 Model Components

The subband power P, in dB, was calculated using 22 fourth-order gammatone filters with equivalent rectangular bandwidths. The center frequencies were spaced on a third-octave grid, covering a range from 63 Hz to 8 kHz. The Hilbert envelope of each filter output was extracted and low-pass filtered using a first-order Butterworth filter with a cut-off frequency of 150 Hz. The envelopes were down-sampled to a sampling rate of 1050 Hz.

The modulation power P_{mod}, in dB, was obtained using the subband envelope extraction described above, followed by a modulation filterbank consisting of 3 second-order band-pass filters (center frequencies of 4, 8, and 16 Hz) in parallel with one third-order low-pass filter (cut-off frequency of 2 Hz).

A template-matching procedure was applied to predict the responses obtained in experiment 1. Two talker-specific template sets were considered, consisting of all speech tokens from each talker (i.e., three templates for each CV). The templates were mixed with random white noise at the test-signal SNR and compared to the experimental stimuli. The distances between the models' internal representations of the test signals and the templates were obtained using a standard dynamic time warping (DTW) algorithm (Sakoe and Chiba 1978). The template-matching procedure was conducted nine times with newly generated random noise for the templates. The "correct" template always contained the same speech token as the test signal, while the masking noise differed. In each run, the template showing the smallest distance to the test signal was selected. The modeled responses were converted to proportions of responses. The responses obtained in experiment 2 were predicted similarly, considering only the 15 speech tokens used in experiment 2 as templates.

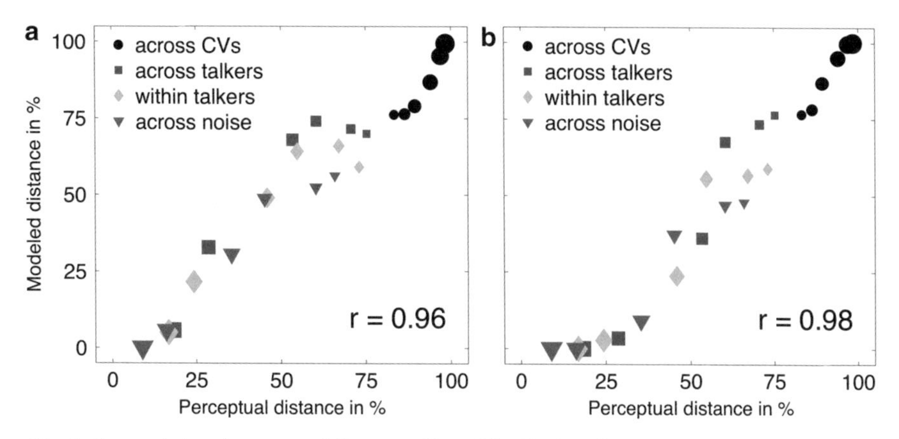

Fig. 3 Source-induced perceptual distances (from Fig. 2) plotted versus corresponding modeled distances obtained using P (*left*) and P_{mod} (*right*). The symbols and colors represent the different distance types, the size of the symbols is proportional to the SNR

5 Simulation Results

5.1 *Sources of Variability*

In accordance with the procedure described in Sect. 2.4, the across-CV, across-talker, within-talker, and across-noise modeled distances were obtained as a function of the SNR from the predicted responses. Figure 3 shows scatter plots of the perceptual distance versus the modeled distances obtained using P (panel a) and P_{mod} (panel b). It can be observed that the perceptual distances were remarkably well-predicted using P as well as P_{mod}, with a Pearson's r of 0.96 and 0.98, respectively.

5.2 *Consonant Recognition and Confusions*

The grand average consonant recognition rates obtained in experiment 1 showed a speech reception threshold (SRT) of −3 dB. The predicted SRTs were overestimated by 2.8 dB using P and by only 0.4 dB using P_{mod}. The token-specific SRTs showed a large spread across speech tokens, which was smaller in both model predictions. However, P_{mod} captured the relative ranking of token-specific SRTs considerably better than P (Spearman's r of 0.4 and 0.04, respectively).

The across-listener average data obtained in experiment 1 were averaged across different speech tokens of the same phonetic identity and across the four lowest SNRs (as most confusions occur for low SNRs). The resulting confusion matrices (CMs) are shown as filled gray circles in both panels of Fig. 4. The model predictions are plotted on top as open black circles. For P (panel a), an underestimation of the recognition for many consonants was observed, indicated by the mismatch

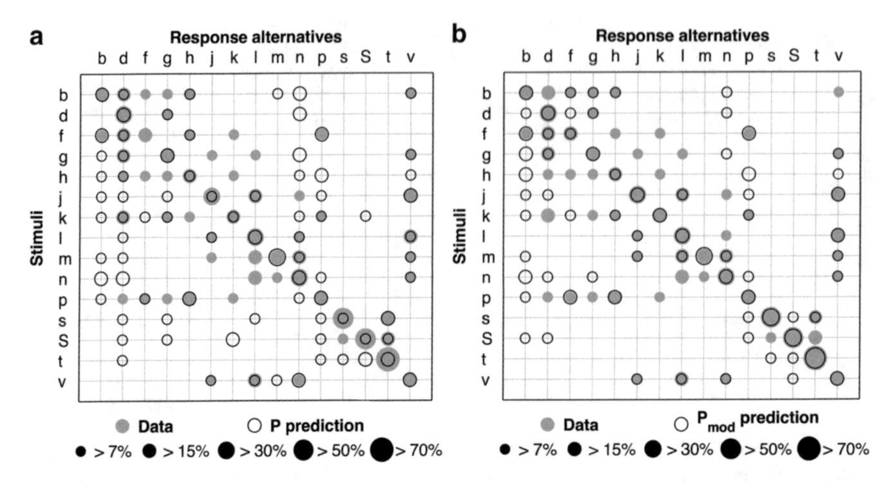

Fig. 4 Confusion matrices obtained in experiment 1, averaged across listeners, speech tokens of the same phonetic identity, and across SNRs of 0, −6, −12, and −15 dB. The perceptual data are shown as filled gray circles in both panels. The model predictions obtained with P (left) and P_{mod} (right) are represented as open *black circles*

of the on-diagonal circles. For P_{mod} (panel b), a good consonant-specific recognition rate match was found. Both models hit most of the confusions, reflected in the proportion of gray off-diagonal circles matched with a black circle. However, there were also many "false alarms", particularly in the P-based predictions (panel a).

6 Summary and Discussion

The investigation of different sources of variability in Sect. 3 indicated that any considered difference in the stimuli produced a measurable effect. The observed perceptual variability across talkers is well established in the related literature (e.g., Phatak et al. 2008); the equally large variability within talkers had not yet been demonstrated. Most remarkably, even a 100-ms time shift in the white masking-noise waveform induced significant perceptual variability, indicating that "steady-state" masking noise should not be considered steady over time in the context of consonant cues. On the receiver side, different NH listeners with identical language background showed substantial differences while individual listeners could fairly reliably reproduce their responses. Averaging consonant perception data (even across NH listeners) thus seems problematic.

The predictions obtained in Sect. 4 with the energy-based (P) and the modulation-based (P_{mod}) pre-processing stages both accounted for the trends in the perceptual data with respect to the considered stimulus-related sources of variability. Consonant recognition was strongly underestimated using P and well-predicted using P_{mod}. An inspection of confusion matrices suggested that both models correctly predicted most of the perceptual confusions, albeit with some “false alarms”. The overall larger predictive power obtained with P_{mod} indicates that slow envelope fluctuations up to 16 Hz are a good predictor for consonant-in-noise perception. This is consistent with the findings by Gallun and Souza (2008).

The perceptual data analysis has implications for the further model design. It was shown that the internal noise increased with decreasing SNR. This could be incorporated in the model using an SNR-dependent random process in the decision stage (instead of SNR-dependent templates). Furthermore, the model predicted responses of a hypothetical “average” NH listener, which is unrealistic given the considerable across-listener variability. It remains a challenge to include listener-specific differences in the model, as it is not clear whether these differences can be accounted for by slight sensitivity differences between the NH listeners, cognitive effects, individual biases, or any combination of these factors. Eventually, an extension of the model towards different types of hearing impairment might be useful to understand the link between individual impairment factors and microscopic speech intelligibility.

Acknowledgments We thank Søren Jørgensen for his contributions to the research underlying this study. This research was supported by the FP7 Marie Curie Initial Training Network INSPIRE.

References

ANSI (1969). ANSI S3.5-1969 American national standard methods for the calculation of the articulation index. Standards Secretariat, Acoustical Society of America

Christiansen TU, Henrichsen PJ (2011). Objective evaluation of consonant-vowel pairs produced by native speakers of Danish. Proceedings of Forum Acusticum 2011.

Gallun F, Souza P (2008) Exploring the role of the modulation spectrum in phoneme recognition. Ear Hear 29(5):800–813

Jürgens T, Brand T (2009) Microscopic prediction of speech recognition for listeners with normal hearing in noise using an auditory model. J Acoust Soc Am 126(5):2635–2648

Li F, Menon A, Allen JB (2010) A psychoacoustic method to find the perceptual cues of stop consonants in natural speech. J Acoust Soc Am 127(4):2599–2610

Miller GA, Nicely PE (1955) An analysis of perceptual confusions among some English consonants. J Acoust Soc Am 27(2):338–352

Phatak SA, Allen JB (2007) Consonant and vowel confusions in speech-weighted noise. J Acoust Soc Am 121(4):2312–2326

Phatak SA, Lovitt A, Allen JB (2008) Consonant confusions in white noise. J Acoust Soc Am 124(2):1220–1233

Sakoe H, Chiba S (1978). Dynamic programming algorithm optimization for spoken word recognition. IEEE Trans Acoust, Speech Signal Proc (ASSP) 26(1), 43–49

Wang MD, Bilger RC (1973) Consonant confusions in noise: a study of perceptual features. J Acoust Soc Am 54(5):1248–1266

On Detectable and Meaningful Speech-Intelligibility Benefits

William M. Whitmer, David McShefferty and Michael A. Akeroyd

Abstract The most important parameter that affects the ability to hear and understand speech in the presence of background noise is the signal-to-noise ratio (SNR). Despite decades of research in speech intelligibility, it is not currently known how much improvement in SNR is needed to provide a meaningful benefit to someone. We propose that the underlying psychophysical basis to a meaningful benefit should be the just noticeable difference (JND) for SNR. The SNR JND was measured in a series of experiments using both adaptive and fixed-level procedures across participants of varying hearing ability. The results showed an average SNR JND of approximately 3 dB for sentences in same-spectrum noise. The role of the stimulus and link to intelligibility was examined by measuring speech-intelligibility psychometric functions and comparing the intelligibility JND estimated from those functions with measured SNR JNDs. Several experiments were then conducted to establish a just meaningful difference (JMD) for SNR. SNR changes that could induce intervention-seeking behaviour for an individual were measured with subjective scaling and report, using the same stimuli as the SNR JND experiment as pre- and post-benefit examples. The results across different rating and willingness-to-change tasks showed that the mean ratings increased near linearly with a change in SNR, but a change of at least 6 dB was necessary to reliably motivate participants to seek intervention. The magnitude of the JNDs and JMDs for speech-intelligibility benefits measured here suggest a gap between what is achievable and what is meaningful.

W. M. Whitmer (✉) · D. McShefferty
MRC/CSO Institute of Hearing Research—Scottish Section, Glasgow, UK
e-mail: bill@ihr.gla.ac.uk

D. McShefferty
e-mail: david@ihr.gla.ac.uk

M. A. Akeroyd
MRC Institute of Hearing Research, Nottingham, UK
e-mail: maa@ihr.mrc.ac.uk

P. van Dijk et al. (eds.), *Physiology, Psychoacoustics and Cognition in Normal and Impaired Hearing,* Advances in Experimental Medicine and Biology 894,
DOI 10.1007/978-3-319-25474-6_47

Keywords Signal-to-noise ratio · Just-noticeable difference · Speech intelligibility · Hearing impairment

1 Introduction

The inability to provide appreciable speech-in-noise benefits can lead to the non-use of hearing aids (e.g., McCormack and Fortnum 2013). Non-use can come from unmet expectations (i.e., benefits without satisfaction; Demorest 1984), which may be at least partly a result of whatever changes the hearing aids provided being either undetectable to the user or not important enough to the user. The current study looked at both of these: how large a change in signal-to-noise ratio (SNR) has to be for it to psychophysically discriminable or clinically meaningful.

This is a curiously under-studied area: previously there has been only one small study related to measuring the just-noticeable difference (JND) in SNR by Killion (2004), in which unspecified participants were at chance discriminating a 2 dB SNR difference for speech in noise. A JND, though, is just a psychophysical benchmark; it is not clear *a priori* that a change of one JND is necessary or sufficient for the change to have any subjective importance to an individual. What is meaningful and clinically significant is key to service-wide treatments, and could be vital for determining provision criteria as well as tempering expectations for the patient. The current study looks at what could induce intervention-seeking behaviour for an individual; that is, how much subjective value do patients ascribe to discriminable speech intelligibility benefits? Here, we use psychophysical methods to rigorously measure (a) the JND for SNR in decibels, (b) the JND for intelligibility in %, and (c) the JND for meaningful benefit, using the same stimuli as examples of pre- and post-benefit situations.

2 Detectable Benefits

2.1 *The JND for Changes in SNR*

The JND for SNR was measured for adults of varying hearing ability at different reference SNRs using adaptive and fixed-level procedures based on the classic level discrimination paradigm (cf. McShefferty et al. 2015). The stimuli were equalised IEEE sentences uttered by a native British English speaker (Smith and Faulkner 2006) in same-spectrum noise, presented over headphones in a sound-attenuated booth. Noises began and ended simultaneous with the speech signal, and there was a 500-ms gap between intervals. Both intervals contained the same randomly chosen sentence, and after both intervals, participants were prompted to respond to which one was clearer. In each interval the levels of both speech and noise were adjusted to maintain a presentation level of 75 dB A; to minimize other cues, the level of each interval was roved independently by a maximum of ±2 dB. In a two-interval/

Table 1 Mean (μ) SNR JND for reference SNRs of −6, 0 and +6 dB pooled across experiments, showing standard deviation (σ) and participant number (*n*) tested at each reference SNR. Median BE4FAs and ages are given with ranges in parentheses

	−6 dB	0 dB	+6 dB
μ (dB SNR)	2.8	2.9	3.7
σ	1.0	1.0	1.4
n	35	99	72
BE4FA (dB HL)	28 (3–56)	29 (−1–85)	32 (−1–71)
Age (yrs)	63 (22–72)	65 (23–76)	64 (22–76)

alternative forced-choice task, participants were presented, in randomized order, with a reference interval at a fixed reference SNR (either −6, 0, or +6 dB), and a target interval at the reference SNR plus an increment (ΔSNR) that was either varied adaptively or presented at fixed, randomly interleaved ΔSNRs. The adaptive procedure estimated 79 % correct from the geometric means of the best two of three three-up/one-down tracks. The fixed-level procedure estimated 79 % correct from fitting a maximum-log-likelihood logistic function to the data. The procedures produced equivalent results to 0.1 dB, so results were combined across procedures for each reference SNR.

The results are shown in Table 1. The SNR JNDs at the three SNR references of −6, 0 and +6 dB SNR were 2.8, 2.9, and 3.7 dB respectively; the first two were statistically no different and both less than the third [$t_{(91.3)}=4.07$; $p=0.0001$ and $t_{(121.4)}=4.11$; $p=0.00007$, respectively].

2.2 *The JND for Changes in Intelligibility*

The JND for intelligibility was estimated by first measuring psychometric functions for IEEE sentences based on keywords at SNRs of −16/−12/−8/−4/0 dB or −8/−4/0/+4/+8 dB. Twenty-four adult participants, median better-ear four frequency average (BE4FA) of 16 dB HL (range −3–49 dB HL) and median age of 57 years (range 27–74), were first presented with sentences at −16 and 0 dB SNR. Fourteen participants could not respond with half of the keywords (5 keywords/sentence) at 0 dB; those participants were tested at the −8–8 dB SNR range. The individual results, averaged across 50 sentences (250 keywords) at each SNR, were fit with a maximum-log-likelihood logistic function. The SNR corresponding to 50 % correct on each individual's psychometric function (SNR50) was calculated and then used as the reference for an adaptive-track JND experiment; the mean SNR50 was 0.20 dB. This gave a mean SNR JND of 3.0 dB (σ = 1.0 dB), which, converted to intelligibility using the slope of the psychometric function, corresponded to a mean Intelligibility JND of 26 % (σ = 7.5 %).

This raises the question as to whether the JNDs initially measured (in Sect. 2.1 above) were intelligibility rather than SNR JNDs (i.e., were the listeners basing their responses on changes in word intelligibility or in changes in signal-to-noise ratio?) We reasoned that by changing the stimuli to ones with a far steeper psychometric function (MacPherson and Akeroyd 2014), namely triple digits, the JND in the appropriate domain would be constant, but the JND in the other domain would change. On testing this hypothesis, we found that the mean SNR JND for digits was 2.5 dB (σ=0.8 dB) and the mean Intelligibility JND estimate for digits was 17.5 % (σ=7.2 %). SNR and Intelligibility JNDs were significantly less using digits in unfiltered random (white) noise than sentences in same-spectrum noise [$t_{(23)}=2.98$; $p=0.007$ and $t_{(23)}=5.00$; $p=0.00005$]. Across stimuli, both the SNR JND differences (mean 0.5 dB) and the Intelligibility JND differences (mean 8.92 dB) were significantly non-zero ($z_{(23)}=2.98$; $p=0.003$, and $z_{(23)}=5.00$; $p\approx 0$). That *both* differences were non-zero was unexpected; it indicates that one cannot be certain whether the 3 dB SNR JND is indeed the JND for SNR or instead the JND for Intelligibility.

3 Meaningful Benefits

To ascertain what is a meaningful benefit to someone requires more than detectability; it requires the subjective input of the participant. The previous method for deriving the SNR JND was hence changed to subjective-comparison tasks using SNR and SNR +ΔSNR pairs as examples of, respectively, pre-benefit and post-benefit situations to measure the "just meaningful difference" (JMD) for SNR. It was not clear in advance to us what query, however, best represents *meaningfulness*. Four tasks are discussed below: a conventional better/worse rating task, a novel conversation-tolerance rating task, a swap-paradigm task and a clinical significance task. For the latter two tasks (Sect. 3.3 and 3.4), the JMD was considered as the ΔSNR where the proportion of affirmative responses (i.e., willingness to swap devices or attend the clinic) were significantly greater than chance (50%).

3.1 Rateable Benefits

The simplest, most direct rating of preference is a better/worse scaling procedure. Thirty six participants (median age: 62 years; range: 31–74) of varying hearing ability (median BE4FA: 21 dB HL; range 3–85 dB HL) heard two intervals of IEEE sentences in same-spectrum noise, one at 0 dB SNR, and the other at 0 dB plus a ΔSNR of 1, 2, 4, 6 or 8 dB. The overall level of each interval was 63 dB A (73 for those with more severe losses); unlike the JND experiments, the level was not roved across intervals. Participants were asked how much better/worse the second interval was on a discrete, signed 11-point scale with −5 marked "much worse" and +5 marked "much better." Each of the ten conditions (5 ΔSNRs × 2 orders) was repeated 12 times in randomized order, resulting in 120 trials completed in three blocks of 40.

The results, including 95 % within-subject confidence intervals, are shown in the left panel of Fig. 1. Ratings increase near linearly as a function of ΔSNR, and are asymmetric: when the +ΔSNR interval preceded the reference, ratings were less than when the +ΔSNR interval followed the reference (e.g., when the second

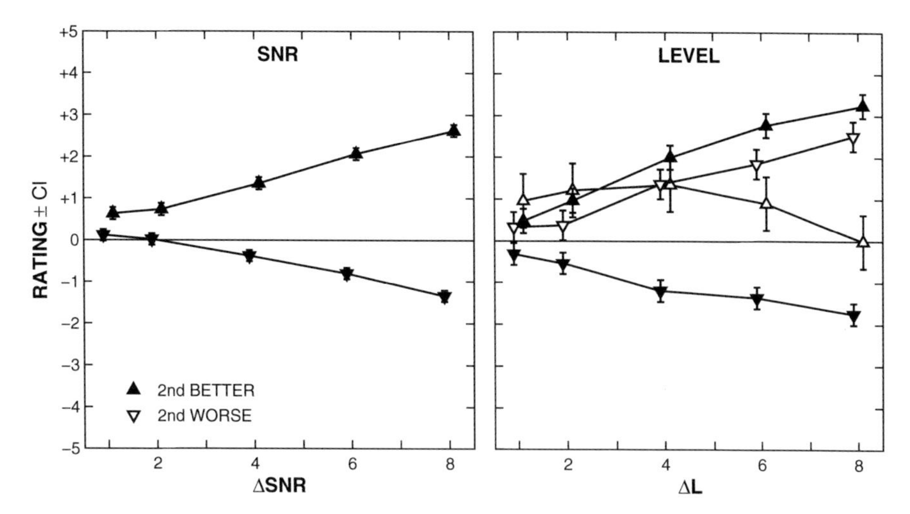

Fig. 1 Better (*upward triangles*) and worse (*downward triangles*) ratings as function of changes in SNR (*left panel*) and overall level (*right panel*). The reference SNR was 0 dB; the reference level was 70 dB A. Open *symbols/lines* (*right panel*) indicate mean ratings for six participants whose max./min. ratings did not occur at the extremities. *Error bars* show 95 % within-subject confidence intervals (based on the analysis of variance subject × condition interaction term)

interval was the better SNR, ratings were greater than one increment at a ΔSNR of 4 dB, but when the second interval was the worse SNR, ratings were greater than one increment only when ΔSNR was 8 dB). This asymmetry was most likely due to an order effect; the second iteration of a sentence has been shown to be more intelligible (e.g., Thwing 1956), hence second-interval deficits would be deflated, and second-interval benefits inflated. Regarding the JMD, the ratings significantly increased only at 4 dB, regardless of order, lending support to the notion of detectability being a requisite for meaningfulness. SNR JNDs for these participants, however, were not well correlated with their individual mean responses at any ΔSNR.

An analogous procedure was used to examine what ratings would apply to a change in level per se as opposed to a change in relative level due to a change in SNR. Thirty-six participants (median age: 67 years; range: 27–77) of varying hearing ability (median BE4FA: 37 dB HL; range: 4–84) heard stimuli at a fixed SNR of 0 dB. The reference level (L) was 70 dB A, and ΔL on any trial was 1, 2, 4, 6 or 8 dB. The level JND was also measured for each participant; this mean was 1.4 dB (σ= 0.4 dB), similar to previous level JNDs for older hearing-impaired adults (Whitmer and Akeroyd 2011).

The results are shown in the right panel of Fig. 1. Six participants exhibited varying negative reactions to the greatest levels, finding a particular non-maximal level rated highest (e.g., 66 and 74 dB A). For the remaining 30 participants, the results are very similar to those for SNR, though ΔLs of 4–8 dB were rated modestly higher than ΔSNRs of 4–8 dB. That is, a change in level without a change in SNR was considered as good as—indeed if not better than—a change in SNR. While this finding does not coincide with their role in intelligibility, it is coincident with detectability of level and SNR changes (JNDs of 1.5 and 3.0 dB, respectively).

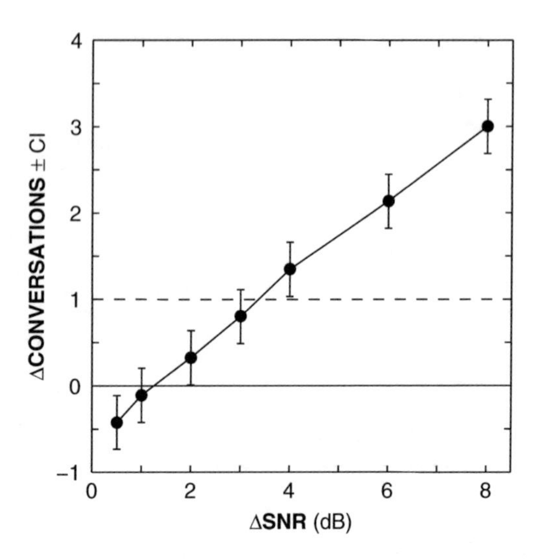

Fig. 2 Mean change in conversations tolerated as a function of the change in SNR. *Error bars* show 95 % within-subject confidence intervals

3.2 Tolerable Benefits

A second, novel method for establishing a meaningful difference was developed based around complaints of not being able to endure noisy conversation. For the conversation tolerance test, listeners heard the same sentence at two SNRs, one at an estimate of their speech-in-noise reception threshold (SNR50) and one at SNR50 plus an increment of 0.5–8 dB. In each of a pair of intervals, the participant was asked, "How many conversations in this situation would you tolerate?" (i.e., a paired single-interval task). Twenty-one adults (median age: 66 years; range: 52–73) with varying hearing ability (median BE4FA: 24 dB HL; range: 4–61) completed the task, repeating each increment in randomized order ten times.

The results are shown in Fig. 2. On average, participants were not prepared to tolerate an additional conversation until the SNR had increased by at least 3 dB, and not prepared to tolerate *more* than one extra conversation until a 4 dB change in SNR, similar to the results in Sect. 3.1. Though a change of one conversation is an arbitrary change, it is arguably a less arbitrary unit of meaningfulness than one point on a better/worse rating scale.

3.3 Swappable Benefits

If another hearing aid offers more speech-in-noise benefit, would someone be willing to trade for it? Using the same stimuli as the earlier rating experiment, 35 participants (median age: 62 years; range: 38–74) of varying hearing ability (median BE4FA: 34 dB HL; range: 4–80) were asked to consider the reference SNR (either −6 or +6 dB) interval as an example of their current device, and the reference SNR + ΔSNR (2, 4, 6 or 8 dB) interval as a different device. They were asked on each trial if they would swap their current for the different device in order to get that change (yes/no). Each condition (reference SNR × ΔSNR) was repeated 30 times in randomized order for a total of 240 trials.

The results are shown in the left panel of Fig. 3. When the initial SNR was −6 dB, the proportion of times that participants, on average, were willing to swap exceeded chance only when the ΔSNR was greater than 4 dB (i.e., 70 % at 6 dB SNR, and 87 % at 8 dB SNR). When the initial SNR was +6 dB, willingness to swap just exceeded chance (56 % willing to swap) at a ΔSNR of 8 dB. The JMD based on the swap paradigm is therefore dependent on the difficulty of the situation (i.e., the reference SNR), but appears to be at least 6 dB SNR.

3.4 Clinically Significant Benefits

While there are statistical bases for a minimal clinically important difference (cf. Jaeschke et al., 1989), the goal here was to develop a benchmark for what speech intelligibility benefit is necessary to motivate an individual to seek intervention. Hence, "clinical significance" was applied in a literal sense: participants were asked whether a positive change in SNR, as an example of what a visit to the clinic would provide, was worth attending the clinic to get that change. Thirty-six participants (median age: 63 years; range: 22–72) of varying hearing ability (median BE4FA: 28 dB HL; range: 2–56) were presented stimuli as in Sect. 3.1 and 0 with 12 trials of each reference SNR (−6 and +6 dB) and each ΔSNR (1, 2, 4, 6 and 8 dB) for a total of 120 trials.

The results are shown in the right panel of Fig. 3. They were similar to the swap experiment (left panel of same figure), though there was a greater tendency towards

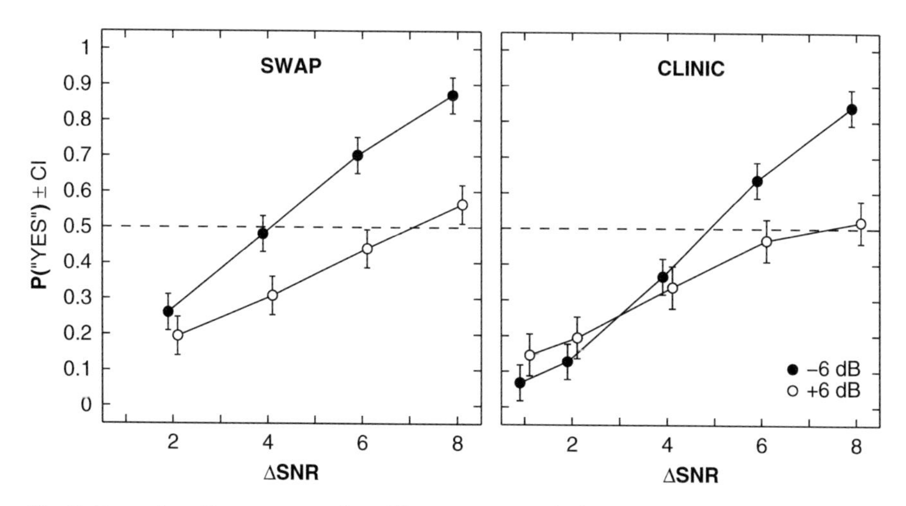

Fig. 3 Proportion of yes responses for willingness to swap devices (*left panel*) and willingness to attend the clinic (*right panel*) as a function of positive change in SNR (ΔSNR) for two reference SNRs: −6 (*filled circles*) and +6 dB SNR (*open circles*). *Error bars* show 95 % within-subject confidence intervals

not attending the clinic compared to not swapping devices at the lowest, sub-JND ΔSNRs. For the −6 dB SNR reference condition, the mean proportion of yes responses only exceeded chance at 6 and 8 dB SNR. For the +6 dB SNR reference condition, mean responses never significantly exceeded chance; furthermore, the function appears asymptotic, so the responses may not have exceeded chance regardless of the change in SNR. If the SNR is already relatively high enough to be clear, then making it clearer still does not appear to induce intervention-seeking behaviour. The JMD derived from clinical significance is the same as the JMD derived from the swap task: at least 6 dB SNR.

4 Conclusions

To determine both detectable and meaningful speech-intelligibility benefits, participants were presented paired examples of speech in noise, one at a reference SNR and the other at a variably higher SNR. The threshold (JND) for a discriminable change was roughly 3 dB SNR. In more advantageous conditions, the SNR JND increased, and decreased for simpler speech (digits vs. sentences; cf. MacPherson and Akeroyd 2014). The threshold (JMD) for a meaningful benefit was at least 6 dB. Curiously, age and hearing loss were not factors well correlated with either individuals' just noticeable or just meaningful differences.

Given the difficulty in achieving SNR benefits greater than 6 dB in realistic environments with current hearing aid processing outwith wireless transmission of the signal (e.g., Whitmer et al. 2011), the evidence here indicates that there is no demonstrable "wow" in speech-intelligibility benefits. Rather, assessing other potential avenues of benefit—such as long-term improvements, attentional/cognitive ease, and psycho-social engagement—could show the advantages of hearing aids despite the benefits they give in SNR being less than the JND or JMD.

Acknowledgments We would like to thank Dr. Michael Stone for supplying some of the stimuli used. The Scottish Section of IHR is supported by intramural funding from the Medical Research Council (grant number U135097131) and the Chief Scientist Office of the Scottish Government.

References

Demorest ME (1984). Techniques for measuring hearing aid benefit through self-report. In: Pickett J (ed) Symposium on hearing aid technology: its present and future. Gallaudet, Washington, DC

Jaeschke R, Singer J, Guyatt GH (1989) Measurement of health status: ascertaining the minimal clinically important difference. Control Clin Trials 10:407–415

Killion MC (2004) Myths about hearing in noise and directional microphones. Hear Rev 11(14–19):72–73

MacPherson A, Akeroyd MA (2014) Variations in the slope of the psychometric functions for speech intelligibility: a systematic survey. Trends Hear 18:1–26

McCormack A, Fortnum H (2013) Why do people with hearing aids not wear them? Int J Audiol 52:360–368

McShefferty D, Whitmer WM, Akeroyd MA (2015) The just-noticeable difference in speech-to-noise ratio. Trends Hear 19:1–9. doi:10.1177/2331216515572316

Smith MW, Faulkner A (2006) Perceptual adaptation by normally hearing listeners to a simulated 'hole' in hearing. J Acoust Soc Am 120:4019–4030

Thwing EJ (1956) Effect of repetition on articulation scores for PB words. J Acoust Soc Am 28:302–303

Whitmer WM, Akeroyd MA (2011) Level discrimination of speech sounds by hearing-impaired individuals with and without hearing amplification. Ear Hear 32:391–398

Whitmer WM, Brennan-Jones CG, Akeroyd MA (2011) The speech intelligibility benefit of a unilateral wireless system for hearing-impaired adults. Int J Audiol 50:905–911

Individual Differences in Behavioural Decision Weights Related to Irregularities in Cochlear Mechanics

Jungmee Lee, Inseok Heo, An-Chieh Chang, Kristen Bond, Christophe Stoelinga, Robert Lutfi and Glenis Long

Abstract An unexpected finding of previous psychophysical studies is that listeners show highly replicable, individualistic patterns of decision weights on frequencies affecting their performance in spectral discrimination tasks—what has been referred to as *individual listening styles*. We, like many other researchers, have attributed these listening styles to peculiarities in how listeners attend to sounds, but we now believe they partially reflect *irregularities* in cochlear micromechanics modifying what listeners hear. The most striking evidence for cochlear irregularities is the presence of low-level spontaneous otoacoustic emissions (SOAEs) measured in the ear canal and the systematic variation in stimulus frequency otoacoustic emissions (SFOAEs), both of which result from back-propagation of waves in the cochlea. SOAEs and SFOAEs vary greatly across individual ears and have been shown to affect behavioural thresholds, behavioural frequency selectivity and judged loudness for tones. The present paper reports pilot data providing evidence that SOAEs and SFOAEs are also predictive of the relative decision weight

J. Lee (✉) · I. Heo · A.-C. Chang · K. Bond · C. Stoelinga · R. Lutfi
Auditory Behavioral Research Laboratory, Communication Sciences and Disorders, University of Wisconsin-Madison, Madison, WI 53706, USA
e-mail: Jungmee.lee@wisc.edu

I. Heo
e-mail: iheo@wisc.edu

A.-C. Chang
e-mail: achang5@wisc.edu

K. Bond
e-mail: kbond3@wisc.edu

C. Stoelinga
e-mail: stoelinga@wisc.edu

R. Lutfi
e-mail: ralutfi@wisc.edu

G. Long
Speech-Language-Hearing Sciences, Graduate Center of City University of New York, New York, NY 10016, USA
e-mail: glong@gc.cuny.edu

P. van Dijk et al. (eds.), *Physiology, Psychoacoustics and Cognition in Normal and Impaired Hearing,* Advances in Experimental Medicine and Biology 894,
DOI 10.1007/978-3-319-25474-6_48

listeners give to a pair of tones in a level discrimination task. In one condition the frequency of one tone was selected to be near that of an SOAE and the frequency of the other was selected to be in a frequency region for which there was no detectable SOAE. In a second condition the frequency of one tone was selected to correspond to an SFOAE maximum, the frequency of the other tone, an SFOAE minimum. In both conditions a statistically significant correlation was found between the average relative decision weight on the two tones and the difference in OAE levels.

Keywords Behavioural decision weights · Level discrimination · Spontaneous otoacoustic emissions · Stimulus frequency otoacoustic emission

1 Introduction

People with normal hearing acuity usually can follow a conversation with their friends at a noisy party, a phenomenon known as the "cocktail party effect" (Cherry 1953). This remarkable ability to attend to target sounds in background noise deteriorates with age and hearing loss. Yet, people who have been diagnosed in the clinic as having a very mild hearing loss or even normal hearing based on their pure tone audiogram (the clinical gold standard for identifying hearing loss) still often report considerable difficulty communicating with others in such noisy environments (King and Stephens 1992). The conventional pure tone audiogram, the cornerstone of hearing loss diagnosis, is not always the best predictor for these kinds of difficulties.

Perturbation analysis has become a popular approach in psychoacoustic research to measure how listeners hear out a target sound in background noise (cf. Berg 1990; Lutfi 1995; Richards 2002). Studies using this paradigm show listeners to have *highly replicable, individualistic patterns* of decision weights on frequencies affecting their ability to hear out specific targets in noise—what has been referred to as *individual listening styles* (Doherty and Lutfi 1996; Lutfi and Liu 2007; Jesteadt et al. 2014; Alexander and Lutfi 2004). Unfortunately this paradigm is extremely time-consuming, rendering it ineffective for clinical use. Finding a quick and an objective way to measure effective listening in noisy environments would provide a dramatic improvement in clinical assessments, potentially resulting in better diagnosis and treatment.

In the clinic, otoacoustic emissions (OAEs) provide a fast, noninvasive means to assess auditory function. Otoacoustic emissions (OAEs) are faint sounds that travel from the cochlea back through the middle ear and are measured in the external auditory canal. Since their discovery in the late 1970s by David Kemp (1978), they have been used in the clinic to evaluate the health of outer hair cells (OHC) and in research to gain scientific insight into cochlear mechanics. Behaviorally they have been shown to predict the pattern of pure-tone quiet thresholds (Long and Tubis 1988; Lee and Long 2012; Dewey and Dhar 2014), auditory frequency selectivity, (Baiduc et al. 2014), and loudness perception (Mauermann et al. 2004).

The effect of threshold microstructure (as measured by OAEs) on loudness perception is particularly noteworthy because relative loudness is also known to be one of the most important factors affecting the decision weights listeners place on different information-bearing components of sounds (Berg 1990; Lutfi and Jesteadt 2006; Epstein and Silva 2009; Thorson 2012; Rasetshwane et al. 2013). This suggests that OAEs might be used to diagnose difficulty in target-in-noise listening tasks through their impact on decision weights. OAEs may be evoked by external sound stimulation (EOAEs) or may occur spontaneously (SOAEs). Stimulus frequency OAEs (SFOAEs), which are evoked using a single frequency sound, are one of the most diagnostic OAEs regarding cochlear function. They show a *highly replicable individualistic pattern* of amplitude maxima and minima when measured with high enough frequency resolution, a pattern called SFOAE fine structure. The level difference between maxima and minima can be as large as 30 dB. Usually SOAEs occur near the maxima of SFOAEs fine structure (Bergevin et al. 2012; Dewey and Dhar 2014). Given that loudness varies with SFOAE maxima and minima and that loudness is a strong predictor of listener decision weights, it is possible that both SFOAEs and SOAEs may be used to predict individual differences in behavioural decision weights.

2 Methods

2.1 Listeners

Data are presented from seven individuals (mean age: 27.42 yrs) with pure tone air-conduction hearing thresholds better than 15 dB HL at all frequencies between 0.5 and 4 kHz, normal tympanograms, and no history of middle ear disease or surgery.

2.2 Measurement and Analysis of Otoacoustic Emissions

SOAEs were evaluated from 3-min recordings of sound in the ear canal obtained after subjects were seated comfortably for 15 min in a double-walled, Industrial Acoustics, sound-attenuation chamber. The signal from the ER10B+microphone was amplified by an Etymotic preamplifier with 20 dB gain before being digitized by a Fireface UC (16 bit, 44100 samples/sec). The signal was then segmented into 1-sec analysis windows (1-Hz frequency resolution) with a step size of 250 ms. Half of the segments with the highest power were discarded in order to reduce the impact of subject generated noise. Then an estimate of the spectrum in the ear canal was obtained by converting the averages of FFT magnitude in each frequency bin to dB SPL. SOAE frequencies were identified as a very narrow peak of energy at least 3 dB above an average of the background level of adjacent frequencies.

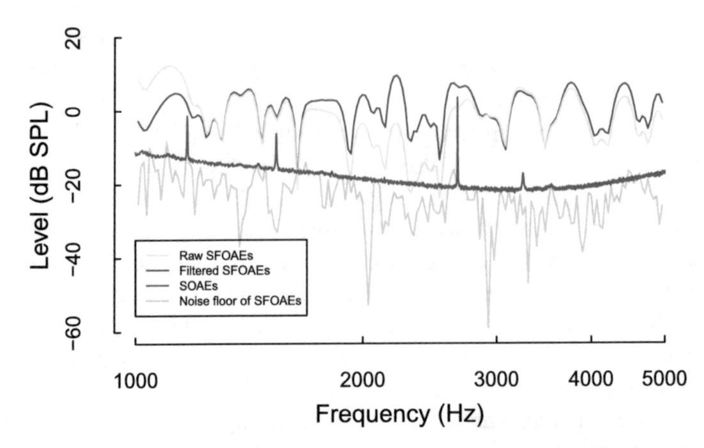

Fig. 1 An example of raw (*light blue*) and the filtered (*blue*) SFOAEs. The raw SFOAE fine structure was high-pass filtered with a high-pass frequency of 1 cycle/octave to reduce the impact of any slow changes in overall SFOAEs level on estimates of fine structure depth. The *red line* represents SOAEs and the *grey line* represents the noise floor

SFOAE measurements were obtained using an adaptation of a swept-tone paradigm (c.f. Long and Talmadge 2007; Long et al. 2008). A probe tone was swept logarithmically at a rate of 2 s per octave from 500 to 8000 Hz at 35 dB SPL (8 s per sweep). The stimuli were delivered to the subjects' ear canal through ER2 tube phones attached to an ER10B + probe microphone used to record the OAEs. The sweep was repeated 32 times to permit reduction of the noise and intermittent artifacts. Each measurement was multiplied by a half-second-long sliding window with a step size of 25 ms, which gives successive windowed signal segments x_m^n where $n = 1, 2, ..., 32$ and m are the indices of repeated number and segments respectively. An artifact-reduced segment $\bar{x}_m$ was obtained by taking a temporal average of the 32 segments having power falling within 75 % of the median, then the magnitude coefficients $\bar{X}_m$ (at frequency $F_m = 500 \cdot 2^{(0.025/2)m}$) were evaluated by LSF procedure to the delayed probe stimulus (Long and Talmadge 1997; Talmadge et al. 1999; Naghibolhosseini et al. 2014). Real-time data processing was controlled by a Fireface UC, using Matlab and Psychtoolbox (Kleiner et al. 2007) in a Windows OS programing environment. The SFOAE fine structure obtained was high-pass filtered with a cutoff frequency of 1 cycle/octave to reduce the impact of any slow changes in overall SFOAEs level (Fig. 1). The levels of maxima (MAX) and minima (MIN) frequencies were determined by the filtered SFOAE fine structure.

2.3 Behavioural Task: Two-Tone Level Discrimination

A two-interval, forced-choice procedure was used: two-tone complexes were presented in two intervals, standard and target, on each trial. All stimuli were presented monaurally at a duration of 300 (in SOAEs experiment) or 400 ms (SFOAEs experiment) with cosine-squared, 5-ms rise/fall ramps. In the target interval, the level of each tone was always 3 dB greater than that in the standard interval. Small independent and random perturbations in level of the individual tones were presented from one presentation to the next. The level perturbations were normally distributed with sigma (σ)=3 dB. The order of standard and target intervals was selected at random on each trial. Listeners were asked to choose the interval in which the target (higher level) sound occurred by pressing a mouse button. Correct feedback was given immediately after the listener's response. Decision weights on the tones for each listener were then estimated from logistic regression coefficients in a general linear model for which the perturbations were predictor variables for the listener's trial-by-trial response (Berg 1990). In experiment 1, the frequencies of the two-tone complex were chosen from SOAE measures from each listener: one at the frequency of an SOAE and the other at a nonSOAE frequency, either lower or higher than the chosen SOAE frequency. The level of each tone in the standard interval was 50 dB SPL. SOAEs usually occur near maxima of SFOAEs fine structure (Bergevin et al. 2012; Dewey and Dhar 2014), but SOAEs are not always detectable at such maxima. Thus, we decided also to measure SFOAE fine structure and select frequencies at the maxima and minima of the fine structure for the behavioural level discrimination task. In experiment 2, two frequencies were chosen from the measured SFOAE fine structure for each listener: one at a maximum of the fine structure and the other at a minimum. The level of the standard stimuli was 35 dB SPL. During a testing session, SOAEs and SFOAEs were recorded prior to and after the behavioural task.

3 Results

Figure 2 shows the mean relative decision weight for the tone at an SOAE frequency as a function of the mean level-difference of the OAE at the SOAE and non-SOAE frequency for six different ears. There was a statistically significant positive correlation between the relative decision weights and the OAE level difference ($r^2=0.759$, $p=0.0148$).

The relative decision weights were compared with the level differences between tones at the maxima and tones at the minima of the fine structure for two listeners (Fig. 3). These listeners have detectable SOAEs, which occur near SFOAE maxima.

The SFOAEs levels obtained at the beginning of the session are associated with the decision weights from the first half of the behavioural trials (filled symbols), and those at the end of session are associated with the decision weights from the second half of the behavioural trials (open symbols). The correlation between the relative decision weight and the level difference between tones near fine-structure maxima and tones near minima is statistically significant ($r^2=0.48$, $p=0.000014$). This outcome suggests that the association between decision weights and OAEs does not depend on detection of SOAEs.

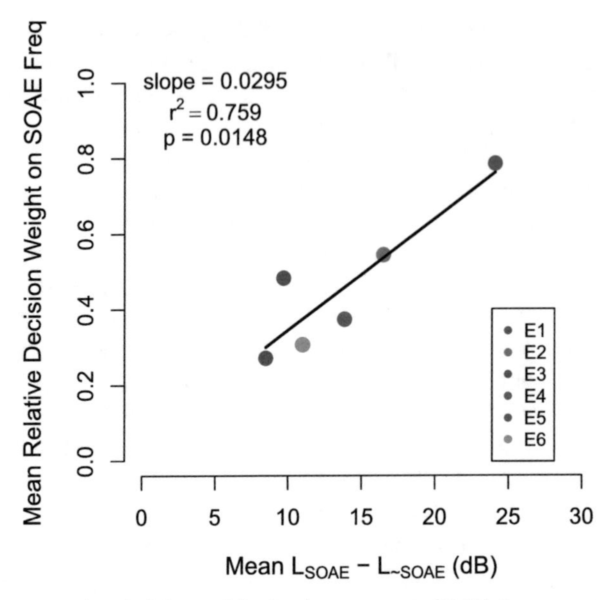

Fig. 2 The mean relative decision weight for the tone at an SOAE frequency as a function of the mean OAE level difference measured at the SOAE and nonSOAE (~SOAE) frequency. Each decision weight was based on at least 400 trials. The relative decision weight was calculated as $W_{SOAE\ tone}/(W_{SOAE\ tone}+W_{nonSOAE\ tone})$, where W is the coefficient from the logistic regression. Each different colour represents a different ear

Fig. 3 The relative decision weights from two listeners responding to tones at either SFOAE maxima or minima as a function of the level difference between the maxima and minima. Each decision weight was based on 1000 trials. The relative decision weight was calculated as $W_{Max\ freq}/(W_{Max\ freq}+W_{Min\ freq})$

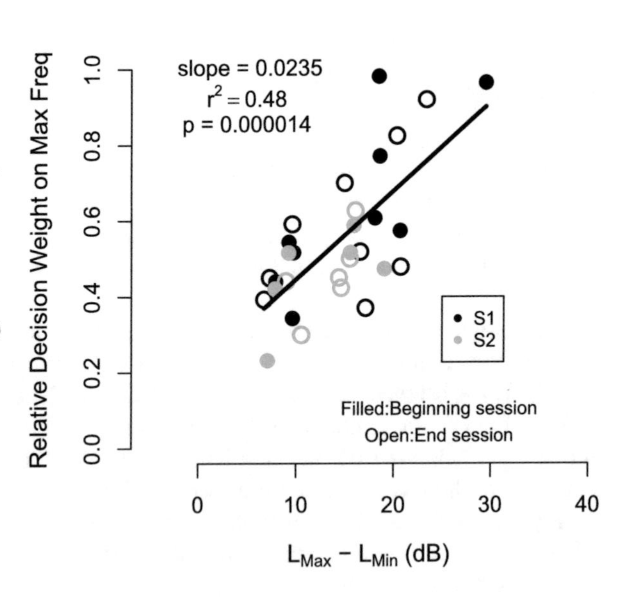

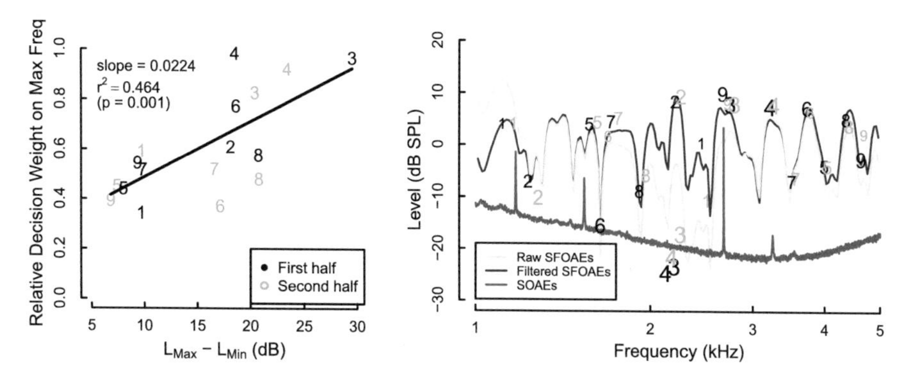

Fig. 4 The left panel represents the relative weight of the tone at SFOAE fine structure maxima as a function of the level difference between SFOAE maxima and minima for subject 1. The frequencies used in each comparison are represented by number pairs (*right panel*). The level difference between the maxima and minima (represented by the same numbers in a pair) is a direct transformation of the relative weights into the level difference using the fitted regression line in the left panel. The rank of the relative weights of frequency pairs is represented by the font size of the numbers in the right panel. Weights from the first half (*black*) and second half (*orange*) of the behavioural trials are both presented

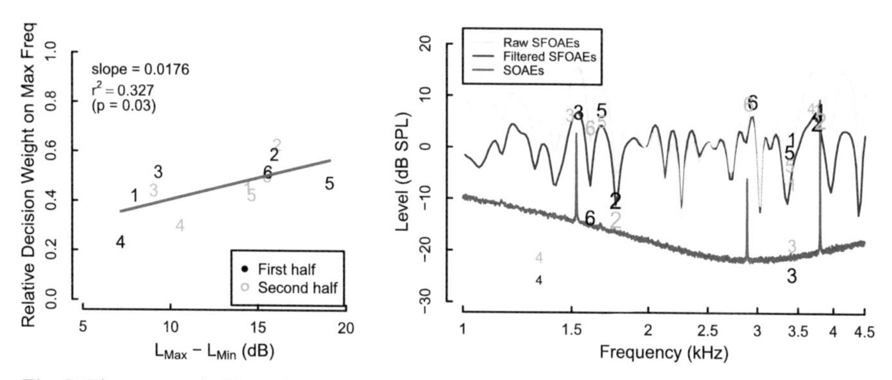

Fig. 5 The same as in Fig. 4 for subject 2

The position of the frequency pairs used for the behavioural task and the decision weight analysis for subject 1 and 2 are shown in Figs. 4 and 5 respectively. The relative weights were transformed into level differences using the fitted function in the right panel and plotted the transformed level differences (an indication of fine structure depth) in the right panel as a function of SFOAE frequency.

4 Discussion

Given that loudness varies with SFOAE maxima and minima and that loudness is a strong predictor of listener decision weights, we hypothesized that both SFOAEs and SOAEs may be used to predict individual differences in decision weights in a level discrimination task. As expected, the data showed a significant positive correlation *between* the level difference of the OAE at the SOAE and non-SOAE frequency *and* the relative decision weights on level discrimination task of two-tone complex (see Fig. 2). Also there was a similar positive correlation with the difference between SFOAE maxima and minima (see Fig. 3). The results suggest that OAE levels might be used to predict individual differences in more complex target-in-noise listening tasks, even possibly in the diagnosis of speech understanding in specific noise backgrounds. For clinical applications swept frequency SFOAEs might provide a better measure of cochlear fine structure inasmuch as they are less time consuming than SOAE measurements and provide a clearer indication of regions of threshold microstructure and variations in loudness.

Acknowledgments This research was supported by NIDCD grant R01 DC001262-21. Authors thank Simon Henin and Joshua Hajicek for providing MATLAB code for the LSF analysis.

References

Alexander JM, Lutfi RA (2004) Informational masking in hearing-impaired and normal-hearing listeners: sensation level and decision weights. J Acoust Soc Am 116(4):2234–2247

Baiduc RR, Lee J, Dhar S (2014) Spontaneous otoacoustic emissions, threshold microstructure, and psychophysical tuning over a wide frequency range in humans. J Acoust Soc Am 135(1):300–314

Berg BG (1990) Observer efficiency and weights in a multiple observation task. J Acoust Soc Am 88:149–158

Bergevin C, Fulcher A, Richmond S, Velenovsky D, Lee J (2012) Interrelationships between spontaneous and low-level stimulus-frequency otoacoustic emissions in humans. Hear Res 285:20–28

Cherry EC (1953) Some experiments on the recognition of speech, with one and two ears. J Acoust Soc Am 25(5):975–979

Dewey JB, Dhar S (2014). Comparing behavioral and otoacoustic emission fine structures. 7th Forum Acusticum, Krakow, Poland

Doherty KA, Lutfi RA (1996) Spectral weights for overall discrimination in listeners with sensorineural hearing loss. J Acoust Soc Am 99(2):1053–1058

Epstein M, Silva I (2009) Analysis of parameters for the estimation of loudness from tone-burst otoacoustic emissions. J Acoust Soc Am 125(6):3855–3810. doi:10.1121/1.3106531

Jesteadt W, Valente DL, Joshi SN, Schmid KK (2014) Perceptual weights for loudness judgments of six-tone complexes. J Acoust Soc Am 136(2):728–735

Kemp DT (1978) Stimulated otoacoustic emissions from within the human auditory system. J Acoust Soc Am 64(5):1386–1391

King K, Stephens D (1992) Auditory and psychological factors in auditory disability with normal hearing. Scand Audiol 21:109–114

Kleiner M, Brainard D, Pelli D (2007). What's new in Psychtoolbox-3. Perception 36:14

Lee J, Long GR (2012) Stimulus characteristics which lessen the impact of threshold fine structure on estimates of hearing status. Hear Res 283:24–32

Long GR, Talmadge CL (1997) Spontaneous otoacoustic emission frequency is modulated by heartbeat. J Acoust Soc Am 102:2831–2848

Long GR, Talmadge CL (2007) New procedure for measuring stimulus frequency otoacoustic emissions. J Acoust Soc Am 122:2969

Long GR, Tubis A (1988) Investigations into the nature of the association between threshold microstructure and otoacoustic emissions. Hear Res 36:125–138

Long GR, Talmadge CL, Jeung C (2008). New Procedure for evaluating SFOAEs without suppression or vector subtraction. Assoc Res Otolaryngol 31

Lutfi RA (1995) Correlation coefficients and correlation ratios as estimates of observer weights in multiple-observation tasks. J Acoust Soc Am 97:1333–1334

Lutfi RA, Jesteadt W (2006) Molecular analysis of the effect of relative tone on multitone pattern discrimination. J Acoust Soc Am 120(6):3853–3860

Lutfi RA, Liu CJ (2007) Individual differences in source identification from synthesized impact sounds. J Acoust Soc Am 122:1017–1028

Mauermann M, Long GR, Kollmeier B (2004) Fine structure of hearing threshold on loudness perception. J Acoust Soc Am 116(2):1066–1080

Naghibolhosseini M, Hajicek J, Henin S, Long GR (2014). Discrete and swept-frequency sfoae with and without suppressor tones. Assoc Res Otolaryngol 37(73)

Rasetshwane DM, Neely ST, Kopun JG, Gorga MP (2013) Relation of distortion-product otoacoustic emission input-output functions to loudness. J Acoust Soc Am 134(1):369–315. doi:10.1121/1.4807560

Richards VM (2002) Effects of a limited class of nonlinearities on estimates of relative weights. J Acoust Soc Am 111:1012–1017

Talmadge CL, Long GR, Tubis A, Dhar S (1999) Experimental confirmation of the two-source interference model for the fine structure of distortion product otoacoustic emissions. J Acoust Soc Am 105:275–292

Thorson MJ, Kopun JG, Neely ST, Tan H, Gorga MP (2012) Reliability of distortion-product otoacoustic emissions and their relation to loudness. J Acoust Soc Am 131(2):1282–1214. doi:10.1121/1.3672654

On the Interplay Between Cochlear Gain Loss and Temporal Envelope Coding Deficits

Sarah Verhulst, Patrycja Piktel, Anoop Jagadeesh and Manfred Mauermann

Abstract Hearing impairment is characterized by two potentially coexisting sensorineural components: (i) cochlear gain loss that yields wider auditory filters, elevated hearing thresholds and compression loss, and (ii) cochlear neuropathy, a noise-induced component of hearing loss that may impact temporal coding fidelity of supra-threshold sound. This study uses a psychoacoustic amplitude modulation (AM) detection task in quiet and multiple noise backgrounds to test whether these aspects of hearing loss can be isolated in listeners with normal to mildly impaired hearing ability. Psychoacoustic results were compared to distortion-product otoacoustic emission (DPOAE) thresholds and envelope-following response (EFR) measures. AM thresholds to pure-tone carriers (4 kHz) in normal-hearing listeners depended on temporal coding fidelity. AM thresholds in hearing-impaired listeners were normal, indicating that reduced cochlear gain may counteract how reduced temporal coding fidelity degrades AM thresholds. The amount with which a 1-octave wide masking noise worsened AM detection was inversely correlated to DPOAE thresholds. The narrowband noise masker was shown to impact the hearing-impaired listeners more so than the normal hearing listeners, suggesting that this masker may be targeting a temporal coding deficit. This study offers a window into how psychoacoustic difference measures can be adopted in the differential diagnostics of hearing deficits in listeners with mixed forms of sensorineural hearing loss.

Keywords Hearing impairment diagnostics · Cochlear neuropathy · Amplitude modulation detection · EFR · DPOAE

S. Verhulst (✉) · P. Piktel · A. Jagadeesh · M. Mauermann
Medizinische Physik and Cluster of Excellence Hearing4all, Department of Medical Physics and Acoustics, Oldenburg University, Carl-von-Ossietzky Strasse 9-11, 26129, Oldenburg, Germany
e-mail: Sarah.verhulst@uni-oldenburg.de

P. Piktel
e-mail: Patrycja.piktel@uni-oldenburg.de

A. Jagadeesh
e-mail: anoop.jagadeesh@uni-oldenburg.de

M. Mauermann
e-mail: Manfred.mauermann@uni-oldenburg.de

P. van Dijk et al. (eds.), *Physiology, Psychoacoustics and Cognition in Normal and Impaired Hearing,* Advances in Experimental Medicine and Biology 894,
DOI 10.1007/978-3-319-25474-6_49

1 Introduction

Listeners with impaired audiograms likely suffer from a combination of pathologies that may interact to affect their speech intelligibility in adverse listening conditions. The well-studied cochlear gain loss aspect of hearing loss due to outer-hair-cell deficits is known to impact audibility of sound and yields reduced frequency selectivity as well as cochlear compression loss. Cochlear neuropathy is associated with a reduction in the number and types of auditory nerve (AN) fibers responsible for robust afferent transmission of sound (Kujawa and Liberman 2009). Whereas cochlear gain loss affects the whole dynamic range of sound intensities, cochlear neuropathy is thought to affect sound encoding at supra-threshold levels as high-threshold AN fibers are most sensitive to noise exposure (Furman et al. 2013). Indeed, a recent study has demonstrated that listeners with normal hearing thresholds can show supra-threshold hearing deficits (e.g., in envelope ITD, AM detection threshold tasks) that are related to temporal coding fidelity in the auditory brainstem while being uncorrelated to audiometric or distortion-product (DPOAE) thresholds (Bharadwaj et al. 2015). DPOAE thresholds offer an objective and purely peripheral correlate to the hearing threshold (Dorn et al. 2001) that is not influenced by AN deficits in afferent transmission.

Even though a temporal coding deficit may influence auditory perception, it is currently not known how cochlear neuropathy interacts with the cochlear gain loss aspect of hearing loss, or whether it is equally important for auditory perception. On the one hand, AM detection is expected to *improve* when cochlear compression is reduced (Moore et al. 1996), while cochlear neuropathy may *degrade* temporal coding fidelity to temporal envelopes (Bharadwaj et al. 2014, 2015). To study the interaction between these components, and to quantify contributions in listeners that may suffer from both aspects of hearing loss, we tested amplitude-modulation detection (100 Hz) in quiet and in the presence of noise maskers. To force the system to rely on redundancy of coding within a single auditory filter, we determined AM thresholds in a fixed-level narrowband noise masker (NB; 40 Hz) condition. The difference between the AM threshold in quiet and with the NB noise masker might be a metric that is free from cochlear compression and sensitive to the temporal coding fidelity aspect of hearing loss. A second differential measure uses a fixed level-broadband noise masker (BB, 1 octave) to test whether auditory filter widening due to cochlear gain loss is more detrimental than temporal coding fidelity in processing temporal modulations in the presence of background noise. To help separating hearing deficits in the psychoacoustic measures, AM thresholds were compared to DPOAE thresholds (cochlear gain loss) and EFR measures targeting temporal coding fidelity.

2 Methods

The test population was formed with 10 subjects (4 male, 6 female), aged from 20 to 32 (mean: 25.9), that had near normal hearing thresholds (<15 dB HL, flat), 4 subjects (3 male, 1 female) aged from 25 to 35 (mean: 25.9), with a normal 4 kHz threshold, but a slightly sloping audiogram in the higher frequencies, and 7 subjects (3 male, 4 female) aged from 39 to 75 (mean: 60.7), with a near 25 dB dB loss at 4 kHz and a mild hearing loss at the higher frequencies. We chose listeners with mild hearing losses to ascertain they would have measurable DPOAEs.

Sound delivery for OAEs, EFRs, and AM detection threshold measurements was provided by ER-2 insert earphones attached to a TDT-HB7 headphone driver and Fireface UCX soundcard. Stimuli were generated in Matlab and calibrated using a B&K type 4157 ear simulator and sound level meter. OAEs were recorded using the OLAMP software and an ER10B + microphone, EFRs were recorded using a 32-channel Biosemi EEG amplifier using a custom built triggerbox, and analyzed using the ANLFFR and Matlab software.

2.1 Amplitude Modulation Detection Thresholds

AM detection thresholds for 4-kHz, 500-ms pure-tone carriers were obtained using a 3AFC method (1-up, 2 down) in quiet and in masking noise. Thresholds were obtained from the last six reversals at the smallest step size and 4 repetitions were measured of which the first run was discarded. The initial modulation depth value was -6 dB ($20 \cdot \log_{10}(m)$, $m=50\,\%$), and varied adaptively with stepsizes of 10, 5, 3 and 1 dB. The modulation frequency was set to 100 Hz to target auditory brainstem processing based on the shorter EFR group delays reported for modulation frequencies above ~80 Hz (Purcell et al. 2004), while providing the modulation within a single equivalent rectangular bandwidth (ERB). Target levels were 60 dB SPL for the pure tone and NH listeners, and adjusted to a level corresponding 25 categorical loudness units (approx. 45–50 dB SL) to allow for equal sensation in the very slight and mild hearing-impaired group. The two on-frequency maskers were presented at the stimulus level—15 dB SPL spectral level (narrowband; NB) 2212 $-$ 50 dB SPL spectral level for the broadband condition (BB). The spectral level calibration refers to a method in which the noise levels were calibrated relative to the level of the stimulus in the frequency spectrum rather than from the rms of the time-domain waveform. Noise bandwidths were 40 Hz (NB) and 1 octave (BB) centered around 4 kHz.

2.2 Envelope Following Responses

EFRs were obtained for 4 kHz 1-octave wide noise carriers presented at 70 dB SPL for a modulation frequency of 120 Hz and modulation depths ($20\log_{10}(m)$) of -8, -4 and 0 dB relative to $m=100\,\%$. 600 repetitions of 600 ms modulated epochs were recorded on 32 channels, and after filtering (60–600 Hz), artefact rejection (100 μV), epoching and baseline correction, the FFT of the averaged epochs in each of the 32 channels was calculated. EFR strength (in dB) was determined as the spectral level at the frequency of the modulator and averaged across the 32 channels. EFR strength of the 100 % modulated condition was measured along with the slope of the EFR strength as a function of modulation depth reduction. The latter slope measure was proposed to reflect temporal coding fidelity to supra-threshold sounds (Bharadwaj et al. 2015). Slopes were only considered when at least the 0 and −4 dB EFR levels were above the noise floor; if the −8 dB EFR was not present, the level of the noise floor was used as the level of the −8 dB EFR strength. The slope was thus calculated using a linear fit to 3 datapoints: 0, −4 and −8 dB.

2.3 *DPOAE Thresholds*

DPOAE based hearing thresholds were derived from 2 f_1-f_2 DPOAE I/O measurements using the DPOAE sweep method (Long et al. 2008). The primary frequencies were exponentially swept up (2 s/octave) over a 1/3 octave range around the geometric mean of 4 kHz at a constant frequency ratio f_2/f_1 of 1.2. Using a sufficiently sharp least squared fit filter (here ca. 2.2 Hz), the distortion component (DCOAE) can be extracted from the DPOAE recording (Long et al. 2008). The DCOAE is generated around the characteristic site of f_2 and thus predominantly provides information about the f_2 site without being influenced by DPOAE fine structure that is known to affect I/O functions unwantedly (Mauermann and Kollmeier 2004). DCOAE I/O functions were computed as average over 34 DCOAE I/O functions across the measured frequency range. A matched cubic function $L_{DP} = a + \left(\frac{1}{q} \cdot (L_2 - b)^3\right)$ with parameters a, b, and q was fit to the data points. DPOAE thresholds were determined as the level of L_2 at which the extrapolated fitting curve would reach a level of −25 dB SPL (~0 Pa).

3 Results

Figure 1 depicts correlations between the DPOAE threshold and the EFR measures for NH (blue circles), very slight (black circle) and mild (red squares) HI listeners. There were no significant correlations between the DPOAE measures and the EFR measures indicating that the EFR measures at 70 dB SPL reflect more aspects of hearing loss than captured by auditory threshold measures alone. Note that the audiometric hearing threshold and the DP threshold were highly correlated ($p < 1e^{-6}$) suggesting they both reflect the perceived threshold of hearing. Likely a combination of cochlear gain loss, temporal coding fidelity, along with potential head-size differences affects the 100 % modulated EFR strength. Because the EFR slope measure was not correlated to EFR strength in the same listeners, it may be that the differential slope measure to modulation depth reduction is more sensitive to temporal coding fidelity as earlier suggested in Bharadwaj et al. (2015).

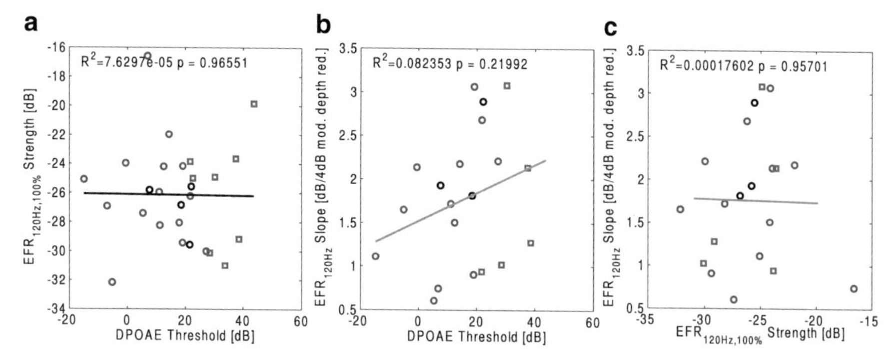

Fig. 1 Correlations between DPOAE thresholds and EFR strength (m = 100 %) (panel A) and EFR slope measures as a function of modulation depth reduction (panel B). Panel C shows that the EFR slopes and strength measure do not correlate indicating they reflect different aspects of auditory coding. The *blue circles* indicate NH listeners, the *black circles* represent NH listeners with a very slight sloping high-frequency hearing loss, and the mild HI listeners (*red squares*) had elevated hearing thresholds at 4 kHz. Because these figures reflect general correlations between objective metrics, data from additional listeners (Verhulst et al. 2015) were added to this analysis

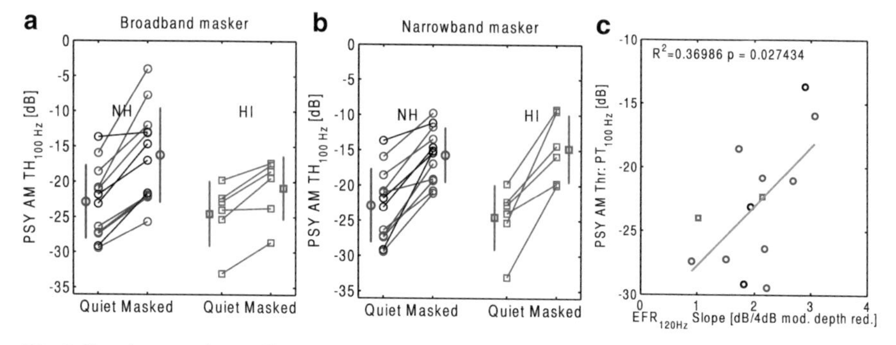

Fig. 2 Psychoacoustic amplitude modulation detection thresholds for the NH (*blue* and *black*) and mild HI listeners in the quiet condition and in the presence of a fixed level broadband (panel A) and narrowband (panel B) noise masker. The *black symbols* reflect those NH listeners that had normal hearing thresholds but a slightly sloping audiogram at the high frequencies. Panel C shows the relation between EFR slope measure and the AM detection thresholds in quiet, indicating that temporal coding fidelity predicts performance in this task for listeners with normal hearing thresholds

Figure 2 shows the psychoacoustic AM detection thresholds in quiet and in the presence of broadband (panel A) and narrowband noise (panel B). Average AM detection thresholds were similar for NH and mild HI listeners supporting the observation that temporal modulation detection is not determined by the hearing threshold (Moore et al. 1996). In fact, the variation in thresholds for the NH listeners was significantly correlated to the EFR slope metric (panel C) demonstrating that temporal coding fidelity predicts performance when hearing thresholds are normal (see also Bharadwaj et al. 2015). For the mild HI listeners, this relationship is more complex as the reduced temporal coding fidelity in those listeners (i.e., only 2 out of

6 listeners had EFR responses down to −4 dB AM depth) would predict worse AM detection thresholds. The observation that HI listeners had normal AM performance despite their reduced temporal coding fidelity suggests that cochlear gain loss can compensate for temporal coding deficits to yield normal AM detection thresholds.

AM detection thresholds in background noise worsened for all listeners. Whereas the broadband noise had a variable effect on the NH listeners, the mild HI listeners were only mildly affected (Fig. 2a). In contrast, the narrowband noise impacted AM detection performance in HI listeners significantly more strongly than did broadband noise (Fig. 2a, 2b). This difference between the masker conditions was absent for the NH listeners.

Degradation in AM detection performance due to the presence of background noise is depicted in Fig. 3 for the broadband (panels A&C) and narrowband noise conditions (panels B&C) and compared to the objective DPOAE and EFR measures.

The amount with which AM detection thresholds worsened for the NH listeners was significantly correlated to the DPOAE threshold indicating that those listeners with the widest auditory filters and steepest compression slopes were less impacted by the addition of the noise. Because this correlation happened for both the NB and BB noise conditions, perhaps cochlear compression and to a lesser extent the width of the auditory filters might be responsible for this result. The EFR slope metric was not significantly correlated to the AM detection threshold reductions in the NH listeners. Because the EFR slope was correlated to the AM detection performance in quiet for the NH listeners, the degradation measure plotted in Fig. 3 may have factored out its influence. Unfortunately, despite the normal EFR strengths to 100 % modulation depths, the tested HI listeners had poor EFR strength for the −4 and −8 dB modulation depths making EFR slope estimates and associated correlations for this subgroup impossible.

4 Discussion

The present study offers a window into how cochlear neuropathy and cochlear gain loss interact to affect perception of fast temporal modulations important for speech perception in adverse conditions. AM detection thresholds and EFR strength to 100 % modulated stimuli were normal in the mild HI listeners we tested. These findings are in line with other studies that show normal AM detection thresholds (Moore et al. 1996) and EFR strengths (Zhong et al. 2014) for subjects with elevated hearing thresholds. However, interactions between hearing deficits were apparent from correlating the psychoacoustic results with objective measures in the same listeners. Whereas AM detection for NH listeners was correlated to their EFR slopes as a measure for temporal coding fidelity, cochlear gain loss was shown to compensate for reduced temporal coding in the HI listeners to yield near normal AM detection thresholds.

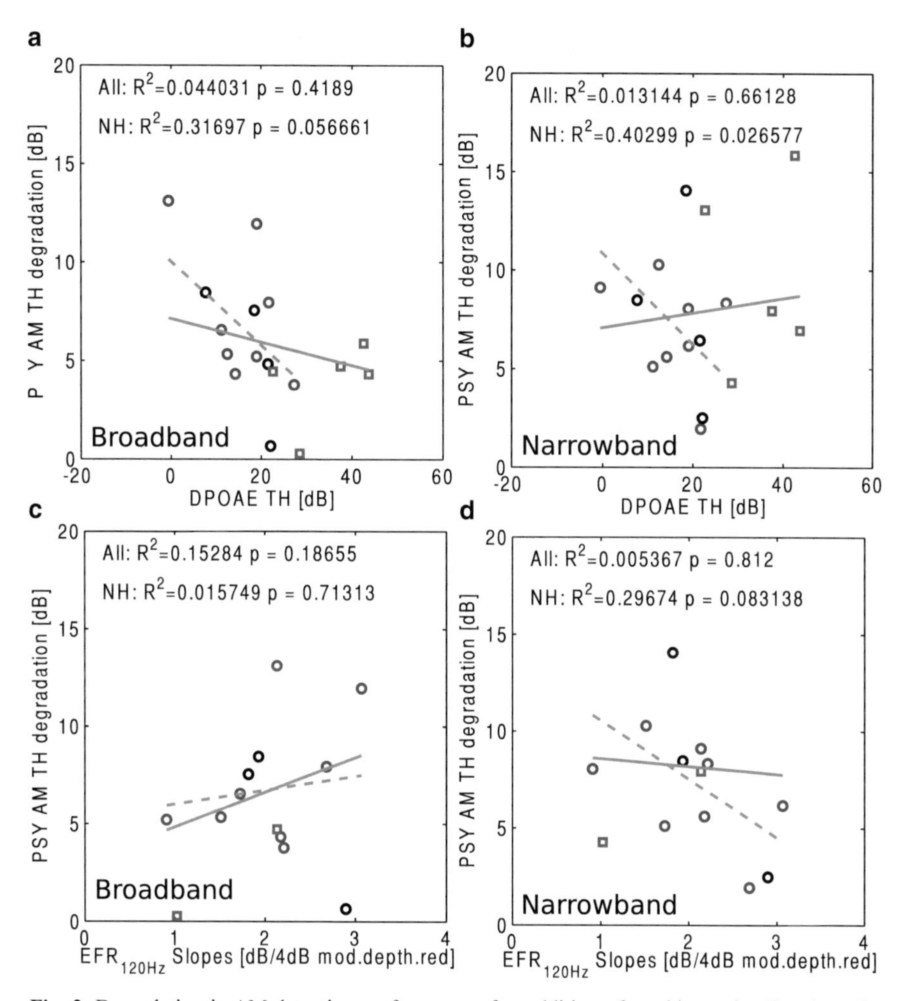

Fig. 3 Degradation in AM detection performance after addition of masking noise (i.e. degradation = AM threshold in noise—AM threshold in quiet) and its relation to objective measures in the same listeners: DPOAE thresholds and EFR slopes. Correlations were calculated for the whole population (*solid*; All) and for the listeners with normal hearing thresholds (*dashed*; NH)

Because AM detection thresholds did not correlate to hearing threshold measures and reflected interactions between temporal coding fidelity and cochlear gain, it is informative to use differential metrics to tease apart different subcomponents of sensorineural hearing loss. Adding a fixed-level broadband noise masker is expected to impact AM detection performance in two ways: (i) wider auditory filters would pass through more noise and degrade performance accordingly, and (ii) AM information within the auditory filter would be more noisy in each coding channel (e.g., in each auditory nerve fiber), such that a sufficient number of AN fibers needs to be present to perform well in the task. Because the bandwidth of the NB masker fell within an ERB, this condition was expected to be more detrimental to listeners that suffer from reduced temporal coding fidelity irrespective of the width of their auditory filters. An important third factor that could influence all the psychoacoustic results is the individual amount of cochlear compression that is present at the tested frequency. Because cochlear compression loss would enhance perceived modulation depth both in the quiet and noise masking conditions, it was assumed that the differential metric would be able to parse out this effect.

Comparison between AM detection thresholds in quiet and noise demonstrated that the mild HI listeners were heavily impacted by the NB noise. This finding is in line with the idea that the NB noise targets temporal coding fidelity within a single auditory filter, especially because 4 out of 6 HI listeners did not have EFRs at the −8 dB modulation depth. Also the NH listeners showed a large variability in how the NB noise impacted their performance with a trend ($p=0.09$) towards degraded AM detection performance in NB noise for those listeners with steeper EFR slopes. Even though the NB condition was designed to target within auditory filter aspects of temporal coding fidelity, it is possible that another mechanism could also explain these results. For example, because the temporal envelope of the NB noise waveform is much more fluctuating than the BB noise envelope, it is possible that the quality of a modulation coding mechanism in the brainstem and not the numbers of auditory-nerve fibers responsible for a robust coding of temporal envelopes could also explain the NB results.

Lastly, it is interesting to observe that AM detection performance in the mild HI listeners was significantly more impacted by the NB than the BB noise. This difference was not observed for the NH listeners. If the NB condition reflects a temporal coding deficit, then it appeared not to dictate performance in the BB condition for the HI listeners, suggesting that perhaps the overall loss of compression due to cochlear gain loss would enhance modulation sensitivity such that a fixed-level BB noise did not degrade performance substantially, despite the reduced temporal coding observed from the EFR measures. It is too early to make strong conclusions regarding the underlying mechanisms based on the present dataset, as additional data and additional metrics that reflects the individual listeners cochlear compression should be added to further tease apart the psychoacoustic results. In this respect, both categorical loudness scaling metrics and DPOAE compression slope estimates could be included.

To conclude, differential psychoacoustic and EFR methods methods form a promising candidate to separate different aspects of hearing loss in listeners with mixed sensorineural pathologies. Differential diagnostic metrics separating cochlear gain loss from temporal coding deficits are necessary to understand contributions of different interacting pathologies to the perceptual performance.

Acknowledgments The authors thank Frauke Ernst and Anita Gorges for assistance with EEG measurements and subject recruitment and Steven van de Par for discussions related to this work. Work supported by the DFG Cluster of Excellence EXC 1077/1 "Hearing4all".

References

Bharadwaj H, Verhulst S, Shaheen L, Liberman MC, Shinn-Cunningham B (2014) Cochlear neuropathy and the coding of supra-threshold sound. Front Syst Neurosci 8:26

Bharadwaj H, Masud S, Mehraei G, Verhulst S, Shinn-Cunningham B (2015) Individual differences reveal correlates of hidden hearing deficits. J Neurosci 35(5):2161–2172

Dorn PA, Konrad-Martin D, Neely ST, Keefe DH, Cyr E, Gorga MP (2001) Distortion product otoacoustic emission input/output functions in normal-hearing and hearing-impaired human ears. J Acoust Soc Am 110(6):3119–3131

Furman AC, Kujawa SG, Liberman MC (2013) Noise-induced cochlear neuropathy is selective for fibers with low spontaneous rates. J Neurophysiol 110(3):577

Kujawa SG, Liberman MC (2009) Adding insult to injury: cochlear nerve degeneration after "temporary" noise-induced hearing loss. J Neurosci 29:14077

Long GR, Talmadge CL, Lee J (2008) Measuring distortion product otoacoustic emissions using continuously sweeping primaries. J Acoust Soc Am 124:1613–1626

Mauermann M, Kollmeier B (2004) Distortion product otoacoustic emission (DPOAE) input/output functions and the influence of the second DPOAE source. J Acoust Soc Am 116:2199

Moore BC, Wojtczak M, Vickers DA (1996) Effect of loudness recruitment on the perception of amplitude modulation. J Acoust Soc Am 100:481–489

Purcell DW, John SM, Schneider BA, Picton TW (2004) Human temporal auditory acuity as assessed by envelope following responses. J Acoust Soc Am 116(6):3581

Verhulst S, Ernst F, van de Par S (2015). In search for the cochlear neuropathy component of audiometric hearing loss. Association for research in otolaryngology midwinter meeting, poster

Zhong Z, Henry KS, Heinz MG (2014) Sensorineural hearing loss amplifies neural coding of envelope information in the central auditory system of chinchillas. Hear Res 309:55–62

Frequency Tuning of the Efferent Effect on Cochlear Gain in Humans

Vit Drga, Christopher J. Plack and Ifat Yasin

Abstract Cochlear gain reduction via efferent feedback from the medial olivocochlear bundle is frequency specific (Guinan, Curr Opin Otolaryngol Head Neck Surg 18:447–453, 2010). The present study with humans used the Fixed Duration Masking Curve psychoacoustical method (Yasin et al., J Acoust Soc Am 133:4145–4155, 2013a; Yasin et al., Basic aspects of hearing: physiology and perception, pp 39–46, 2013b; Yasin et al., J Neurosci 34:15319–15326, 2014) to estimate the frequency specificity of the efferent effect at the cochlear level. The combined duration of the masker-plus-signal stimulus was 25 ms, within the efferent onset delay of about 31–43 ms (James et al., Clin Otolaryngol 27:106–112, 2002). Masker level (4.0 or 1.8 kHz) at threshold was obtained for a 4-kHz signal in the absence or presence of an ipsilateral 60 dB SPL, 160-ms precursor (200-Hz bandwidth) centred at frequencies between 2.5 and 5.5 kHz. Efferent-mediated cochlear gain reduction was greatest for precursors with frequencies the same as, or close to that of, the signal (gain was reduced by about 20 dB), and least for precursors with frequencies well removed from that of the signal (gain remained at around 40 dB). The tuning of the efferent effect filter (tuning extending 0.5–0.7 octaves above and below the signal frequency) is within the range obtained in humans using otoacoustic emissions (Lilaonitkul and Guinan, J Assoc Res Otolaryngol 10:459–470, 2009; Zhao and Dhar, J Neurophysiol 108:25–30, 2012). The 10 dB bandwidth of the efferent-effect filter at 4000 Hz was about 1300 Hz (Q_{10} of 3.1). The FDMC method can be used to provide an unbiased measure of the bandwidth of the efferent effect filter using ipsilateral efferent stimulation.

Keywords Auditory · Fixed-duration masking curves · Forward masking · Compression · Gain · Efferent · Precursor · Frequency-range

---

V. Drga (✉) · I. Yasin
Ear Institute, University College London (UCL), 332 Grays Inn Road, WC1X 8EE London, UK
e-mail: v.drga@ucl.ac.uk

I. Yasin
e-mail: i.yasin@ucl.ac.uk

C. J. Plack
School of Psychological Sciences, The University of Manchester, M13 9PL Manchester, UK
e-mail: Chris.Plack@manchester.ac.uk

P. van Dijk et al. (eds.), *Physiology, Psychoacoustics and Cognition in Normal and Impaired Hearing,* Advances in Experimental Medicine and Biology 894,
DOI 10.1007/978-3-319-25474-6_50

1 Introduction

In addition to ascending (afferent) neural pathways, the mammalian auditory system contains descending (efferent) neural projections from higher to lower levels of the auditory system [ipsilateral and contralateral efferent systems of cat (Liberman 1988; Huffman and Henson 1990) and human (Guinan 2006)]. The mammalian auditory system includes a brainstem-mediated efferent pathway from the superior olivary complex by way of the medial olivocochlear bundle (MOCB) which reduces the cochlear gain applied over time to the basilar membrane (BM) response to sound [electrical stimulation of the olivocochlear bundle (OCB) in guinea-pigs (Murugasu and Russell 1996)]. The human MOCB response (as measured using otoacoustic emissions: OAEs) has an onset delay of between 25 and 40 ms and rise and decay time constants in the region of 280 and 160 ms, respectively (Backus and Guinan 2006). Recordings from guinea pigs show the greatest reduction in BM vibration due to efferent stimulation when the stimulating tone is close to or above the characteristic frequency (CF) associated with the recording site (Russell and Murugasu 1997). Similarly, suppression of human otoacoustic emissions via efferent activation is more effective for stimulating frequencies slightly above and below the test probe frequency (Maison et al. 2000).

The advantage of psychoacoustical methods to infer the efferent effect is that they can be used in cases of both normal and mildly impaired hearing (unlike OAEs which can be eliminated by a mild hearing loss). Psychoacoustical studies suggest the efferent effect appears to decrease as the precursor frequency is set higher or lower in frequency than the subsequently presented masker (Bacon and Viemeister 1985; Bacon and Moore 1986). The present study uses a psychoacoustical forward-masking technique [fixed-duration masking curve (FDMC) method (Yasin et al. 2013a, b, 2014)] in which a precursor sound is presented to elicit the efferent response. The FDMC method avoids the confounds of previous forward-masking methods used to measure the efferent effect by using a combined duration of the masker-plus-signal stimulus of 25 ms, which is within the efferent onset delay. Hence, the representation of the signal should be unaffected by efferent activity elicited by the masker. The effect of efferent activation on cochlear gain can then be separately studied by presenting a precursor sound (to activate the efferent system) prior to the onset of the combined FDMC masker-signal stimulus. The present study used the FDMC method to estimate the frequency range of efferent-mediated cochlear gain reduction in humans.

2 Method

2.1 Listeners

Three normal-hearing listeners participated (L1 (author), L2 and L3). All listeners had absolute thresholds better than 20 dB HL for sound frequencies between 0.25

and 8 kHz (ANSI 1996) and were practised in related psychoacoustical experiments. Listeners undertook between 4 and 10 h of training on subsets of the stimuli used for the main experiment. L2 and L3 were paid £6 /h for participation.

2.2 *Stimuli*

The signal was a 4-kHz sinusoid and the sinusoidal masker was either on-frequency (4 kHz) or off-frequency (1.8 kHz); both signal and masker always began in sine phase (Yasin et al. 2013a, b, 2014). The signal had a total duration of 6-ms (3-ms onset and offset ramps, 0-ms steady-state). Absolute thresholds for this signal, presented to the left ear, were 18.2, 20.1 and 23.5 dB SPL for listeners L1, L2 and L3. The masker had a total duration of 19 ms (2-ms onset and offset ramps, 15-ms steady-state). The frequency range of the efferent effect on cochlear gain was studied by presenting a bandlimited precursor sound centred at different frequencies prior to the onset of the combined FDMC masker-signal stimulus (see Fig. 1). The ipsilateral precursor, when present, was a 160-ms (5-ms onset and offset ramps, 150-ms steady-state), 200-Hz-wide (brickwall filtered) noiseband presented at 60 dB SPL. The precursor was presented at centre frequencies of 2.5, 3.0, 3.5, 3.75, 4.0, 4.25, 4.5, 5.0 or 5.5 kHz at a precursor offset-masker onset silent interval of 10 ms. A silent interval of 10 ms was chosen to reduce potential confusion effects when the precursor frequency was close to the signal frequency. Stimuli were presented via the left channel of a pair of Sennheiser HD 600 headphones.

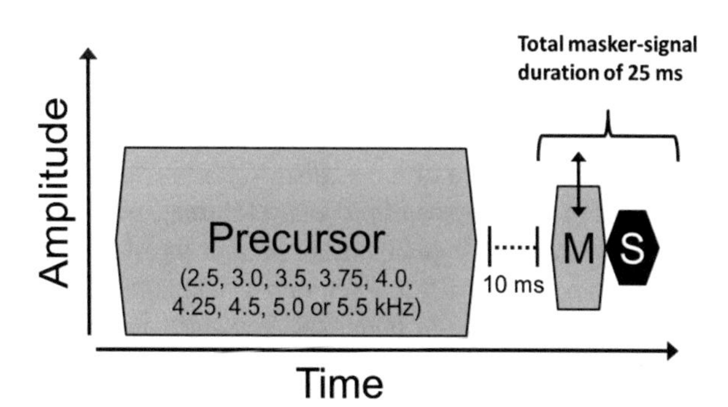

Fig. 1 Schematic of the stimuli used in phase 2 of the study. Masker (M) level at threshold for the signal (S) was measured using the FDMC method. Off-frequency or on-frequency masker-plus-signal total duration was 25 ms and masker-signal temporal interval was 0 ms. A precursor was presented prior to the combined masker-signal stimulus at frequencies of 2.5, 3.0, 3.5, 3.75, 4.0, 4.25, 4.5, 5.0 or 5.5 kHz at a level of 60 dB SPL. The silent temporal interval between precursor offset and masker onset was 10 ms. Double-headed arrow indicates that the masker level was adaptively varied in each stimulus trial

2.3 Procedure

The experiment consisted of two phases (Yasin et al. 2014). In phase 1, for each listener, signal level at threshold was measured per precursor frequency in the presence of a precursor and absence of any tonal masker, to take the forward masking of the precursor into account. In phase 2, the signal level was set at 15 dB SL (based on phase 1 results), a tonal masker was presented as well as the precursor, and masker level at threshold was measured for each precursor frequency and masker frequency. Phase 1 and phase 2 thresholds were also measured for a no-precursor condition. Thresholds were measured using a two-interval, two-alternative forced choice adaptive tracking procedure using a 2-down 1-up rule (phase 1), or a using a 2-up 1-down rule (phase 2) (Levitt 1971). Trial-by-trial feedback was provided. Per block of trials, the initial adaptive track step size was 5 dB, which reduced to 2 dB after four reversals. Threshold was obtained by averaging stimulus levels for the next eight reversals, but the block was rerun if the standard deviation was greater than 6 dB, or the nominal masker level reached 108 dB SPL or higher (due to soundcard output clipping). Listeners ran in 2 h sessions, taking breaks as needed. Condition order was randomized within sessions. Reported listener threshold values for phase 2 are the mean of 3–6 separate threshold estimates.

3 Results

The mean data obtained in the presence of a precursor are shown in Fig. 3. Masker levels at threshold obtained using a 1.8-kHz masker range from about 60 to 80 dB SPL, and remain roughly constant within this range as precursor frequency is increased from 2.5 to 5.5 kHz. Variability in data from listener L3 accounts for most of the dip seen in this data series. Masker levels at threshold obtained using a 4-kHz masker range from about 25 to 45 dB SPL, with the greatest masker level at threshold (about 45 dB SPL) associated with the presence of a precursor with a frequency close to that of the signal. Comparing masker levels at threshold for on- and off-frequency maskers in the presence of a precursor (elicits efferent activation) with masker levels obtained in the absence of precursor (absence of efferent activation), it can be seen that the greatest change in masker level at threshold (of about 20 dB) occurs when there is an on-frequency masker, rather than an off-frequency masker, and this change is greatest when the precursor frequency is the same as, or just below, the 4-kHz signal.

An estimate of inferred basilar-membrane gain can be obtained from the data as the difference between on- and off-frequency masker level values. The mean gain estimated in the absence of a precursor is 41 dB. The estimated gain (as difference between on- and off-frequency masker levels) in the presence of a precursor of different frequencies is shown in Fig. 3. Maximum gain estimates of about 40 dB indicate little effect of any efferent-mediated cochlear gain reduction, and are asso-

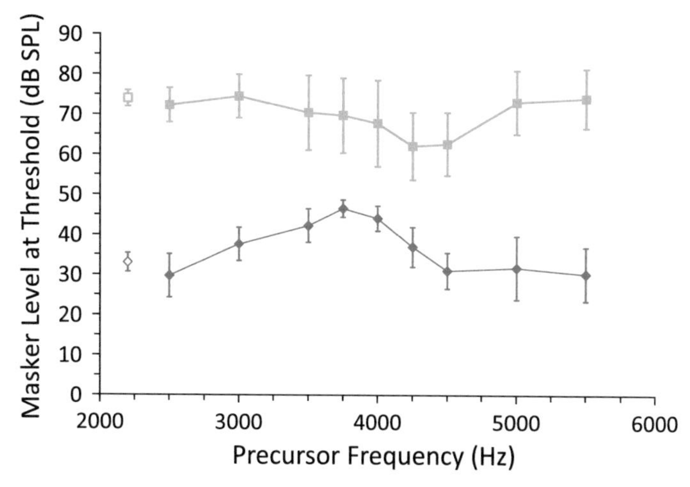

Fig. 2 Mean masker level at threshold (across three listeners) as a function of precursor frequency for on-frequency maskers (4 kHz, lower series) or off-frequency maskers (1.8 kHz, upper series). Open symbols at *left* show masker level at threshold in the absence of a precursor. Error bars show ± 1 standard error.

ciated with precursor frequencies well above and below that of the signal frequency of 4 kHz. Minimum gain estimates of about 20 dB indicate maximal effect of efferent-mediated cochlear gain reduction, and are associated with precursor frequencies close to that of signal frequency of 4 kHz. The estimated 10 dB bandwidth of the efferent-effect filter is 1300 Hz (Q_{10} of 3.1).

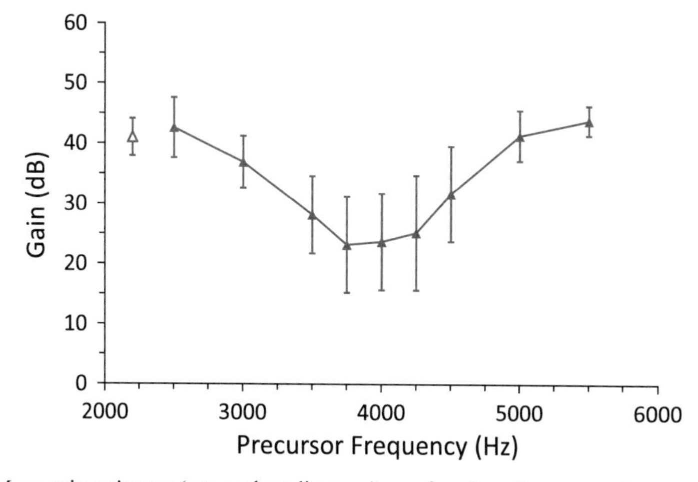

Fig. 3 Mean gain estimates (across three listeners) as a function of precursor frequency. Gain is calculated as the difference between masker levels at threshold obtained for on-frequency (4-kHz) and off-frequency (1.8-kHz) maskers (Fig. 2). Open symbol at *left* shows mean gain value in the absence of a precursor. Error bars show ± 1 standard error.

4 Discussion

The FDMC method was used to obtain an estimate of the efferent effect in the absence (without a precursor) and presence (with a precursor presented at different frequencies) of efferent activation to investigate the frequency specificity of the efferent effect on cochlear gain. There was minimum cochlear gain reduction via efferent-activation when the precursor frequency was well removed from that of the signal frequency of 4-kHz, and maximal cochlear gain reduction (about 20 dB) when the precursor frequency was the same as, or close to, that of the signal. The results suggest that the efferent effect extends for precursors up to about 0.5–0.7 octaves above and below the signal frequency. The frequency range of the efferent effect found in the present study is similar to that observed with the measurement of otoacoustic emissions in humans, using simultaneous or contralateral elicitors. The tuning of the efferent effect can extend 0.5–2 octaves above and below the frequency range of interest (spontaneous OAEs) or probe frequency (stimulus-frequency OAEs), as measured using simultaneous ipsilateral/contralateral tones (Lilaonitkul and Guinan 2009; Zhao and Dhar 2012). In most cases the tone/narrowband noise eliciting the efferent effect has maximal effect above a presentation level of 60 dB SPL, and in some cases the efferent effect has been shown to be more pronounced if the elicitor/precursor sound is about 0.5–1 octave below the probe frequency (e.g., Mott et al. 1989; Harrison and Burns 1993). However, the present study does not appear to show such a distinct asymmetry in the tuning of the efferent effect.

Overall, direct comparisons of gain estimates between OAE- and psychophysical-based measures of efferent activation are problematic, even if measuring SFOAEs with elicitors (analogous to a precursor in psychophysical studies) mainly due to constraints in stimulus design. However, once the effects of OAE stimulus constraints such as the use of a broadband elicitor (presented contralaterally to elicit the greatest efferent response), use of an ongoing long-duration probe-tone stimulus (required for pre- and post-elicitor measurement of SFOAEs), and the possibility of two-tone cochlear suppressive effects with ipsilateral elicitor presentation are taken into account, the reduction in SFOAEs with an elicitor (e.g., Guinan et al. 2003) show gain reduction trends that may be in some cases comparable to those found in the present study. The present findings suggest that the 10 dB bandwidth of the efferent-effect filter at 4000 Hz is about 1300 Hz (Q_{10} of 3.1).

5 Conclusions

1. The FDMC method can be used to provide a measure of the frequency tuning of the efferent effect in humans using narrowband ipsilateral precursors.
2. Cochlear gain reduction is greatest for precursors with frequencies close to that of the signal, and least for precursors with frequencies well removed from the signal frequency.

3. The tuning of the efferent effect filter (0.5–0.7 octaves above and below the signal frequency) is within the range obtained in humans using OAEs (Lilaonitkul and Guinan 2009; Zhao and Dhar 2012).
4. The 10-dB bandwidth of the efferent-effect filter centred at 4000 Hz is about 1300 Hz (Q_{10} of 3.1).

Acknowledgements The research was supported by an EPSRC grant EP/H022732/1 and an International Project grant awarded by Action on Hearing Loss.

References

ANSI (1996) ANSI S3.6-1996 Specification for audiometers. American National Standards Institute, New York

Backus BC, Guinan JJ Jr (2006) Time-course of the human medial olivocochlear reflex. J Acoust Soc Am 119:2889–2904

Bacon SP, Moore BCJ (1986) Temporal effects in simultaneous pure-tone masking: effects of signal frequency, masker/signal frequency ratio, and masker level. Hear Res 23:257–266

Bacon SP, Viemeister NF (1985) The temporal course of simultaneous tone-on-tone masking. J Acoust Soc Am 78:1231–1235

Guinan JJ Jr, Backus BC, Lilaonitkul W, Aharonson V (2003) Medial olivocochlear efferent reflex in humans: otoacoustic emission (OAE) measurement issues and the advantages of stimulus frequency OAEs. J Assoc Res Otolaryngol 4:521–540

Guinan JJ Jr (2006) Olivocochlear efferents: anatomy, physiology, function, and the measurement of efferent effects in humans. Ear Hear 27:589–607

Guinan JJ Jr (2010) Cochlear efferent innervation and function. Curr Opin Otolaryngol Head Neck Surg 18:447–453

Harrison WA, Burns EM (1993) Effects of contralateral acoustic stimulation on spontaneous otoacoustic emissions. J Acoust Soc Am 94:2649–2658

Huffman R, Henson OW (1990) The descending auditory pathway and acousticomotor systems: connections with the inferior colliculus. Brain Res Rev 15:295–323

James AL, Mount RJ, Harrison RV (2002) Contralateral suppression of DPOAE measured in real time. Clin Otolaryngol 27:106–112

Levitt H (1971) Transformed up-down methods in psychoacoustics. J Acoust Soc Am 49:467–477

Liberman MC (1988) Response properties of cochlear efferent neurons: monaural vs. binaural stimulation and the effects of noise. J Neurophysiol 60:1779–1798

Lilaonitkul W, Guinan JJ (2009) Human medial olivocochlear reflex: effects as functions of contralateral, ipsilateral, and bilateral elicitor bandwidths. J Assoc Res Otolaryngol 10:459–470

Maison S, Micheyl C, Andeol G, Gallego S, Collet L (2000) Activation of medial olivocochlear efferent system in humans: influence of stimulus bandwidth. Hear Res 140:111–125

Mott JB, Norton SJ, Neely ST, Warr B (1989) Changes in spontaneous otoacoustic emissions produced by acoustic stimulation of the contralateral ear. Hear Res 38:229–242

Murugasu E, Russell IJ (1996) The effect of efferent stimulation on basilar membrane displacement in the basal turn of the guinea pig cochlea. J Neurosci 16:325–332

Russell IJ, Murugasu E (1997) Efferent inhibition suppresses basilar membrane responses to near characteristic frequency tones of moderate to high intensities. J Acoust Soc Am 102:1734–1738

Yasin I, Drga V, Plack CJ (2013a) Estimating peripheral gain and compression using fixed-duration masking curves. J Acoust Soc Am 133:4145–4155

Yasin I, Drga V, Plack CJ (2013b) Improved psychophysical methods to estimate peripheral gain and compression. In: Moore BCJ, Patterson RD, Winter IM, Carlyon RP, Gockel HE (eds) Basic aspects of hearing: physiology and perception. Springer-Verlag, Heidelberg, pp 39–46

Yasin I, Drga V, Plack CJ (2014) Effect of human auditory efferent feedback on cochlear gain and compression. J Neurosci 34:15319–15326

Zhao W, Dhar S (2012) Frequency tuning of the contralateral medial olivocochlear reflex in humans. J Neurophysiol 108:25–30